Smart Innovation, Systems and Technologies

Volume 54

Series editors

Robert James Howlett, KES International, Shoreham-by-sea, UK
e-mail: rjhowlett@kesinternational.org

Lakhmi C. Jain, University of Canberra, Canberra, Australia;
Bournemouth University, UK;
KES International, UK
e-mails: jainlc2002@yahoo.co.uk; Lakhmi.Jain@canberra.edu.au

About this Series

The Smart Innovation, Systems and Technologies book series encompasses the topics of knowledge, intelligence, innovation and sustainability. The aim of the series is to make available a platform for the publication of books on all aspects of single and multi-disciplinary research on these themes in order to make the latest results available in a readily-accessible form. Volumes on interdisciplinary research combining two or more of these areas is particularly sought.

The series covers systems and paradigms that employ knowledge and intelligence in a broad sense. Its scope is systems having embedded knowledge and intelligence, which may be applied to the solution of world problems in industry, the environment and the community. It also focusses on the knowledge-transfer methodologies and innovation strategies employed to make this happen effectively. The combination of intelligent systems tools and a broad range of applications introduces a need for a synergy of disciplines from science, technology, business and the humanities. The series will include conference proceedings, edited collections, monographs, handbooks, reference books, and other relevant types of book in areas of science and technology where smart systems and technologies can offer innovative solutions.

High quality content is an essential feature for all book proposals accepted for the series. It is expected that editors of all accepted volumes will ensure that contributions are subjected to an appropriate level of reviewing process and adhere to KES quality principles.

More information about this series at http://www.springer.com/series/8767

Simone Bassis · Anna Esposito
Francesco Carlo Morabito
Eros Pasero
Editors

Advances in Neural Networks

Computational Intelligence for ICT

 Springer

Editors
Simone Bassis
Department of Computer Science
University of Milano
Milan
Italy

Anna Esposito
Department of Psychology
Seconda Università di Napoli
Caserta
Italy

and

IIASS
Vietri sul Mare
Italy

Francesco Carlo Morabito
Department of Informatics, Mathematics,
    Electronics and Transportation
Mediterranea University of Reggio Calabria
Reggio Calabria
Italy

Eros Pasero
Dipartimento di Elettronica e
    Telecomunicazioni
Politecnico di Torino
Turin
Italy

ISSN 2190-3018   ISSN 2190-3026   (electronic)
Smart Innovation, Systems and Technologies
ISBN 978-3-319-81591-6   ISBN 978-3-319-33747-0   (eBook)
DOI 10.1007/978-3-319-33747-0

Printed on acid-free paper

This Springer imprint is published by Springer Nature
The registered company is Springer International Publishing AG Switzerland

Preface

Advances in Neural Networks: Computational and Theoretical Issues is a book series dedicated to recent advances in computational and theoretical issues of artificial intelligence methods. Special attention is reserved to information communication technologies (ICT) applications that are of public utility and profitable for a living science that simplify user access to future, remote and nearby social services. In particular, avatars replacing human in high-responsibility tasks, companion agents for elderly and impaired people, robots interacting with humans in extreme, stressful time-critical conditions, wearable sensors and apps for clinical applications. In addition, new mathematical models for representing data, reasoning, learning and interacting are considered as well as new psychological and computational approaches from existing cognitive frameworks and algorithmic solutions. Any book edition will select new topics that report progresses towards the implementation of automaton human levels of intelligence, crucial for developing believable and trustable HCI systems that enhance the end user's quality of life. The content of the book is organized in sections and in each edition new topics are afforded and discussed on the basis of their contribution in the conception of new ICT functionalities and evaluation methods for modelling the concepts of learning, reasoning, interaction and data interpretation. Each edition of the book is related to a long running international conference, International Workshop on Neural Networks (WIRN, currently at the 25th edition), where researchers contributing to the book meet each year to propose new topics, advances and applications in the field of neural networks, machine learning and artificial intelligence.

After the conference, the topics of major interest are selected and researchers proposing these topics are invited to contribute to the book.

This second edition, different from the previous ones, is dedicated to computational intelligence for ICT (the subtitle of the current book), and emphasizes robotic applications, embedded systems, and ICT computational and theoretical methods for psychological and neurological diseases. The last theme is inspired by the European Call for Collaboration "Connect2Sea, Supporting European Union and Southeast Asia ICT strategic partnership and policy dialogue: Connecting ICT

EU-SEA Research, Development and Innovation Knowledge Network" which supported a special session of WIRN organized by the EU Partners and the SEA Partners.

The content of book is organized in sections. Each section is dedicated to a specific topic. The topics accounted for this edition are the following:

1. Introduction
2. Machine learning
3. Artificial neural networks: algorithms and models
4. Intelligent cyberphysical and embedded system
5. Computational intelligence methods for biomedical ICT in neurological diseases
6. Neural networks-based approaches to industrial processes
7. Reconfigurable, modular, adaptive, smart robotic systems for optoelectronics industry: the White'R instantiation

Given the themes afforded, we believe that this book is unique in proposing a holistic and multidisciplinary approach to implement autonomous and complex human–computer interfaces. We would like to thank the contributors and the reviewers.

May 2015   Simone Bassis
                                                                                  Anna Esposito
                                                                     Francesco Carlo Morabito
                                                                                   Eros Pasero

Organization

Executive Committee

Anna Esposito, Seconda Università di Napoli and IIASS, Italy
Marcos Faundez-Zanuy, Pompeu Fabra University (UPF), Spain
Francesco Carlo Morabito, Università Mediterranea di Reggio Calabria, Italy
Francesco A.N. Palmieri, Seconda Università di Napoli, Italy
Eros Pasero, Politecnico di Torino, Italy
Stefano Rovetta, Università di Genova, Italy
Stefano Squartini, Università Politecnica delle Marche, Italy
Aurelio Uncini, Università di Roma "La Sapienza", Italy
Salvatore Vitabile, Università di Palermo (IT)

International Advisory Committee SIREN

Metin Akay, Arizona State University, USA
Pierre Baldi, University of California, Irvine, USA
Piero P. Bonissone, Computing and Decision Sciences, USA
Leon O. Chua, University of California, Berkeley, USA
Jaime Gil-Lafuente, University of Barcelona, Spain
Giacomo Indiveri, Institute of Neuroinformatics, Zurich, Switzerland
Nikola Kasabov, Auckland University of Technology, New Zealand
Vera Kurkova, Academy of Sciences, Czech Republic
Shoji Makino, NTT Communication Science Laboratories, USA
Dominic Palmer-Brown, London Metropolitan University, UK
Witold Pedrycz, University of Alberta, Canada
Harold H. Szu, Army Night Vision Electronic Sensing Dir., USA
Jose Principe, University of Florida at Gainesville, USA
Alessandro Villa, Università Joseph Fourier, Grenoble 1, France

Fredric M. Ham, Florida Institute of Technology, USA
Cesare Alippi, Politecnico di Milano, Italy
Marios M. Polycarpou, Politecnico of Milano, Italy

Program Committee

Alda Troncone
Alessandro Formisano
Alessio Micheli
Amedeo Buonanno
Andrea Tettamanzi
Angelo Ciaramella
Anna Di Benedetto
Anna Esposito
Antonino Staiano
Aurelio Uncini
Bill Howell
Danilo Comminiello
Emanuele Principi
Enea Cippitelli
Eros Pasero
Fabio Antonacci
Fabio La Foresta
F. Carlo Morabito
Francesco Castaldo
Francesco Ferracuti
Francesco Masulli
Francesco Palmieri
Francesco Rugiano
Francesco Trovo
Gennaro Cordasco
Giuliano Armano
Isabella Palamara
Italo Zoppis
Luca Liparulo
Luca Mesin
Luigi Di Grazia
Manuel Roveri
Marco Fagiani
Marco Muselli
Marcos Faundez-Zanuy
Massimo Panella
Maurizio Campolo

Maurizio Fiaschè
Michele Scarpiniti
Nadia Mammone
Pasquale Coscia
Pietro Falco
Roberto Bonfigli
Roberto Serra
Silvia Scarpetta
Simone Bassis
Simone Scardapane
Stefano Rovetta
Stefano Squartini
Valentina Colla

Sponsoring Institutions

International Institute for Advanced Scientific Studies (IIASS) of Vietri S/M (Italy)
Department of Psychology, Second University of Napoli (Italy)
Provincia di Salerno (Italy)
Comune di Vietri sul Mare, Salerno (Italy)
International Neural Network Society (INNS)
University Mediterranea of Reggio Calabria (Italy)

Contents

Part I
Introductory Chapter

Some Notes on Computational and Theoretical Issues in Artificial Intelligence and Machine Learning

Anna Esposito, Simone Bassis, Francesco Carlo Morabito
and Eros Pasero

Abstract In the attempt to implement applications of public utility which simplify the user access to future, remote and nearby social services, new mathematical models and new psychological and computational approaches from existing cognitive frameworks and algorithmic solutions have been developed. The nature of these instruments is either deterministic, probabilistic, or both. Their use depends upon their contribute to the conception of new ICT functionalities and evaluation methods for modelling concepts of learning, reasoning, and data interpretation. This introductory chapter provide a brief overview on the theoretical and computational issues of such artificial intelligent methods and how they are applied to several research problems.

Keywords Neural network models · Probabilistic learning · Artificial intelligent methods

1 Introduction

The idea of developing suitable methodologies for automatic artificial intelligence is very stimulating both from an intellectual and a practical perspective. Intellectually, they hold, for scientists and developers alike, the relevant promise and

A. Esposito (✉)
Dipartimento di Psicologia and IIASS, Seconda Università di Napoli, Caserta, Italy
e-mail: iiass.annaesp@tin.it

S. Bassis
Università di Milano, Milan, Italy
e-mail: bassis@di.unimi.it

F.C. Morabito
Università degli Studi "Mediterranea" di Reggio Calabria, Reggio Calabria, Italy
e-mail: morabito@unirc.it

E. Pasero
Dip. Elettronica e Telecomunicazioni, Politecnico di Torino, Turin, Italy
e-mail: eros.pasero@polito.it

© Springer International Publishing Switzerland 2016
S. Bassis et al. (eds.), *Advances in Neural Networks*, Smart Innovation,
Systems and Technologies 54, DOI 10.1007/978-3-319-33747-0_1

challenge to improve our understanding of the human brain and allow to develop machines with an automaton level of intelligence. From a practical viewpoint, artificial intelligent methods are aimed to solve time and memory unbounded computational problems[1] allowing the development of complex applications that must help in improving the human productivity and their quality of life.

A huge amount of researches has been done in this direction over the last three decades. As early as the 1950s, simple automatic intelligent systems have been built, yielding credible intelligent performance. However, it was soon found that the algorithmic and computational solutions adopted in these systems were not easily adaptable to more sophisticated applications, because several problems emerged that significantly degraded the system's performance.

Researchers discovered that the performance of intelligent computational methods are dependent on the problem to be solved and related application's domains, and that, their modeling and implementation must account of cognitive and behavioral frameworks, as far as they are aimed to solve human daily life problems.

Being aware of this, automatic systems *have been developed with the aim to propose computational and mathematical models, such as neural networks and expert systems, which are able to infer and gain the required knowledge from a set of contextual examples or a set of rules describing the problem at the hand* [1], as well as learning structures, such as Hidden Markov Models, HMM [2] which are able to capture data probability distributions when trained on a statistically significant amount of data examples (training dataset).

2 Template, Feature, and Probabilistic-Based Approaches to Learning Problems

Under the label of "learning problems" can be grouped several diverse problems, such as pattern recognition, target detection, perception, reasoning and data interpretation. Several strategies have been proposed for implementing "learning" algorithms devoted to solve such problems, which can be mainly grouped in (a) template-based, (b) feature-based, and (c) probabilistic-based approaches.

The *template-based* approach is one of the most commonly used in this context. This approach has provided a family of techniques which have advanced the field and relative applications considerably. The underlying idea is quite simple. A collection of prototypical patterns (the templates) are stored as reference patterns, representing the set of "candidate" patterns. The learning is then carried out by matching an unknown pattern which each of these reference templates and selecting the one that best match the unknown input. This technique has been applied to

---

[1]*NP-complete* and *NP-hard* problems, where NP indicates that the problem has a *Non Polynomial* solution either in terms of computational time or of memory occupancy, or both.

different problems, in different fields, such as robotics for grasp selection [3], spoken language learning [4], detection and recognition of objects in images [5, 6], face recognition [7], geostatistical modeling [8], and continuous speech recognition [9] among many. In the last example, some templates representing complete words are constructed, avoiding segmentation or classification errors generated by smaller and acoustically variable speech units, such as phonemes. However, these techniques showed several limitations upon the pattern recognition problem to which they are applied. For example speech recognition remains a computationally intractable problem when using template-based techniques. This is because speech varies considerably in speaking rate and words are compressed or elongated according to the emphasis and intonation the speaker intends to give to her/his speech. Template matching systems must be able to dynamically stretch and compress time axes between test and reference patterns in order to work properly. Even though optimal strategies, called *Dynamic Programming* techniques [10] have been proposed to achieve such time alignment, template-based recognition schemes can have excellent performance only when implemented on small isolated word vocabularies [11]. One of the most serious problems to be faced with the template-based approach is to define distance measurements able to achieve a fine discrimination, and insensitive to irrelevant changes of the input patterns. For example, it has been shown that familiarity with a person's voice facilitate human speech recognition [12] and that familiar visual patterns are better identified than unfamiliar ones [13]. However, Euclidian distances are not able to incorporate such information and similar voices, as well as similar patterns, can result very distant when such metrics are applied. This is particularly true for speech, since a message's semantic meaning may not primarily be conveyed by the verbal information (the produced words) since nonverbal, paralinguistic and indexical (proper of the speaker) information may play the same semantic and pragmatic functions [14, 15].

To face such problems, currently template-based approaches (mostly in the speech domain)) are evolved in episodic, multiple trace, exemplar, example-based, or instance-based approaches [16–21].

The *feature-based* approaches are considered an alternative to template-based approaches. These techniques require to first identify a set of features from the data that must capture relevant signal information and then, exploit this knowledge for developing algorithms to extract such features and implementing classifiers that combine features to carry out a recognition decision. Feature-based approaches require therefore, two processing phases and are more complex and fragile than template-based intelligent methods, since they strongly depend on the goodness of the extracted features. Extracting features depend upon the nature of the problem at the hand, since features can be extracted from images, video and speech data, satellite and GPS data, among many. Appropriate feature extraction algorithms have been proposed in specific research domains. It is worth mentioning among many, some feature extraction algorithms that can be generally used in several application domains, such Principal Component Analysis (PCA), Linear Discriminant Analysis, Linear Predictive Coding (LPC), Multilinear Principal Component Analysis, Fast Fourier Transform (FFT) [22], Independent Component Analysis

[23], Empirical Mode Decomposition [24] as well as Mel Frequency Cepstral Coefficients (MFCC), Perceptual Critical Band Features (PCBF), Mel Bank Filtering (MELB), Perceptual Linear Predictive coefficients (PLP) [25, 26] and their first and second derivatives (Δ and $\Delta\Delta$), mostly adopted in the speech processing domain for speech recognition, speaker verification and others typical problems. In some cases, when it is still unclear which features are of importance for the problem at the hand (as in the case of emotion recognition, see [27], as an example), features are extracted by using some of the above-mentioned algorithms and on all of them, feature selection algorithms are applied (exploiting some criteria of relevance, see the Sequential Floating Forward Selection (SFFS) algorithm—an iterative method to find the best subset of features [28] to select those that best fit the classification problem at the hand [29].

After the extraction of suitable features, automatic recognition or classification is then implemented by exploiting several classifiers, which in literature are reported as Artificial Neural Networks, Deep Learning Architecture, k-Nearest Neighbor, Linear Discriminant Analysis, Support Vector Machine, Complex Networks, Bayesian classifier based on Gaussian Mixture Model, Hidden Markov Models, among many others [30–34]. In this context, Bayesian classifiers and Hidden Markov Models (HMM) are also seen as *probabilistic-based* approaches since they are used to characterize statistical properties of the data and use probability theory in dealing with doubtful and incomplete information. In pattern recognition problems, doubts and incompleteness can be produced by many different causes. For example, in speech recognition incomplete and doubtful information are produced by confusable sounds, contextual effects, variability in the acoustic realization of utterances from different speakers, and homophone words among many other causes. Truthfully, for HMM models the separation among template, feature, and probabilistic-based approaches does not apply, since they use feature based representations of their input data, and recent implementations fuse together both classifiers and HMM models into hybrid pattern matching systems [35, 36]. Conceptually, an HMM is a double embedded stochastic process, with an underlying stochastic process which is *hidden*, and can only be observed through another set of stochastic processes that produce the sequence of observations [2]. As a consequence, a HMM is characterized by a finite-state Markov chain and by a set of probability distributions at chain outputs. Transition parameters in the Markov chain model the time variabilities while the parameters in the output distributions model the spectral variabilities. HMM are popular because many algorithms and many variants of them have been implemented, in particular for speech applications. In particular HMM are considered the probabilistic approach par excellence in dealing with speech problems. This is because HMM had the excellent ability to improve their performance as a function of the size of the database on which are trained. This propriety and the fact that current computer technologies easily allow the storing, handling, as well as, training on large amount of data, is has favored their use, since it provides essentially "a brute force approach" to the curse of data dimensions (some algorithms are not able to run in polynomial on a large amount of data, see [22]. Nevertheless, there are limitation in their implementation. The main one is the assumption that speech sounds and their duration depend only on the current and previous state and that they are independent from all other states. For this reason, they are not able to handle long-span temporal dependencies, which are essential proprieties of some classification and recognition problems, as for speech. In addition speech is described as a sequences of states that simulate essentially the transition from one phoneme to another or one word to another. This assumption is far for being realistic, since phonemes and words are the verbal part of the spoken message and may not contain all of the information needed for it to be understood. For example, the nonverbal and speaker specific information are detached by this representation [21]. In spite of these limitations, these models worked extremely well for several pattern recognition problems such as DNA motif elucidation [37], ion channel kinetics problems [38], unsupervised segmentation of random discrete noisy data [39] and handwriting recognition [40] among many others.

3　Content of This Book

This book is putting emphasis on computational and artificial intelligent methods for learning and their relative applications in robotics, embedded systems, and ICT interfaces for psychological and neurological diseases. The content of book is organized in Parts. Each Part is dedicated to a specific topic, and includes peer reviewed chapters reporting current applications and/or research results in particular research topics. The book is a follow-up of the scientific workshop on Neural Networks (WIRN 2015) held in Vietri sul Mare, Italy, from the 20th to the 22nd of May 2015. The workshop, at its 27th edition become a traditional scientific event that brought together scientists from many countries, and several scientific disciplines. Each chapter is an extended version of the original contribute presented at the workshop, and together the reviewers' peer revisions it also benefits from the live discussion during the presentation. The chapters reported in the book constitute an original work are not published elsewhere.

Part 1 introduces the class of artificial intelligent methods and their relative applications through a short chapter proposed by Esposito and colleagues.

Part 2 is dedicated to *machine learning* methods and current applications of them in different scientific domains such as vision, medicine, graph theory, audio classification tasks, control theory. Machine learning is a research field in artificial intelligence aiming to develop mathematical models and subsequent algorithms learning from examples and able to detect structures in the data by picking up unseen regularities and similarities. The Part includes 10 short chapters respectively on machine learning techniques for Semi-Automatic Brain Lesion Segmentation in Gamma Knife Treatments (proposed by Rundo and colleagues), machine learning based data mining for Milky Way Filamentary Structures Reconstruction (proposed by Riccio and colleagues), 3-D Hand Pose Estimation from Kinect's Point Cloud Using Appearance Matching (proposed by Coscia and colleagues), Frequency-Domain Adaptive Filtering in Hypercomplex Systems (proposed by

Ortolani and colleagues), Bayesian Clustering on Images (proposed by Buonanno and colleagues), Selection of Negative Examples for Node Label Prediction (proposed by Frasca, and Malchiodi), Recent Advances on Distributed Unsupervised Learning (proposed by Rosato and colleagues), a Rule Based Recommender System i.e. an algorithm aimed at predicting user responses to options (proposed by Apolloni and colleagues), an extension of the Fuzzy Inference System (FIS) paradigm [41] to the case where the universe of discourse is hidden to the learning algorithm (proposed by Apolloni and colleagues), Model Complexity Control in Clustering (proposed by Rovetta).

Part 3 affords learning paradigm by artificial neural networks and related algorithms and models. Neural networks (NN) are a class of mathematical learning models inspired by the human brain biological structure. Even though NNs are also a subset of machine learning algorithms, it seems more appropriate deserve a full Part to them. This Part includes 7 short chapters where NNs are applied to most disparate classes of daily life problems and proved to be very successful with respect to traditional algorithms. In particular, application of NNs are reported for the Recognition of Specific Motion Symptoms of the Parkinson's Disease (by Lorenzi and colleagues), Functional Link Expansions for Audio Classification Tasks (by Scardapane and colleagues), a Comparison of Consensus Strategies for Distributed Learning (by Fierimonte and colleagues), the efficiency of the Systolic Hebb Agnostic Resonance Perceptron (SHARP) in implementing Sparse Distributed Storage and Sparse Distributed Code (proposed by Marchese), a class of Multilayer Perceptrons, linked with Lukasiewicz logic, able to describe non-linear phenomena (proposed by Di Nola and colleagues), Bayesian-Based Neural Networks for Solar Photovoltaic Power Forecasting (proposed by Ciaramella and colleagues) and a Pyramidal Deep Architectures for Person Re-Identification (proposed Iodice and colleagues).

Part 4 theme is on "Computational Intelligence Methods for Biomedical ICT in Neurological Diseases". The topic is rooted in the European Project for Collaboration named "Connect2Sea, Supporting European Union and Southeast Asia ICT strategic partnership and policy dialogue: Connecting ICT EU-SEA Research, Development and Innovation Knowledge Networks" which partially supported a special session of the WIRN 2015 organized by the EU Partners and the SEA Partners.

The Part includes 14 short chapters facing the problem to describe, model, and implements computational intelligence techniques for nearby and/or remote Information Communication Technology (ICT) interfaces providing services and supports in biomedicine, and particularly in psychological and neurological diseases. In this context the 14 chapters are dedicated respectively to a pilot study on the decoding of dynamic emotional expressions in Major Depressive Disorders (Esposito and colleagues), A Multimodal Approach for Dysphagia Analysis (López-de-Ipiña and colleagues), Modified Bee Firefly Algorithms for Mammographic Mass Classification (Mazwin and colleagues), Quantifying the Complexity of Epileptic EE (Mammone and colleagues), A Computerized Text Analysis on What Relatives and Caregivers of Children with Type 1 Diabetes Talk About

(Troncone and colleagues), An ICT Based Model for Wellness and Health Care (Mirarchi and colleagues), Implicit and Explicit Attitudes for Predicting Decision-Making (Pace and colleagues), Internet Dependence in High School Student (Sergi and colleagues), Universal Matched Filter Template versus Individualized Template for the Detection of Movement Intentions (Akmal and colleagues), Sparse fNIRS Feature Estimation via Unsupervised Learning for Mental Workload Classification (Thao and colleagues), Correlations Between Personality Disorders and Brain Frontal Functions (Sperandeo and colleagues), Processing Bio-Medical Data with Class-Dependent Feature Selection (Zhou and Wang and colleagues), LQR based Training of Adaptive Neuro-Fuzzy Controller (Rashid and colleagues), Structured versus Shuffled Connectivity in Cortical Dynamics (Apicella and colleagues).

Part 5 is dedicated to Neural Networks-based approaches to Industrial Processes and includes 5 short chapters respectively on Machine Learning Techniques for the Estimation of Particle Size Distribution in Industrial Plants (proposed by Rossetti and colleagues), an algorithm Combining Multiple Neural Networks to Predict Bronze Alloy Elemental Composition (proposed by D'Andrea and colleagues), Acceptability, Advantages and Perspectives in applying Neuro-Fuzzy Techniques to Industrial Problems (proposed by Colla and colleagues), the Importance of Variable Selection for Neural Networks-Based Classification in Industrial (proposed by Cateni and Colla), and a Nonlinear Time Series Analysis of Air Pollutants with Missing (proposed by Albano and colleagues).

The argument of Part 6 is on Intelligent Cyberphysical and Embedded Systems, a class of engineered systems constituted of collaborating computational elements closely interacting with the system's physical entities. Since this is a new research field [42], the 5 chapters included in the Part can be considered as prototyping works. The topics faced by the 5 chapter are respectively on Making Intelligent the Embedded Systems through Cognitive Outlier and Fault Detection (proposed by Roveri and Trovó), Moral Implications Of Human Disgust-Related Emotions Detected Using EEG Based BCI Devices (proposed by Cameli and colleagues), Learning Hardware Friendly Classifiers through Algorithmic Risk Minimization (proposed by Oneto and Anguita), A Hidden Markov Model-Based Approach to Grasping Hand Gestures Classification (proposed by Di Benedetto and colleagues), A Low Cost ECG Biometry System Based on an Ensemble of Support Vector Machine classifiers (proposed by Mesin and colleagues).

Part 7 is dedicated to results recently obtained by partners of the EU funded project Reconfigurable Modular Adaptive Smart Robotic Systems for Optoelectronics Industry: the White'R Instantiation (http://whiterproject.eu/project/). The Part includes 10 chapters respectively on A Regenerative Approach to Dynamically Design the Forward Kinematics for an Industrial Reconfigurable Anthropomorphic Robotic Arm (proposed by Avram and Valente), an ANN based Decision Support System Fostering Production Plan Optimization Through Preventive Maintenance Management (proposed by Cinus and colleagues), Gaussian Beam Optics Model for Multi-emitter Laser Diode Module Configuration Design (proposed by Yu and colleagues), Flexible Automated Island for (Dis)Assembly of a Variety of Silicon

Solar Cell Geometries for Specialist Solar Products (proposed by Dunnill and colleagues), Analysis of Production Scenario and KPI Calculation (proposed by Lai and Tusacciu), Standardization and Green Labelling Procedure Methodology (proposed by Lai and colleagues) A Novel Hybrid Fuzzy Multi-Objective Linear Programming Method (proposed by Fiasché and colleagues), A Production Scheduling Algorithm for Distributed Mini Factories (proposed by Seregni and colleagues), Multi-horizon, Multi-objective Training Planner (proposed by Pinzone and colleagues) Use of Laser Scanners in Machine Tools (proposed by Silvestri and colleagues).

4 Conclusion

The multidisciplinary themes discussed in this volume offer a unique opportunity to face the study of modeling concepts of automatic learning, reasoning, and data interpretation from several perspectives in order to satisfy several needs, particularly to implements many ICT applications. The reader of this book must cope with the topics' diversity, even connected under the goal to model and implement automatic autonomous learning systems. Diversity must be considered here as the key for succeeding in sharing different approaches and in suggesting new theoretical frameworks and new solutions for the same old problems.

References

1. Esposito, A.: The importance of data for training intelligent devices. In: Apolloni, B., Kurfess, C. (eds) From Synapses to Rules: Discovering Symbolic Knowledge from Neural Processed Data, pp. 229–250. Kluwer Academic press (2002)
2. Rabiner, L.R.: A tutorial on hidden markov models and selected applications in speech recognition. Proc. IEEE **77**(2), 257–286 (1989)
3. Herzog, A., Pastor, P., Kalakrishnan, M., et al.: Template-based learning of grasp selection. In: IEEE International Conference on Robotics and Automation (ICRA), pp. 2379– 2384, Saint Paul, MN (USA) (2012)
4. Gutkin, A., King, S.: Inductive string template-based learning of spoken language (2005). https://www.era.lib.ed.ac.uk/handle/1842/932
5. Brunelli, R.: Template Matching Techniques in Computer Vision: Theory and Practice. Wiley (2009). ISBN: 978-0-470-51706-2
6. Cacciola, M., Calcagno, S., Morabito, F.C., Versaci, M.: Computational intelligence aspects for defect classification in aeronautic composites by using ultrasonic pulse. IEEE Trans. Ultrason. Ferroelectr. Freq Control, **55**(4), 870–878 (2008)
7. Yuen, C.T., Rizon, M., San, W.S., Seong, T.C.: Facial features for template matching based face recognition. Am. J. Eng. Appl. Sci. **3**(1) 899–903 (2010)
8. Tahmasebi, P., Hezarkhani, A., Sahimi, M.: Multiple-point geostatistical modeling based on the cross-correlation functions. Comput. Geosci. **16**(3), 779–797 (2012)
9. De Wachter, M., Matton, M., et al.: Template-based continuous speech recognition. IEEE Trans. Audio Speech Lang. Process. **15**(4), 1377–1390 (2007)

10. Sniedovich, M.: Dynamic Programming: Foundations and principles. Taylor & Francis (2010)
11. Rabiner, L.R., Levinson, S.E.: Isolated and connected word recognition: theory and selected applications. IEEE Trans. Commun. **29**(5), 621–659 (1981)
12. Goldinger, S.D.: Words and voices: episodic traces in spoken word identification and recognition memory. J. Exp. Psychol. Learn. Memory Cogn. **33**, 1166–1183 (1996)
13. Jacoby, L., Hayman, C.: Specific visual transfer in word identification. J. Exp. Psychol. Learn. Memory Cogn. **13**, 456–463 (1987)
14. Esposito, A., Esposito, A.M.: On speech and gesture synchrony. In: Esposito, A., et al. (eds.) Analysis of Verbal and Nonverbal Communication and Enactment: The Processing Issue. LNCS vol. 6800, pp. 252–272. Springer, Heidelberg (2011). ISBN: 978-3-642-25774-2
15. Esposito, A., Marinaro, M.: What pauses can tell us about speech and gesture partnership. In: Esposito, A., et al. (eds) Fundamentals of Verbal and Nonverbal Communication and the Biometric Issue. NATO Publishing Series, Human and Societal Dynamics, vol. 18,pp. 45−57. IOS press, The Netherlands (2007)
16. Aradilla, G., Vepa, J., Bourlard, H.: Improving speech recognition using a data-driven approach. In: Proceedings of EUROSPEECH, Lisbon, pp. 3333−3336 (2005)
17. De Wachter, M., Demuynck, K., Wambacq, P., Van Compernolle, D.: A locally weighted distance measure for example based speech recognition. In Proceedings of ICASSP, pp. 181–184, Montreal, Canada (2004)
18. Hawkins.: Contribution of fine phonetic detail to speech understanding. In: Proceedings of 15th International Congress of Phonetic Sciences (ICPhS03), pp. 293–296, Barcelona, Spain (2003)
19. Maier, V., Moore, R.K.: An investigation into a simulation of episodic memory for automatic speech recognition. In: Proceedings of Interspeech-2005, pp. 1245–1248, Lisbon (2005)
20. Matton, M., De Wachter, M., Van Compernolle, D., Cools, R.: Maximum mutual information training of distance measures for template based speech recognition. In: Proceedings of International Conference on Speech and Computer, pp. 511−514, Patras, Greece (2005)
21. Strik, H.: How to handle pronunciation variation in ASR: by storing episodes in memory? In: Proceedings of ITRW on Speech Recognition and Intrinsic Variation (SRIV2006), pp. 33–38, Toulouse, France (2006)
22. Bishop, C.M.: Pattern Recognition and Machine Learning. Springer, Heidelberg (2006)
23. Morabito, F.C.: Independent component analysis and feature extraction for NDT data. Mater. Eval. **58**(1), 85–92 (2000)
24. Labate, D., La Foresta, F., Morabito, G., Palamara, I., Morabito, F.C.: On the use of empirical mode decomposition (EMD) for Alzheimer's disease diagnosis. In: Smart Innovation, Systems and Technologies, vol. 37, pp. 121−128 (2015)
25. Aversano, G., Esposito, A.: Automatic parameter estimation for a context-independent speech segmentation algorithm. In: Sojka, P., et al. (eds) Text Speech and Dialogue, LNAI 2448, 293–300. Springer, Heidelberg (2002)
26. Esposito, A., Aversano, G.: Text independent methods for speech segmentation. In: Chollet, G., et al. (eds) Nonlinear Speech Modeling and Applications. Lectures Notes in Computer Science, vol. 3445, pp. 261–290. Springer, Heidelberg (2005)
27. Atassi, H., Smékal, Z., Esposito, A.: Emotion recognition from spontaneous Slavic speech. In: Proceedings of 3rd IEEE Interernational Conference on Cognitive Infocommunications (CogInfoCom2012), pp. 389–394, Kosice, Slovakia (2012)
28. Pudil, P., Novovicová, J., Kittler, K.: Floating search methods in feature selection. Patt. Recogn. Lett. **15**(11), 1119–1125 (1994)
29. Atassi, H., Esposito, A: Speaker independent approach to the classification of emotional vocal expressions. In: Proceedings of IEEE Conference on Tools with Artificial Intelligence (ICTAI 2008), vol. 1, pp. 487–494, Dayton, OH, USA, 3−5 Nov 2008
30. Bengio, Y.: learning deep architectures for AI. In: Foundations and Trends in Machine Learning, vol. 2, no. 1, pp. 1–127 (2009)
31. Duda, R.O., Hart, P.E., Stork, D.G.: Pattern Classification, 2nd edn. Wiley, New York (2001)

32. Mammone, N., Labate, D., Lay-Ekuakille, A., Morabito, F.C.: Analysis of absence seizure generation using EEG spatial-temporal regularity measures, Int. J. Neural Syst. **22**(6), art. no. 1250024 (2012)
33. Morabito, F.C., Campolo, M., Labate, D., Morabito, G., Bonanno, L., Bramanti, A.: De Salvo, S., Bramanti, P.: A longitudinal EEG study of Alzheimer's disease progression based on a complex network approach. Int. J. Neural Syst. **25**(2), art.no. 1550005 (2015)
34. Schuller, B.: Deep learning our everyday emotions: a short overview. In: Bassis, et al. (eds) Advances in Neural Networks: Computational and Theoretical Issues. SIST Series, vol. 37, pp. 339–346. Springer, Heidelberg (2015)
35. Frankel, J., King, S.: Speech recognition using linear dynamic models. IEEE Trans. Audio Speech Lang. Process. **15**(1), 246–256 (2007)
36. Lee, C.H.: On automatic speech recognition at the dawn of the 21st century. IEICE Trans. Inf. Syst. **E86-D** (3) 377–396 (2003). Special Issue on Speech Information Processing
37. Wong, K.C., Chan, T.M., Peng, C., Li, Y., Zhang, Z.: DNA motif elucidation using belief propagation. Nucleic Acids Res. **41**(16), e153 (2013)
38. Nicolai, C., Sachs, F.: Solving ion channel kinetics with the Qub software. Biophys. Rev. Lett. **08**, 191–211 (2013)
39. Boudaren, M.Y., Monfrini, E., Pieczynski, W.: Unsupervised segmentation of random discrete data hidden with switching noise distributions. IEEE Signal Process. Lett. **19**(10), 619–622 (2012)
40. Fink, GA.: Markov Models for Pattern Recognition—From Theory to Applications. Advances in Computer Vision and Pattern Recognition, pp. 1–253. Springer (2014)
41. Jang, J.S.R.: ANFIS: adaptive-network-based fuzzy inference system. IEEE Trans. Syst. Man Cybern. **23**(3), 665–685 (1993)
42. Khaitan, S.K., McCalley, J.D.: Design techniques and applications of cyberphysical systems: a survey. IEEE Syst. J. **9**(2), 350–365 (2014)

Part II
Machine Learning

Semi-automatic Brain Lesion Segmentation in Gamma Knife Treatments Using an Unsupervised Fuzzy C-Means Clustering Technique

Leonardo Rundo, Carmelo Militello, Salvatore Vitabile,
Giorgio Russo, Pietro Pisciotta, Francesco Marletta,
Massimo Ippolito, Corrado D'Arrigo, Massimo Midiri
and Maria Carla Gilardi

Abstract MR Imaging is being increasingly used in radiation treatment planning as well as for staging and assessing tumor response. Leksell Gamma Knife® is a device for stereotactic neuro-radiosurgery to deal with inaccessible or insufficiently treated lesions with traditional surgery or radiotherapy. The target to be treated with radiation beams is currently contoured through slice-by-slice manual segmentation on MR images. This procedure is time consuming and operator-dependent. Segmentation result repeatability may be ensured only by using automatic/semi-automatic methods with the clinicians supporting the planning phase. In this paper a semi-automatic segmentation method, based on an unsupervised Fuzzy C-Means clustering technique, is proposed. The presented approach allows for the target segmentation and its volume calculation. Segmentation tests on 5 MRI series were performed, using both area-based and distance-based metrics. The following average values have been obtained: $DS = 95.10$, $JC = 90.82$, $TPF = 95.86$, $FNF = 2.18$, $MAD = 0.302$, $MAXD = 1.260$, $H = 1.636$.

Keywords Semi-automatic segmentation · Gamma knife treatments · Unsupervised FCM clustering · MR imaging · Brain lesions

L. Rundo · C. Militello (✉) · G. Russo · P. Pisciotta · M.C. Gilardi
Istituto di Bioimmagini e Fisiologia Molecolare - Consiglio Nazionale delle Ricerche (IBFM-CNR), Cefalù (PA), Italy
e-mail: carmelo.militello@ibfm.cnr.it

S. Vitabile · M. Midiri
Dipartimento di Biopatologia e Biotecnologie Mediche (DIBIMED),
Università degli Studi di Palermo, Palermo, Italy

F. Marletta · M. Ippolito · C. D'Arrigo
Azienda Ospedaliera Cannizzaro, Catania, Italy

© Springer International Publishing Switzerland 2016
S. Bassis et al. (eds.), *Advances in Neural Networks*, Smart Innovation,
Systems and Technologies 54, DOI 10.1007/978-3-319-33747-0_2

1   Introduction

Magnetic Resonance Imaging (MRI) of the brain can provide images of very high quality revealing detailed information especially concerning the extent of abnormalities, and thus has a great potential in radiotherapy planning of brain tumors [1]. Currently, MRI is considered to be superior to Computed Tomography (CT) in determining the extent of tumor infiltration [2]. Leksell Gamma Knife® (Elekta, Stockholm, Sweden) is a device for stereotactic neuro-radiosurgery to deal with inaccessible or insufficiently treated lesions with traditional surgery or radiotherapy [3, 4]. Gamma Knife (GK) surgery does not require the skull to be opened for treatment. The patient is treated in one session and normally can go home the same day. Radioactive beams (generated by cobalt-60 sources) are focused on the target through a metal helmet (Fig. 1a). Before treatment, a neurosurgeon places a stereotactic head frame on the patient, used to accurately locate the target areas and to prevent head movement during imaging and treatment. After treatment planning phases, the actual treatment begins and the head frame is attached to the Gamma Knife metal helmet (Fig. 1b).

Nowadays, the target volume (Region Of Interest, ROI) is identified with slice-by-slice manual segmentation on MR images only, acquired while the patient is already wearing the stereotactic head frame just before treatment. This procedure is time consuming, because many slices have to be manually segmented in a short time. Manual segmentation is also an operator-dependent procedure.

The repeatability of the tumor boundaries delineation may be ensured only by using automatic/semi-automatic methods, supporting the clinicians in the planning phase [5–7]. Consequently, this improves the assessment of the Gamma Knife treatment response. As illustrated in [2], semi-automated approaches provide more reproducible measurements compared to conventional manual tracing. This was proved by quantitatively comparing intra- and inter-operator reliability of the two methods.

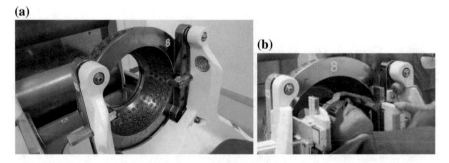

Fig. 1 Leksell Gamma Knife® equipment: **a** metal helmet with collimators; **b** metal helmet with patient positioned, while wearing the stereotactic head frame

In this paper a semi-automatic segmentation method, based on the unsupervised *Fuzzy C-Means (FCM)* clustering technique, is proposed. This approach helps segment the target and automatically calculate the lesion volume to be treated with the Gamma Knife® device. To evaluate the effectiveness of the proposed approach, initial segmentation tests, using both area-based and distance-based metrics, were performed on 5 MRI series.

This manuscript is organized as follows: Sect. 2 introduces an overview of brain lesion segmentation from MR images; Sect. 3 describes the proposed segmentation method; Sect. 4 illustrates the experimental results obtained in the segmentation tests; finally, some discussions and conclusions are provided in Sect. 5.

2 Background

In the current literature there are no works closely related to the segmentation of target volumes for Gamma Knife treatment support. We focused on papers dealing with target segmentation in radiotherapy.

The authors of [8] argue that it is better to use local image statistics instead of a global ones. A localized region-based *Active Contour* semi-automatic segmentation method (by means of *Level Set functions*) was applied on brain MR images of patients to obtain clinical target volumes. A similar method is reported in [9], in which *Hybrid Level Sets (HLS)* are exploited. The segmentation is simultaneously driven by region and boundary information.

In this context, Computational Intelligence techniques are often exploited. Mazzara et al. [10] present and evaluate two fully automated brain MRI tumor segmentation approaches: supervised *k-Nearest Neighbors (kNN)* and automatic *Knowledge-Guided (KG)*. Analogously, an approach based on *Support Vector Machine (SVM)* discriminative classification is proposed in [11]. For these supervised methods, however, a training step is always required. Lastly, Hall et al. [12] extensively compared literal and approximate *Fuzzy C-Means* unsupervised clustering algorithms, and a supervised *computational Neural Network*, in brain MRI segmentation. The comparison study suggested comparable experimental results between supervised and unsupervised learning.

3 The Proposed Segmentation Approach

In this section, a semi-automatic segmentation approach of the brain lesions from axial slices of brain MRI concerning Gamma Knife patients is presented. The proposed method combines existing literature methods in a smart and innovative fashion, resulting in an advanced clustering application for brain lesion segmentation. The user selects only a bounding area containing the tumor zone and no parameter setup is required (Fig. 2). Thereby, operator-dependence is minimized.

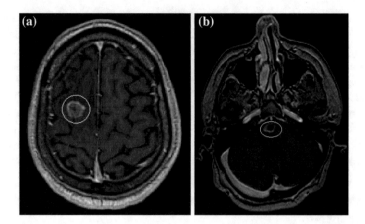

Fig. 2 Example of input MR axial slice with the bounding area (*white ellipsoid*), selected by the user, containing the tumor: **a** lesion without necrosis; **b** lesion with necrosis (*dark area*)

We used an ellipsoidal ROI selection tool, because it is widely provided by commercial medical imaging applications. This pseudocircular selected area is then enclosed and zero-padded into the minimum box in order to deal with a rectangular image during the following processing phases. The flow diagram of the overall segmentation method is shown in Fig. 3.

3.1 Pre-processing

Some pre-processing operations are applied on masked input images after the bounding region selection accomplished by the user, in order to improve segmentation results (Fig. 2a).

First of all, the range of intensity values of the selected part is expanded to extract the ROI more easily. Thereby, a contrast stretching operation is applied by means of a linear intensity transformation that converts the input intensity r values into the full dynamic range $s \in [0, 1]$.

A smoothing operation, using an average filter, is also required to remove the MRI acquisition noise (i.e. due to magnetic field non-uniformities). The application of the pre-processing steps on a cropped input image (Fig. 2a) is shown in Fig. 4b. The image contrast is visibly enhanced.

3.2 ROI Segmentation

Since the *FCM* algorithm classifies input data, the achieved clusters are represented by connected-regions and the corresponding boundaries are closed even when holes

Fig. 3 Flow diagram of the proposed brain lesion segmentation method

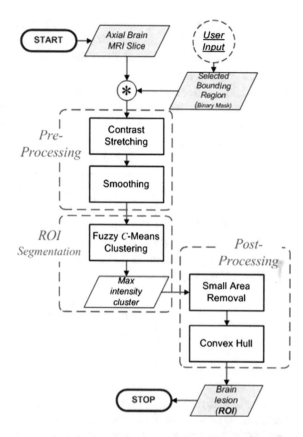

Fig. 4 Pre-Processing phase: **a** detail of the input MR image (part of the area inside the selected bounding region in Fig. 2a containing the tumor); **b** image after pre-processing steps

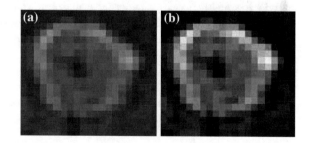

are present. This fact is very important in radiotherapy or radiosurgery treatment planning, because lesion segmentation is a critical step that must be performed accurately and effectively.

FCM cluster analysis is an iterative technique belonging to the realm of unsupervised machine learning [13, 14]. It classifies a dataset (i.e. a digital image) into groups (regions). Mathematically, the goal of this algorithm is to divide a set $X = \{\mathbf{x}_1, \mathbf{x}_2, \ldots, \mathbf{x}_N\}$ of N objects (statistical samples represented as vectors belonging to n-dimensional Euclidean space $\Re^n$.) into C clusters (partitions of the input dataset) [15]. A partition P is defined as a set family $P = \{Y_1, Y_2, \ldots, Y_C\}$.

The crisp version (*K-Means*) states that the clusters must be a proper subset X ($\emptyset \subset Y_i \subset X, \forall i$) and their set union must reconstruct the whole dataset ($\bigcup_{i=1}^{C} Y_i = X$). Moreover, the various clusters are mutually exclusive ($Y_i \cap Y_j = \emptyset, \forall i \neq j$), i.e. each feature vector may belong to only one group. Whereas, a fuzzy partition P is defined as a fuzzy set family $P = \{Y_1, Y_2, \ldots, Y_C\}$ such that each object can have a partial membership to multiple clusters. The matrix $\mathbf{U} = [u_{ik}] \in \Re^{C \times N}$ defines a fuzzy C-partition of the set X through C membership functions $u_i : X \to [0, 1]$, whose values $u_{ik} := u_i(x_k) \in [0, 1]$ are interpreted as membership grades of each element $\mathbf{x}_k$ to the i-th fuzzy set (cluster) Y_i and have to satisfy the constraints in (1). Briefly, the sets of all the fuzzy C-partitions of the input X are defined by: $M_{fuzzy}^{(C)} = \{\mathbf{U} \in \Re^{C \times N} : u_{ik} \in [0, 1]\}$. From an algorithmic perspective, FCM is an optimization problem where an objective function must be iteratively minimized using the least-squares method. The fuzziness of the classification process is defined by the weighting exponent m.

$$
\begin{cases}
0 \leq u_{ik} \leq 1 \\
\sum_{i=1}^{C} u_{ik} = 1, \quad \forall k \in [1, N] \\
0 < \sum_{k=1}^{N} u_{ik} < N, \quad \forall i \in [1, C]
\end{cases}
\tag{1}
$$

The *FCM* algorithm is applied just on pixels included in the selected ellipsoidal bounding region. Therefore, the zero-valued background pixels, added with zero-padding, are never considered. In particular, two clusters are employed ($C = 2$) in order to suitably classify a hyper-intense lesion from the healthy part of the brain. In fact, in contrast-enhanced MR brain scans, metastases have a brighter core than periphery and distinct borders distinguishing them from the surrounding normal brain tissue [16].

As the brain tumors are imaged as hyper-intense regions, during the defuzzification step, the pixels that have the maximum membership with the brightest cluster are selected. Figure 5a and 5b show two segmentation results achieved using the FCM clustering algorithm.

3.3   Post-processing

In some cases, central areas of the lesions could comprise necrotic tissue. For radiotherapeutic purposes, also these necrotic areas must be included in the target volume during the treatment planning phase.

A small area removal operation is used to delete any unwanted connected-components included in the highest intensity cluster. These small regions

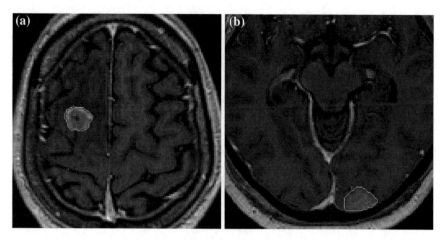

Fig. 5 Brain lesion segmentation using *FCM* clustering: **a** output of the MR image shown in Fig. 2a; **b** result obtained on another input MR image. (2× zoom factor)

may be due to anatomical ambiguities when the lesion is situated near high valued pixels (i.e. cranial bones or the corpus callosum).

Lastly, a *convex hull algorithm* [17] is performed to envelope the segmented lesion in the smallest convex polygon containing the automatically segmented ROI. In fact, brain lesions have a nearly spherical or pseudospherical appearance [16]; this also justifies the choice of an ellipsoidal ROI selection tool for the initial bounding region definition. Moreover, to include any necrosis inside the segmented ROI, this operation comprises hole filling too. An example is shown in Fig. 5b, where both the boundaries with and without the convex hull application are superimposed. Two examples are shown in Fig. 6a and 6b, where both the boundaries without and with the convex hull application are super-imposed.

4   First Experimental Results

To evaluate the accuracy of the proposed segmentation method, several measures were calculated by comparing the automatically segmented ROIs (achieved by the proposed approach) with the target volume manually contoured by a neurosurgeon (considered as our "gold standard"). Supervised evaluation is used to quantify the goodness of the segmentation outputs.

Area-based metrics and distance-based metrics are calculated for all slices of each MR image series, according to the definitions reported in Fenster & Chui [18] and Cárdenes et al. [19]. As evidenced in Figs. 5a and 5b, the proposed segmentation approach achieves correct results even with inhomogeneous input data.

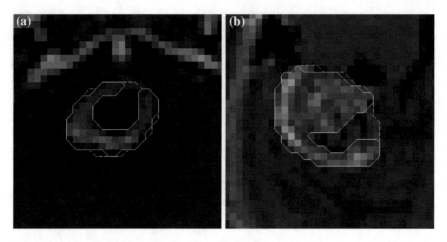

Fig. 6 Brain lesion segmentation without convex hull (*solid line*) and with convex hull application (*dashed line*): **a** output of the MR image shown in Fig. 2b; **b** a further example with necrotic tissue. (8× zoom factor)

4.1 Dataset Composition

The initial dataset is composed of 5 patients affected by brain tumors, who were treated with the Leksell Gamma Knife® stereotactic radio-surgical device. Each MRI series was acquired, just before treatment execution, to be used during the treatment planning phase.

MR images used in the segmentation trials were "T1w FFE" (T1weighted Fast Field Echo) sequences scanned with 1.5 Tesla MRI equipment (Philips Medical Systems, Eindhoven, the Netherlands). Brain lesions appear as hyper-intense zones with, in some cases, a dark area due to necrotic tissue.

4.2 Segmentation Evaluation Metrics

Area-based metrics quantify the similarity between the segmented regions through the proposed method (R_A) and our "gold standard" (R_T). Thus, the region containing "true positives" ($R_{TP} = R_A \cap R_T$) and "false negatives" ($R_{FN} = R_T - R_A$) are defined. In our experimental trials, we used the most important representative evaluation indices: Dice Similarity $DS = 2|R_{TP}|/(|R_A| + |R_T|) \times 100$, Jaccard index $JC = |R_A \cap R_T|/|R_A \cup R_T| \times 100$, Sensitivity aka True Positive Fraction $TPF = |R_{TP}|/|R_T| \times 100$, and Specificity aka False Negative Fraction $FNF = |R_{FN}|/|R_T| \times 100$.

In order to evaluate the contour discrepancy, the distance between the automatically generated boundaries (defined by the vertices $A = \{\mathbf{a}_i : i = 1, 2, \ldots, K\}$) and the manually traced boundaries ($T = \{\mathbf{t}_j : j = 1, 2, \ldots, N\}$) was also estimated.

Table 1 Values of area-based and distance-based metrics obtained on the segmented brain lesions for each patient

Dataset	$DS_\%$	$JC_\%$	$TPF_\%$	$FNF_\%$	MAD	$MAXD$	H
#1	96.29	93.00	96.70	1.87	0.193	0.955	1.471
#2	96.04	92.42	96.12	1.40	0.172	1.000	1.414
#3	94.77	90.15	95.15	2.37	0.290	1.069	1.833
#4	93.78	88.58	96.34	2.88	0.481	1.805	1.828
#5	94.61	89.94	94.97	2.40	0.376	1.472	1.637
AVG ± SD	95.10 ± 1.05	90.82 ± 1.84	95.86 ± 0.76	2.18 ± 0.57	0.302 ± 0.13	1.260 ± 0.37	1.636 ± 0.20

The following distances (expressed in pixels) were calculated: Mean Absolute Difference (*MAD*), Maximum Difference (*MAXD*), and Hausdorff distance (*H*).

In Table 1 the values of both area-based and distance-based metrics obtained in the experimental ROI segmentation tests are shown. High *DS* and *JC* ratios indicate excellent segmentation accuracy. In addition, the average value of *TPF* (sensitivity) proves that the proposed method correctly detects the "true" pathological areas. On the other hand, the very low *FNF* ratio (specificity) involves the ability of not detecting wrong parts within the automatically segmented ROI. Moreover, achieved distance-based indices are coherent with area-based metrics. These findings show the great reliability of the proposed approach even when MR images are affected by acquisition noise and artifacts. Good performance is guaranteed also with poor contrast images or lesions characterized by irregular shapes.

5   Conclusions and Future Works

In this paper an original application of the *FCM* clustering algorithm was presented. The unsupervised learning technique was used to support brain tumor segmentation for Gamma Knife treatments. The proposed semi-automatic segmentation approach was tested on a dataset made up of 5 MRI series, calculating both area-based and distance-based metrics. The achieved experimental results are very encouraging and they show the effectiveness of the proposed approach.

However, it is not always possible to identify the actual tumor extent using the only MRI modality. The combination of multimodal imaging might address this important issue [2]. Surgery, chemotherapy and radiation therapy may alter the acquired MRI data concerning the tumor bed. Therefore, further studies are planned to use a multimodal approach to the target volume selection, which integrates MRI with MET-PET (Methionine—Positron Emission Tomography) imaging [20]. In this way, MR images will highlight the morphology of the lesion volume (Gross Tumor Volume, *GTV*), while PET images will provide information on metabolically active areas (Biological Target Volume, *BTV*) [21].

The proposed approach for lesion segmentation from brain MRI, properly integrated with PET segmentation, would provide a comprehensive support for the precise identification of the region to be treated (Clinical Target Volume, *CTV*) during Gamma Knife treatment planning. The *CTV* will thus be outlined by the physician, through registration and fusion of segmented volumes on PET/MR images.

Acknowledgments This work was supported by "Smart Health 2.0" MIUR project (PON 04a2_C), approved by MIUR D.D. 626/Ric and 703/Ric.

References

1. Beavis, A.W., Gibbs, P., Dealey, R.A., Whitton, V.J.: Radiotherapy treatment planning of brain tumours using MRI alone. British J. Radiol. **71**(845), 544–548 (1998). doi:10.1259/bjr. 71.845.9691900
2. Joe, B.N., Fukui, M.B., Meltzer, C.C., Huang, Q.S., Day, R.S., Greer, P.J., Bozik, M.E.: Brain tumor volume measurement: comparison of manual and semiautomated methods. Radiology **212**(3), 811–816 (1999). doi:10.1148/radiology.212.3.r99se22811
3. Leksell, L.: Stereotact. Radiosurgery. J. Neurol. Neurosurg. Psychiatry **46**, 797–803 (1983). doi:10.1136/jnnp.46.9.797
4. Luxton, G., Petrovich, Z., Jozsef, G., Nedzi, L.A., Apuzzo, M.L.: Stereotactic radiosurgery: principles and comparison of treatment methods. Neurosurgery **32**(2), 241–259 (1993). doi:10. 1227/00006123-199302000-00014
5. Militello, C., Rundo, L., Gilardi, M.C.: Applications of imaging processing to MRgFUS treatment for fibroids: a review. Transl. Cancer Res. **3**(5), 472–482 (2014). doi:10.3978/j.issn. 2218-676X.2014.09.06
6. Salerno, S., Gagliardo, C., Vitabile, S., Militello, C., La Tona, G., Giuffrè, M., Lo Casto, A., Midiri, M.: Semi-automatic volumetric segmentation of the upper airways in patients with pierre robin sequence. Neuroradiol. J. (NRJ) **27**(4), 487–494 (2014). doi:10.15274/NRJ-2014-10067
7. Militello, C., Vitabile, S., Russo, G., Candiano, G., Gagliardo, C., Midiri, M., Gilardi, M.C.: A semi-automatic multi-seed region-growing approach for uterine fibroids segmentation in MRgFUS treatment. In: 7th International Conference on Complex, Intelligent, and Software Intensive Systems, CISIS 2013, art. no. 6603885, pp. 176–182
8. Aslian, H., Sadeghi, M., Mahdavi, S.R., Babapour Mofrad, F., Astarakee, M., Khaledi, N., Fadavi, P.: Magnetic resonance imaging-based target volume delineation in radiation therapy treatment planning for brain tumors using localized region-based active contour. Int. J. Radiat. Oncol. Biol. Phys. **87**(1), 195–201 (2013). doi:10.1016/j.ijrobp.2013.04.049. ISSN: 0360-3016
9. Xie, K., Yang, J., Zhang, Z.G., Zhu, Y.M.: Semi-automated brain tumor and edema segmentation using MRI. Eur. J. Radiol. **56**(1), 12–19. (2005). doi:10.1016/j.ejrad.2005.03. 028. ISSN: 0720-048X
10. Mazzara, G.P., Velthuizen, R.P., Pearlman, J.L., Greenberg, H.M., Wagner, H.: Brain tumor target volume determination for radiation treatment planning through automated MRI segmentation. Int. J. Radiat. Oncol. Biol. Phys. **59**(1), 300–312 (2004). doi:10.1016/j.ijrobp. 2004.01.026. ISSN: 0360-3016
11. Bauer, S., Nolte, L.P., Reyes, M.: Fully automatic segmentation of brain tumor images using support vector machine classification in combination with hierarchical conditional random field regularization. In: Medical Image Computing and Computer-Assisted Intervention—MICCAI 2011. Lecture Notes in Computer Science, vol. 6893, pp. 354–361 (2011). doi:10. 1007/978-3-642-23626-6_44
12. Hall, L.O., Bensaid, A.M., Clarke, L.P., Velthuizen, R.P., Silbiger, M.S., Bezdek, J.C.: A comparison of neural network and fuzzy clustering techniques in segmenting magnetic resonance images of the brain. IEEE Trans. Neural Netw. **3**(5), 672–682 (1992). doi:10.1109/ 72.159057
13. Militello, C., Vitabile, S., Rundo, L., Russo, G., Midiri, M., Gilardi, M.C.: A fully automatic 2D segmentation method for uterine fibroid in MRgFUS treatment evaluation. Computers in Biology and Medicine, **62**, 277–292 (2015). doi:10.1016/j.compbiomed.2015.04.030
14. Chuang, K.S., Tzeng, H.L., Chen, S., Wu, J., Chen, T.J.: Fuzzy c-means clustering with spatial information for image segmentation. Comput. Med. Imaging Graph. **30**(1), 9–15 (2006). doi:10.1016/j.compmedimag.2005.10.001. ISSN: 0895-6111
15. Pal, N.R., Bezdek, J.C.: On cluster validity for the fuzzy c-means model. IEEE Trans. Fuzzy Syst. **3**(3), 370–379 (1995). doi:10.1109/91.413225

16. Ambrosini, R.D., Wang, P., O'Dell, W.G.: Computer-aided detection of metastatic brain tumors using automated three-dimensional template matching. J. Magn. Reson. Imaging **31**(1), 85–93 (2010). doi:10.1002/jmri.22009

17. Zimmer, Y., Tepper, R., Akselrod, S.: An improved method to compute the convex hull of a shape in a binary image. Pattern Recogn. **30**(3), 397–402 (1997) doi:10.1016/S0031-3203(96) 00085-4. ISSN: 0031-3203

18. Fenster, A., Chiu, B.: Evaluation of segmentation algorithms for medical imaging. In: 27th Annual International Conference of the Engineering in Medicine and Biology Society, IEEE-EMBS 2005, pp. 7186–7189 (2005). doi:10.1109/IEMBS.2005.1616166

19. Cárdenes, R., de Luis-García, R., Bach-Cuadra, M: A multidimensional segmentation evaluation for medical image data. Comput. Methods Prog. Biomed. **96**(2), 108–124 (2009). doi:10.1016/j.cmpb.2009.04.009. ISSN: 0169-2607

20. Levivier, M., Wikler Jr., D., Massager, N., David, P., Devriendt, D., Lorenzoni, J., et al.: The integration of metabolic imaging in stereotactic procedures including radiosurgery: a review. J. Neurosurg. **97**, 542–550 (2002). doi:10.3171/jns.2002.97.supplement5.0542

21. Stefano, A., Vitabile, S., Russo, G., Ippolito, M., Sardina, D., Sabini, M.G., et al. A graph-based method for PET image segmentation in radiotherapy planning: a pilot study. In: Petrosino, A. (ed.) Image Analysis and Processing, vol. 8157, pp. 711–720. Springer, Berlin (2013). doi:10.1007/978-3-642-41184-7_72

Machine Learning Based Data Mining for Milky Way Filamentary Structures Reconstruction

Giuseppe Riccio, Stefano Cavuoti, Eugenio Schisano, Massimo Brescia, Amata Mercurio, Davide Elia, Milena Benedettini, Stefano Pezzuto, Sergio Molinari and Anna Maria Di Giorgio

Abstract We present an innovative method called FilExSeC (Filaments Extraction, Selection and Classification), a data mining tool developed to investigate the possibility to refine and optimize the shape reconstruction of filamentary structures detected with a consolidated method based on the flux derivative analysis, through the column-density maps computed from Herschel infrared Galactic Plane Survey (Hi-GAL) observations of the Galactic plane. The present methodology is based on a feature extraction module followed by a machine learning model (Random Forest) dedicated to select features and to classify the pixels of the input images. From tests on both simulations and real observations the method appears reliable and robust with respect to the variability of shape and distribution of filaments. In the cases of highly defined filament structures, the presented method is able to bridge the gaps among the detected fragments, thus improving their shape reconstruction. From a preliminary *a posteriori* analysis of derived filament physical parameters, the method appears potentially able to add a sufficient contribution to complete and refine the filament reconstruction.

Keywords Galaxy evolution · Machine learning · Random forest

1 Introduction

The formation of star clusters is one of the most important events strictly related with the internal evolution of galaxies. Massive stars are responsible for the global ionization of the Interstellar Medium (ISM). About half of the mass in the ISM of

G. Riccio (✉) · S. Cavuoti · M. Brescia (✉) · A. Mercurio
INAF, Astronomical Observatory of Capodimonte, Via Moiariello 16, 80131 Napoli, Italy
e-mail: riccio@na.astro.it

M. Brescia
e-mail: brescia@oacn.inaf.it

E. Schisano · D. Elia · M. Benedettini · S. Pezzuto · S. Molinari · A.M. Di Giorgio
INAF, Institute of Space Astrophysics and Planetology, Via Fosso del
Cavaliere 100, 00133 Roma, Italy

© Springer International Publishing Switzerland 2016
S. Bassis et al. (eds.), *Advances in Neural Networks*, Smart Innovation,
Systems and Technologies 54, DOI 10.1007/978-3-319-33747-0_3

a Galaxy is mainly derived by Molecular Clouds (MC) formed by stars. In such formation scenarios turbulent dynamics produce filamentary structures, where the total amount per unit area of suspended material measured along the length of a column (hereafter column density) agglomerates material from the Interstellar radiation field [7]. As a natural consequence of the cooling process and gravitational instability, the filaments are fragmented into chains of turbulent clumps which may engage the process of star formation. Therefore, the knowledge about the morphology of such filamentary structures is a crucial information to understand the star formation process.

The traditional method, which represents our starting point in the design and application of the presented methodology, carries out filaments detection by thresholding over the image of the multi-directional 2nd derivatives of the signal to identify the spine, the area and the underlying background of the filaments, automatically identifying nodal points and filament branches [9]. From the extracted regions of detected filaments it is possible to estimate several physical parameters of the filament such as the width, length, mean column density and mass per unit length.

In the present work we present a preliminary study of a data mining tool designed to improve the shape definition of filamentary structures extracted by the traditional method. This study is included into the EU FP7 project *ViaLactea*, aimed at exploiting the combination of new-generation Infrared-Radio surveys of the Galactic Plane from space missions and ground-based facilities, by making an extensive use of 3D visual analysis and data mining for the data and science analysis [6].

2   The Data Mining Approach

The presented data mining methodology, called FilExSeC (Filaments Extraction, Selection and Classification), is based on a feature extraction framework, able to map pixels onto a parameter space formed by a dynamical cross-correlation among pixels to produce a variety of textural, signal gradient and statistical features, followed by a Machine Learning (ML) supervised classification and backward feature selection, both based on the known Random Forest model [1]. As shown in Fig. 1, the method is intended as an additional tool inserted into a more complex filament detection pipeline, whose role is to complement traditional consolidated techniques to optimize the overall performance.

FilExSeC is characterized by three main processing steps: (i) encoding of pixel information into a data vector, formed by a set of derived features enclosing the information extracted from textural, flux gradient and statistical cross-correlation with neighbor pixels; (ii) assigning the codified pixels to one of two filament/background classes; and (iii) evaluating the importance of each feature in terms of its contribution to the classification task, with the purpose to detect and remove possible noisy features and to isolate the best subset of them capable to maintain the desired level of performance.

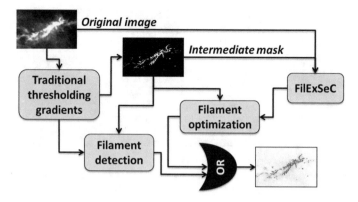

Fig. 1 Layout of the complete filamentary structure detection system

2.1 The Feature Extraction

The proposed method includes a Feature extraction algorithm, whose main target is to characterize pixels of the input image by means of Haar-like, Haralick and statistical features. Haar-like features, [5], are extracted by selecting windows W_I of fixed dimension, where the pixel under analysis lies at the bottom-right vertex of the window. From these areas the integral image $IntIm$ is extracted, by setting the value of each point (x, y) as the sum of all grey levels of pixels belonging to the rectangle having pixel (x, y) and the upper-left pixel as vertices:

$$\text{IntIm}(x, y) = \sum_{x' \le x, y' \le y} W_I(x', y') \tag{1}$$

Features are then calculated by comparing the content of the integral image among different regions, defined as *black* and *white* areas according to different templates specialized to search for different shapes in the image (Fig. 2):

$$f(x, y) = [\text{IntIm}(x, y)]_{black\ area} - [\text{IntIm}(x, y)]_{white\ area} \tag{2}$$

Haralick features, [3], concern the textural analysis that, by investigating the distribution of grey levels of the image, returns information about contrast, homogeneity and regularity. The algorithm is based on the Gray Level Co-occurrence Matrix (GLCM), that takes into account the mutual position of pixels with similar grey levels. The $C_{i,j}$ element of the GLCM, for a fixed direction and distance $\vec{d}$, represents the probability to have two pixels in the image at distance $\vec{d}$ and grey levels Z_i and Z_j respectively. Starting from the GLCM, it is possible to evaluate the most important Haralick features (Fig. 2), calculated within sub-regions with different dimensions and centered on the pixel under analysis, at different distances. Both dimensions and distances are parameters defined during setup.

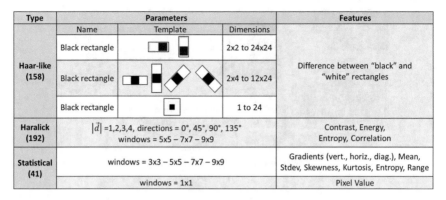

Type	Parameters			Features
	Name	Template	Dimensions	
Haar-like (158)	Black rectangle		2x2 to 24x24	Difference between "black" and "white" rectangles
	Black rectangle		2x4 to 12x24	
	Black rectangle		1 to 24	
Haralick (192)	$\|\vec{d}\|$ =1,2,3,4, directions = 0°, 45°, 90°, 135° windows = 5x5 – 7x7 – 9x9			Contrast, Energy, Entropy, Correlation
Statistical (41)	windows = 3x3 – 5x5 – 7x7 – 9x9			Gradients (vert., horiz., diag.), Mean, Stdev, Skewness, Kurtosis, Entropy, Range
	windows = 1x1			Pixel Value

Fig. 2 Summary of the image parameter space produced by the feature extraction

The statistical features are calculated in sub-regions of varying sizes, centered on the pixel under analysis (Fig. 2).

2.2   The Classification and Feature Selection

The feature extraction produces a list of features for all pixels, representing the input of the next steps.

The pixels belonging to filaments should have correlated values (or trends), although hidden by background noise. These values can be indeed used by a ML algorithm, hereinafter named as classifier, in order to learn how to discriminate the hidden correlation among filament pixels.

The classifier is based on the supervised ML paradigm. It has to be trained on a dataset (training set), where each pixel pattern has associated its known class label (for instance, filament or background). Then the trained model is tested on a blind dataset (test set), which again includes the known class for each pixel, in order to evaluate and validate the training and generalization performance. In the proposed method a Random Forest classifier has been used [1]. After the validation test it is possible to proceed to the next step, i.e. the Feature Selection. In general, the Feature Selection is a technique to reduce the initial data parameter space, by weighting the contribution (information entropy) carried by each feature to the learning capability of the classifier. By minimizing the input parameter space, it is hence possible to simplify the problem and at least to improve the execution efficiency of the model, without affecting its performance. In principle, it is reasonable to guess that some pixel features could be revealed as redundant parameters, by sharing same quantity of information or in some cases by hiding/confusing a signal contribution. Among the most known automated feature selection techniques we decided to use the Backward Elimination [4], a technique starting from the full parameter space available, i.e. initially including all the given features. Then these are evaluated at each iteration in terms of contribution importance, dropped one at a time, starting from the least significant, and the KB is re-fitted until the efficiency of the model begin to get worse. Such technique has revealed to be well suited for image segmentation in other scientific contexts, [11]. At the end of the feature selection phase, a residual subset of features with the higher weights are considered as the candidate parameter space. This subset is then used to definitely train and test the classifier. At the end of this long-time process the trained and refined classifier can be used on new real images.

3 Experiments

The presented method has been tested on the column density maps computed from Herschel [8] observations of the Galactic plane obtained by the Hi-GAL project, . which covers a wide strip of the inner Galactic Plane in five wavebands between 70 and 500 μm, (see [2] for further details). Here we report only one of the most illustrative examples due to problems of available space.

The data sample of the described test consists in a 2973×1001 pixels image of the column density map calculated from Hi-GAL maps in the Galactic longitude range from $l = 217$ to 224° [2]. In the reported example two different experiments are described. Their difference is related to the split of the four tiles to build up training and test sets. As shown in Fig. 3, we assigned two different couples of tiles as training and test sets respectively, by always considering the forth tile as part of training set and the first one as part of the blind test set, while tiles 2 and 3 have been allocated alternately to training and test sets. The reasons of this strategy were on one hand to verify the reliability and robustness of our method with respect to the high variability in the shape and amount of filaments as well as of contaminating background distribution within the region pixels. On the other hand, by taking into account the basic prescription for the machine learning supervised paradigm, particular care has been payed in the selection of the training and test regions, trying to balance the distribution of different levels of signal and background between the two data subsets. It is in fact known that the generalization performance of a classifier strongly depend on the level of homogeneity and coherency between training and test samples. According to this strategy, tiles 1 and 4 are the two samples with the presence of the most relevant filament structures, while the other inner tiles are quite similar in terms of background distribution and low presence of filaments. Therefore, the two presented cases achieve the best balancing between training and test data.

In order to build up the knowledge base required by the supervised ML paradigm, the known target labels associated to the training and test patterns have been derived from an intermediate result of the traditional method, which assigns a binary label to each pixel, by distinguishing between filament or background pixel. Such intermediate binary masks are mainly composed by the central pixels of the filamentary regions. They are obtained by thresholding the eigenvalues map computed by diagonalizing the Hessian matrix of the intensity map of the region [9]. These masks are still partially contaminated by non-filamentary structures (successively removed in

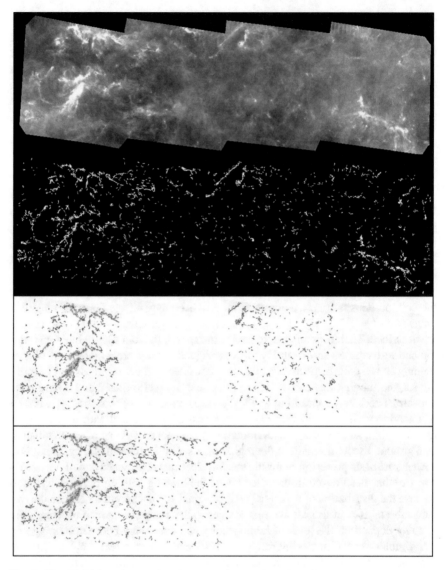

Fig. 3 The Hi-GAL region used in the reported example. It is a 2973 × 1001 pixels image of the column density map composed by 4 tiles. Under the original observed image (*upper side*) the intermediate binary mask is shown. The other two sub-panels in reverse *color* show the output masks of FilExSeC in the two experiments, respectively, by using tiles 1 and 3 as blind test and by replacing tile 3 with the tile 2 in the test set. Here colored pixels are applied on the intermediate binary mask as classified by our method, representing, respectively, NFP (*green*) CFP (*blue*) and UFP (*red*) pixels

the traditional approach by further filtering criteria), but the decision to use an intermediate product minimizes the bias introduced in our method by further filtering steps on the image samples.

The results of the pixel classification is represented in terms of the known confusion matrix [10], in which the pixels are grouped in four categories:

- CFP (Confirmed Filament Pixel), filament pixels correctly recognized;
- UFP (Undetected Filament Pixel), filament pixels wrongly classified;
- NFP (New Filament Pixel), background pixels wrongly classified;
- CBP (Confirmed Background Pixel), background pixels correctly recognized.

Some statistical indicators can be derived by the confusion matrix. For instance, we used the *purity* (precision), i.e. the ratio between the number of correctly classified and total pixels classified as filament, the *completeness* (recall), i.e. the ratio between the number of correctly classified and total pixels belonging to filaments and the *DICE* (F_1-Score), which is a weighted average of purity and completeness. Such statistics have been mainly employed to verify the degree of reliability and robustness of the method with respect to the variability within the real images.

In the case of the two experiments of the presented example the results were, respectively, $\sim 74\%$ of purity, $\sim 52\%$ of completeness and $\sim 61\%$ of DICE in the experiment with test tiles 1 and 3 and $\sim 73\%$ of purity, $\sim 50\%$ of completeness and $\sim 59\%$ of DICE in the experiment with test tiles 1 and 2, for the filament pixels. While the statistics in the case of background pixel class were all enclosed around 98% in both cases. These results appears quite similar in all the performed tests, with very small fluctuations, thus confirming what was expected, and by taking into account the extreme unbalance between the number of filament ($\sim 4\%$) and background ($\sim 96\%$) pixels. Furthermore, the FilExSeC method adds $\sim 16\%$ of new filament pixels (NFP) on average with respect to the intermediate image masks.

4 Discussion and Conclusions

The aim of the innovative method is at improving the shape reconstruction of filamentary structures in IR images, in particular in the outer regions, where the ratio between the signal and the background is lower that in the central regions of the filament. So that the robustness of the method is evaluated for this particular science case, rather than in terms of an absolute detection performance to be compared with other techniques. The most important evaluation here consisted in the verification of an effective capability to refine the shape of filaments already detected by the traditional method, as well as to improve their reconstruction, trying to bridge the occurrences of fragments. We remind in fact, that our method starts from an intermediate result of the traditional method, for instance the intermediate binary masks obtained from the filament spine extraction, slightly enlarged through a method discussed in [9].

The proposed method revealed good reconstruction capabilities in presence of larger filament structures, mostly evident when the classifier is trained by the worst image regions (i.e. the ones with a higher level of background noise mixed to filament signal). This although in such conditions the reconstruction of very thin and short filament structures becomes less efficient (Fig. 3). The global performances have been improved by optimizing Haar-like and statistical parameters as well as by introducing the pixel value as one of the features. Moreover, we have reached the best configuration of the Random Forest. Specific tests performed by varying the number of random trees have revealed the unchanged capability of classification with a relatively small set (1000 trees), thus reducing the complexity of the model.

Furthermore, all the results of performed tests showed how the performance of the method worsen by eliminating the statistical features and that the Haralick features gave a very low contribution to the global efficiency. This was also confirmed by the feature importance evaluation (Fig. 4), in which the textural features (in particular those of Haralick group) were always in the last positions. In this context, the importance of a feature is its relative rank (i.e. depth) used as a decision node of a tree to assess the relevance with respect to the predictability of the target variable. Features that are at the top of the tree induce a higher contribute to the final prediction decision for a larger fraction of the input patterns. The expected fraction of the patterns addressed by these features, can thus be used as an estimate of the relative importance of the features.

Therefore, the Random Forest feature importance analysis highlighted the irrelevance, in terms of information carriage, of Haralick type features, while confirming the predominance of statistical and Haar-Like types, independently from the peculiar

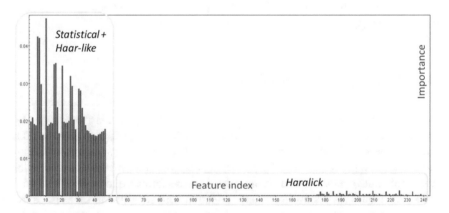

Fig. 4 Histogram of the feature importance as resulted from the feature selection with the Random Forest in the reported example. The highlighted regions show that within the 50 most important features fall those of Haar-Like and statistical types. The Haralick features, located in the second region, resulted with a very low importance. This induced us to exclude them from the parameter space since their contribution is quite negligible, achieving also a benefit in terms of the reduction of computing complexity

aspects of both training and test images. The resulting minimized parameter space, obtained by removing Haralick features, had also a positive impact on the computing time of the workflow, since the 90 % of the processing time was due to Haralick parameters extraction.

Moreover, in case of training and test performed on a same image, like the presented example, it has been concluded that, by enhancing the variability of filament and background distributions in the training set, the method improves the purity and the completeness of filament classification. This can be motivated by the big variety of filament structures and by the extreme variability of background spread over the regions of an image. In fact, understanding where a filamentary structure merges into the surrounding background is the main critical point to be addressed, since it could help to determine its region extent, as well as to realistically estimate the background. The latter is a fundamental condition to obtain a reliable determination of the filament properties [9].

In order to evaluate and validate the results from the physical point of view, it is necessary to estimate the contribution of extended filamentary regions to the calculation of the filament physical parameters on the same regions, through a cross analysis of such physical quantities with the results obtained by the traditional technique. In particular, it is important to evaluate how the physics of filaments is strictly related to its mass and the contribution of the NFP pixels in the calculation of this quantity. The variation of filament integrated column density is due to the contribution of the NFPs. In fact, it must be considered that the change of the distributions of filament/background mass contributions introduced by our method, causes an effect of a variation of the mass when NFPs are introduced and this is one aspect subject to a future further investigation. Furthermore, as also visible in Fig. 3, in many cases our method is able to connect, by means of NFPs, filaments originally tagged as disjointed objects. In such cases, by considering two filaments as a unique structure, both total mass and mass per length unit change, inducing a variation in the physical parameters of the filamentary structure. However, an ongoing further analysis is required to verify the correctness of the reconstruction of interconnections between different filaments, and to better quantify the contribution of FilExSeC to the overall knowledge of the physics of the filaments.

Acknowledgments This work was financed by the 7th European Framework Programme for Research Grant FP7-SPACE-2013-1, *ViaLactea - The Milky Way as a Star Formation Engine*.

References

1. Breiman, L.: Random Forests. Mach. Learn. **45**(1), 25–32 (2001)
2. Elia, D., Molinari, S., Fukui, Y., et al.: ApJ **772**, 45 (2013)
3. Haralick, R.M.: Proc. IEEE **67**(5), 786–804 (1979)
4. Kohavi, R., John, G.H.: Artif. Intell. **97**, 273–324 (1997)
5. Lienhart, R., Maydt, J.: An extended set of Haar-like features for rapid object detection. In: ICIP02, pp. 900–903 (2002)

6. Molinari, S., et al.: Protostars and Planets VI, pp. 125–148. University of Arizona Press, Tucson (2014)
7. Padoan, P., Nordlund, A.: ApJ **576**, 870 (2002)
8. Pilbratt, G., Riedinger, J.R., Passvogel, T., et al.: A&A **518**, L1 (2010)
9. Schisano, E., et al.: ApJ **791** 27 (2014)
10. Stehman, S.V.: Remote Sens. Environ. **62**(1), 77–89 (1997)
11. Tangaro, S., Amoroso, N., Brescia, M., et al.: Comput. Math. Methods Med. 814104 (2015)

3-D Hand Pose Estimation from Kinect's Point Cloud Using Appearance Matching

Pasquale Coscia, Francesco A.N. Palmieri, Francesco Castaldo
and Alberto Cavallo

Abstract We present a novel appearance-based approach for pose estimation of a human hand using the point clouds provided by the low-cost Microsoft Kinect sensor. Both the free-hand case, in which the hand is isolated from the surrounding environment, and the hand-object case, in which the different types of interactions are classified, have been considered. The pose estimation is obtained by applying a modified version of the Iterative Closest Point (ICP) algorithm to the synthetic models. The proposed framework uses a "pure" point cloud as provided by the Kinect sensor without any other information such as RGB values or normal vector components.

Keywords Hand pose · Kinect · Point cloud · RANSAC · DBSCAN · ICP

1 Introduction

Hand pose estimation, by means of one or more cameras, is a desired feature of many information systems in several applications, both in navigation and manipulation. For example in robotic applications, where robots are equipped with dexterous hands, it is important to provide feedback to the robot itself and/or to qualify how well it may be performing human-like grasps. In manipulators, efficient processing for analyzing hand motions may be an important part of a system specially when a device has to control complex machinery with many degrees of freedom (DoFs).

P. Coscia · F.A.N. Palmieri (✉) · F. Castaldo · A. Cavallo
Seconda Universitá di Napoli (SUN), Dipartimento di Ingegneria Industriale e
dell'Informazione, via Roma 29, 81030 Aversa (CE), Italy
e-mail: francesco.palmieri@unina2.it

P. Coscia
e-mail: pasq.coscia@gmail.com

F. Castaldo
e-mail: francesco.castaldo@unina2.it

A. Cavallo
e-mail: alberto.cavallo@unina2.it

© Springer International Publishing Switzerland 2016
S. Bassis et al. (eds.), *Advances in Neural Networks*, Smart Innovation,
Systems and Technologies 54, DOI 10.1007/978-3-319-33747-0_4

The possibility to use depth information with a relatively low-cost sensors has given new perspectives to the hand tracking [1, 2]. Among all, the Kinect sensor has become one of the most used depth sensor. The output of depth sensors is typically a point cloud, i.e. a data structure containing a set of multi-dimensional points expressed within a given coordinate system. In this work only the xyz-coordinates of the point clouds have been used. The point clouds have been stored at approximately 30 Hz and then processed.

The paper is organized as follows: Sect. 2 presents in detail each step of the proposed framework and their relative outputs. Section 3 shows experimental results for a number of real test sequences. The last section discusses the results and provides suggestions for improvements.

2 Our Approach

Figure 1 shows the block diagram of the proposed framework. The first step is background subtraction followed by clustering and matching steps in order to estimate the hand pose. Since the output of the matching step is fed back for the next frame, there is no real tracking. The starting point of the search, at each frame, is based on the results of the previous one. The pose is represented as a frame-dependent rotation matrix $\mathbf{R}[k]$ and a translation vector $\mathbf{t}[k]$ that update the position of the models' point clouds $\mathbf{M}_i$ in the space: $\mathbf{M}_i[k] = \mathbf{R}[k] \cdot \mathbf{M}_i[k-1] + \mathbf{t}[k]$.

The point clouds have been captured using the Simulink Support for Kinect [3] and all the algorithms have been written in MATLAB. Each point cloud, before the processing, contains approximately 200k 3-D points. Our case study consists of a scene in which the hand is located in front of the sensor and has a plane background. For grasping tasks, we have considered only one object at a time, situated on a table. The hand and the objects are placed at a fixed distance. Therefore there is no need to change the scale of the models.

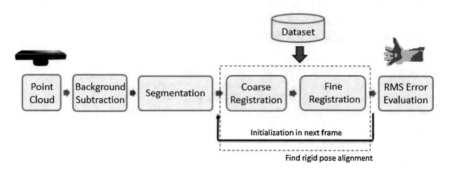

Fig. 1 Overview of the proposed framework for 3-D hand pose estimation

2.1 Background Subtraction

The first step in processing the data consists in removing the background, that in our case was roughly a plane (a wall). To perform the subtraction, we used the RANSAC algorithm. RANSAC stands for "Random Sample Consensus" and it is an iterative method for a robust parameters estimation of a mathematical model from a set of data points containing outliers. The algorithm is described in detail in [4]. In our case the model is represented by a plane, while outliers are the hand and the objects. In this work, we used the function written by Kovesi [5], that uses the RANSAC algorithm to robustly fit a plane to a set of 3-D data points.

Figure 2 shows the result of the RANSAC algorithm during a grasping task. The background is completely removed without affecting the remaining point cloud represented by the hand and the object.

Background subtraction eliminates much of the data points in the cloud, hence it can be managed more easily and quickly for the following steps because the number of 3-D points typically drops by an order of magnitude.

2.2 Segmentation

In order to associate each 3-D point to the corresponding cluster in the scene we have tested three different clustering algorithms: K-means, Euclidean Cluster Extraction and DBSCAN. K-means is a popular clustering algorithm that partitions data in a

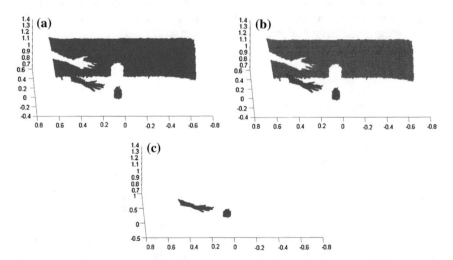

Fig. 2 Results of applying RANSAC plane fitting algorithm: **a** Original point cloud; **b** Points (in *red*) that belong to the plane model determined by RANSAC; **c** Point cloud after background subtraction

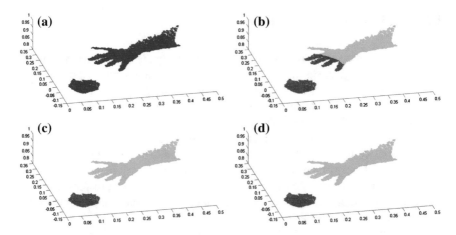

Fig. 3 **a** Original point cloud; Output of the segmentation step by: **b** K-means; **c** Euclidean cluster extraction; **d** DBSCAN

fixed a priori number of clusters. The Euclidean Cluster Extraction is a simple algorithm for clustering a 3-D point cloud using the Euclidean distance proposed by Rusu [6]. DBSCAN (Density based spatial clustering of application with noise) [7] is another technique based on density estimation for arbitrarily shaped clusters in spatial databases.

Figure 3 shows the results of the segmentation step on a point cloud during a grasping task. More generally, if two clusters are close enough, K-means is not able to divide the objects correctly, whereas the DBSCAN and the Euclidean Cluster Extraction algorithms perform much better. In the following steps we have used the results of the DBSCAN algorithm that in our experiments has appeared to show the best performance.

2.3 Synthetic Hand Models Generation

The point clouds prototypes are obtained from the BlenSor software [8], which allows the creation of a point cloud from a 3-D model. A 3-D hand model with 26 DoFs has been built by assembling geometric primitives such as cylinders, spheres and ellipsoids. The shape parameters of each object are set by taking measurements from a real hand. In order to remain within feasible movements, hand and finger motion uses *static* constraints that typically limit the motion of each finger [9], see Fig. 4. In Fig. 5 some of the 3-D models are shown with their associated point clouds that have been used in building the prototypes for our dataset. In particular, we have considered three typical gestures of the hand (palm, fist and gun) and two "interaction" models.

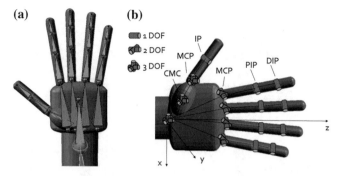

Fig. 4 **a** 3-D hand model and corresponding skeleton that controls its deformations (constrained bones are shown in *green*); **b** DOFs of the hand joints

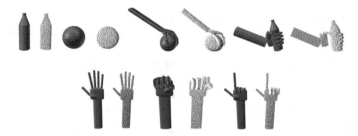

Fig. 5 The 3-D models and their corresponding point clouds generated for our dataset

2.4 *Matching*

In order to perform the matching between the models and the data acquired by the Kinect sensor, the ICP algorithm [10] has been used. The algorithm at each iteration step, selects the nearest points of a cloud (the model, in our case) with respect to the other one (the input data) and computes the transformation, represented by a rotation matrix $\mathbf{R}$ and a translation vector $\mathbf{t}$, that minimize the following equation:

$$E(\mathbf{R}, \mathbf{t}) = \sum_{i=1}^{N_p} \sum_{j=1}^{N_m} w_{i,j} \|\mathbf{p}_i - (\mathbf{R}\mathbf{m}_j + \mathbf{t})\|^2, \tag{1}$$

where $\mathbf{M} = \{m_1, \ldots, m_{N_m}\}$ is the model's point cloud, $\mathbf{P} = \{p_1, \ldots, p_{N_p}\}$ is the input point cloud after the segmentation step and $w_{i,j} \in \{0, 1\}$ represents the point correspondences. A closed form solution for this rigid body transformation can be calculated using a method based on the singular value decomposition (SVD) [11].

Due to many problems, such as local minima, noise and partial overlap, several methods have been proposed to increase the robustness of the ICP algorithm [12, 13].

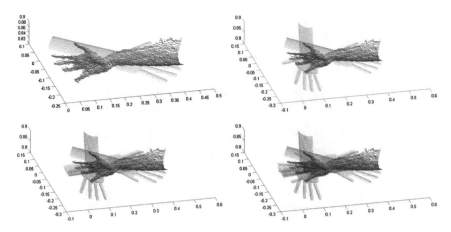

Fig. 6 First four perturbed models obtained by applying the "stochastic" version of the ICP algorithm. Among all perturbed models, the closest one to the data captured by the sensor is chosen

Our approach consists in adding a normally distributed noise and a random rotation matrix to perturb the position of the model's point clouds in order to obtain a good alignment at the first frame (a similar approach was proposed in [14]). More specifically: $\mathbf{m}'_j = \mathbf{R}_{rand}\mathbf{m}_j + \mathbf{s}$, $j = 1, \ldots, N_m$, where $\mathbf{s}$ has zero mean and diagonal covariance matrix $\sigma^2 I$ and $\mathbf{R}_{rand}$ is a random rotation matrix. The parameter σ has been set to 0.1 for the experiments and its value has been determined experimentally. We have used the algorithm proposed by Arvo [15] to generate a random rotation matrix. The "stochastic" version of the ICP algorithm operates as follows: the model $\mathbf{M}$ is perturbed as discussed before, and then the standard version of the ICP algorithm on such perturbed model with respect to the input data $\mathbf{P}$ is applied. This operation is repeated a fixed number of times. Among all perturbed models, the one closest to the data captured by the sensor is chosen. Finally, the standard ICP algorithm is repeated on such model. In Fig. 6 the first four perturbed models obtained by applying the stochastic ICP algorithm are shown.

3   Experimental Results

In order to evaluate the accuracy of the system, we have used the RMS error obtained by considering the Euclidean distance between each point of the selected model and the input data after the clustering step. In the next two sections, the experimental results of our approach are presented. In particular, we have tested the framework in two cases: performing static gestures and grasping an object located on a table. Figure 7 shows the qualitative results in both cases and the corresponding RMS errors.

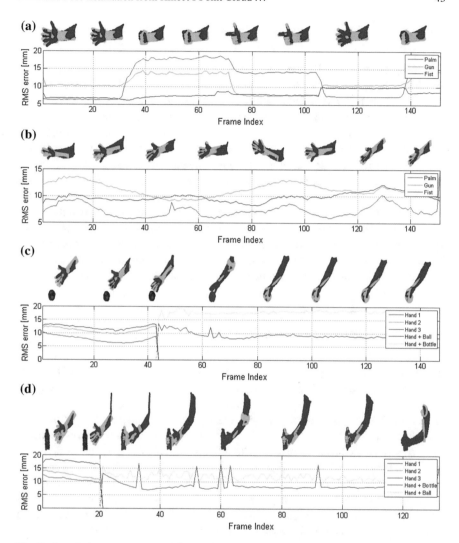

Fig. 7 Sample image sequences of the matching step and the corresponding RMS error: **a** Hand performs static gestures; **b** Hand performs rotations showing the palm; **c** Ball grasping task; **d** Bottle grasping task

Static gestures: The hand is positioned approximately at a distance of 1.8 m from the Kinect sensor. In this case, the system is able to recognize the three models in the dataset. The RMS error is less than 10 mm and it remains constant during all the simulation since the gestures are distinguishable and correspond exactly to the synthetic models. When the hand moves by rotating, the error fluctuates more even if the correct pose is still recognized. In the last part of simulation, the error is greater and the wrong model is selected mainly because of a high rotation of the hand that no longer matches the models in the dataset. The visible part, which is captured by the Kinect, is effectively only the profile of the hand and a greater rotation of the model, even if it is correct, results in an increase of the RMS error.

Grasping objects: Finally, we have tested our system when the hand interacts with an object. In particular, we have considered the grasping of objects easily obtainable in 3-D such as a ball and a bottle. For this experiments, we have replaced the three gestures (palm, gun, fist) with three intermediate poses representing the hand that comes close to the objects. The approach works quite well in the initial phase, recognizing the correct pose of the hand and the objects. As soon as the clustering algorithm returns a single cluster in the scene, only the interaction models are considered. As expected, the approach shows obvious problems in recognizing the correct model, when the clusters begin to join, because they are not completely merged. For the grasping ball task, the RMS error remains less than 10 mm, as in the above experiments. However, for some frames, the correct model is chosen but the estimated pose is wrong, primarily because of a lack of a real tracking. For the grasping bottle task some RMS error peaks are visible due to the presence of the entire arm that increases the surface area that needs to be considered for matching.

4 Conclusions and Future Work

In this work, we have presented a hand pose estimation system which is able to recognize a limited number of hand poses and interaction scenarios. The experimental results are very promising and demonstrate the effectiveness of the proposed approach. The system is able to provide a reasonable 3-D hand pose estimation using predefined models, as confirmed by the RMS error, also considering more complex scenarios in which the hand interacts with different objects. In our future work, possible improvements will be investigated. First of all, the system performance could be improved using parallelized code on GPU or by means of the use of the Point Cloud Library (PCL) in order to obtain a real-time system. Moreover, a larger dataset of typical gestures will be designed. Furthermore, an increase in accuracy could be achieved by implementing a tracking module to support the pose estimation.

References

1. Oikonomidis, I., Kyriazis, N., Argyros, A.: Efficient model-based 3D tracking of hand articulations using Kinect. In: Proceedings. 22nd British Machine Vision Conference, pp. 101.1–101.11 (2011)
2. Melax, S., Keselman, L., Orsten, S.: Dynamics based 3D skeletal hand tracking. In: Proceedings. Graphics Interface Conference, pp. 63–70 (2013)
3. Chikamasa, T.: Simulink support for kinect. http://uk.mathworks.com/matlabcentral/fileexchange/32318-simulink-support-for-kinect
4. Fischler, M.A., Bolles, R.C.: Random sample consensus: a paradigm for model fitting with applications to image analysis and automated cartography. Commun. ACM **24**, 381–395 (1981)

5. Kovesi, P. D.: MATLAB and octave functions for computer vision and image processing. http://www.csse.uwa.edu.au/~pk/research/matlabfns
6. Rusu, R. B.: Semantic 3D Object maps for everyday manipulation in human living environments. Ph.D thesis (2009)
7. Ester, M., Kriegel, H.P., Sander, J., Xu, X.: A density-based algorithm for discovering clusters in large spatial databases with noise. AAAI Press (1996)
8. Gschwandtner, M., Kwitt, R., Uhl, A., Pree, W.: BlenSor: blender sensor simulation toolbox. In: Proceedings of the 7th International Conference on Advances in Visual Computing, pp. 199–208 (2011)
9. Lin, J., Wu, Y., Huang, T.S.: Modeling the constraints of human hand motion. In: Proceedings. Workshop on Human Motion, pp. 121–126 (2000)
10. Besl, P.J., McKay, N.D.: A method for registration of 3-D shapes. IEEE Trans. Pattern Anal. Mach. Intell. **14**(2), 239–256 (1992)
11. Arun, K.S., Huang, T.S., Blostein, S.D.: Least square fitting of two 3-D point sets. IEEE Trans. Pattern Anal. Mach. Intell. **9**(5), 698–700 (1987)
12. Langis, C., Greenspan, M., Godin, G.: The parallel iterative closest point algorithm. In: Proceedings of Third International Conference on 3-D Digital Imaging and Modeling, pp. 195–202 (2001)
13. Li, C., Xue, J., Du, S., Zheng, N.: A fast multi-resolution iterative closest point algorithm. In: Chinese Conference on Pattern Recognition (CCPR), pp. 1–5 (2010)
14. Penney, G.P., Edwards, P.J., King, A.P., Blackall, J.M., Batchelor, P.G., Hawkes, D.J.: A stochastic iterative closest point algorithm (stochastICP). In: 18th IEEE International Conference on Image Processing (ICIP) 2208, pp. 762–769 (2001)
15. Arvo, J.: Fast random rotation matrices. In: Graphics Gems III, pp. 117–120. Academic Press (1992)

Frequency-Domain Adaptive Filtering in Hypercomplex Systems

Francesca Ortolani, Danilo Comminiello, Michele Scarpiniti
and Aurelio Uncini

Abstract In recent years, linear and nonlinear signal processing applications required the development of new multidimensional algorithms. Higher-dimensional algorithms include quaternion-valued filters. One of the drawbacks filter designers have to cope with is the increasing computational cost due to multidimensional processing. A strategy to reduce the computational complexity of long adaptive filters is to implement block algorithms and update the filter coefficients periodically. More efficient techniques embed frequency-domain processing in block algorithms with the use of the Fast Fourier Transform (FFT). Transform-domain adaptive filters in the quaternion field require quaternion-valued transforms. In this paper we also suggest a simple method to obtain a quaternionic DFT/FFT from a complex DFT/FFT. As an example, we propose the Overlap-Save Quaternion Frequency Domain algorithm.

Keywords Adaptive filters · Quaternion · Hypercomplex · Frequency domain · Overlap-Save

1 Introduction

Since the origin of science, men searched for ways to describe the surrounding world. Although maths is tough, it allows an unambiguous representation of nature and its laws. When scientists are called to face the complexity of the problems, the exist-

F. Ortolani · D. Comminiello (✉) · M. Scarpiniti · A. Uncini
Department of Information Engineering Electronics and Telecommunications (DIET),
"La Sapienza" University of Rome, Via Eudossiana 18, 00184 Rome, Italy
e-mail: danilo.comminiello@uniroma1.it

F. Ortolani
e-mail: francesca.ortolani@uniroma1.it

M. Scarpiniti
e-mail: michele.scarpiniti@uniroma1.it

A. Uncini
e-mail: aurel@ieee.org

© Springer International Publishing Switzerland 2016
S. Bassis et al. (eds.), *Advances in Neural Networks*, Smart Innovation,
Systems and Technologies 54, DOI 10.1007/978-3-319-33747-0_5

ing models need to be overcome and the introduction of new numerical systems is peremptory. The most primitive way to quantify objects uses natural numbers, i.e. integer and unsigned numbers. Today we can take advantage of *Hypercomplex* algebras, whose elements consist of multidimensional numbers.

The reasons for increasing the number of dimensions of a numerical system arise from several needs. We may require a number system capable of describing a variety of geometrical objects (points, lines, planes, hyperplanes, spheres, hyperspheres). It would be impossible to draw a hyperplane by means of real numbers, whereas hypercomplex numbers allow describing this kind of object. We may also require to store different information into one single entity. For example, complex numbers carry both the information of amplitude and phase in one number:

$$z = x + \mathbf{i}y \leftrightarrow z = re^{\mathbf{i}\phi} \tag{1}$$

where $r = |z| = \sqrt{x^2 + y^2}$ is the *modulus* of z, representing its amplitude, and $\phi = \arg(z)$ is its *phase*. These needs for working with a higher-dimensional number system are interconnected from a philosophical point of view. We could say, they are geometrical issues. Therefore, a proper numerical infrastructure is necessary for decomposing a physical problem into its own dimensions and spotlighting all its faces, which might remain unseen otherwise. De facto, hypercomplex calculus is also known as *geometric* calculus. Moreover, as written in [12] "Once geometry, in particular differential geometry, is discussed, the presence of physics is unavoidable".

Currently, our activity within the hypercomplex number systems is restricted to *quaternions*, i.e. a four-dimensional hypercomplex subgroup. Specifically, quaternions are a geometric algebra in the sense of Clifford Algebras [4]. At the beginning, quaternions found application in electromagnetism, quantum physics and relativity. Thanks to their compact notation, it is possible to express equations in a more accessible form.

Recent applications, however, include aero-navigation, kinematics, 3D graphics, image processing in general (e.g. color template matching), digital forensics (e.g. DNA sequence matching), weather forecasting and 3D audio processing.

Our effort was to extend frequency-domain adaptive filtering to quaternions. Prior to our work, several quaternion-valued algorithms in the time domain [6, 7, 9, 14] and one algorithm in the frequency domain [8] were proposed. Pioneer in this field was the development of the Quaternionic Least Mean Square algorithm (QLMS) by Mandic et al. [14]. The class of algorithms presented in this paper operates weight adaptation in the frequency domain. There are two main reasons for this choice: fast computation and signal orthogonalization. A low computational cost can be achieved efficiently with the use of the Fast Fourier Transform (FFT), which allows a fast performance of convolutions and crosscorrelations. In addition, the orthogonality properties of the Discrete Fourier Transform (DFT) produce a diagonal input autocorrelation matrix as the transformation length tends to infinity. This makes it possible to get the most out of *power normalization*, a compensation technique intended to have all the algorithm modes converging at the same rate. So, when power normalization is used, a more uniform convergence rate can be obtained easily thanks to the DFT.

Besides the linear filtering algorithms cited above, further algorithm implementations can be found in typical nonlinear environments. Studies concerning the use of quaternion neural networks, amongst many examples, include the definition and application to engineering problems of a quaternionic Multilayer Perceptron (QMLP) model [1, 10, 11] and the development of quaternionic Hopfield-type networks [5, 16]. For what concerns our contribution to the area of neural networks, our idea is to embed our quaternionic algorithms in the conventional learning schemes for neural networks.

2 Introduction to Quaternion Algebra

2.1 Quaternion Math

Quaternions are a hypercomplex normed division algebra consisting of 4 components: one scalar (real part) and 3 imaginary components (vector part):

$$q = q_0 + q_1 \mathbf{i} + q_2 \mathbf{j} + q_3 \mathbf{k} = \text{Re}(q) + \text{Im}(q) \ . \tag{2}$$

A quaternion having zero scalar part is also known as *pure quaternion*. The imaginary units, $\mathbf{i} = (1,0,0)$, $\mathbf{j} = (0,1,0)$, $\mathbf{k} = (0,0,1)$, represent an orthonormal basis in $\mathbb{R}^3$.

The cross product of the versors of the basis gives as results:

$$\mathbf{ij} = \mathbf{i} \times \mathbf{j} = \mathbf{k} \quad \mathbf{jk} = \mathbf{j} \times \mathbf{k} = \mathbf{i} \quad \mathbf{ki} = \mathbf{k} \times \mathbf{i} = \mathbf{j} \tag{3}$$

$$\mathbf{i}^2 = \mathbf{j}^2 = \mathbf{k}^2 = -1 \tag{4}$$

which are the fundamental properties of quaternions.

Quaternion product is non-commutative, i.e. $\mathbf{ij} \neq \mathbf{ji}$, in fact

$$\mathbf{ij} = -\mathbf{ji} \quad \mathbf{jk} = -\mathbf{kj} \quad \mathbf{ki} = -\mathbf{ik} \ . \tag{5}$$

The sum of quaternions is obtained by component-wise addition:

$$\begin{aligned} q \pm p &= \left(q_0 + q_1 \mathbf{i} + q_2 \mathbf{j} + q_3 \mathbf{k}\right) \pm \left(p_0 + p_1 \mathbf{i} + p_2 \mathbf{j} + p_3 \mathbf{k}\right) \\ &= \left(q_0 \pm p_0\right) + \left(q_1 \pm p_1\right) \mathbf{i} + \left(q_2 \pm p_2\right) \mathbf{j} + \left(q_3 \pm p_3\right) \mathbf{k} \ . \end{aligned} \tag{6}$$

The product between quaternions q_1 and q_2 is computed as

$$
\begin{aligned}
q_1 q_2 &= \left(a_0 + a_1\mathbf{i} + a_2\mathbf{j} + a_3\mathbf{k}\right)\left(b_0 + b_1\mathbf{i} + b_2\mathbf{j} + b_3\mathbf{k}\right) \\
&= \left(a_0 b_0 - a_1 b_1 - a_2 b_2 - a_3 b_3\right) + \left(a_0 b_1 + a_1 b_0 + a_2 b_3 - a_3 b_2\right)\mathbf{i} \quad (7) \\
&\quad + \left(a_0 b_2 - a_1 b_3 + a_2 b_0 + a_3 b_1\right)\mathbf{j} + \left(a_0 b_3 + a_1 b_2 - a_2 b_1 + a_3 b_0\right)\mathbf{k} \ .
\end{aligned}
$$

The conjugate of a quaternion is defined as

$$
q^* = w - x\mathbf{i} - y\mathbf{j} - z\mathbf{k} \ . \tag{8}
$$

2.2 Quaternion Discrete Fourier Transform

Frequency domain algorithms require mathematical transformations in order to (pre)process input and output signals. There exists an immediate method that exploits complex DFT/FFT to build up a quaternionic transform (QDFT/QFFT). This method was formerly developed in image processing and presented in [2, 13] with the idea of collecting colour (RGB) or luminance/chrominance signals in one quaternion, rather than treating them as independent vectors. This method is based on the Cayley-Dickson decomposition, according to which quaternions can be thought of as complex numbers whose real and imaginary parts are complex numbers in turn (Cayley-Dickson form):

$$
q = s + p\mathbf{v}_2 = \left(w + x\mathbf{v}_1\right) + \left(y + z\mathbf{v}_1\right)\mathbf{v}_2 = w + x\mathbf{v}_1 + y\mathbf{v}_2 + z\mathbf{v}_3 \ . \tag{9}
$$
$$
\underbrace{\phantom{\left(w + x\mathbf{v}_1\right)}}_{simplex} \ \underbrace{\phantom{\left(y + z\mathbf{v}_1\right)}}_{perplex}
$$

Both the *simplex* and *perplex* parts are isomorphic to field $(\mathbf{i}, \mathbf{j}, \mathbf{k})$, since they are both defined in the same space of the complex operator $\mathbf{v}_1$. Hence a function $f(n)$ can be formulated in terms of an orthonormal basis $\left(\mathbf{v}_1, \mathbf{v}_2, \mathbf{v}_3\right)$:

$$
f(n) = \left[w(n) + x(n)\mathbf{v}_1\right] + \left[y(n) + z(n)\mathbf{v}_1\right]\mathbf{v}_2 \ . \tag{10}
$$

The basis must be chosen in a way that $\mathbf{v}_1 \perp \mathbf{v}_2 \perp \mathbf{v}_3$, $\mathbf{v}_1\mathbf{v}_2 = \mathbf{v}_3$ and $\mathbf{v}_1\mathbf{v}_2\mathbf{v}_3 = -1$.

Thanks to the linearity property of the Fourier transform, the quaternion DFT (11a) and the quaternion *inverse* DFT (11b) can be decomposed into the sum of two complex-valued transforms:

$$
F(u) = \sum_{n=0}^{N-1} \exp\left(-2\pi\mathbf{v}\frac{nu}{N}\right)f(n) = F_s(u) + F_p(u)\mathbf{v}_2 \tag{11a}
$$

$$
f(n) = \frac{1}{N}\sum_{u=0}^{N-1} \exp\left(2\pi\mathbf{v}\frac{nu}{N}\right)F(u) = f_s(n) + f_p(n)\mathbf{v}_2 \tag{11b}
$$

Table 1 Kernel definitions for monodimensional QDFT		Left	Right
	Axis $\boldsymbol{v}$	$e^{-v\omega n}f(\cdot)$	$f(\cdot)e^{-v\omega n}$

where the subscripts s and p denote the simplex and perplex parts, respectively. Both functions $F(u)$ and $f(n)$ are quaternionic functions of N samples of the kind $f(n) = w(n) + x(n)\mathbf{i} + y(n)\mathbf{j} + z(n)\mathbf{k}$. Versor $\boldsymbol{v}$ is an arbitrarily chosen pure unitary quaternion versor and has to meet the constraints $\mathrm{Re}\,[\boldsymbol{v}] \perp \mathrm{Im}\,[\boldsymbol{v}]$, $\|\mathrm{Re}\,[\boldsymbol{v}]\| - \|\mathrm{Im}\,[\boldsymbol{v}]\| = 1$. After executing the two complex DFTs and reassembling the final transformed quaternion, it is sufficient to change back to the original basis by means of inverse change of basis.

Unfortunately, the non-commutativity property of the quaternionic product gives rise to a two-sided monodimensional quaternion transform, i.e. quaternion transforms can be found either in a left- or a right-handed form (transpose with one another) as summed-up in Table 1. For instance, (11a) and (11b) represent the quaternionic Fourier monodimensional left-handed transform (the exponential function is on the left) and its inverse, respectively. Tests conducted during the development of these algorithms revealed that there is no major implication from the existence of a two-sided transform on filtering applications, if the direction of rotation is kept unchanged from the beginning to the end of the algorithm.

3 Overlap-Save Quaternion Frequency Domain Adaptive Filter

We present the OS-QFDAF algorithm, also called *Fast QLMS*. It is a block algorithm, i.e. the filter coefficients $\mathbf{W}_k$ are updated at each block iteration k. A general classification discerns block algorithms by the input block length, the number of the overlapping samples, the type of transform used (for those algorithms working in a transform domain). In OS-QFDAF the usual choice for the filter length is $M_F = M$, where M is the number of the overlapping samples. In order to simplify the implementation, the transform length is chosen as $N = M + L$, where L is the block length.

The algorithm comes with power normalization, a technique intended to improve the convergence to optimum by whitening the input signal, so that the convergence modes are equalized.

Fast convolution in OS-QFDAF is performed using the *Overlap-Save* method.

3.1 Algorithm Overview

Algorithm initialization:

$$\mathbf{W}_0 = \mathbf{0} \ \ (2M\text{-by-}1 \text{ null vector})$$
$$\mu = \mu_0, \ P_0(m) = \delta, \ m = 0, 1, ..., N - 1 \tag{12}$$

$P_k(m)$: power of the m-th frequency bin at block k.
μ_0, δ: initialization constants to be chosen empirically.

Do the following steps for each new input block k:

Compute the QFFT of the filter input samples:

$$\mathbf{X}_k = \text{diag}\left\{ \text{QFFT}\left[\mathbf{x}_{old}^M \ \mathbf{x}_k^L\right]^T \right\} \tag{13}$$

Input block definition:

$$\mathbf{x}_{old}^M = [x(kL - M + 1), \cdots, x(kL - 1)]$$
$$\mathbf{x}_k^L = [x(kL), \cdots, x(kL + L - 1)]$$

Note: diagonalization in (13) allows a formalism similar to Block LMS when the filter output is computed in (14) later on [15].

Compute the filter output in the time domain:

$$\mathbf{Y}_k = \mathbf{X}_k \mathbf{W}_k \tag{14}$$

$$\hat{\mathbf{y}}_k = \left[\text{IQFFT}\left(\mathbf{X}_k \mathbf{W}_k\right)\right]^{\lfloor L \rfloor} \tag{15}$$

$\mathbf{Y}_k$: filter output in the frequency domain.
$\hat{\mathbf{y}}_k$: windowed filter output in the time domain.

Note: since filtering operations are implemented with linear convolution, proper window constraints on data are needed in order to obtain linear convolution from circular convolution.[1]

Error calculation:

$$\hat{\mathbf{d}}_k = \left[d(kL)\ d(kL + 1) \cdots d(kL + L - 1)\right]^T \tag{16}$$

$$\hat{\mathbf{e}}_k = \hat{\mathbf{d}}_k - \hat{\mathbf{y}}_k \tag{17}$$

$$\mathbf{E}_k = \text{QFFT}\left(\left[0_M\ \hat{\mathbf{e}}_k\right]^T\right) \tag{18}$$

$\hat{\mathbf{d}}_k$: desired output vector in the time domain at block k.
$\hat{\mathbf{e}}_k$: error vector in the time domain at block k.
$\mathbf{E}_k$: error vector in the frequency domain at block k.[2]

---

[1]Circular convolution (in the time domain) is provided by the inverse transform of the product of two DFT sequences.
[2]Zero-padding is needed before executing the QFFT.

Update the learning rates (Power Normalization):

$$\mu_k = \mu \cdot \text{diag}\left[P_k^{-1}(0), \ldots, P_k^{-1}(N-1)\right] \tag{19}$$

$$P_k(m) = \lambda P_{k-1}(m) + (1-\lambda)\left|X_k(m)\right|^2 \tag{20}$$

$\lambda \in [0, 1]$: forgetting factor.

Note: Power Normalization is a technique aimed at having all the algorithm modes converging at the same rate. Each filter weight is assigned a step size of its own, so that the disparity between the eigenvalues of the input autocorrelation matrix is reduced.

Update the filter weights:

$$\mathbf{W}_{k+1} = \mathbf{W}_k + \mu_k \mathbf{X}_k^H \mathbf{E}_k \tag{21}$$

$\mathbf{X}_k^H \mathbf{E}_k$: gradient estimation in the frequency domain at step k.

Gradient constraint:

$$\mathbf{W}_{k+1} = \text{QFFT}\left(\begin{bmatrix}[\text{IQFFT}(\mathbf{W}_{k+1})]^{[M]} \\ \mathbf{0}_L\end{bmatrix}\right). \tag{22}$$

If the gradient constraint $[M]$ is neutralized, the product $\mathbf{X}_k^H \mathbf{E}_k$ corresponds to a circular correlation in the time domain and the algorithm is said *Unconstrained QFDAF*. Usually, unconstrained algorithms exhibit a polarized convergence. In order to get convergence to optimum, the filter length M should be increased.

3.2 Computational Cost

Each QFFT requires the execution of 2 complex FFTs. The computation of one complex FFT involves $N\log_2 N$ multiplications. The OS-QFDAF algorithm includes 5 QFFTs, $16N$ multiplications to compute the filter output and $16N$ multiplications to update the filter weights. Given a block of $L = M$ samples, the computational cost for OS-QFDAF is approximately

$$C_{OS-QFDAF} \simeq 2 \cdot 5N\log_2 N + 2 \cdot 16N = 20M\log_2 2M + 64M \tag{23}$$

where $N = L + M = 2M$.

In QLMS, the computation of both the filter output and the cross-correlation in the update equation requires $4 \cdot 4M = 16M$ multiplications for each sample. Overall, for

M samples, the computational cost for QLMS is approximately $C_{QLMS} \simeq 32M \cdot M = 32M^2$. The complexity ratio between OS-QFDAF and QLMS is

$$\frac{C_{OS-QFDAF}}{C_{QLMS}} = \frac{5\log_2 2M + 16}{8M} .$$

(24)

For example, for $M = 64$, OS-QFDAF is about 10 times faster than QLMS.

4 Simulations

The OS-QFDAF algorithm has been tested in a system identification problem. The filter quaternion input was a unit variance colored noise of the form

$$x[n] = bx[n-1] + \frac{\sqrt{1-b^2}}{\sqrt{4}}\eta[n] .$$

(25)

where $\eta[n]$ is a quaternionic zero-mean unit-variance white Gaussian noise.

OS-QFDAF with varying M: Figure 1 shows how choosing too a small block length causes an Excess Mean Square Error (EMSE) in the learning curve, i.e. there is a gap between the steady-state MSE curve and the theoretical MSE bound. As a rule of thumb, too a large M and/or too a large δ decreases the convergence rate of the algorithm.

OS-QFDAF versus QLMS with narrow-band signals: The input signal is now narrow-band (b = 0.99). In such a difficult situation it is possible to outperform the QLMS with the use of the transform domain algorithms as shown in Fig. 2.

Fig. 1 OS-QFDAF with varying M (with power norm.) ($\mu_0 = 0.08$, $\lambda = 0.98, \delta = 4$)

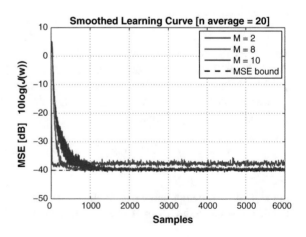

Fig. 2 OS-QFDAF versus QLMS with narrow band signal $b = 0.99$. (Power-norm. OS-QFDAF: $M = 6, \mu_0 = 0.008, \lambda = 0.999, \delta = 50$. Non-power-norm. QLMS and OS-QFDAF $M = 6, \mu = 0.008$)

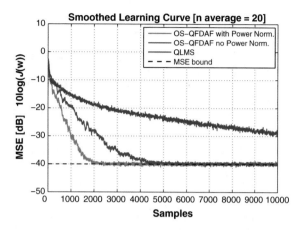

The step size in a power-normalized algorithm is an overall parameter governing the diagonal inverse power matrix. Changing its value does not have a significant effect on the final result. If power normalization is not applied, the step size modifies the algorithm behavior in the same way as in QLMS [14].

5 Conclusions

Quaternions surely have an impact on physics and engineering. Within the hypercomplex number subfields, multidimensional filtering gives the possibility to process data accordingly to their typical space dimensions, thus obtaining robustness and coherence from the results. This is feasible since the correlation between different dimensions is considered and multidimensional data are processed as a whole. The processing of four-dimensional signals increases the computational complexity significantly. For this reason we implemented a quaternionic class of frequency-domain adaptive filters. For instance, in audio signal processing, long impulse response filters require a long memory. In such a case, time-domain processing is not recommended. Block filters in the frequency domain provide a fast computation and improved statistical properties of the signals, as well. Recently, the definition of a quaternionic neuron and the application of quaternion filtering to neural networks have been explored. On this trend, our next efforts are continuing on this route.

References

1. Arena, P., Fortuna, L., Muscato, G., Xibilia, M.: Multilayer perceptrons to approximate quaternion valued functions. Neural Netw. **10**, 335–342 (1997)
2. Ell, T.A., Sangwine, S.J.: Decomposition of 2d hypercomplex fourier transforms into pairs of complex fourier transforms. In: Proceedings of 10th European Signal Processing Conference

3. Haykin, S.: Adaptive Filter Theory. Prentice Hall (1996)
4. Hitzer, E.: Introduction to Clifford's geometric algebra. SICE J. Control, Measur. Sys. Integr. **4**(1), 1–10 (2011)
5. Isokawa, T., Nishimura, H., Matsui, N.: Quaternionic multilayer perceptron with local analyticity. Information **3**, 756–770 (2012)
6. Jahanchahi, C., Took, C., Mandic, D.: A class of quaternion valued affine projection algorithms. Signal Process. **93**, 1712–1723 (2013)
7. Jahanehahi, C., Took, C.C., Mandic, D.P.: The widely linear quaternion recursive least squares filter. In: Proceedings of 2nd International Workshop Cognitive Information Processing (CIP) (2010)
8. Jiang, M., Liu, W., Li, Y.L., Zhang, X.: Frequency-domain quaternion-valued adaptive filtering and its application to wind profile prediction. In: IEEE Region 10 Conference TENCON, pp. 1–5, Oct 2013
9. Kraft, E.: A quaternion-based unscented Kalman filter for orientation tracking. In: Proceedings of 6th International Conference on Information Fusion (ISIF), pp. 47–54 (2003)
10. Kusamichi, H., Isokawa, T., Matsui, N., Y. Ogawa, Y., Maeda, K.: A new scheme for color night vision by quaternion neural network. In: Proceedings of 2nd International Conference Autonomous Robots and Agents, pp. 101–106, Dec 2004
11. Nitta, T.: An extension of the back-propagation algorithm to quaternion. In: 3rd Int. Conference on Neural Information Processing ICONIP, pp. 247–550, Sep 1996
12. Yefremov, A.P.: Number, geometry and nature. Hypercomplex Numbers Geom. Phys. **1**(1), 3–4 (2004)
13. Pei, S.C., Ding, J.J., Chang, J.H.: Efficient implementation of quaternion fourier transform, convolution, and correlation by 2-d complex fft. IEEE Trans. Signal Process. **11**(49), 2783–2797 (2001)
14. Took, C.C., Mandic, D.P.: The quaternion LMS algorithm for adaptive filtering of hypercomplex processes. IEEE Trans. Signal Process. **57**(4), 1316–1327 (2009)
15. Uncini, A.: Fundamentals of Adaptive Signal Processing. Springer (2015)
16. Yoshida, M., Kuroe, Y., Mori, T.: Models of Hopfield-type quaternion neural networks and their energy function. Int. J. Neural Syst. **15**, 129–135 (2005)

Bayesian Clustering on Images with Factor Graphs in Reduced Normal Form

Amedeo Buonanno, Luigi di Grazia and Francesco A.N. Palmieri

Abstract Bayesian clustering implemented on a small Factor Graph is utilized in this work to perform associative recall and pattern recognition on images. The network is trained using a maximum likelihood algorithm on images from a standard data set. The two-class labels are fused with the image data into a unique hidden variable. Performances are evaluated in terms of Kullback-Leibler (KL) divergence between forward and backward messages for images and labels. These experiments reveal the nature of the representation that the learning algorithm builds in the hidden variable.

Keywords Belief propagation · Factor Graph · Image recognition

1 Introduction

Various architectures have been proposed as adaptive Bayesian graphs [1, 2], but a very appealing approach to directed Bayesian graphs for visualization and manipulation, that has not found its full way in the applications, is the Factor Graph (FG) framework and in particular the so-called Normal Form (FGn) [3, 4]. This formulation provides an easy way to visualize and manipulate Bayesian graphs, much like as block diagrams. Factor Graphs assign variables to edges and functions to nodes and in the Reduced Normal Form (FGrn) [5], through the use of replicator units (or equal constraints), the graph is reduced to an architecture in which each variable is connected to two factors at most.

A. Buonanno (✉) · L. di Grazia · F.A.N. Palmieri
Dipartimento di Ingegneria Industriale e dell'Informazione, Seconda Università
di Napoli (SUN), via Roma 29, 81031 Aversa (CE), Italy
e-mail: amedeo.buonanno@unina2.it

L. di Grazia
e-mail: digrazia.luigi88@gmail.com

F.A.N. Palmieri
e-mail: francesco.palmieri@unina2.it

© Springer International Publishing Switzerland 2016
S. Bassis et al. (eds.), *Advances in Neural Networks*, Smart Innovation,
Systems and Technologies 54, DOI 10.1007/978-3-319-33747-0_6

The Bayesian network approach for building artificial system has the advantage of being totally general with respect to the type of data processed. In this framework information coming from different sensors can be easily fused. The information flow is bi-directional via belief propagation and can easily accommodate various kinds of inferences for pattern completion, correction and classification.

In this work we focus on the Latent Variable Model (LVM) [6, 7], also known as Autoclass [8]. The LVM, despite its simplicity, has been extensively used in various tasks. The aim of this work is to gain a better understanding of this model in the FGrn framework as a building block for more complicated multi-layer architectures (an unsupervised multilayer network has been presented in [9]). The great flexibility of the FGrn paradigm can be exploited for introducing supervision in a multi-layer architecture that will be reported elsewhere.

2 Factor Graphs in Reduced Normal Form

In the FGrn framework the Bayesian graph is reduced to a simplified form composed only by *Variables*, *Replicators* (or *Diverters*), *Single-Input/Single-Output (SISO) blocks* and *Source blocks* (Fig. 1). Even though various architectures have been proposed in the literature for Bayesian graphs [4], the FGrn framework is much easier to handle, it is more suitable for defining unique learning equations [5] and consequently for a distributed implementation. Parameter learning, in this representation, can be approached in a unified way because we can concentrate on a unique rule for training any SISO, or Source, factor-block in the system, regardless of its location (visible or hidden) [5]. In our previous work we have also implemented a Simulink library for rapid prototyping of architectures based on this paradigm [10].

Parameters (probabilities) in the SISO block and the source blocks are learned from examples solely on the backward and forward flows available locally. Learning follows an EM (Expectation Maximization) algorithm to maximize global likelihood (see [5] for derivations and comparisons). An extensive review of the paradigm is beyond the scope of this work and the reader can find more details in our recent works [5, 9, 10] or in the classical papers [4, 11].

Fig. 1 FGrn components: **a** a variable branch; **b** a diverter; **c** a SISO block (factor); **d** a source block. The forward and backward messages are explicitly shown

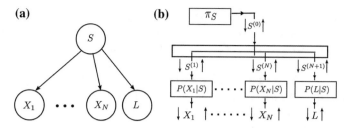

Fig. 2 A $(N + 1)$—tuple with the Latent Variable as a Bayesian graph (**a**) and as a Factor Graph in Reduced Normal Form (**b**)

3 Bayesian Clustering

The *Latent-Variable Model* (LVM) [6], shown in Fig. 2, presents at the bottom N variables X_n, $n = 1 : N$ that belong to a finite alphabet $\mathcal{X} = \{\xi_1, \xi_2, \dots, \xi_{d_X}\}$. In this application they represent the pixels of an images. The variable L, that belongs to the alphabet $\mathcal{L} = \{\lambda_1, \lambda_2, \dots, \lambda_{d_L}\}$, is instead the class label. Generally the complexity of the whole system increases with the cardinality of the alphabets.

The $N + 1$ bottom variables are connected to one Hidden (Latent) Variable S, that belongs to the finite alphabet $S = \{\sigma_1, \sigma_2, \dots, \sigma_{d_S}\}$. The SISO blocks represent $d_S \times d_X$ and $d_S \times d_L$ row-stochastic probability conditional matrices: $P(X_n|S) = [Pr\{X_n = \xi_j|S = \sigma_i\}]_{i=1:d_S}^{j=1:d_X}$ and $P(L|S) = [Pr\{L = \lambda_j|S = \sigma_i\}]_{i=1:d_S}^{j=1:d_L}$.

The system is drawn as a generative model with the arrows pointing down assuming that the source is variable S and the bottom variables are its children. This architecture can be seen also as a Mixture of Categorical Distributions [1]. Each element of the alphabet S represents a "Bayesian cluster" for the $N + 1$ dimensional stochastic vector, $\mathbf{X} = [X_1, X_2, \dots, X_N, L]$. Essentially each bottom variable is independent from the others, given the Hidden Variable [1], and each $N + 1$ dimensional vector $\mathbf{X}$ is assigned to a single cluster. One way to visualize the model is to imagine drawing a sample: for each data point we draw a cluster index $s \in S = \{\sigma_1, \sigma_2, \dots, \sigma_{d_S}\}$ according to the prior distribution π_S. Then for each $n = 1 : N$, we draw $x_n \in \{\xi_1, \xi_2, \dots, \xi_{d_X}\}$ according to $P(X_n|S = s)$ and draw $l \in \{\lambda_1, \lambda_2, \dots, \lambda_{d_L}\}$ according to $P(L|S = s)$. In this model the label variable L is introduced at the same level of the other variables that represent the image pixels. This is different than the Naive Bayes Classifier [2] where the label variable is the conditioning variable S. Here learning is unsupervised and the system builds a representation that fuses the label information when it is available. Information can be injected at any of the bottom nodes and inference can be obtained for each variable using the usual sum-product rule. Note that in the FGrn formulation each SISO block of Fig. 2 uses its own backward message b_{X_n} at the bottom and a different incoming message from the diverter $f_{S^{(n)}}$. Similarly the sum rule produces specific forward messages f_{X_n} at the bottom and different backward messages $b_{S^{(n)}}$ toward the diverter [5].

The Negative Log-Likelihood on observed data (normalized on the number of the data point and observed variables) can be used as a quality index of how much data has been memorized into the system. To reveal the nature of the representation built by the learning algorithm, we focus on the trends of the likelihood associated to the image part $(X_1, ..., X_N)$ and to the label part (L) both independently and globally. The likelihood on the label part (that is equivalent to KL divergence between the real and recognized distributions) is monotonically related to classification, while the likelihood on the image measures the capability of the network to perform "inpainting", i.e. to fill-in the image parts when these may be missing or affected by errors.

4   Simulations

In the following simulations we have used the CIFAR-10 dataset that consists of 60000 32 × 32 color images divided in 10 classes, with 6000 images per class. There are 50000 training images and 10000 test images [12]. To contain the computational effort, we have used here only a small subset of the CIFAR-10 dataset and confined our simulations to 2 classes (Fig. 3).

Pre-processing, Normalization and Quantization: The CIFAR-10 dataset images are converted to grayscale and normalized. The normalization consists in subtracting the mean and in dividing by the standard deviation of each image. For visual data, this corresponds to local brightness and contrast normalization. After normalization each input vector may be optionally whitened [13]. However since we have verified that whitening doesn't have much effect on the results, we have used simply normalized images.

The intensity of each pixel has been quantized uniformly using d_X levels of gray (equispaced intervals). A more advanced minimum variance non uniform quantization can be performed. Furthermore, colors could be included using a global quantization based on a color cube. We have performed simulation on these more elaborated pre-processing schemes, but in this paper we report only results for uniform gray-level quantization and evaluate the effects on likelihood performance when the number of levels changes.

Fig. 3   20 typical grayscale, normalized and quantized to 16 levels images from the data set for the two classes

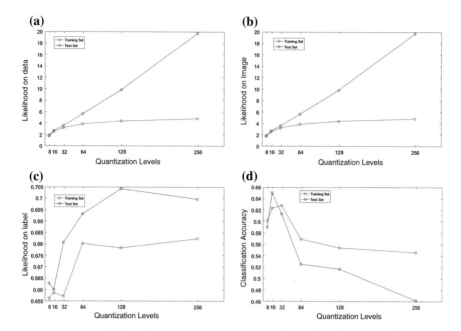

Fig. 4 Effects of increasing the number of quantization level d_X with embedding space size $d_S = 5$. Training Set size: 1000; Test Set size: 600. **a** Average likelihood for all the variables; **b** for the image variables only; **c** for the label variable only; **d** classification accuracy

Performance Evaluation: In Figs. 4 and 5 we show the normalized likelihood when the quantization levels d_X and the embedding space size d_S is varied (in abscissa) both on the training set and the test set (cross-validation). The four plots report respectively: (a) Average likelihood on image and label; (b) Average likelihood only for the image part; (c) Average likelihood only for the label part; (d) Classification accuracy when the max value of the forward distribution on the label is assumed as the final decision. The typical loss of generalization is observed as the two curves tend to diverge both when quantization levels and embedding space size are increased. We also observe in Fig. 4 that increasing the number of gray levels leads to overfitting because the network becomes too confident on the information that it has seen. Notice how generalization is different if we focus on the image variables or on the label variable. The label likelihood benefits from a certain level of overfitting on the image part. $d_X = 16$ levels seems to be a good trade-off between overfitting and quality of the likelihood on the label.

In Fig. 5, where we vary the size d_S of the hidden variable, the trend is similar (here more images have been used). The embedding space for the image should be 4 (in order to avoid the overfitting), but this value is too small to achieve a sufficient classification accuracy for the label.

We see from these results that even though in the model there is nothing that prevents more clusters (larger d_S) to emerge from learning [6], the dimension of the

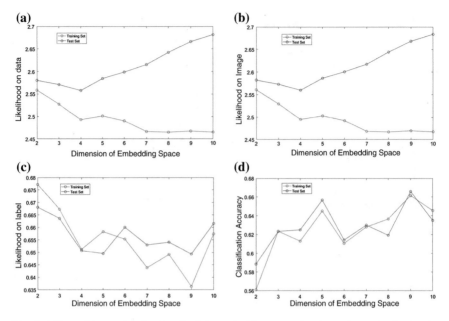

Fig. 5 Effects of increasing the size d_S of the embedding space for $d_X = 16$ quantization levels. Training Set size: 2000; Test Set size: 1200. **a** Average likelihood for all the variables; **b** for the image variables only; **c** for the label variable only; **d** classification accuracy

embedding space that correctly generalizes remains confined to a small number (in the order of the number of classes considered). It is evident that as more dimensions are available in the embedding space, more prototypes are built by the algorithm (see the likelihood curve on the training set), but generalization becomes poor.

To look further into the internal representation of the network that emerges from learning, we show more detailed results for $d_S = 5$ using 1000 images quantized with $d_X = 16$ gray levels. We inject first a set of five delta distributions as forward message f_S on the hidden variable S and obtain at the image terminations five forward distributions f_X whose means are reported on the left of Fig. 6. They are essentially the mean internal representation masks associated to the five embedding components of S. The five learned priors π_S are shown on the second column of Fig. 6 as a bar graph. We note that some of the coordinates of the embedding space are given more importance in the internal representation. For example the fifth mask is the one that is weighted the most out of the five in the generative model. The third column of Fig. 6 shows also the two rows of matrix $P(L|S)$ as a bar graph. The bars represent how much the five masks contribute to each class estimate. For example masks 2 and 4 contribute the most to infer the presence of a ship, while mask 5 is the most important prototype for a car.

We have also probed the learned system with two backward delta distributions at the label termination (b_L). At the image termination we obtain forward distribu-

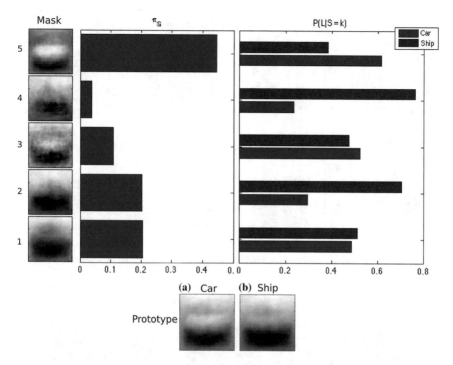

Fig. 6 Masks, prior probabilities, conditional label distributions and prototypes

tions shown in their means as the two images at the bottom of Fig. 6. They could be considered to be the mean prototypes for ships and cars.

Figure 7 shows also some typical results of a Recognition Task. Images here are presented as backward distributions at the image terminations and forward distributions are collected at the label end. We can see how some misclassification errors in (b) and (d) may emerge as the result of maximum likelihood decision when the returned distributions are rather uniform.

Finally in the Fig. 8 the results of a Recall Task are reported. From the original images (Fig. 8a) a certain number of pixels are deleted (Fig. 8b, d) and this information is injected in the network in the form of uniform distributions together with the value of the label. The network responds in the forward messages with a new image reducing the uncertainty on the erased pixels. Figure 8c, e as usual depict the mean distributions in grayscale. The KL divergence computed on the response reflects the improvement. It is interesting to see in Fig. 8e that the system responds on the left part with a shadow, a sort of low resolution image is used to complete the missing information. Clearly increasing the size of the embedding space would have increased the performance on this inpainting process on the training images, but we would have poor generalization.

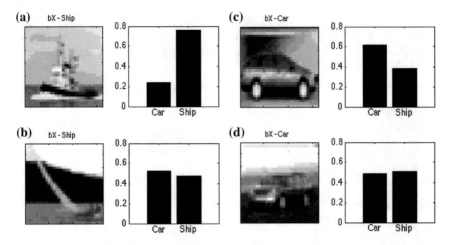

Fig. 7 Recognition Task: Images presented to the network (*left*) and the forward distribution on the label variable (*right*)

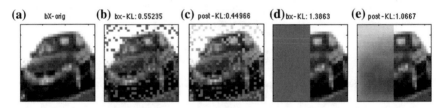

Fig. 8 Recall Task: **a** the original image; **b** the image with 20 % of random erasures; **d** the image with the left half erased; **c** and **e** the respective average forward distributions

5   Conclusions

In this work we have used an LVM in the framework of Factor Graphs in Reduced Normal Form for a joint recognition/recall task on images. Labels and images are treated as equally important variables that depend on one conditioning hidden variable. The results are quite good if we consider that the system is very small in comparison to state-of-the-art deep networks that are based on thousands of parameters, are multi-layer and are trained on very large datasets [14, 15]. The experiment conduced here aimed at revealing the hidden representation that the LVM builds as the result of maximum likelihood learning. The results of this work are very useful in understanding the role of the LVM as a basic building block for more complex architectures. In the FGrn paradigm the supervision can be introduced simply connecting a supervision variable to the diverter. The extendibility of this framework is a crucial characteristic that makes this approach suitable for building supervised multi-layer architectures.

References

1. Koller, D., Friedman, N.: Probabilistic Graphical Models: Principles and Techniques. MIT Press (2009)
2. Barber, D.: Bayesian Reasoning and Machine Learning. Cambridge University Press, (2012)
3. Forney, G.D.: Codes on graphs: normal realizations. IEEE Trans. Inf. Theory **47**, 520–548 (2001)
4. Loeliger, H.A.: An introduction to factor graphs. IEEE Signal Process. Mag. **21**, 28–41 (2004)
5. Palmieri, F.A.N.: A comparison of algorithms for learning hidden variables in normal graphs (2013). arXiv:1308.5576
6. Murphy, K.P.: Machine Learning: A Probabilistic Perspective (Adaptive Computation and Machine Learning series). The MIT Press (2012)
7. Bishop, C.M.: Latent variable models. In: Learning in Graphical Models, pp. 371–403. MIT Press (1999)
8. Cheeseman, P., Stutz, J.: Advances in knowledge discovery and data mining. In: ch. Bayesian Classification (AutoClass): Theory and Results, pp. 153–180. American Association for Artificial Intelligence, Menlo Park, CA, USA (1996)
9. Buonanno, A., Palmieri, F.A.N.: Towards building deep networks with bayesian factor graphs (2015). arXiv:1502.04492
10. Buonanno, A., Palmieri, F.A.N.: Simulink implementation of belief propagation in normal factor graphs. In: Recent Advances of Neural Networks Models and Applications—Proceedings of the 24th Workshop of the Italian Neural Networks Society WIRN 2014, vol. 37 of Smart Innovation, System and Technologies, Springer International Publishing (2015)
11. Kschischang, F.R., Frey, B., Loeliger, H.: Factor graphs and the sum-product algorithm. IEEE Trans. Inf. Theory **47**, 498–519 (2001)
12. Krizhevsky, A., Hinton, G.: Learning multiple layers of features from tiny images, Computer Science Department, University of Toronto. Technical Report **1**(4), 7 (2009)
13. Hyvrinen, A., Hurri, J., Hoyer, P.O.: Natural Image Statistics: A Probabilistic Approach to Early Computational Vision, 1st edn. Springer Publishing Company, Incorporated (2009)
14. Lin, M., Chen, Q., Yan, S.: Network in network. In: Proceedings of International Conference on Learning Representations (ICLR) (2014)
15. Lee, C.-Y., Xie, S., Gallagher, P., Zhang, Z., Tu, Z.: Deeply-supervised nets. In: Proceedings of the Eighteenth International Conference on Artificial Intelligence and Statistics, vol. 38 of JMLR Workshop and Conference Proceedings, pp. 562–570 (2015)

Selection of Negative Examples for Node Label Prediction Through Fuzzy Clustering Techniques

Marco Frasca and Dario Malchiodi

Abstract Negative examples, which are required for most machine learning methods to infer new predictions, are rarely directly recorded in several real world databases for classification problems. A variety of heuristics for the choice of negative examples have been proposed, ranging from simply under-sampling non positive instances, to the analysis of class taxonomy structures. Here we propose an efficient strategy for selecting negative examples designed for Hopfield networks which exploits the clustering properties of positive instances. The method has been validated on the prediction of protein functions of a model organism.

Keywords Negative selection · Fuzzy clustering · Hopfield networks

1 Introduction

In several real-world contexts ranging from text categorization [7] to protein function prediction [8], the notion of "negative" example is not well defined, and the selection of appropriate sets of negative patterns may sensibly improve the predictive capabilities of supervised and semi-supervised methodologies for inductive inference. This area of machine learning, named Positive-Unlabeled (PU) learning, has seen a surge of interest in latest years [8]. Indeed, in different contexts there is no "gold standard" for negative items and only positive instances are the results of accurate studies (negative examples usually are simply "non positive"). Several approaches have been proposed for selecting negative instances, such as randomly sampling items (assuming the probability of getting a false negative to be low) [10], sampling according to positive-negative similarity measures [5], selecting the items positive for the sibling

M. Frasca (✉) · D. Malchiodi
Dipartimento di Informatica, Università degli Studi di Milano, via Comelico 39,
20135 Milan, Italy
e-mail: frasca@di.unimi.it

D. Malchiodi
e-mail: malchiodi@di.unimi.it

© Springer International Publishing Switzerland 2016
S. Bassis et al. (eds.), *Advances in Neural Networks*, Smart Innovation,
Systems and Technologies 54, DOI 10.1007/978-3-319-33747-0_7

and/or ancestral categories of the categories of interest as negative examples [18]. Nevertheless, hierarchical methods cannot be applied in contexts where categories are not structured as a hierarchy (for instance in action video detection [5]). Furthermore, even when parent-child relationships are available for the classes being predicted, strategies based on sibling/ancestral categories may often break down, as some items are annotated to more than one sibling category, and many items have few siblings to use [20].

In this work we propose a novel methodology for graph-based semi-supervised learning which is composed of two main steps: Step (1) a novel strategy for PU learning specific for Hopfield networks (HNs) [11], which can be applied both to structured classes and to hierarchy-less contexts; Step (2) a semi-supervised classifier based on a family of parametric Hopfield networks, which embeds the negative selection performed at Step 1 in the dynamics of network.

At Step 1, the approach for detecting negative instances can be summarized as follows:

1.1. The matrix describing symmetric connections between instances is transformed into a feature matrix, in which each instance is associated with a 3-feature vector, obtained through the application of random walks of length 1, 2 and 3 starting from positive instances.
1.2. The positive points are clustered using a dynamic version of the fuzzy C means algorithm (FCM) [3] exploiting a suitable index [6] in order to decide the optimal number of clusters to summarize data; then, every negative point is assigned a score consisting in the maximum membership it has to the detected clusters of positive items. The low-ranked points according to this score are considered as negatively labeled.
1.3. The remaining points are then further discriminated according to their stability in the Hopfield network constructed from the input graph. In particular, we extend the set of negative instances with those points that locally minimize the energy function.

At Step 2, the Hopfield network is simulated and the items not selected during Step 1 are allowed to change their state along with the items whose label has to be predicted (test set). The final equilibrium state is used to infer the prediction for instances in the test set. In particular, the instances corresponding to neurons fired at equilibrium are considered candidates for the positive class.

We experimentally validated the proposed methodology in the protein function prediction problem, which consists in inferring the biomolecular functions of recently discovered proteins starting from their interactions with already characterized proteins, and in which, apart from rare cases, just the positive annotations are available for the Gene Ontology classes [1]. The comparison with state-of-the-art supervised and semi-supervised label prediction methods and with the "vanilla" Hopfield network shows the effectiveness of our approach. Moreover, the reduced computational complexity, due to the application of clustering techniques only to positive instances (which usually are a large minority of the data set), allows the application of the proposed methodology to contexts characterized by large-size data.

The paper is organized as follows: Sect. 2 describes the afforded prediction task, while Sect. 3 explains why in such context it is difficult to negatively label the available data. Sections 4 and 5 respectively describe and test the proposed methodology. Some concluding remarks close the paper.

2 Node Label Prediction (NLP) in Graphs

Consider a weighted graph $G = (V, W)$, where $V = \{1, 2, \dots, n\}$ is the set of vertices, W is a symmetric $n \times n$ weight matrix, where $W_{ij} \in [0, 1]$ indicates the strength of the evidence of co-functionality between vertices i and j.

For a given class to be predicted, the vertices in V are labeled with $\{+, -\}$, and the labeling is known only for the subset $S \subset V$, whereas it is unknown for $U = V \setminus S$. Furthermore, labeled vertices are partitioned into positive S_+ and negative S_- vertices.

The *Node Label Prediction on graphs* (*NLP*) problem consists in determining a bipartition (U_+, U_-) of vertices in U, such that vertices in U_+ can be considered as candidates for the positive class.

3 The Problem of Selecting Negative Examples for NLP

The problem of selecting negative examples for classification tasks arises in those contexts in which items are classified for the properties they possess, and not for those properties they do not possess. For instance, in text classification it is not practical to label all the topics a document does not contain; hence just the topics a document contains are available [14]. Or in protein function prediction, where the classes to be predicted are the bio-molecular protein functions, proteins are rarely labeled with the functions they do not possess (negative annotation), thus most of all annotations are positive [1]. In our context, we consider unlabeled those vertices for which we want to infer the labels (i.e. U), whereas genes in S which are not positive for the current class (i.e. S_-) are in principle considered as negative. Nevertheless, there is no evidence that vertices in S_- are not positive; some of these instances may be not positive just because not enough studies and analyses have been carried out yet.

To take into account these issues, we assume that there exists a subset $S^p_- \subset S_-$ containing negative instances that are more likely to be classified as positive in future, but such subset is not known. Thus, nodes in $S_- \setminus S^p_-$ can be considered in turn as reliable negative instances. Clearly, the subset S^p_- may vary according to the class to be predicted. In the following we propose a novel method to detect such subset specifically designed for a family of classifiers based on parametric Hopfield networks.

4 Methods

In this section we first describe an algorithm for the *NLP* problem which exploits the properties of a parametric Hopfield model, then we present an extension that embeds in the model a procedure for detecting the negative instances candidates for the set S_-^p.

4.1 An Algorithm for NLP Problem

COSNet, COst-Sensitive neural Network [2, 9], is a semi-supervised algorithm recently proposed for *NLP* problems characterized by unbalanced labelings. *COSNet* is based on parametric HNs $H = \langle W, k, \rho \rangle$, where k is the neuron activation threshold and ρ is a real number in $(0, \frac{\pi}{2})$, that determines the two different values $\{-\cos\rho, \sin\rho\}$ for neuron activation. The distinction between neuron labels and neuron activation values allows the method to automatically determine the activation levels for positive and negative neurons in order to counterbalance the prevalence of labels in the majority class (positive or negative class). The optimal parameters $(\hat{k}, \hat{\rho})$ of the sub-network restricted to labeled nodes are learned so as to move the state determined by the bipartition (S_+, S_-) "as close as possible" to an equilibrium state. The authors have shown that the learned activation values move the state provided by know labels closer to a global minimum of the network restricted to S than the classical activation values $\{-1, 1\}$ (see [9] for details). We denote by $\hat{l}$ the state of labeled network after learning, where $\hat{l}_i = -\cos\hat{\rho}$ if $i \in S_-$, and $\hat{l}_i = \sin\hat{\rho}$ when $i \in S_+$.

Then, the network restricted to unlabeled nodes U is simulated by adopting $\{-\cos\hat{\rho}, \sin\hat{\rho}\}$ as activation values and $\hat{k}$ as unique activation threshold. Neurons in S are not updated. The initial state is set to the null vector, and by assuming that $U = \{1, 2, \cdots, h\}$ and $S = \{h+1, h+2, \cdots, n\}$ (up to a permutation), the network evolves according to the following asynchronous update rule:

$$u_i(t) = \begin{cases} \sin\hat{\rho} & \text{if } \sum_{j=1}^{i-1} W_{ij} u_j(t) + \sum_{j=i+1}^{h} W_{ij} u_j(t-1) - \theta_i > 0 \\ -\cos\hat{\rho} & \text{if } \sum_{j=1}^{i-1} W_{ij} u_j(t) + \sum_{j=i+1}^{h} W_{ij} u_j(t-1) - \theta_i \leq 0 \end{cases} \tag{1}$$

where $u_i(t)$ is the state of neuron $i \in U$ at time t, and $\theta_i = \hat{k} - \sum_{j=h+1}^{n} W_{ij} \hat{l}_j$ is the activation threshold of node i. The state of the network at time t is $\boldsymbol{u}(t) = (u_1(t), u_2(t), \ldots, u_h(t))$, and the main feature of a HN is that it admits a Lyapunov function of the dynamics. In particular, consider the following quadratic state function (energy function):

$$E(\boldsymbol{u}) = -\frac{1}{2} \sum_{\substack{i,j=1 \\ j \neq i}}^{h} W_{ij} u_i u_j + \sum_{i=1}^{h} u_i \theta_i \qquad (2)$$

During the dynamics this function is not increasing and the dynamics converges to an equilibrium state $\hat{\boldsymbol{u}}$, which corresponds to a local minimum of the energy function [11]. The motivation of this approach is that minimizing (2) means maximizing the weighted sum of consistent edges, that is edges connecting neurons at the same state, so as to maximize the coherence with the prior information coded in W and in the labeling $\hat{\boldsymbol{l}}$. The final solution (U_+, U_-) is obtained by setting $U_+ = \{i \in U \mid \hat{u}_i = \sin \hat{\rho}\}$ and $U_- = \{i \in U \mid \hat{u}_i = -\cos \hat{\rho}\}$. In [9] the authors showed that although parameter optimization and label inference are carried out separately, Step 2 and 3 of *COSNet* preserve convergence and optimization properties of the whole HN H.

4.2 A Strategy for Negative Selection

The strategy we propose, able to efficiently identify the subset S^p_-, is composed of three main steps, that we describe in detail in the following.

Node projection onto a feature space. The n by n connection matrix W is transformed into a n by 3 feature matrix F, where the i-th row $F_i = (p^1_i, p^2_i, p^3_i)$ is the feature vector associated with node $i \in V$. The j-th feature p^j_i is the probability that a random walk of length j starting from positive instances ends at node i. This choice comes from previous studies, which have proven that such a feature matrix suffices to propagate information coded in the graph labels [17], that is the information coded in the features corresponding to random walks length k, with $k > 3$, is negligible.

Scoring non positive points through fuzzy clustering. Let $\mathcal{F} = \{F_i | i \in V\}$ be the set of projected points, we denote by $\mathcal{F}_+ = \{F_i | i \in S_+\}$ and $\mathcal{F}_- = \{F_i | i \in S_-\}$ the sets of positive and negative projected points. Points in $\mathcal{F}_-$ are scored according to their relation w.r.t. a fuzzy clustering of the set $\mathcal{F}_+$ of positive points. Such clustering is computed through repeated applications of the FCM algorithm [3]. In order to automatically adjust the number of clusters, at each execution of FCM the number c of clusters is changed, aiming at maximizing the *fuzzy silhouette* index [6], a fuzzy extension of the crisp silhouette index [12] expressly considering the fuzzy nature of the membership functions provided by FCM. Being the space dimension fixed to 3, the computational complexity of each execution of the FCM algorithm (including the assessment through the fuzzy silhouette index) will be $\mathcal{O}(Ic^2|S_+|)$, denoting by I the number of iterations of the clustering algorithm. Since the number of positive instances in our context is very small, *a fortiori* also the number of clusters will be low, thus the overall step has a low computational complexity.

Once an optimal clustering of size c has been found, we assign each point $x \in \mathcal{F}_-$ a *score* $\phi(x)$, obtained considering the maximum membership value of x to the various clusters, that is:

$$\phi(x) = \max_{1 \leq k \leq c} \left\{ \left(\sum_{j=1}^{c} \left(\frac{d(x, v_k)}{d(x, v_j)} \right)^{\frac{2}{a-1}} \right)^{-1} \right\} \tag{3}$$

where $v_1, \ldots, v_c$ are the cluster centroids, while α and d respectively denote the fuzzification parameter and the distance function used by the FCM algorithm. Therefore, a non-positive point having a score lower than a fixed threshold $\tau > 0$ cannot be reasonably attributed to any cluster grouping positive points, whereas nodes corresponding to points $\mathcal{F}_{-,\tau} = \{x \in \mathcal{F}_- | \phi(x) \geq \tau\}$ are good candidates for the set S^p_-.

Even this step can be efficiently computed: the distance d of 3-feature vectors can be computed in $\mathcal{O}(1)$ time and the time complexity of computing $\phi(x)$ $\forall x \in \mathcal{F}_-$ is $\mathcal{O}(|\mathcal{F}_-|)$, since $c \leq |S_+|$ and $|S_+|/|S_-| \ll 1$.

Selecting negative instances through local equilibrium. We denote by $S_{-,\tau} = \{i \in S_- | F_i \in \mathcal{F}_{-,\tau}\}$ the set of negative nodes corresponding to points selected at the previous step. We consider the *COSNet* HN $H = \langle W, k, \rho \rangle$ and estimate the optimal parameters $\hat{k}, \hat{\rho}$. Then, nodes in $S_{-,\tau}$ are further skimmed according to their stability in the labeled sub-network $H_S = \langle W_S, \hat{k}, \hat{\rho} \rangle$, where W_S is the sub-matrix of connections between nodes in S. In particular, a node $i \in S_-$ is *at equilibrium* if $\beta_S(i) = \sum_{j \in S} W_{ij} \hat{l}_j - \hat{k} \leq 0$. Accordingly, we define $S^p_- = \{i \in S_{-,\tau} | \beta_S(i) > 0\}$. Nodes in S^p_- are thereby such that a possible network update of the sub-network H_S would change their state from negative to positive, thus our choice is coherent with both prior information (W and $\hat{l}$) and network local stability.

4.3 Embedding Negative Selection into the Hopfield Model

After selecting the set S^p_-, we run the dynamics of *COSNet* with labeled and unlabeled nodes defined as $\overline{S} = S \setminus S^p_-$ and $\overline{U} = U \cup S^p_-$, respectively. Labels of nodes in S^p_- are thereby not utilized in computing activation thresholds θ_i in (1); nevertheless, as such information has been considered by the learning procedure, to preserve consistency with the learning criterion, the initial state of neurons in S^p_- is set to $-\cos \hat{\rho}$. This choice allows to avoid the parameter relearning, since $-\cos \hat{\rho}$ is the value assigned to nodes in S^p_- by the learning procedure (see [9] for more details).

Finally, it is worth pointing out that, although the dynamics involves also nodes in S^p_-, the inference is still performed solely on nodes in U.

5 Experimental Validation

We validate our approach in predicting the Gene Ontology (GO) functions (release 23-3-13) of the whole genome of the *S.cerevisiae* (a yeast) model organism. In order to predict more specific and unbalanced terms in the ontology, we selected GO Molecular Functions (MF) terms with 30–300 positive annotated genes. The connection network has been obtained by unweighted sum integration on the union genes of 16 networks downloaded from the GeneMANIA website[1] and covering different types of data, ranging from co-expression to physical interactions. The final network has a total of 5775 yeast genes and 127 MF terms.

5.1 Results

We compared our inference methodology with the state-of-the-art supervised and semi-supervised algorithms proposed in the literature for the *NLP* problem. In particular, we considered the *Support Vector Machine* (SVM) algorithm, largely applied in computational biology, and more precisely its probabilistic version [13]; the *Random Forest* (RF) method [4]; a *Label Propagation* (LP) algorithm based on Gaussian random fields and harmonic functions [21]; the classical *random walk* (RW) algorithm without restart with at most 100 steps [15]. Moreover, we also considered the original version of *COSNet*, without negative selection, to evaluate the impact of our negative selection strategy on the performance of the model.

The generalization performances have been estimated through a 5-fold cross-validation procedure, and the performances have been assessed using the *Area Under the Precision-Recall Curve* (AUPRC), the *F-measure* (harmonic mean of precision and recall) and Area Under the ROC Curve (AUC). The AUPRC to some extent represents how close the classifier is to a perfect oracle, which would predict no false positives and have an AUC of 1. Indded, the AUC can be interpreted as the probability that a randomly chosen true positive will be ranked higher by the classifier than a randomly chosen true negative. Finally, the F measures the effectiveness of retrieval by taking into account both the precision and the recall of a classifier, that is the probability that a positive predicted is a true positive, and the proportion of true positives the classifier predicts, respectively. Usually, the F-measure is adopted to evaluate the performance in classifying positive instances in contexts where positives are rare.

The results averaged across the 127 GO terms are reported in Table 1. Our method achieves the best results in terms of all the adopted measures, with improvements statistically significant according to the Wilcoxon signed-rank test [19], except for the AUC results. The semi-supervised methods LP and RW have competitive performance in terms of AUC and AUPRC, where they perform as the second and the third

---

[1]http://www.genemania.org.

Table 1 Average results

Algorithm	AUC	AUPRC	F
LP	0.947	0.525	0.033
RW	0.941	0.473	0.395
RF	0.602	0.075	0.059
SVM	0.576	0.052	0.025
COSNet-neg	**0.949**	**0.529**	**0.582**

COSNet-neg is our method. The best results are highlighted in bold, whereas methods that are significantly better than all the others according to the Wilcoxon signed-rank sum test ($\alpha = 0.05$) are underlined

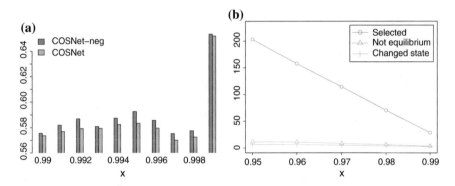

Fig. 1 **a** Comparison of *COSNet* and *COSNet-neg* in terms of average F. *x* is the selected percentile. **b** *Selected* corresponds to $|\mathcal{F}_{-,\tau}|$ averaged across selected classes, *Not equilibrium* to average $|S^p_-|$ and *Changed state* is the average number of nodes in S^p_- which have changed their initial state after the dynamics of the network

best method respectively. Very poor results are obtained by the supervised algorithms SVM and RF, likely due to the fact that they are not well-suited for *NLP* problems with unbalanced labelings.

Finally, the results in terms of F show that our method largely outperforms the compared methodologies as classifier, where only the RW algorithm does not poorly perform w.r.t. remaining methods. F values close to 0 simply means that very few true positives have been predicted.

In order to evaluate the contribution of the negative selection strategy to the performance of the model, in Fig. 1a we report the comparison in terms of F between *COSNet-neg* and *COSNet* (for AUC and AUPRC we have a similar behavior). We selected the parameter τ as the *x*-th percentile of $\phi(\mathcal{F}_-)$, and vary *x* from 0.99 to 0.999 with step 0.001. The negative selection improves the performance for every choice of τ, whereas the average results considerably vary with *x*, because the comparison is restricted to classes with at least one negative instance not at equilibrium (i.e. $|S^p_-| > 0$), and clearly the number of such classes varies with the selected percentile *x*. We chose just these classes for the comparison because when all the neurons in $\mathcal{F}_{-,\tau}$ are at equilibrium, *COSNet* and *COSNet-neg* are identical. After a tuning on a small subset of labeled data, in our experiments we set $x = 0.995$.

Finally, to further analyze the negative selection procedure, in Fig. 1b we report, for different values of x, the average number of negative instances whose score (3) is greater than τ, the average number of those not at equilibrium (average $|S_-^p|$), and the average number of neurons in S_-^p whose initial and final state differ. As expected, the value of $|\mathcal{F}_{-,\tau}|$ increases linearly when τ decreases. Nevertheless, among neurons in $|\mathcal{F}_{-,\tau}|$, just a small proportion is not at equilibrium in the labeled sub-network, and such proportion decreases quickly when τ decreases, showing on the one hand that the non-equilibrium condition for belonging to S_-^p is highly discriminating, on the other hand that the method is quite robust to variations of parameter τ. Furthermore, among neurons not at equilibrium, a small subset changes the initial state during the dynamics of the unlabeled network, and this may be due to: (1) the unlabeled network tends to predict the proportion of positive instances in the training data, and since such proportion is small, it is expected that a relevant part of neurons in S_-^p is predicted as negative; (2) nodes in S_-^p are not at equilibrium when considering just labeled data, but when we complete the information with the unlabeled part, their initial condition of being negative is restored; (3) some of the negative instances selected for the set S_-^p can be associated to feature vectors characterized by noise, since prior information in this context is still affected by both experimental and biological noise [16].

6 Conclusions

In this work we propose a novel strategy for selecting negative examples in contexts where just positive associations can be considered as reliable. In particular, our methodology is designed explicitly for algorithms based on Hopfield networks, and can be applied in context where classes to be predicted are both structured in a hierarchy or hierarchy-less. The negative selection strategy, based on a fuzzy clustering procedure with low computational impact, has been embedded in the dynamics of a Hopfield network. Its effectiveness has been assessed through the comparison with state-of-the-art approaches in predicting the biological functions of the whole proteome of *S.cerevisiae* organism. The promising results encourage to further extend this approach, in particular for improving the negative selection by exploiting also the hierarchy of functional classes.

References

1. Ashburner, M., Ball, C.A., Blake, J.A., Botstein, D., Butler, H., Cherry, J.M., Davis, A.P., Dolinski, K., Dwight, S.S., Eppig, J.T., et al.: Gene ontology: tool for the unification of biology. Nat. Genet. **25**(1), 25–29 (2000)
2. Bertoni, A., Frasca, M., Valentini, G.: Cosnet: A cost sensitive neural network for semi-supervised learning in graphs. In: Machine Learning and Knowledge Discovery in Databases—European Conference, ECML PKDD 2011, Athens, Greece, 5–9 September 2011. Proceedings, Part I. LNAI, vol. 6911, pp. 219–234. Springer-Verlag (2011)

3. Bezdek, J.C., Ehrlich, R., Full, W.: Fcm: the fuzzy c-means clustering algorithm. Comput. Geosci. **10**(2), 191–203 (1984)
4. Breiman, L.: Random forests. Mach. Learn. **45**(1), 5–32 (2001)
5. Burghouts, G.J., Schutte, K., Bouma, H., den Hollander, R.J.M.: Selection of negative samples and two-stage combination of multiple features for action detection in thousands of videos. Mach. Vis. Appl. **25**(1), 85–98 (2014)
6. Campello, R.J.G.B., Hruschka, E.R.: A fuzzy extension of the silhouette width criterion for cluster analysis. Fuzzy Sets Syst. **157**(21), 2858–2875 (2006)
7. Fagni, T., Sebastiani, F.: On the selection of negative examples for hierarchical text categorization. In: Proceedings of the 3rd Language & Technology Conference (LTC07). pp. 24–28 (2007)
8. Ferretti, E., Errecalde, M.L., Anderka, M., Stein, B.: On the use of reliable-negatives selection strategies in the PU learning approach for quality flaws prediction in wikipedia. In: 2014 25th International Workshop on Database and Expert Systems Applications (DEXA), pp. 211–215 (2014)
9. Frasca, M., Bertoni, A., Re, M., Valentini, G.: A neural network algorithm for semi-supervised node label learning from unbalanced data. Neural Netw. **43**, 84–98 (2013)
10. Gomez, S.M., Noble, W.S., Rzhetsky, A.: Learning to predict protein-protein interactions from protein sequences. Bioinformatics **19**(15), 1875–1881 (2003)
11. Hopfield, J.J.: Neural networks and physical systems with emergent collective compatational abilities. Proc. Natl. Acad. Sci. **79**(8), 2554–2558 (1982)
12. Kaufman, L., Rousseeuw, P.J.: Finding Groups in Data. Wiley, New York (1990)
13. Lin, H.T., Lin, C.J., Weng, R.C.: A note on Platt's probabilistic outputs for support vector machines. Mach. Learn. **68**(3), 267–276 (2007)
14. Liu, B., Dai, Y., Li, X., Lee, W.S., Yu, P.S.: Building text classifiers using positive and unlabeled examples. In: Third IEEE International Conference on Data Mining, 2003. ICDM 2003. pp. 179–186 (2003)
15. Lovász, L.: Random walks on graphs: A survey. In: Combinatorics, Paul Erdős is Eighty. pp. 353–397 (1993)
16. Marshall, E.: Getting the noise out of gene arrays. Science **306**(5696), 630–631 (2004)
17. Mostafavi, S., Goldenberg, A., Morris, Q.: Labeling nodes using three degrees of propagation. PLoS ONE **7**(12), e51947 (2012)
18. Mostafavi, S., Morris, Q.: Using the gene ontology hierarchy when predicting gene function. In: Proceedings of the twenty-fifth conference on uncertainty in artificial intelligence. pp. 419–427 (2009)
19. Wilcoxon, F.: Individual comparisons by ranking methods. Biom. Bull. **1**, 80–83 (1945)
20. Youngs, N., Penfold-Brown, D., Drew, K., Shasha, D., Bonneau, R.: Parametric bayesian priors and better choice of negative examples improve protein function prediction. Bioinformatics **29**(9), tt10–98 (2013)
21. Zhu, X., Ghahramani, Z., Lafferty, J.: Semi-supervised learning using gaussian fields and harmonic functions. In: ICML. pp. 912–919 (2003)

Recent Advances on Distributed Unsupervised Learning

Antonello Rosato, Rosa Altilio and Massimo Panella

Abstract Distributed machine learning is a problem of inferring a desired relation when the training data is distributed throughout a network of agents (e.g. sensor networks, robot swarms, etc.). A typical problem of unsupervised learning is clustering, that is grouping patterns based on some similarity/dissimilarity measures. Provided they are highly scalable, fault-tolerant and energy efficient, clustering algorithms can be adopted in large-scale distributed systems. This work surveys the state-of-the-art in this field, presenting algorithms that solve the distributed clustering problem efficiently, with particular attention to the computation and clustering criteria.

Keywords Unsupervised learning · Distributed clustering · Decentralized and diffused computing

1 Introduction

The goal of this paper is to provide a comprehensive list of state-of-the-art algorithms regarding distributed unsupervised learning for clustering [8]. There are many applications of clustering in different fields, first of all in computer vision where there is need of efficient algorithms in applications such as object recognition, facial recognition, environment mapping, etc. Unsupervised learning can also be used for finance, bioinformatics and data mining applications, where it is very useful to implement unsupervised learning algorithms especially in big data environments. Other applications are in the field of signal processing, physics, statistics and computational intelligence more in general.

A. Rosato · R. Altilio · M. Panella (✉)
Department of Information Engineering, Electronics and Telecommunications (DIET),
University of Rome "La Sapienza", Via Eudossiana 18, 00184 Rome, Italy
e-mail: massimo.panella@uniroma1.it

A. Rosato
e-mail: antonello.rosato@uniroma1.it

R. Altilio
e-mail: rosa.altilio@uniroma1.it

© Springer International Publishing Switzerland 2016
S. Bassis et al. (eds.), *Advances in Neural Networks*, Smart Innovation,
Systems and Technologies 54, DOI 10.1007/978-3-319-33747-0_8

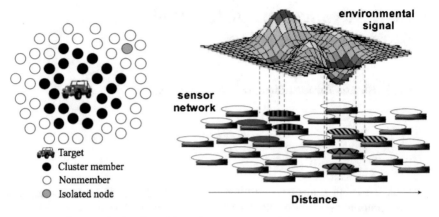

Fig. 1 Visualization of a distributed clustering environment [2]

There are two different solutions for unsupervised learning algorithms regarding the structure of their computation. The first one is the *centralized* architecture: in a pure centralized system there is a central agent (node) where all the learning takes place. In this case the serialisability of the algorithm is always preserved but the computation may take a lot of cycles. In some cases, the centralized architecture is often realized by using multiple agents in a *parallel* architecture where each agent sends the partial results of the computation to the central node via a network of some sort. In this case it is up to the central agent to preserve serialization and fuse the partial results to have a global output.

The other solution is the *distributed* architecture, which is depicted in Fig. 1. In a pure distributed architecture there is no hierarchy and all agents are peers. In this case the data is partitioned among the agents (equally or proportionally to the cal-culation power) and each agent learns only from its own dataset. In practice, also in the distributed architecture there is a network used to exchange information between the agents. This is done to ensure that local results are all coherent.

In this work we will not consider scalable and parallel algorithms over a network of agents starting from a centralized system, since they refer to computational meth-ods only. Instead, we will select the works that present an intrinsically distributed fashion of the learning algorithm.

2   Distributed Clustering Algorithms

2.1   *Algorithms Based on Gaussian Mixture Models*

This class of algorithms is based on Gaussian Mixture Models (GMM); it is assumed that each node retrieves data from an environment that can be described by a proba-bility density function (PDF) as a mixture of elementary conditions [18]. The mixture parameters are commonly estimated in an iterative way through the Expectation-Maximization (EM) algorithm, which is a maximum-likelihood estimator [27]. EM algorithms in general are more computationally affordable and guarantee convergence to a local optimum or a saddle point of the maximum-likelihood [28]. However, in a distributed scenario, transmitting all data to a centralized unit for executing the standard EM algorithm is not an efficient and robust approach.

An affordable solution in these cases is to perform the EM steps followed by a consensus step. The consensus is used to solve the problem of reconciling clustering information about the same data set coming from different sources or from different runs of the same algorithm. Different consensus criteria can be used; the most significant ones in distributed EM clustering algorithms are presented in the following.

In [24] a fully distributed EM algorithm is proposed. This algorithm, called DEM, focus on eliminating the message passing process in the parallel implementation of EM. This is done by cycling through the network and performing incremental E-step and M-step at each node. It uses only the local data and summary statistics passed from the previous node in the cycle. Because of its incremental form, DEM often converges much more rapidly than the standard EM algorithm. The DEM can be modified to be much more effective in terms of power consumption, especially when used in wireless sensor network (WSN) applications. This can be done by doing multiple steps at each node thus reducing the number of cycles through the network and removing the need for extensive communication, this variation of the algorithm is called DEMM. Other variations are related to the organization of dense networks.

The EM algorithms presented in [22] are focused instead on finding the best consensus criteria. Three approaches are proposed: Iterative Voting Consensus (IVC), Iterative Probabilistic Voting Consensus (IPVC) and Iterative Pairwise Consensus (IPC). In IVC the clusters' centres are computed and then every point is reassigned to the closest cluster centre. IPVC is similar to IVC with a refined metric that considers how much each feature of every point differs from those of points in a cluster of other nodes. IPC is different from the other two because it uses a pairwise similarity approach using a similarity matrix computed from the preceding iteration; namely, it can be seen as a variation to the similarity matrix of the K-means method. The authors in [5] present also a consensus-based EM algorithm in which the main novelty compared to the other algorithms cited above is that class-conditional PDFs are allowed to be non-Gaussian. This algorithm (denoted as CB-DEM) relies on a subset of nodes in the network called *bridge sensors*, which impose consensus on the estimated parameters among the set of neighbouring (one-hop) sensors. By using bridge sensors, CB-DEM does not need to find a path among all the nodes in the network (as DEM does).

In the standard EM algorithm for the Gaussian mixtures, the local sufficient statistics can be calculated by only using local data in each sensor node in the E-step. Since the global sufficient statistics are usually required in the M-step, in [10] a distributed EM algorithm is proposed that estimates the global sufficient statistics using local information and neighbours' local information. First, it calculates the local sufficient statistics in the E-step as usual; then, it estimates the global sufficient statistics. Finally, it updates the parameters in the M-step using the estimated global sufficient

Fig. 2 The log-likelihood
function for different
algorithms [30]

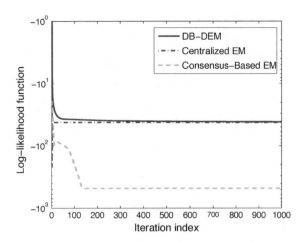

statistics. The last step is done using a consensus filter that acts like a low-pass filter.
One major drawback is that this algorithm requires at each node to know the number
of Gaussian mixtures beforehand.

In [4] a generic algorithm is provided that can solve the EM clustering problem
with a variety of different settings. The novelty is that an adaptive in-network compression is employed and also, a different approach for the M-step is used in [33].
The diffusion adaptation strategy is based on a Newton's recursion instead of using a
steepest-descent argument for optimization. If compared with consensus-based solutions, the diffusion strategy does not require different agents to converge to the same
global statistics; the individual agents are allowed flexibility through adaptation and
through their own assessment of local information.

Since the log-likelihood function is exactly quadratic, Newton's method is suitable because it uses a quadratic approximation to derive the optimal update. Using
diffusion strategies, a distributed EM estimator has been derived in [30], where the
diffusion of the information is embedded in the update of the parameters. In this
Diffusion-Based Distributed EM (DB-DEM) the M-step can be viewed as a two-
time-scale operation. As shown in Fig. 2, the DB-DEM has a behaviour similar to
the centralized one and, requiring only one averaging operation at each M-step, it
is better than the previously considered CB-EM, which is not able to estimate the
parameters accurately with a same number of averaging iterations.

2.2 Algorithms Based on K-Means

K-Means is one of the simplest approaches to clustering [11]. The number of clusters
is fixed a priori and a pattern is assigned a cluster based on a suited metric measuring
the distance of that pattern from each cluster centroid. Usually, a Euclidean distance

is minimized and centroids are updated iteratively on a basis of pattern assignments to clusters, until they do not change location at a local convergence. The difference from GMM assumption is that K-means depends usually on the L2 norm and it is not based on a statistical expectation. Many works deal with a parallel version of K-Means, these algorithms are not actually distributed because they just split the calculation between different subsystems. In this case each subsystem firstly does the computation and then it sends the results to a central node that takes decisions.

Authors in [15] developed a distributed clustering algorithm called K-DMeans, it has a master-slave architecture focused on Ethernet networks, but can be considered a real distributed version. Once the data is divided into subsets, each agent calculates its central point and informs the other agents of its central point. After that, each agent calculates the distances of patterns of its local subset from each central point and clusters the data. Data not belonging to centroids of the local subset are transferred to another agent associated with that centroid. This process iterates until a chosen discriminative function, such as variance, becomes stable. Unfortunately, each iteration causes high computation and communication costs. In [41] there is an improved version (DK-means) where each sub site needs to send to the main site the central point and the number of patterns only, thus reducing such costs. The problem is that the number of clusters has to be known in advance and also, a central node boosts the clustering performances. A purely distributed version can be found in [19], where the clustering is implemented using the peer-to-peer (P2P) network model using no central nodes.

In [32] is described a distributed version of the k-windows algorithm, which fits well the distributed computing paradigm. The k-windows algorithm has the ability to endogenously determine the number of clusters. This is a fundamental issue in cluster analysis, independent of the particular technique applied. The proposed distributed version of the algorithm is able to determine the number of clusters present in the dataset with a satisfactory accuracy, without setting a priori the number of clusters. In this algorithm no data exchange is allowed among the local nodes and, even with this highly restricting hypothesis, its performance is comparable to other algorithms.

An interesting version of distributed K-means is implemented in [25]. In this work the inherent serialisability of the clustering operations is preserved in the distributed implementation by using the Optimistic Concurrency Control (OCC) approach. With OCC the number of messages exchanged between local agents is highly reduced. This stems from the fact that the conflicts of partial clustering results are resolved much more efficiently than in a standard parallelised K-means. Other versions are also implemented, in particular a distributed DP-Means where the number of clusters is not fixed a priori.

The problem of distributed clustering where the data is distributed across nodes, whose communication is restricted to the edges of an arbitrary graph, is studied in [1] where an algorithm with small communication costs and suited quality of clustering results is presented. The technique for reducing communication in general graphs is based on the construction of a small set of points, which act as a proxy for the entire data set. This algorithm is revised in [20] where Principal Component Analysis (PCA) is used to improve the performance over large networks. When combined with the distributed approach, this leads to an algorithm whose communication costs (in terms of the number of transmitted/shared patterns) are independent of the size and dimension of the original data.

2.3 Algorithms Based on Density Models

In distributed clustering algorithms based on density models, the clustering process iterates until a neighbour's density exceeds a given threshold. In this case, the density refers to the number of objects or data points to be clustered. The most representative work for density-based clustering is [3] where an algorithm called Density Based Spatial Clustering of Applications with Noise (DBSCAN) is presented. This algorithm is designed to solve the clustering problem in a large spatial database with noise. The DBSCAN can resolve clusters of any shape, and the clustering result is independent of the data input order.

There are distributed algorithms based on different extensions of DBSCAN. Xu brought up a parallel version of DBSCAN in [39]; this algorithm firstly uses R*-tree to organize data in the central site, then stores the preprocessed data in each local agent, which communicates with other agents by exchanging messages. In [13] the authors proposed a distributed version of DBSCAN algorithm called DBDC, where there is a central node that organizes the clustering results and resolves conflicts. A more refined version in terms of scalability and performance, denoted as DBDC, is presented in [14]. Ni et al. in [23] put forward a distributed clustering algorithm denoted as LDBDC, which is based on local density estimation. Other algorithms to be considered in this context are: KDEC algorithm for distributed clustering based on kernel density estimation preserving the privacy of data [17]; forward Distributed Density Based Clustering (DDC) algorithm for very large and distributed heterogeneous datasets [16]; Ordering Point To Identify Clustering Structure (OPTICS) algorithm in a distributed environment [9].

2.4 Algorithms Based on Fuzzy Logic

In this section some fuzzy clustering algorithms are presented. In fuzzy clustering, every cluster is considered as a fuzzy set and every point has a degree of membership to clusters, rather than belonging exclusively to a cluster or not. For example, points on the edge of a cluster are not hardly assigned to the nearest cluster but may be assigned with a lower degree than points in the centre of cluster. Most of the works in this field pertain to parallelisation of serial algorithms. A very comprehensive analysis and overview of parallel fuzzy clustering can be found in [34].

The only algorithm that can be efficiently implemented in a purely distributed fashion is based on the well-known Fuzzy C-Means (FCM), and its extensions in the joint input-output space for data regression [26]. In [29] an early approach was proposed with particular attention to image segmentation. The sequential FCM algorithm is computationally intensive and has significant memory requirements. In fact, for many applications such as medical image segmentation and geographical image analysis, which deal with large size images, sequential FCM is very slow. The parallel FCM algorithm, by dividing the computations among the processors and minimizing the need for accessing secondary storage, enhances the performance and efficiency of image segmentation tasks.

In [35] authors present an extension of the consensus-based algorithm to distributed (DFCM). A procedure to automatically estimate the number of clusters with the DFCM clustering algorithm is also provided. This algorithm merges FCM with the consensus criteria previously illustrated, giving interesting performance in applications where sites have data from multiple sources. The study in [42] focuses on the data clustering in distributed P2P networks and proposes a novel collaborative clustering algorithm. The proposed method reduces and balances the communication overhead among peers. The focus of the algorithm is P2P networking and the fuzzy clustering is developed using a collaborative approach between local agents. Experiments on several synthetic and real-world datasets show that the proposed algorithms yield comparative or better performance compared with other clustering methods.

2.5 Other Algorithms

In this section some other algorithms are presented, which cannot be fitted into the categories viewed above. These distributed clustering methods are different from the others for a variety of reasons.

A class of algorithms to mention is *Ensemble Clustering* [31]. They are characterized by two main phases: (i) to generate a number of individual clustering solutions, and (ii) to find a final consensus solution with these clustering solutions. In [21] a graph-based approach is used. In [37] a Bayesian approach (denoted as BCE) is presented, it provides a way to combine clustering results avoiding cluster label correspondence problems that are encountered in graph based approaches. In [12] local agents exchange the clusters' centroids only. On the other hand, in [36] the focus is on keeping the privacy of data.

The authors in [6, 38] presented the so called 'Samarah' method for distributed learning. Each local learner uses it's own strategy (K-Means, Cobweb, etc.) and its own data, and a corresponding function between objects from each pair of datasets must be defined. The distributed consensus step is implemented in a collaborative way: clustering methods collaborate to find an agreement about the clustering of dataset. In [7] an application of Samarah to remote sensing image is discussed.

Other good unsupervised learning algorithms are the ones designed for sensor networks where the main goal is not to cluster data but to classify and group nodes of a network. Once the nodes are clustered, the information between them can be exchanged more efficiently. One example of this class of algorithms can be found in [40], where authors propose a new energy-efficient approach for clustering nodes

Table 1 Analysed techniques

Algorithms	Pros	Cons
Gaussian mixtures	Scalable and robust, low communication costs	Necessity of bridge sensors, given number of Gaussian components
K-Means	Low processing costs, less exchange of information	Given number of clusters, necessity of a central node
Density models	Discover clusters of any shape and size	Master-slave architecture or local representatives
Other algorithms	Scalable and robust	Necessity of a central node, same number of clusters at each node

in ad-hoc sensor networks. In Table 1 we have summarized the Pros and Cons of the considered algorithms. Most of them use a partition of the original dataset and so, the placement of data will influence the result. In this regard, the ideal partition should satisfy the basic requirements of load balancing, minimum communication cost and distributed data access.

3 Conclusions

Recent approaches to purely distributed unsupervised learning are presented and reviewed in this paper. These algorithms have been selected based on the criteria underlying their implementation. Namely, we have analysed algorithms based on: EM criteria, K-means algorithm, density/diffusion principle, and fuzzy logic. Other algorithms based on different principles are also presented and briefly discussed. All the considered algorithms are all very useful in a big data environment and where resources are limited in a network of agents and peers. Overall, it is evident that the technical literature is still lacking of a comprehensive implementation on distributed clustering for unsupervised learning. The works mentioned herein should represent a good basis of study to develop future approaches, considering the currently hot direction towards EM criteria combined with consensus and ensemble learning, because of their highly scalable properties and good performance.

References

1. Balcan, M., Ehrlich, S., Liang, Y.: Distributed k-means and k-median clustering on general topologies. Adv. Neural Inf. Process. Syst. **26**, 1995–2003 (2013)
2. Charalambous, C., Cui, S.: A bio-inspired distributed clustering algorithm for wireless sensor networks. In: Proceedings of the 4th Annual Int. Conf. on Wireless Internet (WICON'08) (2008)

3. Ester, M., Kriegel, H., Sander, J., Xu, X.: A density-based algorithm for discovering clusters in large spatial databases with noise. In: Proceedings of International Conference on Knowledge Discovery and Data Mining (KDD), vol. 96, pp. 226–231 (1996)

4. Eyal, I., Keidar, I., Rom, R.: Distributed data clustering in sensor networks. Distrib. Comput. **24**(5), 207–222 (2010)

5. Forero, P., Cano, A., Giannakis, G.: Consensus-based distributed expectation-maximization algorithm for density estimation and classification using wireless sensor networks. In: Proceedings of ICASSP. pp. 1989–1992 (2008)

6. Forestier, G., Gançarski, P., Wemmert, C.: Collaborative clustering with background knowledge. Data Knowl. Eng. Arch. **69**(2), 211–228 (2010)

7. Gançarski, P.: Remote sensing image interpretation. http://omiv2.u-strasbg.fr/imagemining/documents/IMAGEMINING-Gancarski-Multistrategy.pdf. Accessed 04 April 2015

8. Ghahramani, Z.: Unsupervised Learning. In: Lecture Notes in Computer Science, vol. 3176, pp. 72–112. Springer (2004)

9. Ghanem, S., Kechadi, T., Tari, A.: New approach for distributed clustering. In: Proceedings of IEEE International Conference on Spatial Data Mining and Geographical Knowledge Services (ICSDM). pp. 60–65 (2011)

10. Gu, D.: Distributed EM algorithm for Gaussian mixtures in sensor networks. IEEE Trans. Neural Netw. **19**(7), 1154–1166 (2008)

11. Hartigan, J., Wong, M.: Algorithm AS 136: a k-means clustering algorithm. J. R. Stat. Soc. Series C (Appl. Stat.) **28**(1), 100–108 (1979)

12. Hore, P., Hall, L., Goldgof, D.: A scalable framework for cluster ensembles. Pattern Recognit. **42**(5), 676–688 (2009)

13. Januzaj, E., Kriegel, H., Pfeifle, M.: Towards effective and efficient distributed clustering. In: Proceedings of Workshop on Clustering Large Data Sets (ICDM). pp. 49–58 (2003)

14. Januzaj, E., Kriegel, H., Pfeifle, M.: Scalable density-based distributed clustering. In: Proceedings of European Conference on Principles and Practice of Knowledge Discovery in Databases (PKDD). pp. 231–244 (2004)

15. Kantabutra, S., Couch, A.: Parallel k-means clustering algorithm on NOWs. MedTec Tech. J. **1**(6), 243–248 (2000)

16. Khac, N., Aouad, L., Kechadi, T.: A new approach for distributed density based clustering on grid platform. In: Lecture Notes in Computer Science, vol. 4587, pp. 247–258. Springer (2007)

17. Klusch, M., Lodi, S., Moro, G.: Distributed clustering based on sampling local density estimates. In: Proceedings of the Int. Joint Conference on Artificial Intelligence (IJCAI'03). pp. 485–490 (2003)

18. Laird, N., Dempster, A.P., Rubin, D.: Maximum likelihood from incomplete data via the EM algorithm. J. R. Stat. Soc. Series B **39**(1), 1–38 (1977)

19. Li, L., Tang, J., Ge, B.: K-DmeansWM: an effective distributed clustering algorithm based on P2P. Comput. Sci. **37**(1), 39–41 (2010)

20. Liang, Y., Balcan, M., Kanchanapally, V.: Distributed PCA and k-means clustering. In: The Big Learning Workshop at NIPS (2013)

21. Mimaroglu, S., Erdil, E.: Combining multiple clusterings using similarity graph. Pattern Recognit. **44**, 694–703 (2011)

22. Nguyen, N., Caruana, R.: Consensus clustering. In: Proceedings of IEEE International Conference on Data Mining. pp. 607–612 (2006)

23. Ni, W., Chen, G., Wu, Y.: Local density based distributed clustering algorithm. J. Softw. pp. 2339–2348 (2008)

24. Nowak, R.: Distributed EM algorithms for density estimation and clustering in sensor networks. IEEE Trans. Signal Process. **51**(8), 2245–2253 (2003)

25. Pan, X., Gonzalez, J., Jegelka, S., Broderick, T., Jordan, M.: Optimistic concurrency control for distributed unsupervised learning. In: Proceedings of 27th Annual Conference on Neural Information Processing Systems. pp. 1403–1411 (2013)

26. Panella, M.: A hierarchical procedure for the synthesis of ANFIS networks. Adv. Fuzzy Syst. **2012**, 1–12 (2012)

27. Panella, M., Rizzi, A., Martinelli, G.: Refining accuracy of environmental data prediction by MoG neural networks. Neurocomputing **55**(3–4), 521–549 (2003)
28. Parisi, R., Cirillo, A., Panella, M., Uncini, A.: Source localization in reverberant environments by consistent peak selection. In: Proceedings of ICASSP. vol. 1, pp. I–37–I–40 (2007)
29. Rahmi, S., Zargham, M., Thakre, A., Chhillar, D.: A parallel fuzzy c-mean algorithm for image segmentation. In: Proceedings of NAFIPS'04. vol. 1, pp. 234–237 (2004)
30. Silva-Pereira, S., Pages-Zamora, A., Lopez-Valcarce, R.: A diffusion-based distributed EM algorithm for density estimation in wireless sensor networks. In: Proceedings of ICASSP. pp. 4449–4453 (2013)
31. Strehl, A., Ghosh, J.: Cluster ensembles a knowledge reuse framework for combining multiple partitions. J. Mach. Learn. Res. **3**, 583–617 (2002)
32. Tasoulis, D., Vrahatis, M.: Unsupervised distributed clustering. Parallel Distrib. Comput. Netw. pp. 347–351 (2004)
33. Towfic, Z., Chen, J., Sayed, A.: Collaborative learning of mixture models using diffusion adaptation. In: Proceedings of IEEE International Workshop on Machine Learning for Signal Processing (MLSP). pp. 1–6 (2011)
34. Vendramin, L.: Estudo e desenvolvimento de algoritmos para agrupamento fuzzy de dados em cenarios centralizados e distribuidos. http://www.teses.usp.br/teses/disponiveis/55/55134/tde-10092012-163429/publico/LucasVendramin.pdf. Accessed 04 April 2015
35. Vendramin, L., Campello, R., Coletta, L., Hruschka, E.: Distributed fuzzy clustering with automatic detection of the number of clusters. In: Proceedings of International Symposium on Distributed Computing and Artificial Intelligence, Advances in Intelligent and Soft Computing. vol. 91, pp. 133–140 (2011)
36. Wang, H., Li, Z., Cheng, Y.: Distributed and parallelled EM algorithm for distributed cluster ensemble. In: Proceedings of Pacific-Asia Workshop on Computational Intelligence and Industrial Application (PACIIA'08). vol. 2, pp. 3–8 (2008)
37. Wang, H., Shan, H., Banerjee, A.: Bayesian cluster ensembles, machine learning and knowledge discovery. In: Lecture Notes in Computer Science, vol. 6323, pp. 435–450. Springer (2009)
38. Wemmert, C., Gançarski, P., Korczak, J.: A collaborative approach to combine multiple learning methods. Int. J. Artif. Intell. Tools **9**(1), 59–78 (2000)
39. Xu, X., Jager, J., Kriegel, H.: A fast parallel clustering algorithm for large spatial databases. Data Min. Knowl. Discov. **3**(3), 263–290 (1999)
40. Younis, O., Fahmy, S.: Distributed clustering in ad-hoc sensor networks: a hybrid, energy-efficient approach. IEEE Trans. Mob. Comput. **3**(4), 366–379 (2004)
41. Zhen, M., Ji, G.: DK-means, an improved distributed clustering algorithm. J. Comput. Res. Dev. **44**(2), 84–88 (2007)
42. Zhou, J., Chen, C.P., Chen, L., Li, H.X.: A collaborative fuzzy clustering algorithm in distributed network environments. IEEE Trans. Fuzzy Syst. **22**(6), 1443–1456 (2014)

A Rule Based Recommender System

**Bruno Apolloni, Simone Bassis, Marco Mesiti,
Stefano Valtolina and Francesco Epifania**

Abstract We introduce a new recommending paradigm based on the *genomic* features of the candidate objects. The system is based on the tree structure of the object metadata which we convert in acceptance rules, leaving the user the discretion of selecting the most convincing rules for her/his scope. We framed the deriving recommendation system on a content management platform within the scope of the European Project NETT and tested it on the Entree UCI benchmark.

Keywords Recommender system · Decision trees · Genomic features

1 Introduction

In recent years we have seen a remarkable proliferation of recommender systems (RSs) in most disparate fields, ranging from movies, music, books, to financial services and live insurances. Usually implemented as web applications, they constitute a class of algorithms aimed at predicting user responses to options, by generating meaningful recommendations to a collection of users for items that might be of their interest.

B. Apolloni · S. Bassis (✉) · M. Mesiti · S. Valtolina · F. Epifania
Dipartimento di Informatica, Università degli Studi di Milano,
Via Comelico 39/41, 20135 Milan, Italy
e-mail: bassis@di.unimi.it

B. Apolloni
e-mail: apolloni@di.unimi.it

M. Mesiti
e-mail: mesiti@di.unimi.it

S. Valtolina
e-mail: valtolina@di.unimi.it

F. Epifania
Social Things S.r.l., Via de Rolandi Battista n. 1, 20126 Milano, Italy
e-mail: epifania@di.unimi.it

© Springer International Publishing Switzerland 2016
S. Bassis et al. (eds.), *Advances in Neural Networks*, Smart Innovation,
Systems and Technologies 54, DOI 10.1007/978-3-319-33747-0_9

A RS generates and provides recommendations following three phases: (1) the user provides input to the system which is explicit (e.g. feedback or ratings of resources) and/or implicit (e.g. resources visited and time spent on them); (2) the inputs are processed in order to obtain a representation that allows one to infer user interests and preferences, that is, to build a "user model". This representation could be given either as a simple matrix rating of products (texts, lessons, movies), or in terms of more complex data structures that combine rating and content information; (3) the system processes suggestions using the user model, assigning to the recommendation a suitable confidence level.

While initially RSs were based on standard statistical techniques [2], such as correlation analysis, in recent years predictive modeling, machine learning and data mining techniques were engaged in this involving challenge. Mainly, this happened when both the scientific community and practitioners recognized the many issues such methods may conceal, in order to produce meaningful and high-quality results in a wide spectrum of real-world scenarios. From then on, RSs have been investigated under the hat of various well-known branches of machine learning and data mining, such as classification, clustering, and dimensionality reduction techniques, specializing them to the peculiar aspects of RSs.

Traditionally RSs have been classified into three main categories:

- Collaborative Filtering (CF) Systems: users are recommended items based on the past ratings of all users collectively. CF can be further subdivided into neighborhood-based (computing a prediction through a k-nearest neighbor (kNN) approach from a weighted combination of neighbors rating, where neighboring users are those having highest similarity with the questioned user [3]) and model-based approaches (treating the recommendation problem as a parametric estimation task [18]). In the latter group, latent variable models [11] have gained much popularity in recent years: here, to compute the similarity between users and items, some hidden lower-dimensional structure in the data is discovered through either numerical linear algebra and statistical matrix analysis [13], or more sophisticated manifold learning techniques [10]. Current research in CF systems is directed toward the ensemble of different models and their enhancements through factorization techniques [17].
- Content-Based (CB) Systems: only those preferences of the user waiting for suggestions are involved in the recommendation process. These systems try to suggest products/resources that are similar to those that the user has liked in the past. CB systems are usually subdivided into Information Retrieval [1] (where the content associated with the user preferences is treated as a query, and the unrated documents are scored with relevance/similarity to it) and classification tasks (where each example represents the content of an item, and a user past ratings are used as labels for these examples). Algorithms such as Naive Bayes classifiers [16], k-nearest neighbor, decision trees, and neural networks [18] represent the state-of-the-art in the field.
- Hybrid Recommendation Systems: these hybrid systems aim to achieve the advantages of content-based and collaborative filtering systems, combining both

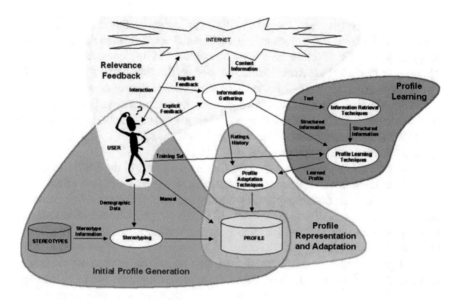

Fig. 1 A synopsis of recommender systems

approaches in order to mitigate the limitations associated with the use of one or the other type of system. To cite a few examples, ensemble methods, such as boosting techniques [14], are used to aggregate the information provided by CF and CB systems; alternatively, such information may be combined under a single probabilistic framework in the form of a generative model [20].

Less known but likewise relevant, we mention two more families having received the attention of the scientific community in the last years (see Fig. 1 taken from [15]):

- Demographic recommender systems, based on demographic classes. Personal information is gathered around user stereotypes. Such demographic groups are used, for instance, to suggest a range of products and services [12]. The classes may be identified by short surveys, machine learning methods, or correlation techniques. The benefit of a demographic approach is that it may not require a history of user ratings which is needed, on the contrary, by collaborative and content-based techniques.
- Utility-based and knowledge-based RSs, both rooted on the evaluation of the match between a user need and the set of available options. Thus, the user profile is the utility function, and the system employs constraint satisfaction techniques to locate the best match. Scientific research has concentrated in discovering the best techniques for identifying and exploiting the user-specific utility function [7].

Technically, the recommendation task is currently afforded as a combination of learning algorithms, statistical tools, and recognition algorithms—the areas ascribed to the study of computational intelligence. Facing a huge base of data, a common action to exploit them is their compression, either through statistical methods, such as independent components analysis, or through logical methods, such as cluster analysis and decision trees. Both families are aimed at extracting relevant features, possibly by a simple selection, mostly by a generation of new ones. Roughly speaking, we may consider the former methods routed on the phenotypic features and the latter on the genotypic ones—the seeds of most learning procedures. Our RS too passes through this genotypic compression, with the methodological option of offering the data genotypes as a former suggestion to the user, and a subsequent screening of the available resources filtered through the genotypes selected by the user, exactly like it happens in DNA microarray analysis [21].

With reference to the above taxonomy, our method falls in the CB Systems category, where in terms of recommender strategies we propose entropic utility functions. These are particularly suited for the application instance we will consider, within the scope of the European project NETT. However, given the adimensionality features of the entropy-based reasoning, we expect the method to prove efficient in many operational fields.

The paper is organized as follows. In Sect. 2 we explain the method and the implementation tools, while in Sect. 3 we frame the proposed RS in the NETT benchmark and discuss some preliminary experiments, providing some forewords as well.

2   The Method and Its Implementation Tools

Consider a set of N records, each consisting of m fields characterizing objects of our interest. Call these fields *metadata* μ_is and the objects *payloads* π_is. We distinguish between qualitative (nominal/ordinal) and quantitative (continuous/discrete) metadata, where the latter are suitably normalized in [0, 1]. Moreover each record is affected by a rank ρ, typically normalized in [0, 1] as well.

The RS goal is to understand the structure of the feature set corresponding to records ranked from a given threshold on. Note that we are not looking for a metadata–rank regression. Rather, we prefer considering a binary attribute (acceptable, not acceptable) and ask the user to select among acceptable genomes.

In essence, our goal is to infer a set of rules in the Horn clauses format, made of some antecedents and one consequent. The consequent is fixed: "π_i is good". The antecedents are Boolean conditions c_j (true/false) concerning sentences of two kinds: (i) "$\mu_i \lesseqgtr \theta$", where θ stands for any symbolic (for nominal metadata) or numeric constant (for quantitative variables); and (ii) "$\mu_i \in A$", with A a suitable set of constants associated with enumerated, hence qualitative, metadata. Hence the format is the following:

$$\textbf{if } c_1 \textbf{ and } c_2 \ \dots \ \textbf{ and } c_k \quad \textbf{then } \pi_i \textbf{ is good} \tag{1}$$

We may obtain these rules starting from one of the many algorithms generating decision trees dividing good from bad items, where the difference between the various methods stands in the entropic criteria and the stopping rules adopted to obtain a tree, and in the further pruning heuristics used to derive rules that are limited in

Table 1 A set of two candidate rules

Id	Rule
R_1	*skill_required* Communication Skill in Marketing Information Management = low **and** *language* = Italian ⇒ *good_course*
R_2	*skill_acquired* Communication Skill in Marketing Information Management = medium-high **and** *skill_acquired* Communication Skill in Communications Basic = high **and** *age* = teenager-adult ⇒ *good_course*

number, short in length (number of antecedents), and efficient as for classification errors. In particular we use RIPPER*k*, a variant of the Incremental Reduced Error Pruning (IREP) proposed by Cohen [5] to reduce the error rate, guaranteeing in the meanwhile a high efficiency on large samples, and in particular its Java version JRip available in the WEKA environment [8]. This choice was mainly addressed by computational complexity reasons, as we move from the cubic complexity in the number of items of the well known C4.5 [19] to the linear complexity of JRip. Rather, the distinguishing feature of our method is the use of these rules: not to exploit the classification results, rather to be used as hyper-metadata of the questioned items. In our favorite application field, the user, in search of didactic material for assembling a course on a given topic, will face rules like those reported in Table 1. Then, it is up to her/him to decide which rules characterize the material s/he's searching for. The second phase of the procedure is rather conventional. Once the rules have been selected, the system extracts all payloads π_is (i.e. didactic resources in our favorite scenario) satisfying them, which will then be further explored via keywords, describing the topics of each didactic resource. The latter have been previously extracted by

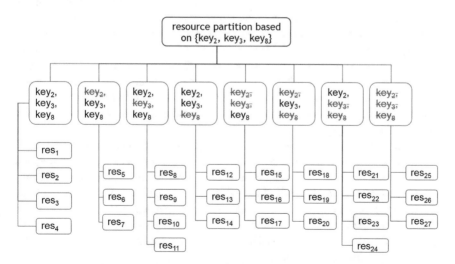

Fig. 2 The keyword subset selection process. The number of resources reduces uniformly, independently from the choice of the user (branch in the picture)

Table 2 Some restaurant paths

192	N	192	N	175	R	175	R	166						
175	R	175	R	166	Q	166	Q	44						
333	N	333	N	460	P	460	P	222	M	222	M	34	M	34	M	99
17	N	17	N	107	N	107	N	192	L	469	L	469			
53	M	53	M	166	M	166	M	166	L	397	L	397			
407	N	407	N	407	Q	407	Q	407	S	407	S	496	L	635	L	635

each π_i so as to constitute their labels. Thus the algorithm looks for the best keyword selection in terms of the ones providing the highest entropy partition of the extracted π_is (see Fig. 2). With this strategy, we are guaranteed that the number of selected resources still reduces uniformly at an exponential rate for whatever keyword subset chosen by the user.

3 Numerical Results

Is genotypic recommendation more efficient that the phenotypic one?

Our target application is a RS supporting the NETT platform[1] in guiding the user exploration to assemble a new course on Entrepreneurship Education. Given the relatively novelty of this discipline, the platform collects didactic materials—books, articles, slides, and didactic projects: didactic objects for short—to be consulted by teachers in order to build up a new course in the field. Given the social nature of the platform, on the one hand we may expect its repository to be populated by a huge number of didactic objects π_is (order of ten thousand). On the other hand, we may expect a teacher to be available to deeply examine some dozens of them. Hence the RS must help her/him to extract the most suitable objects from the repository. We realized this system, while its user experience is still under test. Rather, in this paper, we refer to the restaurant recommendation case study, initially considered by Burke [4], in order to state some quantitative considerations on the performance of our procedure.

The Entree Chicago recommendation benchmark is available in the UCI KDD repository [9]. Apart from ancillary data, it consists of sequences of pairs ⟨restaurant id, next action label⟩ as shown in Table 2. The former refers to Chicago restaurants; they are proposed by a webservice in response to a user solicitation. The first id is the suggestion of the service as a result of some options selected by the user, concerning kind of food, glamor, cost, etc. Next ids are suggestions again in response to the action requested by the user through the labels in the second part of the pair—whose meaning is reported in Table 3.

---

[1] http://siren.laren.di.unimi.it/nett/mnett/.

Table 3 The possible actions performed by the user in search of a better restaurant

Label	Description
L	Browse (move on the next restaurant in a list of recommendations)
M	Cheaper (search for a restaurant like this one, but cheaper)
N	Nicer (search for a restaurant like this one, but nicer)
O	Closer (closer to my house)
P	More traditional (serving more traditional cuisine)
Q	More creative (serving more creative cuisine)
R	More lively (with a livelier atmosphere)
S	Quieter (with a quieter atmosphere)
T	Change cuisine (serving a different kind of food)

The last pair in a sequence has the second element empty, thus denoting that no further action is required since the last restaurant is considered satisfactory.

We processed these records, in number of 50, 672 through JRip and PART [6], an algorithm present in the WEKA environment which builds a partial C4.5 decision tree in each iteration and makes the best leaf into a rule. From a preliminary survey of the results we decided to focus on JRip, looking at the main benefit that a great part of records is discriminated by a short number of rules. Namely, of the 15, 493 records leading to a good restaurant selection, 90 % are recognized by 20 % of the rules. These quantities have to be considered as an average behavior. Indeed, first of all we must decide which records are good and which bad. We decree as bad records those having a percentage of L labels greater than 40 or 50 % (*glimpse* values). Then, another questionable parameter is the truncation of the records to be submitted to the rule generator. Recognizing that 90 % of records have a length up to 8, we decided to generate rules by submitting records truncated to a number of pairs ranging between 5 and 8 (*trunc* values). Thus, our consideration comes from mediating 8 different rule generation scenarios.

A further preprocessing of the data comes from the will of removing noise, that we assume coinciding both with rules satisfied by very few records, for instance 0.3 % of the total, and with restaurant visited, as starting point or ending point of the records, a very small number of times, for instance up to 2 or 3. With this shrinking of the database, we came to the following conclusions:

- The rules well separate the records, with an average overlapping rate less than 20 %, reducing to 10 % in the best case (see Table 4).
- The basins of attraction of these rules is very limited. Namely, starting from the first restaurant suggested by the Entree webservice, a rule brings unequivocally to one or two ending restaurants (few exceptions apart concerning the less-frequent/noise restaurant—see above) independently from the action taken by the user. On the opposite direction, a given restaurant is reached by a limited number of starting ones, with some exceptions for a few restaurants which gather much many sequences. This is shown by the two histograms in Fig. 3.

Table 4 The paths overlapping table when *trunc* value = 7 and *glimpse* value = 40

	r_1	r_2	r_3	r_4	r_5	r_6	r_7	r_8	r_9	r_{10}	r_{11}	r_{12}	r_{13}	r_{14}	r_{15}	r_{16}	r_{17}
r_1	9110	0	0	0	0	0	0	0	0	0	0	0	0	0	0	0	0
r_2	0	423	0	0	0	0	0	0	98	25	35	16	0	0	1	0	0
r_3	0	0	117	0	0	0	0	0	11	4	8	8	0	0	0	0	0
r_4	0	0	0	1838	0	0	0	0	0	0	0	0	0	0	0	0	0
r_5	0	0	0	0	130	0	0	0	4	8	18	4	0	0	1	0	0
r_6	0	0	0	0	0	120	0	0	7	13	5	11	0	0	1	0	0
r_7	0	0	0	0	0	0	1604	0	0	0	0	0	0	0	0	0	0
r_8	0	0	0	0	0	0	0	52	4	4	1	7	0	0	0	0	0
r_9	0	98	11	0	4	7	0	4	210	0	0	0	0	0	0	3	0
r_{10}	0	25	4	0	8	13	0	4	0	80	0	0	1	0	0	0	4
r_{11}	0	35	8	0	18	5	0	1	0	0	91	0	0	0	0	2	1
r_{12}	0	16	8	0	4	11	0	7	0	0	0	68	0	0	0	1	0
r_{13}	0	0	0	0	0	0	0	0	0	1	0	0	6	0	0	0	0
r_{14}	0	0	0	0	0	0	0	0	0	0	0	0	0	19	2	0	0
r_{15}	0	1	0	0	1	1	0	0	0	0	0	0	0	2	10	0	0
r_{16}	0	0	0	0	0	0	0	0	3	0	2	1	0	0	0	12	0
r_{17}	0	0	0	0	0	0	0	0	0	4	1	0	0	0	0	0	10

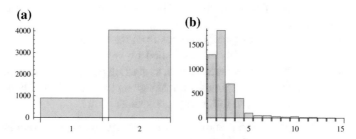

Fig. 3 Histograms of the number of different origins for a same destination (**a**) and vice versa (**b**) in the restaurant paths. The value 1 in **a** shows that around 900 paths start from restaurant visited less than 2 times

These simple experiments allow us to conclude that the JRip rules we used may represent a distinguishing *genomic* trait of the users. These traits, paired with the good starting points suggested by the Entree webservice, allow to associate each people with her/his favorite restaurant without ambiguities. The meaningful of the genomic trait stands also in the restaurant selectivity. Apart from a few exceptions, restaurant receives people with at most 4 different ⟨rule, starting point⟩ characterizations.

Acknowledgments This work has been supported by the European Project NETT.

References

1. Balabanovic, M., Shoham, Y.: Fab: content-based, collaborative recommendation. Commun. ACM **40**, 66–72 (1997)
2. Billsus, D., Pazzani, M.: Learning collaborative information filters. In: 15th International Conference on Machine Learning, pp. 46–54. Morgan Kaufmann (1998)
3. Breese, J., Heckerman, D., Kadie, C.: Empirical analysis of predictive algorithms for collaborative filtering. In: 14th Conference on Uncertainty in Artificial Intelligence (1998)
4. Burke, R.: The Wasabi personal shopper: a case-based recommender system. In: 11th National Conference on Innovative Applications of Artificial Intelligence, pp. 844–849 (1999)
5. Cohen, W.W.: Fast effective rule induction. In: Twelfth International Conference on Machine Learning, pp. 115–123. Morgan Kaufmann (1995)
6. Frank, E., Witten, I.H.: Generating accurate rule sets without global optimization. In: Shavlik, J. (ed.) 15th International Conference on Machine Learning, pp. 144–151. Morgan Kaufmann (1998)
7. Guttman, R.H., Moukas, A.G., Maes, P.: Agent-mediated electronic commerce: a survey. Knowl. Eng. Rev. **13**(2), 147–159 (1998)
8. Hall, M., Frank, E., Holmes, G., Pfahringer, B., Reutemann, P., Witten, I.H.: The WEKA data mining software: an update. SIGKDD Explor. **11**(1) (2009). http://www.cs.waikato.ac.nz/ml/weka
9. Hettich, S., Bay, S.D.: The UCI KDD Archive (1999). http://kdd.ics.uci.edu
10. Hofmann, T.: Latent semantic analysis for collaborative filtering. AACM Trans. Inf. Syst. (2004)
11. Koren, Y., Bell, R., Volinsky, C.: Matrix factorization techniques for recommender systems. IEEE Comput. **42**, 30–37 (2009)
12. Krulwich, B.: Lifestyle finder: intelligent user profiling using large-scale demographic data. Artif. Intell. Mag. **18**(2), 37–45 (1997)
13. Lee, D.D., Seung, H.S.: Learning the parts of objects by non-negative matrix factorization. Nature **788**, 401 (1999)
14. Melville, P., Mooney, R.J., Nagarajan, R.: Content-boosted collaborative filtering for improved recommendations. In: 18th National Conference on Artificial Intelligence, pp. 187–192 (2002)
15. Montaner, M., López, B., De La Rosa, J.L.: A taxonomy of recommender agents on the Internet. Artif. Intell. Rev. **19**(4), 285–330 (2003)
16. Mooney, R.J., Roy, L.: Content-based book recommending using learning for text categorization. In: 5th ACM Conference on Digital Libraries, pp. 187–192 (2000)
17. Pan, R., Scholz, M.: Mind the gaps: weighting the unknown in large-scale one-class collaborative filtering. In: 15th ACM SIGKDD Conference on Knowledge Discovery and Data Mining (KDD) (2009)
18. Pazzani, M.J., Billsus, D.: The identification of interesting web sites. Mach. Learn. **27**, 313–331 (1997)

19. Quinlan, J.R.: C4.5: Programs for Machine Learning. Morgan Kaufmann Publishers Inc., San Francisco, CA, USA (1993)
20. Schein, A., Popescul, A., Ungar, L., Pennock, D.M.: Methods and metrics for cold-start recommendations. In: 25th Annual International ACM SIGIR Conference on Research and Development in Information Retrieval, pp. 253–260. ACM (2002)
21. Shalon, D., Smith, S., Brown, P.: A DNA microarray system for analyzing complex DNA samples using two-color fluorescent probe hybridization. Genome Res. 6(7), 639–645 (1996)

Learning from Nowhere

Bruno Apolloni, Simone Bassis, Jacopo Rota, Gian Luca Galliani, Matteo Gioia and Luca Ferrari

Abstract We extend the Fuzzy Inference System (FIS) paradigm to the case where the universe of discourse is hidden to the learning algorithm. Hence the training set is constituted by a set of fuzzy attributes in whose correspondence some consequents are observed. The scenario is further complicated by the fact that the outputs are evaluated exactly in terms of the same fuzzy sets in a recursive way. The whole works arose from everyday life problems faced by the European Project Social&Smart in the aim of optimally regulating household appliances' runs. We afford it with a two-phase procedure that is reminiscent of the *distal learning* in neurocontrol. A web service is available where the reader may check the efficiency of the assessed procedure.

Keywords Fuzzy inference systems · Fuzzy rule systems · Distal learning · Two-phase learning

B. Apolloni · S. Bassis (✉) · J. Rota · G.L. Galliani · M. Gioia · L. Ferrari
Dipartimento di Scienze dell'Informazione, Università degli Studi di Milano
Via Comelico 39/41, 20135 Milan, Italy
e-mail: bassis@di.unimi.it

B. Apolloni
e-mail: apolloni@di.unimi.it

J. Rota
e-mail: jacopo.rota@studenti.unimi.it

G.L. Galliani
e-mail: galliani@di.unimi.it

M. Gioia
e-mail: gioia@di.unimi.it

L. Ferrari
e-mail: ferrari@di.unimi.it

© Springer International Publishing Switzerland 2016   97
S. Bassis et al. (eds.), *Advances in Neural Networks*, Smart Innovation,
Systems and Technologies 54, DOI 10.1007/978-3-319-33747-0_10

1   Introduction

Since their introduction by Zadeh [15], fuzzy sets have been intended as a rigorous way of dealing with non formalized knowledge falling in the sphere of the experience and intuition of the designers. Rather than *explaining why*, membership functions are a clear way of *describing the granularity* of the information owned by them. Fuzzy rule systems constitute a dynamical version of the fuzzy sets paradigm, hence a favorite tool for solving system control problem when fuzziness affects the description of their dynamics [7]. The general expression of this fuzzy rule system is the following one:

$$
\begin{aligned}
&\textbf{if } x_1 \text{ is } A_{11} \text{ AND } \ldots \text{ AND } x_n \text{ is } A_{1n} \textbf{ then } o \text{ is } B_1 \\
&\textbf{if } x_1 \text{ is } A_{21} \text{ AND } \ldots \text{ AND } x_n \text{ is } A_{2n} \textbf{ then } o \text{ is } B_2 \\
&\quad \vdots \qquad\qquad \vdots \qquad\qquad \vdots \\
&\textbf{if } x_1 \text{ is } A_{k1} \text{ AND } \ldots \text{ AND } x_n \text{ is } A_{kn} \textbf{ then } o \text{ is } B_k
\end{aligned}
\tag{1}
$$

where each rule consists of an antecedent prefixed by **if** followed by a consequent prefixed by **then**; A_{ij} and B_i, for all $i = 1, \ldots, k$ (k—the number of rules), $j = 1, \ldots, n$ (n—the number of conditions), are fuzzy sets defined in the corresponding input and output spaces, while x_i and o are variables corresponding to respectively the i-th condition and conclusion.

Within the European project SandS [1, 2] (as a contraction of Social&Smart[1]) aimed at optimally and adaptively ruling in remote the microcontrollers of our household appliances, we faced peculiar fuzzy rules that concerning, for instance, a bread maker sound as:

- **if** the loaf is less crusty AND soggy **then** increase rising time
- **if** the loaf is very crusty AND crunch **then** decrease rising time

These rules may be easily framed in the Sugeno reasoning model [8] where the consequents are crisp variables to be computed through a weighted mixture of functions depending directly on the input variables; in formula:

$$
o = \sum_{i=1}^{k} \overline{w}_i f_i(\boldsymbol{x}, \boldsymbol{s}) = \frac{\sum_{i=1}^{k} w_i f_i(\boldsymbol{x}, \boldsymbol{s})}{\sum_{i=1}^{k} w_i}
\tag{2}
$$

where f_i defines the activation function (for short *Sugeno function*) of the i-rule, whose shape, arguments and free parameters $\boldsymbol{s}$ depend on the chosen model, and w_i denotes the satisfaction degree of the premise of the rule. Finally, $\boldsymbol{x}$ and o refer to respectively the crisp input variables and the overall output of the system.

Figure 1 gives a schematic idea of the system in relation to the two aforementioned rules. Their distinguishing features come from the fact that when the user says that "the loaf is very crusty", s/he has no reference metric variable to set to a

---

[1]http://www.sands-project.eu.

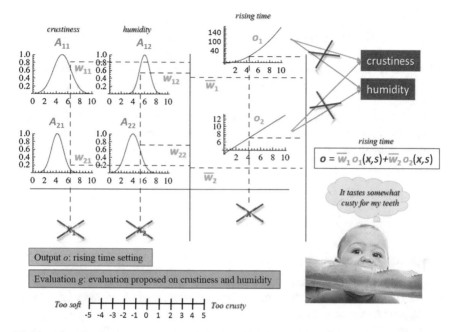

Fig. 1 A learning from nowhere instance. In the picture, x_1 and x_2 refer respectively to the fuzzy variables crustiness and humidity; the fuzzy sets A_{11} and A_{12} refer to the linguistic variables "less crusty" and "soggy" (i.e. "very humid"), while A_{21} and A_{22} to "very crusty" and "crunch" (i.e. "less humid"), respectively; the output o_j of each Sugeno function describes the numeric contribution of the j-th rule to the rising time. All contributions are gathered in the output of the system through (2)

specific value, even though s/he distinguishes different kinds of crustiness that gathers in the same quantifier "very crusty". This denotes that variables x_js exist; but some of them are hidden to the user her/himself. Nevertheless, to identify a suitable value of o (both to compute the satisfaction degree of the premise and to instantiate the Sugeno function in the consequent) we must estimate their values (hence the acronym *learning from nowhere*)—a problem not far from the one of identifying the state transition matrix in the Hidden Markov Models [11].

The second difficulty is connected to f_i in two respects. On the one hand we hazard functional forms such as linear and quadratic ones to interpret the consequent "increase/decrease" of the rising time. But this is a common plague of the Sugeno approach [3]. On the other hand, we don't know the relation between the operational parameter (the rising time) and its effect on the crustiness and humidity appreciation on the part of the user. This entails a non trivial identification problem to be faced in a way reminiscent of distal learning in neurocontrol [6] with a two-phases (identification and control) algorithm. In particular, the operational parameters, being directly controlled by the learner, play the role of *proximal* variables in the distal learning framework. Conversely, user judgments, for which target values are assumed to be available, are *distal* variables that the learner controls indirectly through the intermediary of the proximal variables. In order to tune the system to meet user requirements, the first task afforded by the learner is to identify it by discovering, through a classical supervised learning task, the mapping between operational parameters and user judgments (hence the term *identification* phase). After having identified the system, the learner switches to a *control* phase, aimed at regulating the system so as to satisfy user needs.

In this paper we afford this problem for the actual assessment of the SandS system on a bread maker. The paper is organized as follows. In Sect. 2 we formalize the training problem and specify the adopted options. In Sect. 3 we discuss the numerical results. Conclusions are drawn in Sect. 4.

2   The Formal Framework

The proposed fuzzy rule system is characterized by:

- n_c crisp variables y_is, associated to the operational parameters;
- n_f fuzzy variables g_is, each described by r linguistic quantifiers, associated to the task execution evaluation;
- k rules, grouped in n_c clusters each containing $k_c = r^{n_f+1}$ rules, where each cluster is responsible for the update of the single operational parameter, and each rule consists in principle of $n_f + 1$ antecedents (the n_f fuzzy variables plus the operational parameter the cluster refers to).

Framing the lead example in the above context, Fig. 1 shows two out of the k_c rules constituting the cluster responsible for the rising time, each composed of $n_f = 2$ fuzzy variables g_1 and g_2 (loaf crustness and humidity).[2] In our experiments we fixed $r = 3$ as a suitable compromise between richness of the model and manageability in operational terms. For instance, the three fuzzy sets associated to loaf crustness have been chosen as: *less*, *properly*, and *very* crusty.

With reference to the standard ANFIS architecture [4], one of the well-known neuro fuzzy inference system (FIS) implementing the Sugeno reasoning model, the system architecture is composed of 6 layers (see Fig. 2), as follows:

- **layer 0**, where we enter the $n = n_c + n_f$ input variables $x = (y, g)$, is complicated by the hidden features of those underlying the fuzzy sets (namely, the judgments g). This resolves in the addition of learned shift values δs to initial ones, where the latter may be either the nominal values of the fuzzy numbers used by the tester or completely random numbers (in a reasonable range) computed by the system. The output of this layer is a vector $\widetilde{x}$, where:

$$\widetilde{x}_i = \begin{cases} x_i & \text{if } x_i \text{ is a crisp variable } y_i \\ x_i + \delta_i & \text{if } x_i \text{ is a fuzzy variable } g_i \end{cases} \quad (3)$$

---

[2]In the experiments each rule receives in input also a crisp variable corresponding to the operational parameter, i.e. the rising time in the example, outputted by the FIS at the previous iteration.

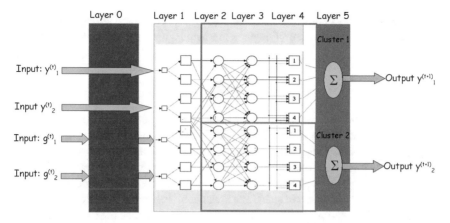

Fig. 2 The proposed neuro-fuzzy system architecture

Coming back to the lead example and using as initial values for the fuzzy sets their nominal values, the fuzzy variable $g_1 \equiv$ *very crusty* translates in the first rule as the crisp value obtained by adding to the mode of the fuzzy set A_{11} the learned shift parameter δ_1. In turn the mode is learned as well by the neuro-fuzzy procedure;

- in the fuzzification **layer 1**, where the membership functions (m.f.s) are introduced, we focus only on triangular functions $\mu_{A_{\{a,b,c\}}}(\tilde{x})$ and asymmetric Gaussian-like functions $\mu_{A_{\{v,\sigma_l,\sigma_r\}}}(\tilde{x})$ which require only three parameters to be specified; namely:

$$\mu_{A_{\{a,b,c\}}}(\tilde{x}) = \begin{cases} \frac{\tilde{x}-a}{(b-a)} & \text{if } \tilde{x} \in [a,b] \\ \frac{c-\tilde{x}}{(c-b)} & \text{if } \tilde{x} \in (b,c] \; ; \\ 0 & \text{otherwise} \end{cases} \qquad \mu_{A_{\{v,\sigma_l,\sigma_r\}}}(\tilde{x}) = \begin{cases} e^{-\frac{(\tilde{x}-v)^2}{\sigma_l^2}} & \text{if } \tilde{x} \leq v \\ e^{-\frac{(\tilde{x}-v)^2}{\sigma_r^2}} & \text{if } \tilde{x} > v \end{cases} \quad (4)$$

Each membership value reflects how much each input variable matches with the corresponding fuzzy set;

- in **layer 2**, we compute the satisfaction degree w_j of the premise of the j-th rule (the conjunction of the antecedents) as the product of the membership degrees of the metric variables, i.e. using the product T-norm. Remembering that A_{ji} refers to the m.f. fuzzifying the i-th variable in the j-rule, we have:

$$w_j = \prod_{i=1}^{n} \mu_{A_{ji}}(\tilde{x}_i) \qquad (5)$$

The weight of the j-rule reflects how much the variables in input to that rules satisfy its premises.

- in **layer 3**, the normalized satisfaction degree $\overline{w}_j$ is computed as usual by dividing a single degree by the sum of degrees in a same cluster c consisting of k_c rules:

$$\overline{w}_j = \frac{w_j}{\sum_{z=1}^{k_c} w_z} \tag{6}$$

The in-cluster-based normalization is a powerful tool considerably contributing to inject nonlinearities in the overall system, in the meanwhile preserving the comparable role held by each operational parameter in the task execution;

- in **layer 4**, Sugeno functions f_i are computed as a function of the crisp variables $\widetilde{x}$ computed in layer 0 (both y and g, after having learned the shift values modulating the latter). In particular, we opted for a linear combination of suitable (possibly different) scalar functions ς of each input variable (for instance the identity function, power law, log, exp, etc.) each one depending on a vector s of free parameters to be suitably learned. In the lead example, each Sugeno function provides the numerical contribution of the corresponding rule to the rising time, reflecting the action to be taken when its premises are satisfied. Finally,

- in **layer 5**, the outputs of the Sugeno functions are summed in each cluster c with a weight equal to the normalized satisfaction degree computed at layer 3:

$$o_c = \sum_{j=1}^{k_c} \overline{w}_j f_j(\widetilde{x}, s_j) \tag{7}$$

In other words, the more a rule is compatible with the input instance, and the higher will be its contribution to the overall output, which reads as the value assigned to the rising time during the next task execution in the lead example.

To sum up, each cluster c in the system, in input the value of both the operational parameters $y^{(t)}$ and the task execution evaluation $g^{(t)}$ (or suitable subsets of them, depending on the operational task to be afforded), computes as output the operational parameter $o_c^{(t+1)} \equiv y_c^{(t+1)}$ at the next time step, which in turn receives user evaluation in terms of the fuzzy variables $g^{(t+1)}$. For the sake of conciseness, we will drop the temporal index in the notation whenever no ambiguities arise. Moreover, due to the twofold role played by the operational variables y_c in terms of both input and output of the system, we still refer to the latter with o_c.

We afford the entire training task in terms of a back-propagation algorithm [13], hence a long derivative chain as the following:

$$\frac{\partial E}{\partial \theta} = \frac{\partial E}{\partial g} \cdot \frac{\partial g}{\partial o} \cdot \frac{\partial o}{\partial \theta} \tag{8}$$

where we used scalar symbols to lighten the notation. Namely, the error E is the canonical mean square error between original signal τ (the target) and reconstructed signal o in the identification phase, whereas it is assumed to be the square of the task execution evaluation—the judgment g—in the control phase (see Table 1). This

Table 1 The analytics of the two-phase learning procedure

Ident. phase	Control phase		
	Analytic		Numeric
$E = (o - \tau)^2$	$E = g^2$		
$\frac{\partial E}{\partial o} = 2(o - \tau)$	$\frac{\partial E}{\partial g} = 2g$		
	$\frac{\partial g}{\partial o} = \left(\frac{\partial o^{(t+1)}}{\partial g^{(t+1)}} \right)^{-1}$		$\approx \frac{g^{(t+1)} - g^{(t)}}{o^{(t+1)} - o^{(t)}}$
$\frac{\partial o}{\partial \theta} =$ canonical learning update rule			

choice is motivated by the fact that the set point of g, i.e. the judgment correspond-ing to user satisfaction, is 0, while positive/negative deviations from it encompass direction and magnitude of user dissatisfaction. Note that in the identification phase g is a dummy variable, so that $\frac{\partial E}{\partial g} \cdot \frac{\partial g}{\partial o}$ contracts in $\frac{\partial E}{\partial o}$, with o as in (7), whereas in the control phase $\frac{\partial E}{\partial g}$ is g itself, modulo a constant. The last derivatives of o w.r.t. all the underlying parameters (identifying fuzzy sets and Sugeno functions) are computed as usual [8]. Note that, since the fuzzy set supports are hidden, we added another parameter, namely the shift δ to the nominal value of the fuzzy number g (i.e. the center b of the triangular m.f.s and the mean v of the asymmetric Gaussian-like m.f.s in (4)), whose derivative computation is analogous to the one performed on fuzzy set parameters.

Moreover, according to the *principle of justifiable information granularity* [10], which states that a fuzzy set should reflects (or matches) the available experimental data to the highest extent, in the meanwhile being specific enough to come with a well-defined semantic, we further rearrange the inputs $\widetilde{x}_i$s in order to make them well framed in the fuzzy sets.[3] In case of triangular m.f.s, we simply rescale them in one-shot (at each learning cycle) so that the γ and $1 - \gamma$ quantiles coincide with the maxmin and minmax of the fuzzy numbers, respectively, for a proper γ (say equal to 0.1). In case of Gaussian-like m.f.s, we preferred working with an incremental procedure, so as to avoid abrupt oscillations in the training process. Namely, having introduced for each linguistic variable two translation parameters β, γ and a scaling parameter α, we applied the affine transformation:

$$\rho(\widetilde{x}) = \alpha(\widetilde{x} - \beta) + \gamma \tag{9}$$

to the shifted input value of each linguistic variable modulating the fuzzy number g_i it refers to (e.g. bread crunchiness) so as to match both position and spread of all the currently learned m.f.s (i.e. soggy, crunch). In particular, we considered a mixture $\mathcal{M}$ of asymmetric Gaussian random variables [5], differing from the companion set of m.f.s only in the normalization factor, and computed its mean $v_{\mathcal{M}}$ and standard

---

[3] Actually, for each fuzzy variable g_i, the output (3) of layer 0 is further processed so as to include as last step this transformation.

deviation $\sigma_{\mathcal{M}}$ [12]. At each system evaluation, through a gradient-descent procedure, we gradually modified translation and scaling parameters in (9) in order to minimize the sum of the mean square errors between the two aforementioned statistics $v_{\mathcal{M}}$ and $\sigma_{\mathcal{M}}$, and their sample realizations computed on the observed data.

In conclusion, focusing on asymmetric Gaussian-like m.f.s, our inference concerns the parameter vector θ, with:

$$\theta = \left(v_{ji}, \sigma_{lji}, \sigma_{rji}, \delta_i, s_{cji} \right) \tag{10}$$

where v, σ_l and σ_r denote respectively mean, left and right standard deviation of the Gaussian m.f. as in (4), δ is the aforementioned shift of any fuzzy input variable, s reads as the generic parameter of the Sugeno function, while c, j and i are indexes coupled respectively with the output units, the rules, and the input variables. In the identification phase their increments are ruled as shown in Table 2.

Rather, in the control phase the error function reads:

$$E = \sum_i^{n_f} \left(g_i^{(t+1)} \right)^2 \tag{11}$$

where $g_i^{(t+1)}$ is the i-th evaluation provided by the user at time $t+1$.

Here, the derivative $\frac{\partial g^{(t+1)}}{\partial o^{(t+1)}}$ is the most critical part of the chain rule (8), finding either:

Table 2 Back-propagation-like equations ruling the identification phase of the proposed FIS

$\Delta s_{cji} = \eta \left(\tau_c - o_c \right) \overline{w}_j \cdot \varsigma_{cj}(\widetilde{x}_i)$	(12.1)
$\Delta v_{ji} = \begin{cases} \eta \sum_c^{n_c} \left(\tau_c - o_c \right) \left(s_{cji} - o_c \right) \overline{w}_j \left(\widetilde{x}_i - v_{ji} \right) / \sigma_{lji}^2 & \text{if } \widetilde{x}_i \leq v_{ji} \\ \eta \sum_c^{n_c} \left(\tau_c - o_c \right) \left(s_{cji} - o_c \right) \overline{w}_j \left(\widetilde{x}_i - v_{ji} \right) / \sigma_{rji}^2 & \text{if } \widetilde{x}_i > v_{ji} \end{cases}$	(12.2)
$\Delta\sigma_{lji} = \begin{cases} \eta \sum_c^{n_c} \left(\tau_c - o_c \right) \left(s_{cji} - o_c \right) \overline{w}_j \left(\widetilde{x}_i - v_{ji} \right)^2 / \sigma_{lji}^2 & \text{if } \widetilde{x}_i \leq v_{ji} \\ 0 & \text{if } \widetilde{x}_i > v_{ji} \end{cases}$	(12.3)
$\Delta\sigma_{rji} = \begin{cases} 0 & \text{if } \widetilde{x}_i \leq v_{ji} \\ \eta \sum_c^{n_c} \left(\tau_c - o_c \right) \left(s_{cji} - o_c \right) \overline{w}_j \left(\widetilde{x}_i - v_{ji} \right)^2 / \sigma_{rji}^2 & \text{if } \widetilde{x}_i > v_{ji} \end{cases}$	(12.4)
$\Delta\delta_i = \sum_{j=1}^{k_c} \begin{cases} \eta \sum_c^{n_c} \left(\tau_c - o_c \right) \left(\left(s_{cji} - o_c \right) \overline{w}_j \left(-\dfrac{\widetilde{x}_i - v_{ji}}{\sigma_{lji}^2} \right) + \overline{w}_j \dfrac{\partial s_{cji}}{\partial \widetilde{x}} \right) & \text{if } \widetilde{x}_i \leq v_{ji} \\ \eta \sum_c^{n_c} \left(\tau_c - o_c \right) \left(\left(s_{cji} - o_c \right) \overline{w}_j \left(-\dfrac{\widetilde{x}_i - v_{ji}}{\sigma_{rji}^2} \right) + \overline{w}_j \dfrac{\partial s_{cji}}{\partial \widetilde{x}} \right) & \text{if } \widetilde{x}_i > v_{ji} \end{cases}$	(12.5)

Parameter η denotes the learning rate. The only input variables $\widetilde{x}_i$ coupled with the shift δ_i in (12.5) are the fuzzy judgments g_i

- an analytical solution in the reciprocal of the derivative $\frac{\partial o^{(t+1)}}{\partial g^{(t)}}$, as it emerges from the identification of $o^{(t+1)}$, or
- a numerical solution in the ratio of the differences at consecutive time steps of the two quantities, i.e. $\frac{g^{(t+1)}-g^{(t)}}{o^{(t+1)}-o^{(t)}}$, to bypass identification errors;

giving rise, respectively, to the expressions:

$$\frac{\partial E}{\partial \theta} = \eta \sum_{i}^{n_c} \sum_{i'}^{n_f} 2g_{i'}^{(t+1)} \left(\frac{\partial o_i^{(t+1)}}{\partial g_{i'}^{(t)}} \right)^{-1} \frac{\partial o_i^{(t+1)}}{\partial \theta} \tag{13}$$

$$\frac{\partial E}{\partial \theta} = \eta \sum_{i}^{n_c} \left(o_i^{(t+1)} - \tau_i \right)^2 \sum_{i'}^{n_f} 2g_{i'}^{(t+1)} \frac{g_{i'}^{(t+1)} - g_{i'}^{(t)}}{o_i^{(t+1)} - o_i^{(t)}} \frac{\partial o_i^{(t+1)}}{\partial \theta} \tag{14}$$

The former indeed suffers from the typical plague of being identified in a domain that may be far from the one where o will be applied during the control phase—a drawback typically overcome by alternating the two phases [14]. The second solution may deserve overfitting, and suffers from the cold start problem; operatively we may lessen these drawbacks by storing all the judgment and signal histories (namely $\{g^{(t)}\}_t$ and $\{o^{(t)}\}_t$), and using the value $g^{(t')}$ nearest to the actual $g^{(t)}$, with $t' < t$ (and analogously with o), in the computation of $\frac{\partial g^{(t+1)}}{\partial o^{(t+1)}}$ instead of consecutive ones.

3   Numerical Results

Moving to our recipe generation task, we applied our two-phase procedure to the bread maker experiment dataset.[4] In this case we have 10 parameters (time and temperature in the 5 baking phases: first and second leavening, pre-cooking, cooking, and browning) and 4 evaluations as well (fragrance, softness, baking, crustiness). Clusters of rules are used to singularly generate the operational parameters. Each cluster consists of $k_c = 3^5$ rules, where $r = 3$ is the number of levels of the adopted fuzzy quantifiers and 5 is the number of antecedents in the rules: the $n_f = 4$ evaluations plus the questioned parameter computed in the previous run. Namely, for homogeneity with the other antecedents we decided to group the values assumed by the single questioned parameter in the training set in three clusters and specialize the rules for each fuzzy quantifier and for each cluster as well. As first training instance, we omitted training the FIS on parameters 6, due to its scarce variability, and 10, because temperature settings above the security threshold (set to 150 in the microcontroller) are dummy. Rather, the identification on the remaining parameters (after proper normalization) proved enough satisfactory (see Fig. 3).

---

[4]The database is available at http://ns3366758.ip-37-187-78.eu/exp/excel.

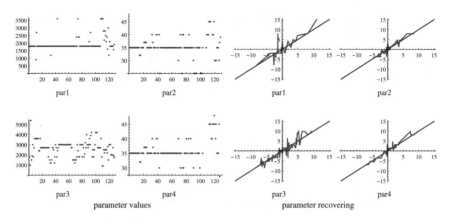

Fig. 3 Plots of the first four tested operational parameter values and their recovering in the bread-making experiment. Analogous trends have been observed for the four remaining parameters

However, many records in the training set have been collected during a sort of users' training phase, so that their suggestion on the new recipe, which constitutes the FIS target on the identification phase, may prove definitely misleading. Hence we decided doping the training set with a set of examples made up of the original instances as for the premises and of the optimal parameter settings (that we learned during the experimental campaign) as for the consequences. This expedient was profitable, as confirmed by the mean square error curves reported in Fig. 4.

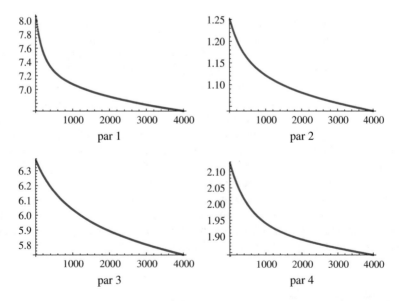

Fig. 4 Mean square error descent on the first four operational parameters of the doped training set

Essentially the system is pushed toward the optimal parameters in the consequents, profiting of a well biased noise represented by the original examples. With
this FIS configuration the system provides good suggestion on new inputs (hence in
generalization). However, to render the system really adaptive to the judgments of
the single user so as to learn how to get a better evaluation from her/him, we needed
to train the FIS following the single user interactions. Indeed, we had to embed the
training phase inside an overall cycle where: (i) the user asks for a task execution;
(ii) the social network computes the related recipe as a result of previous recipes and
related evaluations, and dispatches it to the appliance; (iii) the appliance executes
the recipe; and (iv) the user issues an evaluation on the executed task, so closing the
loop. The problem is that step iii is definitely time consuming (generally, in the order
of hours), thus slowing the entire procedure down. Hence we devised the following
expedient. After a good system identification (performed as before), we may assume
that FIS computes the correct parameters in the doped part of the training set, so that
the evaluation given in correspondence of these parameters properly applies to the
computed parameters as well. Hence we may use this part of the training set to train
the system on the evaluations as well (the control phase). Rather, we adopted an intermediate strategy by pairing the identification error with the control error through a
mutual weighting based on a linearly convex composition. Namely, with reference
to Table 1, the error becomes:

$$E = \xi \sum_i^{n_f} g_i^2 + (1 - \xi) \sum_i^{n_c} (\tau_i - o_i)^2 \qquad (15)$$

for proper (small enough) $\xi \in [0, 1]$. Now that the FIS has been weaned, it is ready to
work on-line. In this case, the error expression remains the same, where the second
term acts as a regularization term, since the target τ now is the previous parameter
value and ξ is more or less big, depending on the evaluation values. The criterion is:
if the overall evaluation is close to 0, then refrain FIS to produce new parameters that
are far from the current ones. *Vice versa* do not hesitate moving far away from them
if the current evaluation is poor (away from 0). Table 3 reports a log of these on-line
interactions leading to a gentle reduction of the leavening and cooking parameters in
response to the user feeling, expressed in terms of slight defects of softness, baking,
and crustness.

Table 3 Log of on-line user interactions with the system

1st Leaven.		2nd Leaven.		Pre cooking		Cooking		Browning		Evaluation			
Time	Temp.	Time	Temp.	Time	Temp.	Time	Temp.	Time	Temp.	Fragr.	Softn.	Baking	Crust
1821	33	2788	35	575	59	3885	118	101	126	+3	+1	+1	+1
1792	33	2813	35	569	59	3872	118	100	126	0	+2	+2	+2
1820	33	2787	35	570	59	3885	118	97	126	0	+1	+1	+1
1766	33	2754	34	560	59	3889	117	93	124	0	+2	+2	+2

4 Conclusions

In this paper we introduce a new learning framework that we call *learning from nowhere*, which frames between the two cases investigated so far: the one where crisp variables are fuzzified to enter a fuzzy rule, and the other where fuzzy sets are dealt with as a whole within a fuzzy rule. The related learning procedure is a rather obvious extension of the common back-propagation procedure, inheriting both its robustness as for numerical results and its weakness as for theoretical guarantees. As it is, we do not claim comparative efficiency w.r.t. other approaches to the problem. Rather we stress its suitability to solve in concrete a problem currently considered unfeasible by the white goods manufacturers: optimally conducting household appliances in a true fuzzy framework where the principal inputs come from the user feeling. Hence we consider all operational aspects of the related training problem, having as proof of evidence of its solution a record of people daily preparing his loaf of bread according to the suggestions of the trained system. Future work will be devoted to generalize the framework to wide families of personal appliances and devise Key Performance Indicator (KPI) tools [9].

Acknowledgments This work has been supported by the European Project FP7 317947 Social&Smart.

References

1. Apolloni, B., Fiaschè, M., Galliani, G.L., Zizzo, C., Caridakis, G., Siolas, G., Kollias, K., Grana Romay, G., Barriento, F., San Jose, S.: Social things—the SandS instantiation. In: Second IEEE Workshop on the Internet of Things: Smart Objects and Services, IoT-SoS 2013 (2013)
2. Apolloni, B., Galliani, G.L., Gioia, M., Bassis, S., Rota, J., Ferrari, L.: Social things: now we can. Intelligenza Artificiale **9**(2) (2015)
3. Cococcioni, M., Lazzerini, B., Marcelloni, F.: Estimating the concentration of optically active constituents of sea water by Takagi-Sugeno models with quadratic rule consequents. Pattern Recogn. **40**(10), 2846–2860 (2007)
4. Jang, J.S.R.: ANFIS: adaptive-network-based fuzzy inference system. IEEE Trans. Syst. Man Cybern. **23**(3), 665–685 (1993)
5. John, S.: The three-parameter two-piece normal family of distributions and its fitting. Commun. Stat. Theor. Methods **11**(8), 879–885 (1982)
6. Jordan, M.I., Rumelhart, D.E.: Forward models: supervised learning with a distal teacher. Cogn. Sci. **16**(3), 307–354 (1992)
7. Miu, S., Hayashi, Y.: Neuro-fuzzy rule generatioion: survey in soft computing framework. IEEE Trans. Neural Netw. **11**, 748–768 (2000)
8. Nauck, D., Klawonn, F., Kruse, R.: Foundations of Neuro-Fuzzy Systems. Wiley, New York, NY, USA (1997)
9. Parmenter, D.: Key Performance Indicators: Developing, Implementing, and Using Winning KPIs, 2nd edn. Wiley, Hoboken (2010)
10. Pedrycz, W.: The principle of justifiable granularity and an optimization of information granularity allocation as fundamentals of granular computing. J. Inf. Process. Syst. **7**(3), 397–412 (2011)

11. Rabiner, L., Juang, B.: An introduction to hidden Markov models. IEEE Acoust. Speech Sig. Process. Mag. **3**, 4–16 (1986)
12. Robertson, C.A., Fryer, J.G.: Some descriptive properties of normal mixtures. Scand. Actuar. J. **1969**(3–4), 137–146 (1969)
13. Werbos, P.J.: The Roots of Backpropagation. From Ordered Derivatives to Neural Networks and Political Forecasting. Wiley, New York, NY (1994)
14. Wu, Q., Hogg, B., Irwin, G.: A neural network regulator for turbogenerators. IEEE Trans. Neural Netw. **3**(1), 95–100 (1992)
15. Zadeh, L.: Fuzzy sets. Inf. Control **8**, 338–353 (1965)

Model Complexity Control in Clustering

Stefano Rovetta

Abstract This work deals with model complexity in clustering. Methods to control complexity in unsupervised learning are reviewed. A method that decouples the number of clusters from clustering model complexity is presented and its properties are discussed with the help of experiments on benchmark data sets.

Keywords Data clustering · Model complexity · Spectral clustering

1 Introduction

Unsupervised learning, learning representations of the data as opposed to learning a map between data domains, is at the core of most successful machine learning approaches, including for instance deep learning.

Data clustering is one of the most common representations used to summarize the data, and very often central clustering, the representation of data by means of a reduced set of centroids or reference vectors or prototypes, is the model of choice. Both problems of central clustering and vector quantization are cast as minimization of nearest-centroid mean squared distance

$$D = \frac{1}{n} \sum_{l=1}^{n} \|\mathbf{x}_l - \mathbf{y}(\mathbf{x}_l)\|^2 \tag{1}$$

where $\mathbf{x}_l \in \{\mathbf{x}_1, \ldots, \mathbf{x}_n\} \subset \mathbb{R}^d$ are the input vectors and $\mathbf{y}(\mathbf{x}) \in \{\mathbf{y}_1, \ldots, \mathbf{y}_c\} \subset \mathbb{R}^d$ are centroids, with $\mathbf{y}(\mathbf{x}_l)$ the nearest centroid to $\mathbf{x}_l$. However, while data clustering requires c to represent the number of separate regions in the data distribution, at the possible expense of approximation quality, in vector quantization c is usually higher than the number of regions, and is mainly limited by resource constraints according to rate-distortion theory [6, 21].

S. Rovetta (✉)
DIBRIS – University of Genova, Via Dodecaneso 35, 16146 Genoa, Italy
e-mail: stefano.rovetta@unige.it

© Springer International Publishing Switzerland 2016
S. Bassis et al. (eds.), *Advances in Neural Networks*, Smart Innovation,
Systems and Technologies 54, DOI 10.1007/978-3-319-33747-0_11

This work focuses on another clustering paradigm, and shows how the prime example of similarity-based clustering, namely, spectral clustering, in its approximated form can be enriched with complexity control. This combination results in the ability to learn cluster shapes whose geometric complexity can be controlled independently of their number.

2   Model Complexity in Unsupervised Learning

Model complexity in a supervised setting can be defined in many ways, but the core idea is providing a measure of how flexible a model is in representing a target concept. For instance, the Vapnik-Chervonenkis dimension [25] measures the maximum shatterable sample size, where shattering refers to solving all possible problems resulting from all possible label assignments. A model that is more flexible than necessary implies a higher probability of overfitting, i.e., fitting the data rather than their underlying distribution. When dealing with unsupervised learning, however, the goal of learning is to describe the data, or some interesting property of the data, as in clustering where "natural" groupings are searched.

We can refer to the following three mutually related concepts:

1. *Model size* is the size of a representation of a model, for instance measured by the number of parameters, including the number of clusters.
2. *Model complexity* is the ability of a model to represent a rich set of target concepts, for instance by reaching a low objective function value; clearly a larger model size usually implies higher complexity.
3. *Complexity measure* is an index that is directly related to model complexity, used to express it quantitatively and compare it across models. Since complexity is not, in itself, a well-defined concept, several different indices have been proposed.

The unsupervised learning literature provides many examples of studies relating some suitable complexity measure directly to model size. In fact, this is not necessarily the case; it is well known that for supervised learning model capacity, as measured for instance by the Vapnik-Chervonenkis dimension or as estimated by the empirical Rademacher complexity, can be low even in the presence of many parameters, or high for very simple examples (the Vapnik-Chervonenkis dimension of $f(x) = \sin \alpha x$ is infinite even if f has only one parameter).

This work does not focus on the problem measuring model complexity, but on its modulation *independently of the final number of clusters*.

Many popular clustering methods have a complexity that is directly related to the number of clusters: for instance the only parameters in k means are centroids, so there is a one-to-one relationship between model complexity and size. We should note that is also possible to modulate complexity by acting on other representation aspects, for instance numerical precision of centroid coordinates or centroid position (lattice quantization) [27]. However, these techniques are not very useful in clustering since they provide a solution to an approximation (vector quantization) rather than explanation (clustering) problem, precisely because c is not required to reflect some intrinsic property of the data distribution.

If we focus on fuzzy clustering [2], another way to modulate complexity is available: the fuzziness can be varied by changing a model parameter, in one form or another present in all fuzzy clustering methods. Even in this case the modulation has a direct effect on model size, i.e., the effective number of clusters, since when the fuzziness parameter grows some centroids will collapse into one. This effect was explicitly studied and exploited in the Deterministic Annealing framework [18].

A dual approach is provided in the works by Buhmann [3, 4], who proposes approaches that, instead of modulating distortion to find the most meaningful number of clusters, emphasize representation performance by modulating the number of clusters, in this being closer to vector quantization. Similar criteria were applied to Neural Gas learning by the author [16].

Other complexity control methods are more general, although they can be and have been applied to unsupervised learning. These include the Kolmogorov-Chaitin algorithmic complexity framework [13] and Rissanen's stochastic complexity framework [17].

In contrast with central clustering, another class of clustering methods uses similarity (or distance) matrices. Examples of this class include single-linkage and full-linkage agglomerative clustering [24]. The peculiarity of this approach lies in the fact that cluster shape is not biased toward some a-priori assumption (in fact, cluster shape might not even be defined if similarities do not correspond to a geometric representation of data in some space). Therefore they offer the theoretical possibility to decouple model *complexity* from model *size*. And yet, this possibility seems to be largely unexplored. A study on model complexity in similarity-based clustering was presented in [12] where the clustering method under examination is Shepard and Arabie's ADCLUS [22] and complexity is related to accuracy of similarity measures and therefore their statistical uncertainty, so it is measured by the Bayesian Information Criterion (BIC) [20].

Spectral clustering in its various flavours is a prominent, modern example of similarity-based clustering. In this work I show how to modulate model complexity in the online spectral clustering model presented in [19], focusing on the effect of changing the internal representation, which allows for richer or smoother cluster shapes independently of their number.

3 Approximated Spectral Clustering

Many approaches and variations to spectral clustering have been proposed. All of them rely on the square similarity matrix $W = K(X)$ where $K()$ is a suitable similarity function between data items, for instance the heat kernel $e^{\|\mathbf{x}_i - \mathbf{x}_j\|^2 / \sigma^2}$ with scale parameter *sigma*, and $K(X)_{i,j} = K(X, X)_{i,j} = K(\mathbf{x}_i, \mathbf{x}_j)$ (this compact formalism is available in our software implementation). W is a Gram matrix in the Hilbert space with inner product $< \mathbf{x}_i, \mathbf{x}_j >= K(\mathbf{x}_i, \mathbf{x}_j)$.

This similarity matrix is then interpreted as the connection weights of a symmetric graph, which is analyzed by means of concepts from spectral graph theory. Specifically, it turns out that the *Laplacian matrix L* of a bipartite graph (see below for definitions of *L*) has a null smallest eigenvalue, corresponding to an eigenvector with constant components, while the second smallest eigenvalue (called the *algebraic connectivity* of the graph) corresponds to an eigenvector called the *Fiedler vector* [9] which is an indicator for the connected components of the graph [5]. When there are $k > 2$ connected components it is not possible to directly associate eigenvectors to clusters, but the subspace spanned by eigenvectors number $2 \ldots k + 1$ has a much clearer structure than the original data space and can efficiently be clustered with a simple method, e.g., *k*-means, especially if row-normalized. These properties still hold, albeit approximately, in case of a graph that has weakly connected, as opposed to isolated, components, as is usually the case when a graph is obtained from a similarity measure. I refer to [26] for a survey on spectral clustering.

To compute the Laplacian there are three main possible approaches. In all cases, first the degree matrix D is computed, a diagonal matrix $D_{ii} = \sum_{j=1}^{n} V_{ij}$; then L, the Laplacian matrix is defined as either of the following:

Definition	Description
$L_u = D - V$	Unnormalized Laplacian
$L_{rw} = I - VD^{-1}$	Asymmetrically normalized Laplacian, analogy with random walks theory [23]
$L_{sym} = I - D^{-1/2}VD^{-1/2}$	Symmetrically normalized Laplacian [14]

Reference [26] illustrates very clearly the properties of these three possible choices.

In [19] I have introduced an approximate form of spectral clustering, needed for online operation. Since the rows of W can be viewed as an embedding of the rows of X in terms of similarities from the rows of X itself, the idea is to use $c < n$ landmark points $\mathbf{y}_1, \ldots, \mathbf{y}_c$ in data space, that quantize the data distribution, and to approximate the Gram matrix $W_{ij} = K(\mathbf{x}_i, \mathbf{x}_j)$ with a rectangular matrix computed as $\hat{W} = K(X, Y)$, i.e., $\hat{W}_{ij} = K(\mathbf{x}_i, \mathbf{y}_j)$. The Laplacian matrix is then built, according to one of the possible formulations, starting from the square matrix $V = \hat{W}\hat{W}'$ (the prime indicates transposition) which is known to have eigenvectors given by the left singular vectors of $\hat{W}$ and eigenvalues given by the squared eigenvalues of $\hat{W}$. The spectral embedding is therefore a low-rank approximation of that obtained from W. Note that V is a real, square, symmetric matrix, so it has real eigendecomposition just like W. As a consequence, a legitimate Laplacian can be built and studied exactly as in full spectral clustering methods.

However, in this case it is possible to modulate the rank of this approximation by suitably selecting the number of landmarks c. This offers the ability to control the representational capacity of this spectral embedding. The consistency of this kind of approximation, ultimately grounded in the general topic of low-rank matrix approximation [8], has been studied in the context of Nystrom approximation of W

[10], which consists in the particular choice of Y as a subset of the rows of X and a method for estimating the parts of W corresponding to the rows of X not in Y.

It has been verified, both in the course of this research and in other comparable cases from the literature [7], that the actual position of landmarks is not very critical, as long as it is able to act as a proxy to the original data distributions. Since the task is to approximate X with Y, two options are randomly subsampling of X to obtain Y, and training Y as a vector quantization codebook for X. The first option is adopted in [10]. Here the second approach has been selected to limit the inter-experiment variance resulting from random sampling.

Algorithm 1 illustrates the required steps for the Approximate Spectral Clustering (ASC) procedure. In the following some experimental results are shown to demonstrate the technique and the effect of model complexity control.

Algorithm 1 ASC - The Approximate Spectral Clustering procedure

---

select suitable $K()$, c, k generate landmarks Y by performing k-means clustering on X
with c centroids
compute similarity matrix $\hat{W}$ between data X and landmarks Y using $K()$
compute correlation matrix $V = \hat{W}\hat{W}'$
compute degree matrix $D = \mathrm{diag}(V)$
compute the desired form of Laplacian L from V and D (unnormalized, sym, rw)
compute E, first $k + 1$ (column) eigenvectors of L
discard first eigenvector by removing first column from E
perform k-means clustering on rows of E obtaining a cluster attribution for each row E_i
attribute pattern x_i to the cluster of row E_i

---

4 Experiments

Some experimental verifications were performed to highlight the properties of the proposed technique. It should be noted that, depending on the specific $K()$, data dimensionality is irrelevant since we are working on mutual similarities. In general, evaluation of results is made by external validation on labelled data sets. This is done by means of two indices. The first is cluster purity, the fraction of points belonging to the majority class in their respective clusters

$$p = \frac{1}{n} \sum_{h=1}^{k} \max_{j:1...k} q_{hj} \tag{2}$$

where q_{hj} is the number of points in cluster h having label j. The second is the (unadjusted) Rand index [15]

$$r = \frac{2}{n(n-1)}(q_{00} + q_{11}) \tag{3}$$

where q_{00} is the number of pairs of data points not in the same cluster AND not in the same class, while q_{11} is the number of pairs of data points in the same cluster AND in the same class. The unadjusted version has been used because, when clusterings and labellings are very similar ($r > 0.8$), it is more or less equivalent to the adjusted version [11], but it is simpler.

Both indices can be used to compare either a clustering with a labeling, or a candidate clustering with a ground-truth clustering.

4.1 A Toy Example

A data set was generated by sampling two 2-dimensional Gaussian distributions, with equal standard deviation $\sigma = 0.5$, centred in $(-1, -1)$ and $(1, 1)$ respectively, obtaining two well-separated clusters with 50 points per cluster. The experiment involves applying ASC with $c = 100, 50, 10, 2$.

Figure 1 illustrates the result with different symbols for data points (crosses) and landmarks (circles, squares) labelled according to the output of ASC. Note that $c = 100$ is equivalent to FSC, while $c = 2$ is not mathematically equivalent to k-means with $k = 2$, but gives results with the same expressivity, being only able to provide linear separability. In all cases the two clusters are efficiently separated.

4.2 Two Spirals

The behaviour of ASC as a function of c is further illustrated by a test at the other end of the spectrum w.r.t. the previous one, the "Two Spirals" data set. In this case the clustering problem is artificially complicated. The interspersed spirals (refer to Fig. 2) are sampled with 60 points each. Figure 2a presents the ground truth with circles and crosses indicating the two spirals respectively. In Fig. 2b, c, d the same symbols indicate the output labeling from ASC.

Again we can appreciate that for $c = 2$ (Fig. 2b) only a linear separation is obtained. For $c = 20$ (Fig. 2c) the number of landmarks is still not sufficient to achieve a suitable separation, although the obtained clusterings are richer in shape. In the final case, $c = 90$ (Fig. 2d) and the separation is perfect.

Figure 3 is a plot of cluster purity versus c. It can be appreciated that there is a threshold behaviour: below $c = 80$ purity stays essentially constant at the value of a random guess, while starting from that value purity steeply grows and rapidly reaches 1. This analysis points out the presence of a *clustering intrinsic dimension*, the degree of complexity that is necessary to correctly recognize clusters.

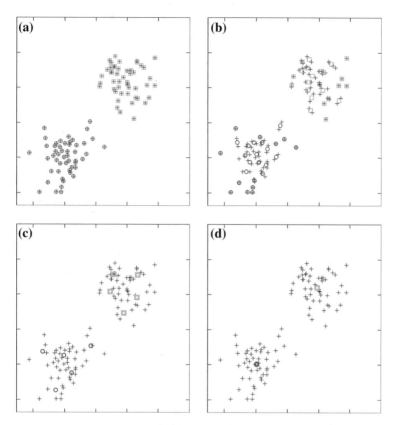

Fig. 1 Toy example: Gaussian data (*crosses*) with decreasing numbers of landmarks (*circles* and *squares*): **a** 100, one per data point; **b** 50; **c** 10; **d** 2

4.3 Semeion

The Semeion data set is a sample of 1593 handwritten decimal digit images from about 80 writers. Each image is black-and-white, 16×16 and contains an individual digit stretched to fit the square. It was provided to the UCI Machine Learning Repository [1] by Semeion Research Center of Sciences of Communication, via Sersale 117, 00128 Rome, Italy and Tattile Via Gaetano Donizetti, 1-3-5,25030 Mairano (Brescia), Italy. Figure 4 shows some sample images from the data set.

The experiments were made with a subset of digits (namely, digits 0, 1, and 5) and the results are reported in Table 1. It is known that the Semeion data are linearly separable. Here we can also observe that classes are also well isolated and separated, giving rise to fairly good clusters, since even k-means attains a very good result (albeit worse than the other methods). As is to be expected, FSC provides

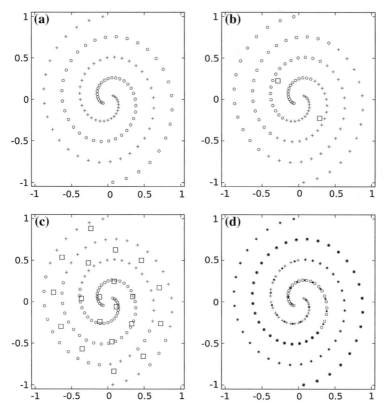

Fig. 2 Two Spirals: **a** dataset with ground truth; **b** clustering result with 2 landmarks, **c** with 20 landmarks, **d** with 90 landmarks. Landmarks are marked by squares. Purity is 1 or 100 % in the last example

Fig. 3 Purity of clusters on the Two Spirals data set versus number of landmarks

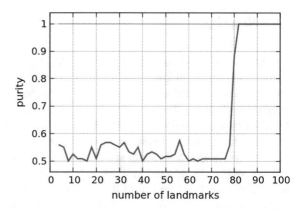

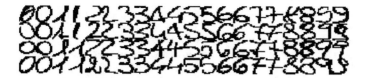

Fig. 4 A sample of digits from the Semeion data set

Table 1 Results on the Semeion data set, digits 0, 1, 5

Method	Rand index	Purity
ASC ($c = 3$)	0.938	0.950
FSC	0.940	0.952
k-means	0.925	0.940

the best result, both in terms of Rand index and of purity, with respect to the class labels. However ASC is very close to FSC, with a very low complexity (the number of landmarks is $c = 3$), which confirms that the problem is not difficult even in an unsupervised setting.

5 Conclusion

A method for controlling model complexity in spectral clustering, as opposed to measuring complexity (on the one hand) and controlling model size (on the other hand), has been presented and discussed with the help of some experiments.

Future work includes the use of this technique in representation learning for supervised problems.

References

1. Asuncion, A., Newman, D.J.: UCI machine learning repository (2007)
2. Bezdek, J.C.: Pattern Recognition with Fuzzy Objective Function Algorithms. Kluwer Academic Publishers, Norwell, MA, USA (1981)
3. Buhmann, J., Tishby, N.: Empirical risk approximation: a statistical learning theory of data clustering. NATO ASI series. Series F Comput. Syst. Sci. 57–68 (1998)
4. Buhmann, J.M.: Information theoretic model validation for clustering. In: 2010 IEEE International Symposium on Information Theory Proceedings (ISIT), pp. 1398–1402. IEEE (2010)
5. Chung, F.R.K.: Spectral Graph Theory (CBMS Regional Conference Series in Mathematics, No. 92). American Mathematical Society (1997)
6. Cover, T.M., Thomas, J.A.: Elements of Information Theory. Wiley Series in Telecommunications and Signal Processing. Wiley, New Jersey, USA (2006)
7. De Silva, V., Tenenbaum, J.B.: Sparse multidimensional scaling using landmark points. Technical report, Stanford University (2004)

8. Eckart, C., Young, G.: The approximation of one matrix by another of lower rank. Psychometrika **1**(3), 211–218 (1936)
9. Fiedler, M.: Algebraic connectivity of graphs. Czech. Math. J. **23**(2), 298–305 (1973)
10. Fowlkes, C., Belongie, S., Chung, F., Malik, J.: Spectral grouping using the nystrom method. IEEE Trans. Pattern Anal. Mach. Intell. **26**(2), 214–225 (2004)
11. Hubert, L., Arabie, P.: Comparing partitions. J. Classif. 193–218 (1985)
12. Lee, M.D.: On the complexity of additive clustering models. J. Math. Psychol. **45**(1), 131–148 (2001)
13. Li, M., Vitányi, P.M.: An Introduction to Kolmogorov Complexity and Its Applications, 3rd edn. Springer, New York (2008)
14. Ng, A.Y., Jordan, M.I., Weiss, Y.: On spectral clustering: analysis and an algorithm. In: Dietterich, T.G., Becker, S., Ghahramani, Z. (eds.) Advances in Neural Information Processing Systems 14. MIT Press, Cambridge, MA (2002)
15. Rand, W.: Objective criteria for the evaluation of clustering methods. J. Am. Stat. Assoc. **66**, 846–850 (1971)
16. Ridella, S., Rovetta, S., Zunino, R.: Plastic algorithm for adaptive vector quantization. Neural Comput. Appl. **7**(1), 37–51 (1998)
17. Rissanen, J.: Stochastic Complexity in Statistical Inquiry, vol. 511. World Scientific Singapore (1989)
18. Rose, K., Gurewitz, E., Fox, G.: A deterministic annealing approach to clustering. Pattern Recogn. Lett. **11**, 589–594 (1990)
19. Rovetta, S., Masulli, F.: Online spectral clustering and the neural mechanisms of concept formation. In: Apolloni, B., Bassis, S., Esposito, A., Morabito, F. (eds.) WIRN2014—Proceedings of the 24th Italian Workshop on Neural Networks. Springer (2014)
20. Schwarz, G., et al.: Estimating the dimension of a model. The Ann. Stat. **6**(2), 461–464 (1978)
21. Shannon, C.E.: A mathematical theory of communication. ACM SIGMOBILE Mob. Comput. Commun. Rev. **5**(1), 3–55 (2001)
22. Shepard, R.N., Arabie, P.: Additive clustering: representation of similarities as combinations of discrete overlapping properties. Psychol. Rev. **86**(2), 87 (1979)
23. Shi, J., Malik, J.: Normalized cuts and image segmentation. IEEE Trans. Pattern Anal. Mach. Intell. **22**(8), 888–905 (2000)
24. Sneath, P.H.A., Sokal, R.R.: Numerical Taxonomy: The Principles and Practice of Numerical Classification. W.H. Freeman, San Francisco (1973)
25. Vapnik, V.N.: Statistical Learning Theory. Wiley, New York (1998)
26. Von Luxburg, U.: A tutorial on spectral clustering. Stat. Comput. **17**(4), 395–416 (2007)
27. Zamir, R., Feder, M.: On lattice quantization noise. IEEE Trans. Inf. Theor. **42**(4), 1152–1159 (1996)

Part III
Artificial Neural Networks: Algorithms and Models

Using Neural Networks
for the Recognition of Specific Motion
Symptoms of the Parkinson's Disease

Paolo Lorenzi, Rosario Rao, Giulio Romano, Ardian Kita,
Martin Serpa, Federico Filesi, Matteo Bologna, Antonello Suppa,
Alfredo Berardelli and Fernanda Irrera

Abstract It is proposed a wearable sensing system based on Inertial Measurement Unit which uses Artificial Neural Networks (ANNs) for the detection of specific motion disorders typical of the Parkinson's disease (PD). The system is made of a single inertial sensor positioned on the head by the ear. It recognizes noticeable gait disorders potentially dangerous for PD patients and can give an audio feedback. The algorithm of recognition based on ANNs is extremely versatile and correctly operating for any individual gait feature. It provides robust and reliable detection of the targeted kinetic features and requires fast and light calculations. The final headset system will be extremely energy efficient thanks to its compactness, to the fact that the ANN avoids computational energy wasting and to the fact that the audio feedback to the patient does not require any wired/wireless connection. This improves the system performance in terms of battery life and monitoring time.

1 Introduction and Hard System

The advancement of sensing technologies allows today to develop smart systems to monitor in a quantitative and objective way the humans activities [1]. In particular, the availability of highly accurate inertial sensors opened to a wide range of applications where the 3D orientation, linear and angular velocity and acceleration

P. Lorenzi · R. Rao · G. Romano · A. Kita · M. Serpa · F. Filesi · F. Irrera (✉)
Department of Information Engineering, Electronics and Telecommunications DIET,
"Sapienza" University of Rome, via Eudossiana 18, 00184 Rome, Italy
e-mail: irrera@diet.uniroma1.it

P. Lorenzi
e-mail: lorenzi@diet.uniroma1.it

G. Romano
e-mail: romano@diet.uniroma1.it

M. Bologna · A. Suppa · A. Berardelli
Department of Neurology and Psichiatry, Piazzale Aldo Moro 1, 00184 Rome, Italy

© Springer International Publishing Switzerland 2016
S. Bassis et al. (eds.), *Advances in Neural Networks*, Smart Innovation,
Systems and Technologies 54, DOI 10.1007/978-3-319-33747-0_12

had to be determined. In the field of the assistance of unhealthy people, a significant portion of the research refers to the Parkinson's disease (PD) [2]. Most common symptoms of the PD involve the motion sphere, including muscular rigidity, tremors, bradykinesia, hypokinesia/akinesia, postural instability. A system for long-time monitoring of PD patients during their daily life is strongly desired by doctors to customize the therapy and control the stage of the disease. Also, this would be crucial in the prevention of catastrophic events as falls, which are often the consequence of motion disorders and freezing of gait (FOG). FOG is a paroxysmal block of movement. From the gait point of view, it typically results in a progressive step shortening (which we call pre-freezing state) and a stop, during which patients refer that their feet are "stuck to the ground" (FOG state). During the FOG patients make attempts to complete the step, oscillating and thrusting forward the trunk, which can cause falls. It has been demonstrated that rhythmic auditory stimulation (RAS) as a metronome can release the involuntary stop condition and getting patients out of the FOG state avoid falls [3–5].

We propose the realization of a Smart HEadset Sensing System (SHESS) able to recognize in real time specific kinetic features associated to motion disorders typical of (but not limited to) the PD. SHESS is thought to be used outdoor, during the normal activity of the monitored person. It is composed by a single IMU (3 axis accelerometers and 3 angle gyroscopes) to be positioned laterally on the head, close to the ear. The algorithm of recognition is based on ANNs and the simplicity of its structure makes it extremely light and fast in terms of computing. Furthermore, the algorithm is very versatile and totally independent on the individual gait feature.

Other systems for the FOG detection have been proposed in the past few years, based on the use of inertial sensors positioned on the patient body, to be used outdoor [6–8] and indoor [9–11]. All of them require using a smartphone or a tablet or a PC for signal processing. Respect to the other systems proposed in literature, SHESS has major energetic, computational (and managing) benefits. In fact, the electronic hardware is compacted in a single package and an eventual audio feedback would not require any wired/wireless (energy wasting) connection (is an headset). Second, the recognition based on ANN is extremely versatile and correctly operating for any individual gait feature, at the same time providing very light and fast calculations, robust and reliable detection of the targeted kinetic feature [12, 13]. Third, the real time signal processing does not require using a smartphone or other external device. At this stage of the research, calculations were performed out-board with a PC. Successful recognition of irregular steps, trunk oscillations and stop state is presented here. It was obtained using a board designed for collecting, storing, processing signals in real-time and eventually wireless transmitting them. It is shown in Fig. 1 [14, 15]. It uses STMicroelectronics components: a wireless AHRS containing ultra-low power STM32L1 microcontroller and a LSM9DS0 inertial unit, a 256 KB non-volatile memory, a bluetooth module (not useful in the proposed system). The board is a few square centimeters large.

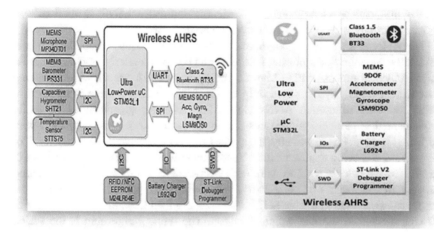

Fig. 1 Sketch of the wireless AHRS and of the board

2 The Soft System Operation

The system has just one IMU positioned on the head by the ear. The reference frame is depicted in Fig. 2. The vertical direction is the y-axis, the gait direction is the x-axis. While walking, the mass center of the sensor oscillates in the y direction between the two extreme positions (*walk state*), and when the person stops (*stop state*) the mass center stops as well. The acceleration along the x direction (A_x) and along the y direction (A_y) vary in the two states. As an example, A_{ymax} is outlined. Typical curves of A_y and A_x are drawn in Fig. 2: the lower (blu) line is A_x, the continuous (red) line is A_y. In that case, the person was first in a stop state, then he started walking and made 10 steps. The portions of signals associated to the steps look quite different from the portion associated to the stop state, and 10 peaks of

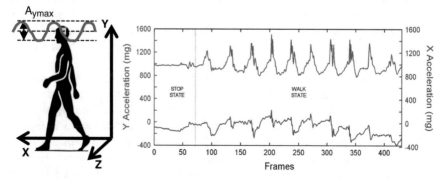

Fig. 2 *Left* Sketch of the reference framework of the headset sensor. *Right* typical curves of raw data of Ay (*upper curve*) and Ax (*lower curve*) during a regular walk and in the stop state

acceleration can be clearly distinguished. The innovative ANN method of gait pattern recognition proposed here uses accelerations raw data.

The ANN is a powerful data-driven, self-adaptive, flexible computational tool having the capability of capturing nonlinear and complex underlying characteristics of any physical process with a high degree of accuracy. It has the ability to learn how to recognize patterns based only on the data given, and there is no need to build any kind of model. ANNs algorithms have the advantage to be easily implemented in parallel architectures (i.e. in multicore processors or systems with GPUs). That reduces drastically the processing time compared to other kind of algorithms, achieving similar results. Neural net computation may be carried out in parallel so we can design hardware taking advantage of this capability. A neural network of low degree of complexity, with only two layers (the hidden and the output layer), is used in this work. The network is a feed forward type consisting of 10 neurons with a sigmoid weight function. The training algorithm is the scaled conjugate gradient backpropagation (implemented in Matlab) [16, 17]. The 80 % of the samples is used for training and 20 % for validation. The cross entropy is chosen as performance function [18]. Ten epochs are sufficient to train the ANN in any studied case (discussed in the following), which indicates that the algorithm is very light and fast.

The approach consists of a few main steps. First of all, we consider a signal containing a known number of reference patterns with a known size. The known signal is partitioned in subsequences having the same size of the reference pattern and the reference pattern is compared with the subsequences. To improve flexibility, the Dynamic Time Warping (DTW) technique is used, since it allows to compare *similar* patterns rather than just *one specific* pattern in the time subsequence [19]. DTW is a nonlinear time normalization technique based on dynamic programming. Given two time series of different duration, a cost function can be calculated [20]. The cost function represents the cumulative distance between two time series, based on the warping function. The warping function is the path which links the beginning and the ending nodes of the time series and exhibits the minimum cumulative distance between them. A threshold of the cost function is set, which determines the degree of similarity between the reference signal and the specific subsequence. The threshold is tuned as to achieve the known number of reference patterns (threshold optimization). Sequences having a cost function below the optimized threshold are positive inputs to the ANN (1). In the other cases the ANN input is negative (0).

Training: The ANN training procedure for step recognition is now described. The flow diagram is reported in Fig. 3. The starting point is the A_y signal associated to a known walk. The step reference pattern is selected in it. An example of a step reference pattern is selected in the walk region of Fig. 1 and is shown in Fig. 4a (with arbitrary origin). Apart from the amplitude, the reference pattern is characterized by the size (number of frames), which is linearly related to the step time. In the case of Fig. 4a the size is about 37 frames. Then, the signal is partitioned in subsequences and the subsequences are compared with the reference pattern using the DTW. An example of the cost function of the DTW is shown in Fig. 4 (right).

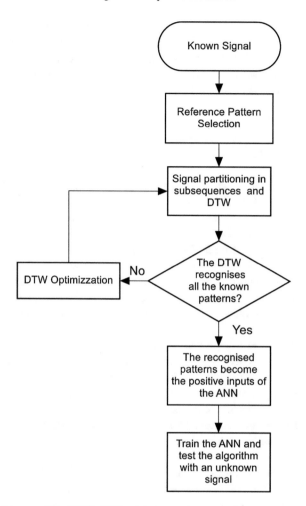

Fig. 3 Flow diagram of the DTW-ANN training procedure

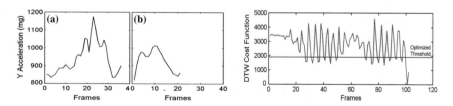

Fig. 4 *Left* **a** Reference pattern associated to a regular step. **b** Reference pattern associated to a short step. *Right* Cost function of the DTW and optimized threshold

When the DTW recognizes the reference pattern in a subsequence then the corresponding ANN input is positive. On the contrary, if the known steps are not all recognized, the size of the reference pattern and/or the threshold of the cost function are changed (DTW optimization) and the DTW is run again. The optimized cost function threshold is outlined in Fig. 4.

3 Experimental Results with the ANN

The ANN is now tested using *unknown* signals. We report results of three tests of interest for PD: stop (typical of a FOG event), step shortening (typical of a pre-freezing phase), trunk fluctuations (which corresponds to an increased risk of fall).

First test: The raw A_y signal of an *unknown* walk is plotted in Fig. 5 (upper red curve). Four intervals can be distinguished: interval I is intuitively associated to a stop state (the A_y value keeps constant at 1000 mg), intervals II and IV are clearly associated to a periodic movement, interval III refers to something intermediate between a periodic movement and a stop state. The ANN was trained to recognize the stop state. The reference signal in this case was selected in interval I (it was an almost straight line, not shown for brevity) and the result is the lower dotted (ciano) curve. As expected, the ANN output is 1 in the I interval, is 0 during intervals II and III and assumes values between 0 and 1in the IV state, as whether short steps were present. The presence of regular steps was investigated using the reference pattern of Fig. 4a. The outputs of the ANN in this case are shown in Fig. 5 with the upper (blue) curve. As one can see, nine steps were recognized in region II and nine steps in region IV (ANN outputs close to 1). No regular steps were identified in region III. We can conclude that interval III was recognized as a not walk state and a not stop state.

Second test: The second *unknown* Ay signal is shown in Fig. 6 (upper red curve). In this test, the ANN had to recognize the short steps and distinguish them from regular ones. Therefore, in this experiment the ANN was trained using a

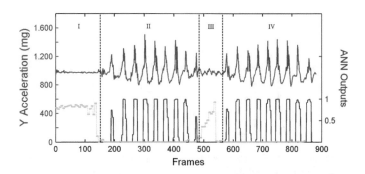

Fig. 5 Raw Ay signal of the first *unknown* test signal (*upper curve*) composed by stop state and walk state. ANN output associated to stop (*lower dotted curve*) and to walk (*lower solid curve*)

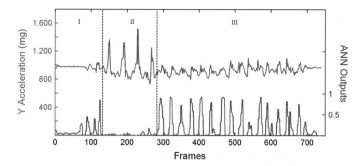

Fig. 6 Raw Ay signal of the second *unknown* test signal (*upper curve*) composed by stop, walk state and irregular short steps. ANN output associated to the irregular short steps (*lower curve*)

reference pattern selected in region III. The new reference pattern is displayed in Fig. 4b with arbitrary origin. (The comparison with the reference pattern of a regular step, reported in Fig. 4a, outlines different shapes, amplitudes and sizes). The ANN outputs are shown in Fig. 6 (lower blue curve). Although steps were irregular and feature variable length, the ANN recognized the short steps in interval III, where just one of the fifteen was regarded as uncertain (step # 10). Furthermore, a couple of irregular steps were also detected, when passing from region I to region II and from region II to region III.

Third test: In this test, the ANN had to recognize trunk fluctuations in the x-y plane (referring to Fig. 2). In this experiment, legs were motionless and only the trunk oscillated pivoting on the pelvis. This situations is of particular interest since during a freezing of gait PD patients feel that their feet are stuck to the ground and they try repeatedly to make a step thrusting out and overbalancing. This is clearly associated to an increased risk of fall.

In this case, the fact that the sensor is positioned on the head guarantees the maximum sensitivity to the movement. In this experiment, the angle respect to the vertical axis varied in the range ±20°. Again, the Ay raw signal was analyzed and the curve is shown in Fig. 7 (upper red curve). As expected, the trunk oscillations are very well characterized (region III). Regions I and II are associated,

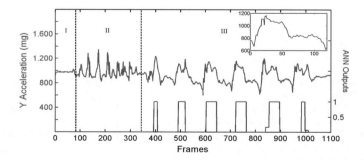

Fig. 7 Raw Ay signal of the third *unknown* test signal (*upper curve*) composed by stop state, walk state and trunk oscillations. ANN output associated to the trunk oscillations (*lower curve*)

respectively, to a stop state and a walk state. The ANN was trained to recognize trunk fluctuations using a reference pattern selected in region III. It is shown in the inset. The ANN outputs are displayed in Fig. 7 (lower blue curve). Recognition was excellent and all the trunk oscillations yielded ANN = 1.

4 Conclusions

A smart wireless sensing system designed for long time monitoring of specific movement features in the Parkinson's disease (PD) has been proposed. The system is composed by a single inertial sensor to be positioned laterally on the head, close to the ear, for headset applications. The classification of the motion features is performed using an ANN, starting from the raw signals of the sensor. The simplicity of this recognition method reflects in extremely light and fast computing. This fact in conjunction with the fact that an eventual audio feedback does not need any wired/wireless connection makes the final headset system extremely energy efficient, impacting positively the battery life and, definitely, the monitoring time.

The ANN recognition method is extremely versatile and totally independent on the individual gait feature. It has been proved to be very flexible obtaining excellent results without the need of large number of samples. Experimental tests have been performed to check the validity of the ANN approach. ANN results were (successfully) compared with results obtained by empirical algorithms. The tests regarded some motion conditions of particular interest for the recognition of specific disorders associated to the Parkinson's disease. In particular, the system was trained to recognize the stop state (typical of the freezing of gait), short and irregular steps (one of the pre-freezing state symptoms), trunk fluctuations (occurring when the patient experiences a freezing of gait and tries to make steps, potentially very dangerous for the patient). As a final result, the incipit of the freezing of gait can be identified. This is an excellent result because rhythmic auditory stimulations (audio-feedback) provided to the PD patient by the headset just at the onset of the freezing would prevent the patient from catastrophic events as falls.

Acknowledgments Authors wish to thank ST Microelectronics for providing the IMUs.

References

1. Mukhopadhyay, S.: Wearable sensors for human activity monitoring: a review. Sens. J. IEEE. **15**, 1321–1330 (2015)
2. Hirtz, D., Thurman, D., Gwinn-Hardy, K., Mohamed, M., Chaudhuri, A., et al.: How common are the "common" neurologic disorders? Neurology. **68**, 326–337 (2007)
3. Plotnik, M., Shema, S., Dorfman, M., Gazit, E., Brozgol, M., et al.: A motor learning-based intervention to ameliorate freezing of gait in subjects with Parkinson's disease. J. Neurol. **261**, 1329–1339 (2014)

4. Arias, P., Cudeiro, J.: Effect of rhythmic auditory stimulation on gait in parkinsonian patients with and without freezing of gait. PLoS ONE **5** (2010)
5. Arias, P., Cudeiro, J.: Effects of rhythmic sensory stimulation (auditory, visual) on gait in Parkinson's disease patients. Exp. Brain Res. **186**, 589–601 (2008)
6. Moore, S.T., MacDougall, H.G., Gracies, J.-M., Cohen, H.S., Ondo, W.G.: Long-term monitoring of gait in Parkinson's disease. Gait Posture. **26**, 200–207 (2007)
7. Lorenzi, P., Rao, R., Suppa, A., Kita, A., Parisi, R., Romano, G., Berardelli, A., Irrera, F.: Wearable wireless inertial sensors for long-time monitoring of specific motor symptoms in parkinson's disease. In: BIODEVICES 2015, BIOSTEC 2015, Lisbon (P), pp. 168–173
8. Bachlin, M., Plotnik, M., Roggen, D., Giladi, N., Hausdorff, J., et al.: A wearable system to assist walking of Parkinson s disease patients. Methods Inf. Med. **49**, 88–95 (2010)
9. Moore, S., Yungher, D., Morris, T., Dilda, V., MacDougall, H., et al.: Autonomous identification of freezing of gait in Parkinson's disease from lower-body segmental accelerometry. J. NeuroEng. Rehabil. **10**, 19 (2013)
10. Sijobert, B., Azevedo, C., Denys, J., Geny, C.: IMU based detection of freezing of gait and festination in parkinson's disease. In: IFESS Malaysia (2014)
11. Moore, S.T., MacDougall, H.G., Ondo, W.G.: Ambulatory monitoring of freezing of gait in Parkinson's disease. J. Neurosci. Methods **167**, 340–348 (2008)
12. Tu, J.V.: Advantages and disadvantages of using artificial neural networks versus logistic regression for predicting medical outcomes. J. Clin. Epidemiol. **49**, 1225–1231 (1996)
13. Kohonen, T., Barna, G., Chrisley, R.: Statistical pattern recognition with neural networks: Benchmarking studies. In: IEEE International Conference on Neural Network, pp. 61–68 (1988)
14. Comotti, D., Galizzi, M., Vitali, A.: neMEMSi: one step forward in wireless attitude and heading reference systems. In: Inertial Sensors and Systems (ISISS), International Symposium on, pp. 1–4 (2014)
15. Caldara, M., Comotti, D., Galizzi, M., Locatelli, P., Re, V., et al.: A novel body sensor network for parkinson's disease patients rehabilitation assessment. wearable and implantable body sensor networks (BSN). In: 2014 11th International Conference on. IEEE, pp. 81–86 (2014)
16. Bishop, C.M., et al.: Neural Networks for Pattern Recognition. Clarendon, Oxford (1995)
17. Chau, T.: A review of analytical techniques for gait data. Part 2: neural network and wavelet methods. Gait Posture. **13**, 102–120 (2001)
18. Kline, D.M., Berardi, V.L.: Revisiting squared-error and cross-entropy functions for training neural network classifiers. Neural Comput. Appl. **14**, 310–318 (2005)
19. Wang, K., Gasser T., et al.: Alignment of curves by dynamic time warping. Ann. Stat. **25** 1251–1276 (1997)
20. Keogh, E., Ratanamahatana, C.A.: Exact indexing of dynamic time warping. Knowl. Inf. Syst. **7**, 358–386 (2005)

Benchmarking Functional Link Expansions for Audio Classification Tasks

Simone Scardapane, Danilo Comminiello, Michele Scarpiniti,
Raffaele Parisi and Aurelio Uncini

Abstract Functional Link Artificial Neural Networks (FLANNs) have been extensively used for tasks of audio and speech classification, due to their combination
of universal approximation capabilities and fast training. The performance of a
FLANN, however, is known to be dependent on the specific functional link (FL)
expansion that is used. In this paper, we provide an extensive benchmark of multiple
FL expansions on several audio classification problems, including speech discrimination, genre classification, and artist recognition. Our experimental results show
that a random-vector expansion is well suited for classification tasks, achieving the
best accuracy in two out of three tasks.

Keywords Functional links · Audio classification · Speech recognition

1 Introduction

Music information retrieval (MIR) aims at efficiently retrieving songs of interest
from a large database, based on the user's requirements [5]. One of the most important tasks in MIR is automatic music classification (AMC), i.e. the capability of
automatically assigning one or more labels of interest to a song, depending on its
audio characteristics. Examples of labels are the genre, artist, or the perceived mood.

S. Scardapane (✉) · D. Comminiello · M. Scarpiniti · R. Parisi · A. Uncini
Department of Information Engineering, Electronics and Telecommunications (DIET),
"Sapienza" University of Rome, Via Eudossiana 18, 00184 Rome, Italy
e-mail: simone.scardapane@uniroma1.it

D. Comminiello
e-mail: danilo.comminiello@uniroma1.it

M. Scarpiniti
e-mail: michele.scarpiniti@uniroma1.it

R. Parisi
e-mail: raffaele.parisi@uniroma1.it

A. Uncini
e-mail: aurel@ieee.org

© Springer International Publishing Switzerland 2016
S. Bassis et al. (eds.), *Advances in Neural Networks*, Smart Innovation,
Systems and Technologies 54, DOI 10.1007/978-3-319-33747-0_13

Clearly, being able to classify a song in a sufficiently large number of categories is extremely helpful in answering a specific user's query.

The problem of AMC can be divided in two components, namely, the choice of a suitable musical feature extraction procedure, and of an appropriate classifier. This latter choice is worsened by the intrinsic difficulties associated to AMC, among which we can list a generally large number of features (to provide a comprehensive overview of the audio content of a signal), an equally large number of class labels, and the necessary subjectivity in assigning such labels to each song. In the literature, classical choices for music classification include support vector machines (SVM) [5], gaussian mixture models (GMM) [16], and multilayer perceptrons (MLP) [11].

A particularly promising line of research stems for the use of functional-link (FL) networks [8]. These are two-layered networks, where the first layer, known as the expansion block, is composed of a given number of *fixed* non-linearities [2, 3, 8]. Due to this, the overall learning problem of an FL network can be cast as a standard linear least-square problem, which can be solved efficiently even in the case of very large datasets. FL-like networks have been shown to achieve comparable results to standard MLP, while requiring a smaller computational time [11, 13, 15]. It is known, however, that the performance of an FL network is dependent on the choice of the expansion block [8]. This in turn depends on the specific application and on the audio signals involved in the processing. Generally speaking, basis functions must satisfy universal approximation constraints and may be a subset of orthogonal polynomials, such as Chebyshev, Legendre and trigonometric polynomials, or just approximating functions, such as sigmoid and Gaussian functions. In this last case, the parameters of the approximating functions are generally assigned stochastically from a predefined probability distribution, and equivalent models have been popularized under the name of extreme learning machine (ELM) [12].

In this paper, we investigate the problem of choosing a suitable expansion block in the case of music classification. To this end, we compare four standard non-linear expansions when applying FL network to three music classification tasks, including genre and artist classification, and music/speech discrimination. This extends and complements our previous work, in which we performed a similar analysis in the case of audio quality enhancement subject to non-linear distortions of a reference signal [1]. Differently from the results analyzed in [1], our experiments show that, in the case of audio classification, random vector expansions provide the best performance, obtaining the lowest classification error in two out of three tasks, and a comparable performance in the last case.

The rest of the paper is organized as follows. Section 2 details the basic mathematical formulation of FL networks. Section 3 presents the 4 functional expansions considered in this work. Then, Sect. 4 introduces the datasets that are used for comparison, together with the basic parameter optimization procedure for the networks. Results are discussed in Sect. 5, while Sect. 6 concludes the paper.

2 Functional Link Neural Networks

Given an input vector $\mathbf{x} \in \mathbb{R}^d$, the output of an FL network is computed as:

$$f(\mathbf{x}) = \sum_{i=1}^{B} \beta_i h_i(\mathbf{x}) = \beta^T \mathbf{h}(\mathbf{x}), \tag{1}$$

where each $h_i(\cdot)$ is a fixed non-linear term, denoted as functional-link or base, $\beta = [\beta_1, \ldots, \beta_B]^T$, while the overall vector $\mathbf{h}(\cdot)$ is denoted as expansion block. Possible choices for $\mathbf{h}(\cdot)$ are detailed in the subsequent section. In a music classification task, the input $\mathbf{x}$ is a suitable representation of a music signal, which may include descriptions of its temporal behavior, frequency and cepstral terms, or meta-information deriving from a user's labeling [5]. We are given a dataset of N pairs song/class for training, denoted as $T = \{(\mathbf{x}_1, y_1), \ldots, (\mathbf{x}_L, y_N)\}$. Let:

$$\mathbf{H} = \begin{bmatrix} h_1(\mathbf{x}_1) & \cdots & h_B(\mathbf{x}_1) \\ \vdots & \ddots & \vdots \\ h_1(\mathbf{x}_N) & \cdots & h_B(\mathbf{x}_N) \end{bmatrix} \tag{2}$$

and $\mathbf{y} = [y_1, y_2, \ldots, y_N]^T$ be the hidden matrix and the output vector respectively. The optimal weights β^* of the FL are obtained as the solution of the overdetermined regularized least-squares problem:

$$\min_{\beta} \frac{1}{2} \|\mathbf{H}\beta - \mathbf{y}\|^2 + \frac{\lambda}{2} \|\beta\|^2, \tag{3}$$

where $\lambda > 0$ is a regularization factor. Solution to Eq. (3) can be expressed in closed form as:

$$\beta^* = (\mathbf{H}^T \mathbf{H} + \lambda \mathbf{I})^{-1} \mathbf{H}^T \mathbf{y}. \tag{4}$$

The previous discussion extends straightforwardly to the case of multiple outputs [14]. This is necessary for classification with K classes, where we adopt the standard 1-of-K encoding, associating a K-dimensional binary vector $\mathbf{y}_i$ to each pattern, where $y_{ij} = 1$ if the pattern i is of class j, 0 otherwise. In this case, the predicted class can be extracted from the K-dimensional FL output $\mathbf{f}(\mathbf{x}_i)$ as:

$$g(\mathbf{x}_i) = \arg\max_{i \in \{1, \ldots, K\}} \mathbf{f}(\mathbf{x}_i). \tag{5}$$

3   Functional Link Expansions

In this section we introduce the most commonly used functional link expansions, that we subsequently compare in our experiments. The first three are deterministic expansions computed from each element of the input, while the fourth is composed of stochastic sigmoid expansions.

3.1   Chebyshev Polynomial Expansion

The Chebyshev polynomial expansion for a single feature x_j of the pattern $\mathbf{x}$ is computed recursively as [8]:

$$h_k\left(x_j\right) = 2x_j h_{k-1}\left(x_j\right) - h_{k-2}\left(x_j\right) , \tag{6}$$

for $k = 0, \ldots, P - 1$, where P is the *expansion order*. The overall expansion block is then obtained by concatenating the expansions for each element of the input vector. In (6), initial values (i.e., for $k = 0$) are:

$$h_{-1}\left(x_j\right) = x_j ,$$
$$h_{-2}\left(x_j\right) = 1. \tag{7}$$

The Chebyshev expansion is based on a power series expansion, which is able to efficiently approximate a nonlinear function with a very small error near the point of expansion. However, far from it, the error may increases rapidly. With reference to different power series of the same degree, Chebyshev polynomials are computationally cheap and more efficient although, when the power series converges slowly, the computational cost dramatically increases.

3.2   Legendre Polynomial Expansion

The Legendre polynomial expansion is defined for a single feature x_j as:

$$h_k\left(x_j\right) = \frac{1}{k}\left\{(2k - 1)x_j h_{k-1}\left(x_j\right) - (k - 1)h_{k-2}\left(x_j\right)\right\} \tag{8}$$

for $k = 0, \ldots, P - 1$. Initial values in Eq. (8) are set as (7). In the case of signal processing, the Legendre functional links provide computational advantage with respect to the Chebyshev polynomial expansion, while promising better performance [9].

3.3 Trigonometric Series Expansion

Trigonometric polynomials provide the best compact representation of any nonlinear function in the mean square sense [10]. Additionally, they are computationally cheaper than power series-based polynomials. The trigonometric basis expansion is given by:

$$h_k(x_j) = \begin{cases} \sin\left(p\pi x_j\right), & k = 2p - 2 \\ \cos\left(p\pi x_j\right), & k = 2p - 1 \end{cases}, \tag{9}$$

where $k = 0, \ldots, B$ is the functional link index and $p = 1, \ldots, P$ is the expansion index, being P the expansion order. Note that the expansion order for the trigonometric series is different from the order of both Chebyshev and Legendre polynomials. Cross-products between elements of the pattern $\mathbf{x}$ can also be considered, as detailed in [3].

3.4 Random Vector Expansion

The fourth functional expansion type that we consider is the random vector (RV) functional link [6, 8]. The RV expansion is parametric with respect to a set of internal weights, that are stochastically assigned. A RV functional link (with sigmoid nonlinearity) is given by:

$$h_k(\mathbf{x}) = \frac{1}{1 + e^{(-\mathbf{a}\mathbf{x}+b)}}, \tag{10}$$

where the parameters $\mathbf{a}$ and b are randomly assigned at the beginning of the learning process. It is worth noting that the sigmoid function is just one of the possible choices to apply a nonlinearity to the vector $\mathbf{x}$. Unlike the previous expansion types, the RV expansion does not involve any expansion order. Additionally, the overall number B of functional links is a free parameter in this case, while in the previous expansions it depends on the expansion order. Convergence properties of the RVFL model are analyzed in [6].

4 Experimental Setup

We tested the FL expansions described in the previous section on three standard audio classification benchmarks, whose characteristics are briefly summarized in Table 1. To compute the average misclassification error, we perform a 3-fold cross-validation on the available data. We optimize the models by performing a grid search procedure, using an inner 3-fold cross-validation on the training data to compute the validation performance. In particular, we search the following intervals:

Table 1 General description of the datasets

Dataset name	Features	Instances	Task	Classes	Reference
Garageband	49	1856	Genre recognition	9	[7]
Artist20	30	1413	Artist recognition	20	[4]
GTZAN	13	120	Speech/Music discrimination	2	[16]

- The exponential interval $2^i, i \in \{-5, -4, \ldots, 4, 5\}$ for the regularization coefficient λ in Eq. (4).
- The set $\{1, 2, \ldots, 8, 9\}$ for the expansion order P in Eqs. (6)-(8)-(9).
- The set $\{50, 100, \ldots, 300\}$ for the expansion block size L when using the random-vector expansion.

Additionally, we extract the parameters **a** and b from the uniform probability distribution in $[-1, +1]$. In all cases, input features were normalized between -1 and $+1$ before the experiments. Finally, we repeat the 3-fold cross-validation 10 times to average out statistical effects due to the initialization of the partition and the random-vector expansion. The code is implemented using MATLAB R2013a on an Intel Core2 Duo E7300, @2.66 GHz and 2 GB of RAM.

5   Results and Discussion

The average misclassification error and training time are reported in Table 2, together with one standard deviation. Best results in each dataset are highlighted in bold-face. Out of three datasets, the Legendre expansion obtains a lower error in the

Table 2 Final misclassification error and training time for the four functional expansions, together with standard deviation

Dataset	Algorithm	Error	Time [s]
Garageband	TRI-FL	0.415 ± 0.013	0.156 ± 0.030
	CHEB-FL	0.407 ± 0.0126	$\mathbf{0.055 \pm 0.001}$
	LEG-FL	$\mathbf{0.404 \pm 0.0140}$	0.072 ± 0.026
	RV-FL	0.411 ± 0.017	0.090 ± 0.015
Artist20	TRI-FL	0.410 ± 0.016	0.084 ± 0.013
	CHEB-FL	0.401 ± 0.021	$\mathbf{0.037 \pm 0.001}$
	LEG-FL	0.442 ± 0.020	0.040 ± 0.001
	RV-FL	$\mathbf{0.375 \pm 0.018}$	0.070 ± 0.014
GTZAN	TRI-FL	0.316 ± 0.073	$\mathbf{0.001 \pm 0.002}$
	CHEB-FL	0.317 ± 0.066	0.004 ± 0.001
	LEG-FL	0.334 ± 0.071	0.004 ± 0.001
	RV-FL	$\mathbf{0.222 \pm 0.062}$	0.005 ± 0.002

Best results are shown in boldface

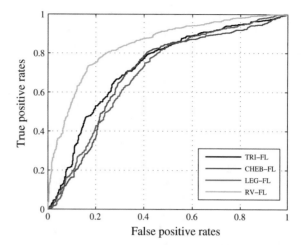

Garageband dataset, while the random-vector expansion has the the lowest error in the remaining two datasets. It is interesting to note, however, that in the first case the differences between the four algorithms are small (roughly 1 % of error), while in the second and third case RV-FL strongly outperforms the other two, with a 2.5 % and 9 % decrease in error respectively, with respect to the second best. Additionally, despite its stochastic nature, the variance of RV-FL is always comparable to the other three models. To comment more on this discrepancy, we show in Fig. 1 the ROC curve for the GTZAN dataset, which is the only binary classification dataset in our experiments. It can be seen that RV-FL (shown with a green line) strongly dominates the other three curves, showing the superior performance of the random-vector expansion in this case.

We show in Table 3 the optimal parameters found by the grid search procedure. As a rule of thumb, we can see that for large datasets (Garageband and Artist20), the Legendre expansion requires a larger expansion order with respect to the trigonometric and Chebyshev expansions. Similarly, the random vector FL requires a large expansion block (around 250 hidden nodes), while for smaller datasets (the GTZAN), the optimal expansion decreases to around 100 hidden nodes. The trigonometric and Chebyshev expansions also requires a large regularization coefficient, which can be kept much smaller for the random-vector expansion (in the larger datasets), and for the Legendre one.

With respect to training time, all four expansions are performing comparably. In particular, the gap between the slowest algorithm and the fastest one never exceeded more than 0.1 seconds. Generally speaking, the trigonometric expansion is the most expensive one to compute, except for datasets with a low number of features, such as GTZAN. In the other cases, the Chebyshev expansion is the fastest one, followed closely by the Legendre expansion, with the random-vector in the middle.

Based on this analysis, we can conclude that, for audio classification tasks, the random expansion seems to outperform other common choices, such as the trigono-

Table 3 Optimal parameters found by the grid search procedure, averaged across the runs

Dataset	Algorithm	Reg. Coeff. λ	Exp. ord. p / B
Garageband	TRI-FL	15.52	1.53
	CHEB-FL	12.14	3
	LEG-FL	9.48	4.27
	RV-FL	0.75	263.3
Artist20	TRI-FL	8.92	1
	CHEB-FL	6.89	3
	LEG-FL	0.30	3.4
	RV-FL	0.33	270
GTZAN	TRI-FL	16.55	3.8
	CHEB-FL	17.93	3.07
	LEG-FL	1.18	7.4
	RV-FL	10.13	113.3

The value in the fourth column is the expansion order p for TRI-FL, CHEB-FL and LEG-FL, and the hidden layer's size B for RV-FL

metric one. This is an interesting result, since it shows a difference between the performance of FL networks with random vector expansions in dynamic audio modeling tasks [1], and static classification tasks. We hypothesize this is due to its higher capacity of extracting non-linear features from the original signal.

6   Conclusions

FL networks are common learning models when considering audio classification tasks. In this paper, we presented an analysis of several functional expansion blocks, considering three different tasks, including genre and artist recognition. Our experimental results suggest that the random vector expansion outperforms other common choices, while requiring a comparable training time.

References

1. Comminiello, D., Scardapane, S., Scarpiniti, M., Parisi, R., Uncini, A.: Functional link expansions for nonlinear modeling of audio and speech signals. In: Proceedings of the International Joint Conference on Neural Networks (2015)
2. Comminiello, D., Scardapane, S., Scarpiniti, M., Parisi, R., Uncini, A.: Online selection of functional links for nonlinear system identification. In: Smart Innovation, Systems and Technologies, Springer International Publishing AG, **37**, pp. 39–47 (2015)
3. Comminiello, D., Scarpiniti, M., Azpicueta-Ruiz, L.A., Arenas-García, J., Uncini, A.: Functional link adaptive filters for nonlinear acoustic echo cancellation. IEEE Trans. Acoust. Speech Signal Process. **21**(7), 1502–1512 (2013)

4. Ellis, D.P.W.: Classifying music audio with timbral and chroma features. In: Proceedings of the 8th International Conference on Music Information Retrieval, pp. 339–340. Austrian Computer Society (2007)
5. Fu, Z., Lu, G., Ting, K.M., Zhang, D.: A survey of audio-based music classification and annotation. IEEE Trans. Multimedia **13**(2), 303–319 (2011)
6. Igelnik, B., Pao, Y.H.: Stochastic choice of basis functions in adaptive function approximation and the functional-link net. IEEE Trans. Neural Netw. **6**(6), 1320–1329 (1995)
7. Mierswa, I., Morik, K.: Automatic feature extraction for classifying audio data. Mach. Learn. **58**(2–3), 127–149 (2005)
8. Pao, Y.H.: Adaptive Pattern Recognition and Neural Networks. Addison-Wesley, Reading, MA (1989)
9. Patra, J.C., Chin, W.C., Meher, P.K., Chakraborty, G.: Legendre-FLANN-based nonlinear channel equalization in wireless communication system. In: Proceedings of the IEEE International Conference on Systems, Man and Cybernetics (SMC), Singapore, pp. 1826–1831, Oct 2008
10. Patra, J.C., Pal, R.N., Chatterji, B.N., Panda, G.: Identification of nonlinear dynamic systems using functional link artificial neural networks. IEEE Trans. Syst. Man Cybern. Part B **29**(2), 254–262 (1999)
11. Scardapane, S., Comminiello, D., Scarpiniti, M., Uncini, A.: Music classification using extreme learning machines. In: 8th International Symposium on Image and Signal Processing and Analysis (ISPA), Trieste, Italy, pp. 377–381, Sep 2013
12. Scardapane, S., Comminiello, D., Scarpiniti, M., Uncini, A.: Online sequential extreme learning machine with kernels. IEEE Trans. Neural Netw. Learn. Syst. **26**(9), 2214–2220 (2015). doi:10.1109/TNNLS.2014.2382094
13. Scardapane, S., Fierimonte, R., Wang, D., Panella, M., Uncini, A.: Distributed music classification using random vector functional-link nets. In: Proceedings of the International Joint Conference on Neural Networks (2015)
14. Scardapane, S., Wang, D., Panella, M., Uncini, A.: Distributed learning for random vector functional-link networks. Inf. Sci. **301**, 271–284 (2015)
15. Turnbull, D., Elkan, C.: Fast recognition of musical genres using RBF networks. IEEE Trans. Knowl. Data Eng. **17**(4), 580–584 (2005)
16. Tzanetakis, G., Cook, P.: Musical genre classification of audio signals. IEEE Trans. Speech Audio Process. **10**(5), 293–302 (2002)

A Comparison of Consensus Strategies for Distributed Learning of Random Vector Functional-Link Networks

Roberto Fierimonte, Simone Scardapane, Massimo Panella and Aurelio Uncini

Abstract Distributed machine learning is the problem of inferring a desired relation when the training data is distributed throughout a network of agents (e.g. robots in a robot swarm). Multiple families of distributed learning algorithms are based on the decentralized average consensus (DAC) protocol, an efficient algorithm for computing an average starting from local measurement vectors. The performance of DAC, however, is strongly dependent on the choice of a weighting matrix associated to the network. In this paper, we perform a comparative analysis of the relative performance of 4 different strategies for choosing the weighting matrix. As an applicative example, we consider the distributed sequential algorithm for Random Vector Functional-Link networks. As expected, our experimental simulations show that the training time required by the algorithm is drastically reduced when considering a proper initialization of the weights.

Keywords Consensus · Distributed machine learning · Random vector functional-link

R. Fierimonte · S. Scardapane (✉) · M. Panella · A. Uncini
Department of Information Engineering Electronics
and Telecommunications (DIET), "Sapienza" University of Rome,
Via Eudossiana 18, 00184 Rome, Italy
e-mail: simone.scardapane@uniroma1.it

R. Fierimonte
e-mail: roberto.fierimonte@gmail.com

M. Panella
e-mail: massimo.panella@uniroma1.it

A. Uncini
e-mail: aurel@ieee.org

© Springer International Publishing Switzerland 2016
S. Bassis et al. (eds.), *Advances in Neural Networks*, Smart Innovation,
Systems and Technologies 54, DOI 10.1007/978-3-319-33747-0_14

1 Introduction

In the last decade, the problem of distributed machine learning, i.e. the problem of learning by a set of data originated in a distributed system, has become an extensive reserched topic, since a large number of real-world applications can be modeled in the form of a distributed learning problem (e.g. inference in Wireless Sensor Networks [2], decentralized databases and datacenters [1], music classification over P2P networks [10], and several others).

Different algorithms were developed to deal with the problem, including works on distributed support vector machines [4], distributed neural networks [5, 11], and applications of distributed optimization techniques [2]. One recently developed strategy is the 'learning by consensus' (LBC) strategy [5]. LBC allows to transform any centralized iterative learning algorithm into a fully distributed protocol. It consists in the alternative application of local update rules and distributed averaging steps, implemented via the distributed average consensus (DAC) protocol [6, 7]. The use of DAC allows to obtain a fully decentralized algorithm, without the need for a fusion center, and with only local communication between neighboring nodes. In [10], we extended the LBC framework to the decentralized sequential setting, where data arrives continuously at every node. In particular, we presented a fully distributed, sequential algorithm for a particular class of neural networks known as Random Vector Functional-Links (RVFLs) [8]. RVFLs are composed of a fixed layer of non-linearities, followed by an adaptable linear layer. Due to this, the resulting training algorithms can be formulated in the form of linear regression problems, with strong capabilities of scaling to large training sets, as demonstrated in several previous works [11].

The DAC protocol computes the average in an iterative fashion, by locally weighting the estimations of each neighbor at every node. Hence, its convergence behavior is strongly dependent on the particular choice of mixing parameters. Among the many choices that guarantee global convergence, several strategies for choosing them have been proposed in the literature, each with its own asymptotic behavior and computational requirements [12]. Currently, a thorough analysis and comparison on this topic is missing. In this paper, we begin this investigation by comparing the performance of 4 different strategies for the DAC protocol, using the algorithm presented in [10] as a benchmark application. The strategies that we analyze vary from choosing a fixed value for every coefficient, to more complex choices satisfying strong optimality conditions. Our experimental results show that the performance of the DAC protocol, and by consequence the performance of any distributed training algorithm based on its application, can improve significantly with proper choices of the mixing parameters.

The rest of the paper is organized as follows. In Sect. 2 we present the DAC protocol, together with the 4 strategies that we investigate. Then, in Sect. 4 we briefly describe the distributed algorithm for RVFLs presented in [10]. Experimental results are then provided in Sect. 5, while Sect. 6 concludes the paper.

2 Decentralized Average Consensus

Distributed average consensus, or simply Consensus, is a totally distributed iterative protocol designed to compute the average of a measurements vector within a
network. We assume the network in the form of a graph $G(V, E)$ with nodes V and
edges E, wherein connectivity is known a priori and can be expressed in the form of
an adjacency matrix $\mathbf{A}$ whose elements are such that:

$$A_{ij} = \begin{cases} 1 & \{i,j\} \in E \\ 0 & \{i,j\} \notin E \end{cases}. \tag{1}$$

For the sake of simplicity, we consider only connected graphs, i.e. graphs where
exists a directed path between each pair of nodes $u, v \in V$. Let $\boldsymbol{\beta}_i(t)$ be the vector of
measurements associated with the ith node of the network at instant t, and $N = |V|$,
the task of the protocol is for all the nodes to converge to the average of the initial
values of the measurements:

$$\hat{\boldsymbol{\beta}} = \frac{1}{N} \sum_{i=1}^{N} \boldsymbol{\beta}_i(0). \tag{2}$$

For discrete-time distributed systems the DAC protocol is defined by a set of linear
updating equations in the form of:

$$\boldsymbol{\beta}_i(t+1) = \sum_{j=1}^{N} w_{ij} \boldsymbol{\beta}_j(t), \tag{3}$$

which can be reformulated compactly as a linear system:

$$\boldsymbol{\beta}(t+1) = \mathbf{W} \boldsymbol{\beta}(t). \tag{4}$$

The matrix $\mathbf{W}$ is named weights matrix, and the value of its generic element w_{ij}
denotes the strength of the connection between nodes i and j, or alternatively, the
confidence that node i assigns to the information coming from node j. We denote
with $\mathscr{W}$ the set of the admissible weights matrices:

$$\mathscr{W} = \{\mathbf{W} \in \mathbb{R}^{N \times N} : w_{ij} = 0 \quad \text{if} \quad i \neq j, \{i,j\} \notin E\}. \tag{5}$$

For suitable choices of the weights matrix $\mathbf{W} \in \mathscr{W}$, the iterations defined in Eq.
(4) converge locally to the average (2). A large number of modifications for the basic
DAC protocol detailed here were proposed in literature, including the case where
the agreement is a weighted average of the initial values [2], and applications to
problems with dynamic topologies and time delays [7]. In this paper, we focus on
the simplest case, which represents fixed, undirected network topologies.

3 Consensus Strategies

Different strategies for the DAC protocol correspond to different choices of the weights matrix. Clearly, the choice of a particular weight matrix depends on the available information at every node about the network topology, and on their specific computational requirements.

3.1 Max-Degree

The first strategy that we consider is the max-degree weights matrix, which is a common choice in real-world applications, and is defined entry-wise by:

$$
w_{ij} = \begin{cases} 1/(d+1) & i \neq j, \{i,j\} \in E \\ 1 - d_i/(d+1) & i = j \\ 0 & i \neq j, \{i,j\} \notin E \end{cases} , \tag{6}
$$

where d_i is the degree of the ith node, and d is the maximum degree of the network.

3.2 Metropolis-Hastings

An even simpler choice is the Metropolis-Hastings weights matrix:

$$
w_{ij} = \begin{cases} 1/(\max\{d_i, d_j\} + 1) & i \neq j, \{i,j\} \in E \\ 1 - \sum_{j \in \mathcal{N}_i} 1/(\max\{d_i, d_j\} + 1) & i = j \\ 0 & i \neq j, \{i,j\} \notin E \end{cases} , \tag{7}
$$

where $\mathcal{N}_i$ is the set of nodes' indexes directly connected to node i. Differently from the max-degree strategy, the Metropolis-Hastings strategy does not require the knowledge of global information (the maximum degree) about the network topology, but requires that each node knows the degrees of all its neighbors.

3.3 Minimum Asymptotic

The third matrix strategy considered here corresponds to the optimal strategy introduced in [12], wherein the weights matrix is constructed to minimize the asymptotic convergence factor $\rho(\mathbf{W} - \mathbf{1}\mathbf{1}^{\mathrm{T}}/N)$, where $\rho(\cdot)$ denotes the spectral radius operator. This is achieved by solving the constrained optimization problem:

$$\begin{aligned} \text{minimize} \quad & \rho(\mathbf{W} - \mathbf{1}\mathbf{1}^{\mathrm{T}}/N) \\ \text{subject to} \quad & \mathbf{W} \in \mathscr{W}, \quad \mathbf{1}^{\mathrm{T}}\mathbf{W} = \mathbf{1}^{\mathrm{T}}, \quad \mathbf{W}\mathbf{1} = \mathbf{1} \end{aligned} \tag{8}$$

Problem (8) is non-convex, but it can be shown to be equivalent to a semidefinite programming (SDP) problem [12], solvable using efficient ad-hoc algorithms.

3.4 Laplacian Heuristic

The fourth and last matrix considered in this work is an heuristic approach [12] based on constant edge weights matrix:

$$\mathbf{W} = \mathbf{I} - \alpha\mathbf{L}, \tag{9}$$

where $\alpha \in \mathbb{R}$ is a user-defined parameter, and $\mathbf{L}$ is the Laplacian matrix associated to the network [2]. For weights matrices in the form of (9), the asymptotic convergence factor satisfies:

$$\begin{aligned} \rho(\mathbf{W} - \mathbf{1}\mathbf{1}^{\mathrm{T}}/N) &= \max\{\lambda_2(\mathbf{W}), -\lambda_n(\mathbf{W})\} \\ &= \max\{1 - \alpha\lambda_{n-1}(\mathbf{L}), \alpha\lambda_1(\mathbf{L}) - 1\}, \end{aligned} \tag{10}$$

where $\lambda_i(\mathbf{W})$ denotes the ith eigenvalue associated to $\mathbf{W}$. The value of α that minimizes (10) is given by:

$$\alpha^* = \frac{2}{\lambda_1(\mathbf{L}) + \lambda_{N-1}(\mathbf{L})} . \tag{11}$$

4 Data-Distributed RVFL Network

Let us consider an artificial neural network with a single hidden layer and a single output node, obtained as a weighted sum of B nonlinear transformations of the input:

$$f_\omega(\mathbf{x}) = \sum_{m=1}^{B} \beta_m h_m(x; \omega_m) = \boldsymbol{\beta}^{\mathrm{T}}\mathbf{h}(\mathbf{x}; \omega_1, \dots, \omega_m), \tag{12}$$

where $\mathbf{x} \in \mathbb{R}^d$, and each function $h_i(\cdot)$ is parameterized by a real-valued vector ω_i. The resulting model is named Functional-Link Artificial Neural Network (FLANN) and the functions $h_i(\cdot)$ are named basis functions, or functional links [3]. A particular instance of FLANN is the Random Vector Functional-Link (RVFL) network, in which the internal parameters of the B basis functions $\{\boldsymbol{\omega}_i\}_{i=1\ldots B}$ are randomly generated from a fixed probability distribution before the learning process [11]. In [8] it was proven that, if the dimensionality of the functional expansion is adequately high, RVFLs possess universal approximation capabilities for a wide range of basis functions $\mathbf{h}(\cdot)$.

Let $T = \{(\mathbf{x}_1, y_1), \ldots, (\mathbf{x}_L, y_L)\}$ be the set of data available for the training, define the hidden matrix as

$$\mathbf{H} = \begin{bmatrix} h_1(\mathbf{x}_1) & \cdots & h_m(\mathbf{x}_1) \\ \vdots & \ddots & \vdots \\ h_1(\mathbf{x}_L) & \cdots & h_m(\mathbf{x}_L) \end{bmatrix} \tag{13}$$

and let $\mathbf{y} = [y_1, y_2, \ldots, y_l]^{\mathrm{T}}$ be the output vector. Notice that we omitted the parameterization with respect to the parameters $\boldsymbol{\omega}_i$ for better readability. The optimal weights $\boldsymbol{\beta}^*$ of the RVFL are obtained as the solution of the regularized least-squares problem:

$$\min_{\boldsymbol{\beta}} \frac{1}{2}\|\mathbf{H}\boldsymbol{\beta} - \mathbf{y}\|^2 + \frac{\lambda}{2}\|\boldsymbol{\beta}\|^2, \tag{14}$$

where $\lambda > 0$ is a regularization factor. Solution to Eq. (14) can be expressed in closed form as:

$$\boldsymbol{\beta}^* = (\mathbf{H}^{\mathrm{T}}\mathbf{H} + \lambda\mathbf{I})^{-1}\mathbf{H}^{\mathrm{T}}\mathbf{y}. \tag{15}$$

In a sequential scenario, at each time step we receive a portion of the overall dataset T, i.e. $T = \bigcup_{i=1}^{P} T_i$, for a given number P of batches. Let us define $\mathbf{H}_i$ and $\mathbf{y}_i$ as the hidden matrix and output vector computed over the ith batch, respectively. The optimal solution in Eq. (15) can be computed recursively by means of the recursive least-square (RLS) algorithm [9]:

$$\mathbf{P}(n+1) = \mathbf{P}(n) - \mathbf{P}(n)\mathbf{H}_{n+1}^{T}\mathbf{M}_{n+1}^{-1}\mathbf{P}(n), \tag{16}$$

$$\boldsymbol{\beta}(n+1) = \boldsymbol{\beta} + \mathbf{P}(n+1)\mathbf{H}_{n+1}^{T}\left[\mathbf{y}_{n+1} - \mathbf{H}_{n+1}\boldsymbol{\beta}(n)\right], \tag{17}$$

where $\mathbf{P}(n)$ is an auxiliary state matrix, and we defined:

$$\mathbf{M}_{n+1} = \mathbf{I} + \mathbf{H}_{n+1}\mathbf{P}(n)\mathbf{H}_{n+1}^{T}. \tag{18}$$

When considering a distributed scenario, we suppose that training data is partitioned throughout a network of agents, such that at every time step each agent receives a new batch, and the local batches are mutually independent. A distributed algorithm for RVFL for the case of a single batch is presented in [11], and later extended to the more general setting in [10]. The algorithm presented in [10] is based on alternating local

Table 1 Description of the datasets

Name	Features	Instances	Task
G50C	50	550	Gaussian of origin (classification)
CCPP	4	9568	Plant output (regression)

updates in Eqs. (16)–(17), with global averaging steps over the output weights, based on the DAC protocol. Although we will choose this algorithm as an experimental benchmark for the 4 consensus strategies herein considered, a detailed analysis of it goes beyond the scope of the paper. For the interested reader, we refer to [10, 11] and references therein.

5 Experimental Results

We compare the performance of the 4 different strategies illustrated in Sect. 3 in terms of number of iterations required to converge to the average and speed of convergence. In order to avoid that a particular network topology compromises the statistical significance of the experiments, we perform 25 rounds of simulation. In each round, we generate a random topology for an 8-nodes network, according to the so-called Erdős-Rényi model, such that every pair of nodes is connected with a fixed probability p. The only global requirement is that the overall topology is connected. In the experiments, we set $p = 0.5$. We consider 2 public available datasets, whose overview is given in Table 1. The G50C is an artificial classification dataset (see for example [11]), while the CCPP is a regression dataset taken from the UCI repository.[1] At each round, datasets are subdivided in batches following the procedure detailed in [10]. Since in real applications the value of the average is not available to the nodes, in order to evaluate the number of iterations, we consider that all the nodes reached consensus when $\|\beta_i(t) - \beta_i(t-1)\|^2 \leq 10^{-6}$ for any value of i.

In Fig. 1 we show the average number of iterations required by the DAC protocol, averaged over the rounds. The x-axis in Fig. 1 shows the index of the processed batch. As expected, the number of DAC iterations shows a decreasing trend as the number of processed training batches grows, since the nodes are slowly converging to a single RVFL model (see [10]). The main result in Fig. 1, however, is that a suitable choice of the mixing strategy can significantly improve the convergence time (and hence the training time) required by the algorithm. In particular, the optimal strategy defined by Eq. (8) achieves the best performance, with a reduction of the required number of iterations up to 35 and 28 % when compared with max-degree and Metropolis-Hasting strategies respectively. On the other side, the strategy based on constant edge matrix in Eq. (9) shows different behaviors for the 2 datasets, probably due to the heuristic nature of this strategy.

---

[1] https://archive.ics.uci.edu/ml/datasets/Combined+Cycle+Power+Plant.

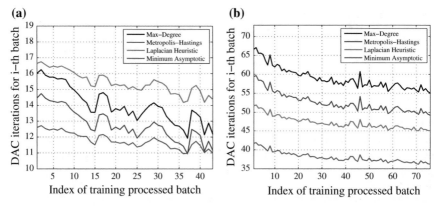

Fig. 1 Evolution of the DAC iterations required by the considered strategies to converge to the average, when processing successive amounts of training batches. **a** Dataset: G50C **b** Dataset: CCPP

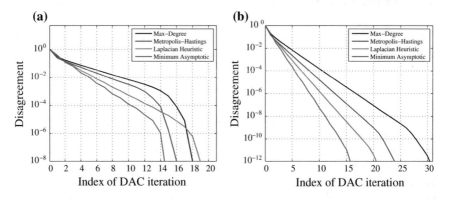

Fig. 2 Evolution of the relative network disagreement for the considered strategies as the number of DAC iterations increases. The y-axis is shown with a logarithmic scale. **a** Dataset: G50C **b** Dataset: CCPP

The second experiment, whose results are shown in Fig. 2, is to show the speed of convergence for the considered strategies. This is made by evaluating the trend of the relative network disagreement:

$$RND(t) = \frac{1}{N} \sum_{i=1}^{N} \frac{\|\boldsymbol{\beta}_i(t) - \hat{\boldsymbol{\beta}}\|^2}{\|\boldsymbol{\beta}_i(0) - \hat{\boldsymbol{\beta}}\|^2}, \tag{19}$$

as the number of DAC iterations increases. The value of $\hat{\boldsymbol{\beta}}$ in Eq. (19) is the true average given in Eq. (2). The y-axis in Fig. 2 is shown with a logarithmic scale. Results show that the "optimal" strategy has the fastest speed of convergence, as expected, while it is interesting to notice how, when compared to max-degree and Metropolis-Hastings weights, the heuristic strategy achieves a rapid decrease in disagreement in the initial iterations, while its speed tends to become slower in the end (this is notice-able in Fig. 2a). This may help to explain the lower performance of this strategy in Fig. 1a.

6 Conclusions

In this paper we have conducted an empirical comparison on the performance of dif-ferent strategies for the DAC protocol. In particular, we compared 4 time-invariant strategies for undirected topologies. Moreover, we focused on the application of the DAC protocol to a recently proposed distributed training algorithm for RVFL net-works.

Experimental results show how an appropriate choice of the weights matrix can lead to considerable improvements both in the number of iterations required by the protocol to converge to the average, and in the speed of convergence. In particular, when compared to other strategies, an "optimal" choice of the weights matrix can save up to 30 % in time.

This work can be set in the development of efficient distributed machine learning algorithms. Although we have focused on a specific learning model, nodes can be trained locally using different models than RVFL networks. Moreover, we made the assumption of a fixed, undirected topology. This assumption is common in literature, but limits the applicability of the model to more complex problems and real-world applications [10]. Future works will extend the analysis to time-varying topologies [7] and strategies involving communication constraints (e.g. on the time and energy required for data exchange).

References

1. Baccarelli, E., Cordeschi, N., Mei, A., Panella, M., Shojafar, M., Stefa, J.: Energy-efficient dynamic traffic offloading and reconfiguration of networked datacenters for big data stream mobile computing: review, challenges, and a case study. IEEE Netw. Mag. (2015)
2. Barbarossa, S., Sardellitti, S., Di Lorenzo, P.: Distributed detection and estimation in wireless sensor networks. In: Chellapa, R., Theodoridis, S. (eds.) E-Reference Signal Processing, pp. 329–408. Elsevier (2013)
3. Comminiello, D., Scarpiniti, M., Azpicueta-Ruiz, L., Arenas-Garcia, J., Uncini, A.: Functional link adaptive filters for nonlinear acoustic echo cancellation. IEEE Trans. Audio, Speech, Lang. Process. 21(7), 1502–1512 (2013)
4. Forero, P.A., Cano, A., Giannakis, G.B.: Consensus-based distributed support vector machines. J. Mach. Learn. Res. 11, 1663–1707 (2010)
5. Georgopoulos, L., Hasler, M.: Distributed machine learning in networks by consensus. Neu-rocomputing 124, 2–12 (2014)
6. Olfati-Saber, R., Fax, J.A., Murray, R.M.: Consensus and cooperation in networked multi-agent systems. Proc. IEEE 95(1), 215–233 (2007)
7. Olfati-Saber, R., Murray, R.M.: Consensus problems in networks of agents with switching topology and time-delays. IEEE Trans. Autom. Control 49(9), 1520–1533 (2004)

8. Pao, Y.H., Park, G.H., Sobajic, D.J.: Learning and generalization characteristics of the random vector functional-link net. Neurocomputing **6**(2), 163–180 (1994)
9. Scardapane, S., Comminiello, D., Scarpiniti, M., Uncini, A.: Online sequential extreme learning machine with kernels. IEEE Trans. Neural Netw. Learn. Syst. (2015)
10. Scardapane, S., Fierimonte, R., Wang, D., Panella, M., Uncini, A.: Distributed music classification using random vector functional-link nets. In: Accepted for presentation at 2015 IEEE/INNS International Joint Conference on Neural Networks (IJCNN'15) (2015)
11. Scardapene, S., Wang, D., Panella, M., Uncini, A.: Distributed learning for random vector functional-link networks. Inf. Sci. **301**, 271–284 (2015)
12. Xiao, L., Boyd, S.: Fast linear iterations for distributed averaging. Syst. Control Lett. **53**(1), 65–78 (2004)

Spatial-Temporal Entangled Sparse Distributed Storage (STE-SDS) and Sparse Distributed Code (SDC) in the Systolic Hebb Agnostic Resonance Perceptron (SHARP) Proposed as Hypothetical Model Linking Mini and Macro-Column Scale Functionality in the Cerebral Cortex

Luca Marchese

Abstract This document contains a specific analysis of the Systolic Hebb Agnostic Resonance Perceptron regarding the concepts of Spatial Temporal Entangled Sparse Distributed Storage (STE-SDS) and Sparse Distributed Code (SDC). The paper explains how SHARP implements STE-SDS in a model linking mini and macro-column scale functionality of the cerebral cortex. SHARP has been presented in a previous paper as a neural network that is executed with extreme efficiency on serial computers. This document analyzes the efficiency of the model involving the STE-SDS property and the associated capability to map similar inputs to similar prototypes stored in the network, with a single addressing operation.

Keywords SHARP · SDC · STE-SDS · Neo-cortex · Mini-column · Macro-column

1 Introduction

The SHARP neural network model has been presented as a possible model linking mini and macro-column scale functionality in the cerebral cortex. In the first paper describing this new paradigm [1], the extreme efficiency of its execution on the serial computers has been emphasized. In the same document, the behavior of the analogic neuron models has been analyzed. The extreme efficiency of the digital model on serial computers (Von Neumann) has been demonstrated with software

L. Marchese (✉)
Syn@ptics, 16151 Genoa, Italy
e-mail: luca.marchese@synaptics.org

© Springer International Publishing Switzerland 2016   153
S. Bassis et al. (eds.), *Advances in Neural Networks*, Smart Innovation,
Systems and Technologies 54, DOI 10.1007/978-3-319-33747-0_15

simulations and pattern recognition tests on different databases. Such efficiency originates from two main properties. The first is related to the simple digital emulation of the analogic neuron models. The second and most important property is the intrinsic Sparse Distributed Storage that is implemented in a way that enables the input stimulus to directly address the small portion of synapses and neurons required to perform the recognition process. In this paper, I will deepen the link between the timing and the SDS in the SHARP architecture.

2 An Overview of the SHARP Model

SHARP originates from a research targeted to design a cortical digital processor with very poor computational capability (compare, sum and subtraction of 8 bits/16 bits integers) and very low power consumption. The result has been an algorithm that can be simulated with extreme efficiency on Von-Neumann computers. The basic architecture of the network can be viewed as a three-dimensional matrix where the three dimensions represent respectively, a mini-column, a macro-column, and a cortical area network. Each mini-column is associated with a specific feature of the input stimulus and contains neurons associated with specific values of such a feature. The neurons in the mini-column are competitive modules (CM) working with a WTA (Winner Takes All) mechanism. A macro-column is an ensemble of mini-columns linked with synaptic connections that represent many configurations of the learned input patterns. Figure 1 shows an example of macro-column and the path of synaptic connections required to complete the recognition of a specific pattern. There is one macro-column for each binary element (C neuron) that participates in the representation of a category associated with a specific configuration of the stimulus. This array of macro-columns constitutes the third dimension and completes the cortical area network (Fig. 2). The macro-column has a feed-forward structure indeed the axons originate from each mini-column and reach all the following mini-columns. The term "systolic" is referred to this incremental flow of feed-forward connections. The term "agnostic" is referred to the resonance (with a "doublet" of spikes [1]) of the neurons in the mini-columns. The resonance of these neurons is "agnostic" because it is triggered by the single value of the single feature and not by a complex configuration of features/values. During the supervised learning activity [1] triggered by the SL neuron, the WTA within the mini-column is modulated by an inhibitory factor. In this way, the modulation of the synapses shapes the generalization capability of the network in a Hyper-Cube Influence Field

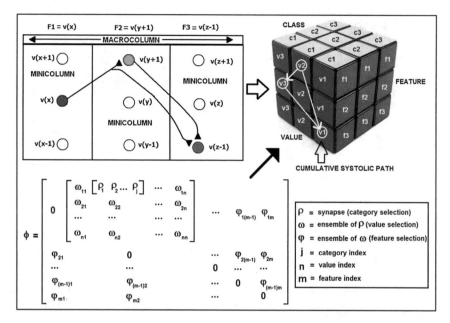

Fig. 1 Three cortical mini-columns in the cortical macro-column. The input pattern F1 = v(x), F2 = v(y + 1), F3 = v(z − 1) is recognized because this pattern has been previously learned and the synapses between the couples of neurons v(x) to v(y + 1), v(x) to v(z − 1), v(y + 1) to v (z − 1) have been reinforced through the Hebb rule. The neuron v(x) emits a spike just because it resonates with the input value. The neuron v(y + 1) emits a spike because it resonates with the input value and receives a consensus spike from the mini-column 1. The neuron v(z − 1) emits a spike because it resonates with the input value and receives two spikes respectively from the mini-column 1 and the mini-column 2. One macro-column exists for any Category neuron. The *cube* represents the basic structure of the SHARP neural network model. The path *v2(f1)*, *v3(f2)*, *v1(f3)* is shown in the macro-column in *c1* (*blue side*). The matrix of all *fx-vx* represents a macro-column. The tridimensional matrix *fx-vx-cx* represents a cortical area network. The fourth dimension involved (time) is not represented here

(Fig. 3). There is one C (Category) neuron and one SL (Supervised Learning) neuron for any macro-column. The activation (firing condition) of a mini-column neuron is:

$$s_{nmc} = 1 \because \left((\omega_m \approx \omega) \wedge \left(\sum_0^{n-1} s_{nc} = n \right) \right); 0 - otherwise \tag{1}$$

$$\omega = doublet\_freq; n = feature\_index; m = value\_index$$

The activation of a category (C) neuron is:

$$s_c = 1 \because \left(\sum_{n=0}^{N-1} s_{nc} = N \right); 0 - otherwise \tag{2}$$

$$N = total\_number\_of\_features; c = category\_index$$

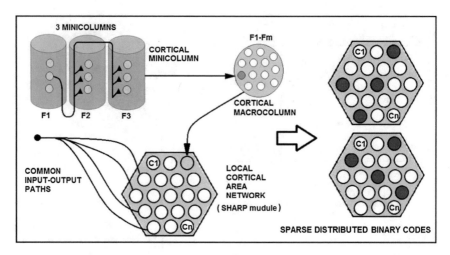

Fig. 2 The picture shows a graphical representation of the SHARP neural network where it is possible to distinguish the components from the mini-columns to the macro-columns and the cortical area network. An ensemble of macro-columns builds a local cortical area network (a SHARP module). Any macro-column is linked with a Category neuron that represents an item of a sparse distributed binary code

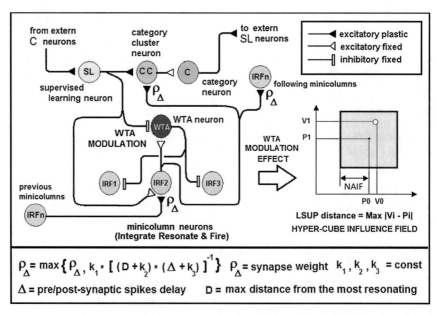

Fig. 3 The external action potentials arriving to the SL (Supervised Learning) neuron activate the resonant mini-column neuron (IRF) and the C neuron (replacing the systolic consensus spikes from the previous mini-columns). The SL neuron partially inhibits the WTA activity through the inhibitory synapse with the WTA neuron. The partial inhibition of the WTA enables the neighborhoods (IRF1, IRF3) to fire and the Hebb rule (STDP) is applied to IRF1, IRF2 and IRF3. The inhibitory modulation on the WTA neuron that is performed by the SL neuron enables the Hyper-Cube Influence Field generalization. IRF stays for "Integrate, Resonate and Fire" [1]. The picture shows only the connectivity of the most resonating neuron. The CC neuron represented in this picture is an option not explained in this context

3   One-Shot Learning and Recognition

The most important characteristic of the SHARP algorithm is the capability to learn in a single step and to recognize a pattern in a timeframe that is independent of the number of the learned patterns. Due to this feature, the algorithm, executed on a Von-Neumann computer, has the same performance of an RBF (Radial Basis Function) network executed on an SIMD (Single Instruction Multiple Data) computer.

The learning activity is triggered by the activation of one or more SL (Supervised Learning) neurons associated with Category neurons [1]. The subsequent activation of the Category neurons triggers the Hebb rule between the resonating neurons in the mini-columns (i.e.: with three mini-columns and R(mcX) = resonating neuron of mini-column X: R(mc1) => R(mc2), R(mc1) => R(mc3), R (mc2) => R(mc3)) and the category neuron (i.e.: R(mc1) => C(y), R(mc2) => C (y), R(mc3) => C(y)). The SL neuron partially inhibits the WTA activity within the mini-columns, enabling the application of the Hebb rule in a range of neighboring neurons to the most resonant neuron in all the mini-columns [1]. If the SL neuron is not activated by an external signal, the network recognizes the input stimuli generating binary patterns with the Category neurons. The following part of this document analyzes the efficiency of the learning and the recognition processes as the consequence of the SDS and the capability to address only the involved synapses through the input stimulus.

4   The Spatial-Temporal Entangled Sparse Distributed Storage (STE-SDS) and the Sparse Distributed Code (SDC)

In a distributed representation of an item of information, multiple elements collectively represent that item and each of those elements can be involved in the representations of other items. Distributed representations are often referred to as population codes. A Sparse Distributed Code (SDC) is a special case of distributed representation in which only a small part of the entire ensemble of representing elements is involved in the representation of any particular item and any element can be involved in the representation of multiple items.

There is an increasing evidence that SDC plays a fundamental role in the cortex and, perhaps, in many other structures of the brain. One important advantage of the SDC over a localist code is that the number of unique patterns that can be stored is much larger than the number of representing neurons. Other models linking mini and macro-columns with interconnections that support the SDC have been presented in other works. I want to mention the work of Rinkus [2–4] that explains with a scientifically relevant approach the biological plausibility of the SDC in the cortex with many references to experimental and theoretical studies. In the model designed by Rinkus [3], the macro-column is proposed to store information in the form of SDC, and the mini-column (specifically, its L2/3 of pyramidals) is proposed as a WTA CM, the purpose of which is to enforce the sparseness of the macro-columnar code. This role of mini and macro-columns matches with the SHARP model. However, this is the only characteristic shared between the SHARP and the model proposed by Rinkus. Rinkus [2], Albright [5] and Földiák [6] deepen the issues related to the biological plausibility of the SDC in the cortex. In this paper, I deepen the computational efficiency of the SDC and explain how this type of information coding is working in the SHARP neural network model. I need to distinguish between the elements that are involved in the storage of the learned patterns and the elements that are involved in the external representation of the category associated with the learned patterns. Therefore I will use respectively the terms SDS (Sparse Distributed Storage) and SDC (Sparse Distributed Code). In the SHARP model, the competing elements are the single neurons inside each mini-column. Indeed, each mini-column is organized in a WTA mode. Therefore, the mini-column contributes to the sparseness of the information distributing the values that a particular feature of the input stimulus can assume, on different neurons. The macro-column distributes the information along the dimension of the features of the input stimulus. Indeed, each mini-column represents a feature, and the entire ensemble of features constitutes a macro-column. The ensemble of macro-columns associated with different Category neurons is proposed as a cortical area network. The representation of an information item is a subset of the ensemble of the synaptic connections between each mini-column and the following mini-columns together with a subset of the synaptic connections between mini-columns and the Category neurons. The extreme efficiency of the SHARP model simulation on a serial computer is originated by the involvement of few neurons that are directly addressed by the values of the features of the input stimulus. The SHARP algorithm checks the completeness of the feed-forward synaptic connections between all the addressed neurons for any specific binary category element, in order to verify if an input pattern is recognized. The check of the synaptic connections must account for their values in order to compute the strength of the recognition. In the model SHARP, the timing of the single spike is meaningful. Therefore, the algorithm must check the synaptic connections in a time frame that is discretized in a certain number of steps representing the delays associated with the synaptic connections. We can see an example of how SHARP recognizes an input stimulus by addressing the correct neurons and checking only the involved synapses. We have the input stimulus composed of three features F1, F2 and F3 and their values are respectively 3, 4 and 7. The previous stimulus was composed of the same features with the values 2, 9 and 6. We set the timeframe composed of two steps (the synapses can have only two delay values). The algorithm checks if one or more category elements exist in which the following three conditions are true:

> 3(F1) is connected with 4(F2) in the category element C at the step(t2),
> 3(F1) is connected with 7(F3) in the category element C at the step(t2),
> 4(F2) is connected with 7(F3) in the category element C at the step(t2).

If all the conditions are true, the strength of the recognition is computed performing a fuzzy AND operation on the synaptic values. If one of the conditions is not true, the previous input stimulus is involved only in the verification of the false condition. Supposing that the false condition is *3(F1) is connected with 7(F3) in the category C at the step(t2)*, the algorithm checks the following conditions:

> 2(F1) is connected with 4(F2) in the category C at the step(t1),
> 2(F1) is connected with 7(F3) in the category C at the step(t1).

If one of the conditions is false, the input stimulus is not recognized. If both the conditions are true, the strength of the recognition is performed computing a fuzzy AND operation on the synaptic values (one value is associated with the true condition at the *step(t2)* and two synaptic values are associated with the true conditions at the *step(t1)*). The process above explained is the functional implementation of the SIASP (Similar Inputs Address Similar Paths) property, in the space and the time domain, and Fig. 4 shows a graphical example. The learning algorithm provides to reduce the strength of the synaptic connections associated with larger delays

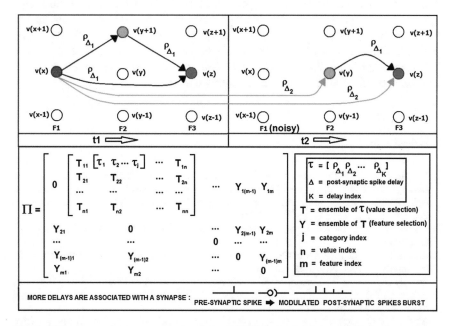

Fig. 4 This picture shows an example of the STE-SDS. At the time step *t1* the input pattern addresses directly a stored pattern. At the successive time step *t2*, the feature *F1* is noisy or missing and the input pattern can be recognized thanks to two delayed synapses. The synaptic values associated with the delays have been modulated during the learning process in the transition between two consequent time steps having these sequences of inputs

(Fig. 3). The presence of different delays in the synapses that interconnect the mini-columns induces a synaptic/dendritic behavior where a presynaptic spike could trigger a postsynaptic burst of decreasing spikes: the SDS is performed in the space and the time domain. I must underline that the distribution of the representation in the space and the time domain is entangled for any mini-column (STE-SDS).

5 Conclusions and Future Work

In this paper, I have deepened some important characteristics of the SHARP neural network model. The most important characteristic is the use of SDC, and I have explained the advantages of the SDC in the artificial neural systems. I have also indicated references of scientific works related to the biological plausibility of the SDC, with the awareness that the SHARP algorithm and the proposed schemes of connectivity have been designed with the target to build brain-inspired intelligent systems and all the details on the biological plausibility are only speculative. I have proposed a model of Sparse Distributed Storage that works in the space domain and the time domain with a continuous entanglement of distribution between the two dimensions. I have called this property STE-SDS (Spatial-Temporal Entangled Sparse Distributed Storage). The analogic version of the SHARP algorithm has been deeply described in a previous document [1]. However the efficient digital software simulation has been only briefly described. I have planned to write a document explaining the algorithm that emulates the behavior of the analogic neurons, described in [1], with simple digital operations.

References

1. Marchese, L.: SHARP (Systolic Hebb—Agnostic Resonance—Perceptron): a bio-inspired spiking neural network model that can be simulated very efficiently on Von Neumann architectures. Am. J. Intell. Syst. **4**(5), 159–195 (2014). SAP (Scientific & Academic Publishing). doi:10.5923/j.ajis.20140405.01
2. Rinkus, G.J.: Quantum computation via sparse distributed representation. NeuroQuantology **10** (2) (2012)
3. Rinkus, G.J.: A cortical sparse distributed coding model linking mini- and macrocolumn-scale functionality. Front. Neuroanat. (2010)
4. Rinkus, G., Lisman, J.: Time invariant recognition of spatiotemporal patterns in a hierarchical cortical model with a caudal-rostral persistence gradient. In: Society for Neuroscience—Annual Meeting, Washington, DC (2005)
5. Albright, T.D., Desimone, R., Gross, C.G.: Columnar organization of directionally selective cells in visual area MT of the macaque. J. Neurophysiol. **51**, 16–31 (1984)
6. Földiák, P.: Sparse coding in the primate cortex. In: Arbib, M.A. (ed.) The Handbook of Brain Theory and Neural Networks, 2nd edn., pp. 1064–1068. MIT Press (2002). ISBN: 0-262-01197-2

Łukasiewicz Equivalent Neural Networks

Antonio Di Nola, Giacomo Lenzi and Gaetano Vitale

Abstract In this paper we propose a particular class of multilayer perceptrons, which describes possibly non-linear phenomena, linked with Łukasiewicz logic; we show how we can name a neural network with a formula and, viceversa, how we can associate a class of neural networks to each formula. Moreover, we introduce the definition of *Łukasiewicz Equivalent Neural Networks* to stress the strong connection between different neural networks via Łukasiewicz logical objects.

Keywords MV-algebras · Riesz MV-algebras · Łukasiewicz Logic · Feedforward Neural Networks · Multilayer Perceptron

1 Introduction

Many-valued logic has been proposed in [1] to model neural networks: it is shown there that, by taking as activation functions ρ the identity truncated to zero and one (i.e., $\rho(x) = (1 \wedge (x \vee 0)))$, it is possible to represent the corresponding neural network as combination of propositions of Łukasiewicz calculus.

In [3] the authors showed that multilayer perceptrons, whose activation functions are the identity truncated to zero and one, can be fully interpreted as logical objects, since they are equivalent to (equivalence classes of) formulas of an extension of Łukasiewicz propositional logic obtained by considering scalar multiplication with real numbers (corresponding to Riesz MV-algebras, defined in [4, 5]).

Now we propose more general multilayer perceptrons which describe not necessarily linear events. We show how we can name a neural network with a formula and,

A. Di Nola · G. Lenzi · G. Vitale (✉)
Department of Mathematics, University of Salerno,
Via Giovanni Paolo II 132, 84084 Fisciano (SA), Italy
e-mail: gvitale@unisa.it

A. Di Nola
e-mail: adinola@unisa.it

G. Lenzi
e-mail: gilenzi@unisa.it

© Springer International Publishing Switzerland 2016
S. Bassis et al. (eds.), *Advances in Neural Networks*, Smart Innovation,
Systems and Technologies 54, DOI 10.1007/978-3-319-33747-0_16

viceversa, how we can associate a class of neural networks to each formula; moreover we introduce the idea of *Łukasiewicz Equivalent Neural Networks* to stress the strong connection between (very different) neural networks via Łukasiewicz logical objects.

2 Multilayer Perceptrons

Artificial neural networks are inspired by the nervous system to process information. There exist many typologies of neural networks used in specific fields. We will focus on feedforward neural networks, in particular multilayer perceptrons, as in [3], which have applications in different fields, such as speech or image recognition. This class of networks consists of multiple layers of neurons, where each neuron in one layer has directed connections to the neurons of the subsequent layer. If we consider a multilayer perceptron with n inputs, l hidden layers, ω_{ij}^h as weight (from the jth neuron of the hidden layer h to the ith neuron of the hidden layer $h + 1$), b_i real number and ρ an activation function (a monotone-nondecreasing continuous function), then each of these networks can be seen as a function $F : [0, 1]^n \to [0, 1]$ such that

$$F(x_1, \dots, x_n) = \rho(\sum_{k=1}^{n^{(l)}} \omega_{0,k}^l \rho(\dots (\sum_{i=1}^{n} \omega_{l,i}^1 x_i + b_i) \dots))).$$

3 Łukasiewicz Logic and Riesz MV-Algebras

MV-algebras are the algebraic structures corresponding to Łukasiewicz many valued logic, as Boolean algebras correspond to classical logic. An MV-algebra is a structure $A = (A, \oplus, ^*, 0)$ that satisfies the following properties:

- $x \oplus (y \oplus z) = (x \oplus y) \oplus z$
- $x \oplus y = y \oplus x$
- $x \oplus 0 = x$
- $x^{**} = x$
- $x \oplus 0^* = 0^*$
- $(x^* \oplus y)^* \oplus y = (y^* \oplus x)^* \oplus x$

The standard MV-algebra is the real unit interval $[0, 1]$, where the constant 0 is the real number 0 and the operations are

$$x \oplus y = min(1, x + y)$$

$$x^* = 1 - x$$

for any $x, y \in [0, 1]$. Another example of MV-algebra is the standard Boolean algebra $\{0, 1\}$ where all the elements are idempotent, i.e. $x \oplus x = x$ and the $\oplus$ is the connective $\vee$ in classical logic. A further class of examples of MV-algebras are M_n (for each $n \in \mathbb{N}$), where the elements are the continuous functions from the cube $[0, 1]^n$ to the real interval $[0, 1]$ which are piecewise linear with integer coefficients. These functions are called *McNaughton functions* and a major result in MV-algebra theory states that M_n is the free MV-algebra with n generators. For more details on the theory of the MV-algebras see [2].

A Riesz MV-algebra (RMV-algebra) is a structure $(A, \oplus,^* , 0, \cdot)$, where $(A, \oplus,^* , 0)$ is an MV-algebra and the operation $\cdot : [0, 1] \times A \to A$ has the following properties for any $r, q \in [0, 1]$ and $x, y \in A$:

- $r \cdot (x \odot y^*) = (r \cdot x) \odot (r \cdot y)^*$
- $r \cdot (x \odot y^*) = (r \cdot x) \odot (r \cdot y)^*$
- $r \cdot (q \cdot x) = (rq) \cdot x$
- $1 \cdot x = x$

where $x \odot y = (x^* \oplus y^*)^*$. We will denote by RM_n the Riesz MV-algebra of the continuous functions from the cube $[0, 1]^n$ to the real interval $[0, 1]$ which are piecewise linear with real coefficients (*Riesz McNaughton functions*). In analogy with the MV-algebra case RM_n is the free Riesz MV-algebra on n generators.

When we talk about a (Riesz) MV-formula, i.e. a syntactic polynomial, we can consider it also as a (Riesz) McNaughton function. Actually, in most of the literature there is no distinction between a (Riesz) McNaughton function and a (Riesz) MV-formula, but it results that, with a different interpretation of the free variables, we can give meaning to MV-formulas by means of other, possibly nonlinear, functions (e.g. we consider generators different from the canonical projections $\pi_1, \ldots, \pi_n$, such as polynomial functions, Lyapunov functions, logistic functions, sigmoidal functions and so on).

4 The Connection Between Neural Networks and Riesz MV-Algebras

Already in the first half of *XXth* century Claude Shannon understood the strong relation between switching circuits and Boolean algebras, and so Boolean algebras were (and they are still) used to describe and analyze circuits by algebraic methods. In an analogous way, in [3] the authors describe the connection between (a particular class of) neural networks and RMV-algebras; for instance the authors define the one-layer neural networks which encode $min(x, y)$ and $max(x, y)$ as follows:

$$min(x, y) = \rho(y) - \rho(y - x)$$

$$max(x, y) = \rho(y) + \rho(x - y)$$

where ρ is the identity truncated to zero and one. In [3] the following theorem was proved.

Theorem 1 *Let the function ρ be the identity truncated to zero and one (i.e., $\rho(x) = (1 \wedge (x \vee 0)))$.*

- *For every l, n, $n^{(2)}$, ..., $n^{(l)} \in \mathbb{N}$, and $\omega_{i,j}^h, b_i \in \mathbb{R}$, the function $F : [0,1]^n \to [0,1]$ defined as*

$$F(x_1, \ldots, x_n) = \rho(\sum_{k=1}^{n^{(l)}} \omega_{0,k}^l \rho(\ldots (\sum_{i=1}^{n} \omega_{l,i}^1 x_i + b_i) \ldots)))$$

is a Riesz McNaughton function;
- *for any Riesz McNaughton function f, there exist l, n, $n^{(2)}$, ..., $n^{(l)} \in \mathbb{N}$, and $\omega_{i,j}^h, b_i \in \mathbb{R}$ such that*

$$f(x_1, \ldots, x_n) = \rho(\sum_{k=1}^{n^{(l)}} \omega_{0,k}^l \rho(\ldots (\sum_{i=1}^{n} \omega_{l,i}^1 x_i + b_i) \ldots))).$$

4.1 Łukasiewicz Equivalent Neural Networks

If we consider particular kinds of functions from $[0,1]^n$ to $[0,1]^n$ (surjective functions) we can still describe non linear phenomena with a Riesz MV-formula, which, this time, does not correspond to a piecewise linear function but rather to a function which can still be decomposed into "regular pieces" (e.g. a piecewise sigmoidal function).

The crucial point is that we can try to apply, with a suitable choice of generators, all the well established methods of the study of piecewise linear functions to piecewise non-linear functions.

As in [3] we have the following definition.

Definition 1 We denote by $\mathcal{N}$ the class of the multilayer perceptrons such that the all activation functions of all neurons coincide with the identity truncated to zero and one.

We can generalize this definition as follows.

Definition 2 We call $\mathcal{LN}$ the class of the multilayer perceptrons such that:

- the activation functions of all neurons from the second hidden layer on is $\rho(x) = (1 \wedge (x \vee 0))$, i.e. the identity truncated to zero and one;
- the activation functions of neurons of the first hidden layer have the form $\varphi \circ \rho(x)$ where φ is a continuous functions from $[0,1]$ to $[0,1]$.

Example

An example of $\varphi(x)$ could be LogSigm, the logistic sigmoid function "adapted" to the interval $[0, 1]$, as showed in the next figure.

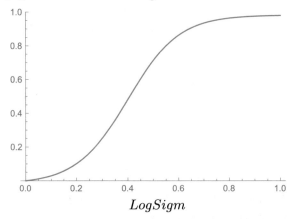

LogSigm

The first hidden layer (which we will call *interpretation layer*) is an interpretation of the free variables (i.e. the input data) or, in some sense, a change of variables.

Roughly speaking we interpret the input variables of the network $x_1, \ldots, x_n$ as continuous functions from $[0, 1]^n$ to $[0, 1]$; so, from the logical point of view, we have not changed the (Riesz) MV-formula which describes the neural network but only the interpretation of the variables.

For these reasons we introduce the definition of *Łukasiewicz Equivalent Neural Networks* as follows.

Definition 3 Given a network in $Ł\mathcal{N}$, the Riesz MV-formula associated to it is the one obtained first by replacing $\varphi \circ \rho$ with ρ, and then building the Riesz MV-formula associated to the resulting network in $\mathcal{N}$.

Definition 4 We say that two networks of $Ł\mathcal{N}$ are *Łukasiewicz Equivalent* iff the two networks have logically equivalent associated Riesz MV-formulas.

5 Examples of Łukasiewicz Equivalent Neural Networks

Let us see now some examples of Łukasiewicz equivalent neural networks (seen as the functions $\psi(\varphi(\bar{x}))$). In every example we will consider a Riesz MV-formula $\psi(\bar{x})$ with many different φ interpretations of the free variables $\bar{x}$, i.e. the activation functions of the interpretation layers.

Example 1

A simple one-variable example of Riesz MV-formula could be $\psi = \bar{x} \odot \bar{x}$. Let us plot the functions associated with this formula when the activation functions of the

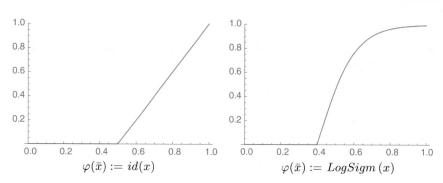

Fig. 1 $\psi(\bar{x}) = \bar{x} \odot \bar{x}$

interpretation layer is respectively the identity truncate function to 0 and 1 and the *LogSigm* (Fig. 1).

In all the following examples we will have (a), (b) and (c) figures, which indicate respectively these variables interpretations:

(a) x and y as the canonical projections π_1 and π_2;
(b) both x and y as *LogSigm* functions, applied only on the first and the second coordinate respectively, i.e. *LogSigm* $\circ$ $\rho(\pi_1)$ and *LogSigm* $\circ$ $\rho(\pi_2)$ (as in the example 1);
(c) x as *LogSigm* function, applied only on the first coordinate, and y as the cubic function π_2^3.

We show how, by changing projections with arbitrary functions φ, we obtain functions (b) and (c) "similar" to the standard case (a), which, however, are no more "linear". The "shape" of the function is preserved, but distortions are introduced.

Example 2: The $\odot$ Operation

We can also consider, in a similar way, the two-variables formula $\psi(\bar{x}, \bar{y}) = \bar{x} \odot \bar{y}$, which is represented in the following graphs (Fig. 2).

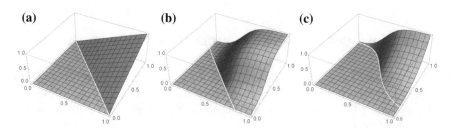

Fig. 2 $\psi(\bar{x}, \bar{y}) = \bar{x} \odot \bar{y}$

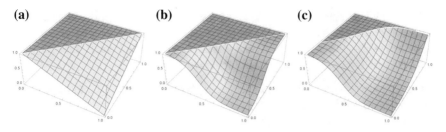

Fig. 3 $\psi(\bar{x}, \bar{y}) = \bar{x} \rightarrow_{\cdot} \bar{y}$

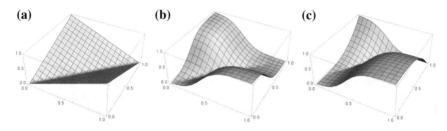

Fig. 4 $\psi(\bar{x}, \bar{y}) = (\bar{x} \odot \bar{y}^*) \oplus (\bar{x}^* \odot \bar{y})$

Example 3: The Łukasiewicz Implication

As in classical logic, also in Łukasiewicz logic we have *implication* ($\rightarrow_{\cdot}$), a propositional connective which is defined as follows: $\bar{x} \rightarrow_{\cdot} \bar{y} = \bar{x}^* \oplus \bar{y}$ (Fig. 3).

Example 4: The Chang Distance

An important MV-formula is $(\bar{x} \odot \bar{y}^*) \oplus (\bar{x}^* \odot \bar{y})$, called *Chang Distance*, which is the absolute value of the difference between x and y (in the usual sense) (Fig. 4).

6 Conclusions and Future Investigations

To sum up we:

1. propose $Ł\mathcal{N}$ as a privileged class of multilayer perceptrons;
2. link $Ł\mathcal{N}$ with Łukasiewicz logic (one of the most important many-valued logics);
3. show that we can use many properties of (Riesz) McNaughton functions for a larger class of functions;
4. propose an equivalence between particular types of multilayer perceptrons, defined by Łukasiewicz logic objects;
5. compute many examples of Łukasiewicz equivalent multilayer perceptrons to show the action of the free variables interpretation.

We think that using (in various ways) the *interpretation layer* it is possible to encode and describe many phenomena (e.g. degenerative diseases, distorted signals, etc.), always using the descriptive power of the Łukasiewicz logic formal language.

In the future investigations we will focus on:

- the composition of many multilayer perceptrons, to describe more complicated phenomena and their relations;
- the implementation of a back propagation model, regarded as a dynamical system.

References

1. Castro, J.L., Trillas, E.: The logic of neural networks. Mathw. Soft Comput. **5**, 23–27 (1998)
2. Cignoli, R., D'Ottaviano, I., Mundici, D.: Algebraic foundations of many valued reasoning, Kluwer (2000)
3. Di Nola, A., Gerla, B., Leustean, I.: Adding real coefficients to Łukasiewicz logic: an application to neural networks. Fuzzy Logic and Applications, pp. 77–85. Springer International Publishing, Switzerland (2013)
4. Di Nola, A., Leustean, I.: Riesz MV-algebras and their logic. In: Proceedings of the 7th Conference of the European Society for Fuzzy Logic and Technology. Atlantis Press, pp. 140–145 (2011)
5. Di Nola, A., Leustean, I.: Łukasiewicz logic and Riesz spaces. Soft Comput. **18**(12), 2349–2363 (2014)

A Bayesian-Based Neural Network Model for Solar Photovoltaic Power Forecasting

Angelo Ciaramella, Antonino Staiano, Guido Cervone and Stefano Alessandrini

Abstract Solar photovoltaic power (PV) generation has increased constantly in several countries in the last ten years becoming an important component of a sustainable solution of the energy problem. In this paper, a methodology to 24 h or 48 h photovoltaic power forecasting based on a Neural Network, trained in a Bayesian framework, is proposed. More specifically, a multi-ahead prediction Multi-Layer Perceptron Neural Network is used, whose parameters are estimated by a probabilistic Bayesian learning technique. The Bayesian framework allows obtaining the confidence intervals and to estimate the error bars of the Neural Network predictions. In order to build an effective model for PV forecasting, the time series of Global Horizontal Irradiance, Cloud Cover, Direct Normal Irradiance, 2-m Temperature, azimuth angle and solar Elevation Angle are used and preprocessed by a Linear Predictive Coding technique. The experimental results show a low percentage of forecasting error on test data, which is encouraging if compared to state-of-the-art methods in literature.

1 Introduction

Solar photovoltaic technology has become one of several renewable energy [1]. It has been receiving global research attention due to its natural abundance, noise pollution free, non-emission of greenhouse gases unlike the fossil powered generation sources which affect the climatic conditions and causing global warming. The technology has

A. Ciaramella · A. Staiano (✉)
Department of Science and Technology, University of Naples "Parthenope",
Isola C4, Centro Direzionale, 80143 Napoli (NA), Italy
e-mail: antonino.staiano@uniparthenope.it

A. Ciaramella
e-mail: angelo.ciaramella@uniparthenope.it

G. Cervone
Penn State University, State College, PA, USA

S. Alessandrini
NCAR, National Centre for Atmospheric Research, Boulder, CO, USA

© Springer International Publishing Switzerland 2016
S. Bassis et al. (eds.), *Advances in Neural Networks*, Smart Innovation,
Systems and Technologies 54, DOI 10.1007/978-3-319-33747-0_17

gained popularity in the establishment of large solar farms in major countries such as the United States of America, Spain, Italy, Japan, China, Australia and in other countries of the world. Mainly for its suitability for power generation in urban and remote isolated rural areas for small scale applications such as water pumping systems and domestic electricity supply for lighting and other uses [6]. In Italy, thanks also to substantial government subsides over the past five years, the annual generation by solar photovoltaic power (PV) has notably increased, from 200 GWh in 2008 to 19418 GWh in 2013, that corresponds to 7 % of the total Italian energy demand [7]. The energy produced by photovoltaic farms has a variable nature depending on astronomical and meteorological factors. The former are the solar elevation and the solar azimuth, which are easily predictable without any uncertainty. The latter, instead, deeply impact on solar photovoltaic predictability. Since the power produced by a PV system depends critically on the variability of solar irradiance and environmental factors, unexpected variations of a PV system output may increase operating costs for the electricity system by increasing requirements of primary reserves, as well as placing potential risks to the reliability of electricity supply.

A priority of a grid operator is to predict changes of the PV system power production, mainly using persistence-type methods, in order to schedule the spinning reserve capacity and to manage the grid operations. In addition to transmission system operators, online power prediction of the PV system is also required by various end-users such as energy traders, energy service providers and independent power producers, to provide inputs for different functions like economic scheduling, energy trading, and security assessment.

In this last years, several researches for forecasting the solar irradiance in different scale times have been made by using Machine Learning and Soft Computing methodologies [14]. In particular, based on Artificial Neural Networks [8, 13], Fuzzy Logic [17] and hybrid system such as ANFIS [18], Recurrent Neural Networks [4] and Support Vector Machines [19]. Most of the works concentrate only on few impact parameters (e.g., only temperature) and they not consider a confidence interval for the estimated prediction.

In this paper a methodology to PV forecasting based on a Neural Network (Multi-Layer Perceptron), trained in a Bayesian framework, is proposed. The Bayesian framework allows obtaining the confidence intervals and to estimate the error bars of the model prediction. In order to build an effective model for PV forecasting, the time series of Global Horizontal Irradiance, Cloud Cover, Direct Normal Irradiance, 2-m Temperature, azimuth angle and solar elevation angle are used. The features of the observed time series are extracted by a Linear Predictive Coding mechanism.

The paper is organized as following. In Sect. 2 the Solar Photovoltaic Power Data are described. In Sect. 3 we introduce the Multi-Layer Perceptron and the Probabilistic Bayesian learning and in Sect. 4 the experimental results are described. Finally, in Sect. 5 we focus on conclusions and future remarks.

2 Solar Photovoltaic Power Data

To maintain grid stability at an effective cost, it has now become crucial to be able to predict with accuracy the renewable energy production which is combined with other more predictable sources (e.g., coal, natural gas) to satisfy the energy demand [11, 12]. In this work, data collected from three PV farms are considered. They are located in Italy, namely Lombardy and Sicily regions, with a nominal power (NP) of 5.21 kW and in Calabria region with a nominal power around 5 MW [16].

3 Multi-Layer Perceptron and the Probabilistic Bayesian Learning

A Neural Network (NN) is usually structured into an input layer of neurons, one or more hidden layers and one output layer. Neurons belonging to adjacent layers are usually fully connected and the various types and architectures are identified both by the different topologies adopted for the connections and by the choice of the activation function. Such networks are generally called Multi-Layer Perceptron (MLP) [2] when the activation functions are sigmoidal or linear. The output of the jth hidden unit is obtained first by forming a weighted linear combination of the d input values, and then by adding a bias to give:

$$z_j = f\left(\sum_{i=0}^{d} w_{ji}^{(1)} x_i\right) \qquad (1)$$

where d is the number of the input, $w_{ji}^{(1)}$ denotes a weight in the first layer (from input i to hidden unit j). Note that $w_{j0}^{(1)}$ denotes the bias for the hidden unit j; and f is an activation function such as the continuous sigmoidal function.

The outputs of the network are obtained by transforming the activation of the hidden units using a second layer of processing elements

$$y_k = g\left(\sum_{j=0}^{M} w_{kj}^{(2)} z_j\right) \qquad (2)$$

where M is the number of hidden unit, $w_{kj}^{(2)}$ denotes a weight in the second layer (from hidden unit j to output unit k). Note that $w_{k0}^{(2)}$ denotes the bias for the output unit k; and g is an activation function of the output units which does not need to be the same function as for the hidden units. The learning procedure is the so called back propagation [2].

Due to its interpolation capabilities, the MLP is one of the most widely used neural architectures. The MLP can be trained also using probabilistic techniques. The Bayesian learning framework offers several advantages over classical ones [2]: (i) it cannot overfit the data; (ii) it is automatically regularized; (iii) the uncertainty in the prediction can be estimated.

In the conventional maximum likelihood approach to training, a single weight vector is found which minimizes the error function; in contrast, the Bayesian scheme considers a probability distribution over the weights. This is described by a prior distribution $p(\mathbf{w})$ which is modified when we observe the data $\mathbf{D}$. This process can be expressed by the Bayes theorem:

$$p(\mathbf{w}|\mathbf{D}) = \frac{p(\mathbf{D}|\mathbf{w})p(\mathbf{w})}{p(\mathbf{D})}. \tag{3}$$

To evaluate the posterior distribution, we need expressions for the prior $p(\mathbf{w})$ and for the likelihood $p(\mathbf{D}|\mathbf{w})$. The prior over weights should reflect the knowledge, if any, about the mapping to be built.

We consider k different sets of weights by using different regularization parameters α_k for each group. To preserve the scaling properties of the network mapping the prior equation can be written as [15]

$$p(\mathbf{w}) = \frac{1}{Z_W(\{\alpha_k\}_k)}e^{-\sum_k \alpha_k E_{W_k}} \tag{4}$$

where k runs over the different weight groups, $Z_W(\{\alpha_k\}_k)$ is a normalization factor, E_{W_k} weight regularizer, α_k is the hyperparameter.

Once the expressions for the prior and the noise model is given, we can evaluate the posterior

$$p(\mathbf{w}|\mathbf{D}, \{\alpha_k\}_k, \beta) = \frac{1}{Z(\{\alpha_k\}_k, \beta)}e^{-\beta E_D - \sum_k \alpha_k E_{W_k}} \tag{5}$$

where β is an hyperparameter, $Z(\{\alpha_k\}_k, \beta)$ is a normalization factor and E_D is an appropriate error function [2].

This distribution is usually very complex and multi-modal, and the determination of the normalization factor is very difficult. Also, the hyperparameters must be integrated out, since they are only used to determine the form of the distributions.

The approach followed is the one introduced by [10], which integrates the parameters separately from the hyperparameters by means of a Gaussian approximation and then finds the mode with respect to the hyperparameters. This procedure gives a good estimation of the probability mass attached to the posterior, in particular way for distributions over high-dimensional spaces [10].

Using a Gaussian approximation the mode of the resulting distribution can be evaluated

$$\{\{\hat{\alpha}_k\}_k, \hat{\beta}\} = \text{argmax}_{\{\alpha_k\}_k, \beta} p(\{\alpha_k\}_k, \beta|\mathbf{D}) = \int p(\{\alpha_k\}_k, \beta, \mathbf{w}|\mathbf{D})d\mathbf{w} \qquad (6)$$

and the hyperparameters so found can be used to evaluate:

$$\hat{\mathbf{w}} = \text{argmax}_{\mathbf{w}} p(\mathbf{w}|\mathbf{D}, \{\alpha_k\}_k, \beta). \qquad (7)$$

The above outlined scheme must be repeated until a self-consistent solution $(\mathbf{w}|\{\alpha_k\}_k, \beta)$ is found.

4 Experimental Results

Data have been collected from the three PV farms previously mentioned. In particular, the data used in the experiments are a combination of model forecasts and observational measurements relative to three stations, namely Lombardy, Calabria and Sicily. The data were available for the following periods: January 2010–December 2011 (Sicily), July 2010–December 2011 (Lombardy), and April 2011–March 2013 (Calabria).

The forecast data is composed of atmospheric forecasts generated using the Regional Atmospheric Modeling System (RAMS), initialized with boundary conditions from European Centre for Medium Range Weather Forecasts (ECMWF) deterministic forecast fields starting at 00 UTC. The forecasted parameters used in the experiments include Cloud Cover (CC), the Global Horizontal Irradiance (GHI) and air Temperature at 2 m above the ground (T2M). Additionally, the data were augmented with computed measurements for solar AZimuth angle (AZ) and solar ELevation angle (EL). The observations include the real quantity of electrical power generated from each of the three solar farms. The data was averaged into hourly values to be consistent with the forecast data.

These predictors are then employed to train an MLP for a 24 h or 48 h ahead prediction. In particular, we consider 24 h of observations to forecast 24 or 48 h of PV concentrations. For each parameter, previously mentioned, a Linear Predictive Coding (LPC) technique [5] is used to extract the main features. LPC is for representing the spectral envelope of a digital signal in compressed form, using the information of a linear predictive model [5]. More in detail, if we consider k LPC coefficients for each parameter (GHI, CC, T2M, AZ, EL, PV), at time t, the NN has $6 \times k$ inputs, $x_i, i = 1, \dots, 6 \times k$ corresponding to the observed 24 h $\mathbf{x} = [x_1, \dots x_{6 \times k}]$. The outputs are 24 or 48 corresponding with the PV observations to forecast. In Fig. 1 the architecture of the NN with 24 outputs is shown. The data set has been divided in training

Fig. 1 MLP for 24 h ahead
prediction

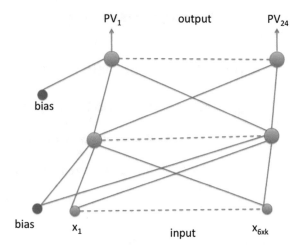

Fig. 2 24 h PV prediction
obtained by MLP NN and
Bayesian framework

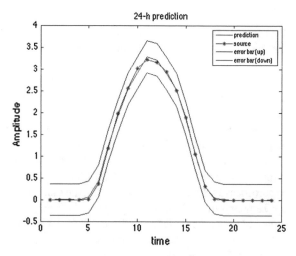

and test sets (70 % and 30 %, respectively). A K-fold cross-validation mechanism is
used to select the optimal number of hidden nodes [3]. In the first experiment we
concentrate on the prediction of PV for 24 h. The NN is composed by 10 hidden
nodes. We obtained a cross-correlation coefficient of 97 % between source and predicted PV sequences on the training set and of 91 % on the test set. Moreover we
use a Root Mean Square Error (RMSE) to calculate the percentage of error and we
obtain 6 and 8 % on the training and test sets, respectively. In Figs. 2 and 3 we report
some predictions obtained by the model on test data. In particular, in Fig. 2 a regular PV concentration is considered and in Fig. 3 an unusual sequence is considered.
In both cases the model permits to predict the concentrations with high accuracy.
Successively, we concentrate on the prediction of PV for 48 h. We obtained a cross-
correlation coefficient of 94 % between source and predicted PV sequences on the

Fig. 3 24 h PV prediction obtained by MLP NN and Bayesian framework

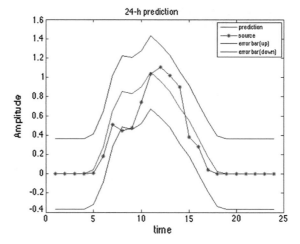

Fig. 4 48 h PV prediction obtained by MLP NN and Bayesian framework

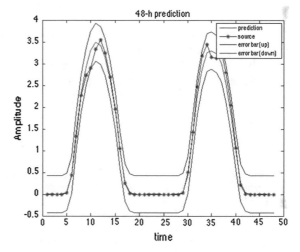

training set and of 89 % on the test set. The RMSEs are of 7 and 9 % on the training and test sets, respectively. In Figs. 4 and 5 we report some predictions obtained by the model on test data. Also in this last example, in both cases the model permits to predict the concentrations with high accuracy.

5 Conclusions

In this work, a methodology to PV forecasting based on a neural network, trained in a Bayesian framework, has been introduced. More specifically, a Multi-Layer Perceptron Neural Network for 24 or 48 h ahead prediction is used whose parameters are

Fig. 5 48 h PV prediction obtained by MLP NN and Bayesian framework

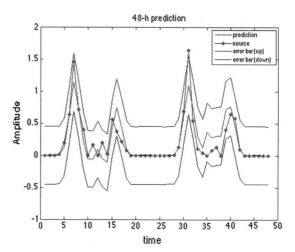

estimated by a probabilistic Bayesian learning technique. The experimental results show a low percentage of error on test data. In the next future the work will be focused on 72 h prediction and the use of other Neural Network based methodologies such as Radial Basis Functions and Recurrent NNs [9].

References

1. Balogun, E.B., Huang, X., Tran, D.: Comparative study of different artificial neural networks methodologies on static solar photovoltaic module. Int. J. Emerg. Technol. Adv. Eng. **4**(10) (2014)
2. Bishop, C.M.: Neural Networks for Pattern Recognition. Oxford University Press (1995)
3. Bishop, C.M.: Pattern Recognition and Machine Learning. Springer (2006)
4. Cao, S., Cao, J.: Forecast of solar irradiance using recurrent neural networks combined with wavelet analysis. Appl. Therm. Eng. **25** 161-172 (2005)
5. Deng, L., Douglas, O.S.: Speech processing: a dynamic and optimization-oriented approach. Marcel Dekker. pp. 41–48 (2003). ISBN: 0-8247-4040-8
6. Deshmukh, R., Bharvirkar, R., Gambhir, A., Phadke, A., Sunshine, C.: Analyzing the dynamics of solar electricity policies in the global context. Renew. Sustain. Energy Rev. **16**(7), pp. 5188-5198 (2012). ISSN: 1364-0321
7. Early data on 2013 electricity demand: 317 billion KWh of demand, −3.4 % compared to 2012 Terna (company) press release. http://www.terna.it/LinkClick.aspx?fileticket=GjQzJQkNXhM3d&tabid=901&mid=154 (2014)
8. Elizondo, D., Hoogenboom, G., Mcclendon, R.W.: Development of a neural network model to predict daily solar radiation. Agric. For. Meteorol. **71**, 115–132
9. Haykin, S.: Neural Networks—A Comprehensive Foundation, 2nd edn. Prentice Hall (1999)
10. MacKay, D.J.C.: Hyperparameters: optimise or integrate out? Maximum entropy and Bayesian methods, Dordrecht (1993)
11. Mahoney, W.P., Parks, K., Wiener, G., Liu, Y., Myers, B., Sun, J., Delle Monache, L., Johnson, D., Hopson, T.: Haupt SE. A wind power forecasting system to optimize grid integration. IEEE Trans. Sustain. Energy Appl. Wind Energy Power Syst. **3**(4), 670–682 (2012). Special issue

12. Marquis, M., Wilczak, J., Ahlstrom, M., Sharp, J., Stern, A., Charles Smith, J., Calvert, S.: Forecasting the wind to reach significant penetration levels of wind energy. Bull. Am. Meteorol. Soc. **92**, 1159–1171 (2011)
13. Mellit, A., Pavan, A.M.: A 24-h forecast of solar irradiance using artificial neural network: application for performance prediction of a grid-connected PV plant at Trieste, Italy. Solar Energy **84**, 807–821 (2010)
14. Mellit, A.: Artificial intelligence techniques for modelling and forecasting of solar radiation data: a review. Int. J. Artif. Intell. Soft Comput. **1**, 52–76 (2008)
15. Neal, R.M.: Bayesian learning for neural networks. Springer, Berlin (1996)
16. Pielke, R.A., Cotton, W.R., Walko, R.L. et al.: A comprehensive meteorological modeling system RAMS, Meteorol. Atmos. Phys. **49**, 69 (1992)
17. Sen, Z.: Fuzzy algorithm for estimation of solar irradiation from sunshine duration, Solar Energy **63**, 39–49 (1998)
18. Sfetsos, A., Coonick, A.H.: Univariate and Multivariate forecasting of hourly solar radiation with artificial intelligence techniques, Solar Energy **68**, 169–178 (2000)
19. Shi, J., Lee, W.-J., Liu, Y., Yang, Y., Wang, P.: Forecasting power output of photovoltaic systems based on weather classification and support vector machines. IEEE Trans. Ind. Appl. **48**(3) (2012)

Strict Pyramidal Deep Architectures for Person Re-identification

Sara Iodice, Alfredo Petrosino and Ihsan Ullah

Abstract We report a strict 3D pyramidal neural network model based on convolutional neural networks and the concept of pyramidal images for person re-identification in video surveillance. Main advantage of the model is that it also maintains the spatial topology of the input image, while presenting a simple connection scheme with lower computational and memory costs than in other neural networks. Challenging results are reported for person re-identification in real-world environments.

1 Introduction

Person re-identification (PRe-ID) is an open problem in computer vision, which tries to solve questions like: "Have I seen this person before?" [18], or "Is this the same person?". In other words, it consists in recognizing an individual in different locations over a set of non-overlapping camera views. This is an important task in computer vision with applications ranging from intelligent video surveillance, like people tracking to behaviour analysis [3, 9–13].

PRe-ID generally requires to match person images captured from surveillance cameras working in wide-angle mode. Therefore, the resolution of these frames is very low (e.g., around 48 × 128 pixels) and lighting conditions are unstable too. Furthermore, the direction of cameras and the pose of people are arbitrary. These factors cause great difficulties in re-identification task, due to two distinctive properties: large intraclass variation and interclass ambiguity. In addition, other aspects such as camera view change, pose variation, non-rigid deformation, unstable illumination condition and low resolution play a key role in making the task challenging.

S. Iodice · A. Petrosino (✉) · I. Ullah
CVPR Lab, Department of Science and Technology,
University of Naples Parthenope, Naples, Italy
e-mail: petrosino@uniparthenope.it; alfredo.petrosino@uniparthenope.it

I. Ullah
Department of Computer Science, University of Milan, Milan, Italy
e-mail: ihsan.ullah@uniparthenope.it

© Springer International Publishing Switzerland 2016   179
S. Bassis et al. (eds.), *Advances in Neural Networks*, Smart Innovation,
Systems and Technologies 54, DOI 10.1007/978-3-319-33747-0_18

The majority of existing methods include two separate phases: feature extraction and metric learning. In general, these features come from separate sources, i.e. colour and texture, some of which are designed by hand, while others are learned. Finally, they are collected together or fused by simple strategies [2, 6, 16].

Recent methods are based on deep learning for extracting more discriminative features. They combine the two separate steps, feature extraction and metric learning, in a unified framework, such as the Siamese Convolutional Neural Network (SCNN) with Deep Metric Learning (DML) [17].

SCNN with DML has an extra edge due to direct learning of similarity metric from image pixels: parameters at each layer are continuously updated guided by a common objective function, which allows to extract more discriminative features, as compared to hand-crafted features in traditional computer vision models. Furthermore, using a multichannel kernel, as in the case of convolutional neural networks (CNN), different kind of features are more naturally combined, clearly obtaining a reasonable superiority over other computer vision fusion techniques, e.g., feature concatenation, sum rule, etc.

CNNs extract features of lower level of discrimination in first layers and then, this level improves in deeper layers. However, one of the limitations related to CNNs models derives from the ambiguity in the feature extraction due to the increase number of feature maps with a low level of discrimination. In contrast, it would be better to use a structure that extracts more features from real information, and then refine them in deeper layers to make them more distinctive. A viable model is the Strict Pyramidal Structure, which we propose in order to refine and reduce ambiguity in the features.

In this context, our main contributions can be summarized as follows:

1. modifying the existing SCNN with DML model following a Strict Pyramidal Structure;
2. providing a deeper analysis of the impact of batch size on performances;
3. demonstrating that unbalanced datasets can be contrasted by using a regularization parameter into the cost function;
4. finally, proving that our proposed pyramid structure model enhances Rank1 performance.

Section 2 of the paper contains overview for the basics of the strict pyramidal model, while Sect. 3 describes the architecture of SCNN model with pyramidal structure. Section 4 reports results on VIPeR dataset, together with specific analysis of the behaviour of proposed architecture.

2   Strictly Pyramidal Structure

To adopt pyramidal structure for decision making as done in brain, we decided to use a strict 3d pyramidal architecture by starting from a big input at first layer and then refining the features in deeper layers until we reach a reduced and most discriminative set of features. This model takes inspiration from early pyramidal neural network model [4, 14] and its more recent version PyraNet that learns parameters from input till output. The objective is to demonstrate that following a strict pyramidal structure performance improves compared to unrestricted models.

2.1 PyraNet

PyraNet model [15] was inspired by the pyramidal neural network (PNN) model reported in [4, 14] with 2D and 1D layers. Differently from the original model in [4] where, the coefficients of receptive fields are adaptive, PyraNet model perform feature extraction 2D layers and classification through 1D layer at the top.

This model is to some extent similar to CNN, however, there are some differences. Firstly, it performs weighted sum operation (or correlation) rather than convolution. Secondly, weights are not in form of a kernel, but each output neuron has a local unique kernel specifically assigned based on input neurons and their respective weights. This leads to a unique kernel for each output neuron. In contrast, an important feature of CNNs is weight sharing concept. This property reduces the large amount of learning parameters, but increases burden on those few parameters.

Finally, Pyranet does not use any pooling layers for reducing the dimension of feature maps; their size is reduced differently through the stride of the kernel in each layer. On the other hand, both architectures utilize back propagation technique for learning parameters. It is important to note that PyraNet their technique achieved 96.3 % accuracy similar to SVM for gender recognition and 5 % more than CNN with same input size images.

2.2 Architecture

Our model takes inspiration by the CNN-DML [17], a siamese neural network, whose structure is well suitable for person re-identification problem. As known, neural networks work in standalone mode, where the input is a sample and the output is a predicted label, as in several pattern recognition problems, i.e. handwritten digit recognition [8], object recognition [5], human action recognition [1], and others, when the training and test set are characterized by the same classes.

On the other hand, the siamese architecture provides a different pattern, where the input is an image pair and the output is binary value, revealing if the two images comes from the same subject, or not.

Another main aspect is that, the weights between the two sub-networks, which are a part of the siamese architecture, can be shared or not. For person re-identification problem, the best choose is clearly sharing parameters to find out the peculiarity of an individual in different pose, or from images acquired by different views.

Our architecture, named Strict Pyramidal DML, is composed by two siamese Strict Pyramidal CNN blocks, $B1$ and $B2$, a connection function, C and a cost function, J. Driven by a common objective function, this model can learn, simultaneously, both the features and an optimal metric to compare them. The Strictly Pyramidal CNN block is composed by 2 convolutional layers, $C1$ and $C3$, 2 max pooling layers, $S2$ and $S4$, and a fully-connected layer F. In particular, $C1$ layer exploits $NF1 = 32$ filters with kernel size 7×7; while $C3$ layer $NF3 = 25$ filters, having size 5×5. Finally, the Strictly Pyramidal CNN block gives in output a vector of $x \in R^{500}$ components, containing the salient features of an individual.

Specifically, the proposal architecture is different from the CNN-DML for:

1. replacing each CNN block by a strictly Pyramidal CNN block;
2. using the full image and not dividing the input image in three parts and giving as input to three differrent CNN's;
3. using simple hyperbolic tangent function instead of ReLU [7] as activation function for each layer;
4. not using cross-channel normalization unit;
5. not padding by zero the input before a CNN block.

In contrast to CNN, the Strictly Pyramidal CNN has a biggest number of filters in first layer, then the number of filters decreases going deeply, layer by layer. This in agreement with our conjecture that, due to its direct interaction with the input, the outer layers require a wide set of filters to extract salient features from the input images; in contrast, going deeply, the number of filters should be decreased to avoid redundant features.

We opt for the architecture based on full image in order to reduce the number of parameters and make the learning process faster. Indeed, using three CNNs the learning process becomes slower due to the higher number of parameters which need to be updated. Furthermore, we choose as activation function the Hyperbolic tangent rather then ReLU, because it provides better results in our experiments. Finally, the cross-channel normalization unit and the padding by zero do not give any improvement in our analysis, thus we avoid to use them, reducing the complexity of whole architecture.

3 Experiments and Analysis

In this section, the results obtained on the state-of-art VIPeR dataset will be explained. The VIPeR dataset contains 632 pedestrian image pairs taken from arbitrary viewpoints under varying illumination conditions. The data were collected in an academic setting over the course of several months. Each image is scaled to 128×48 pixels. We took 11 divisions by randomly selecting 316 disjoint subjects for training and the remaining 316 for testing set. The first split called (Dev. view) is used for parameter tuning, such as the number of training epoch, kernels size, number of kernel to use for each layer etc.; while, the other 10 splits (Test view) for reporting the

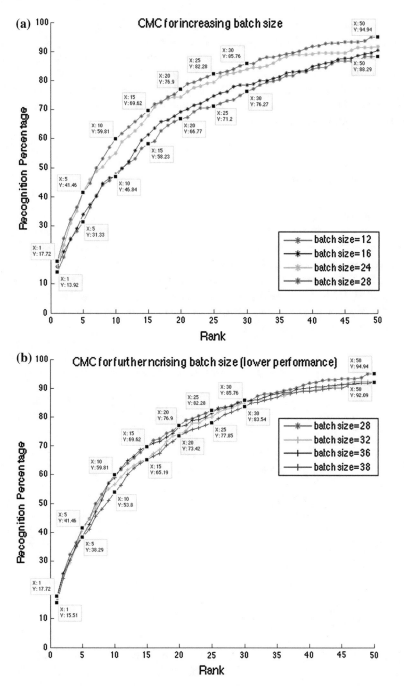

Fig. 1 Figure **a** CMC for increasing batch size; **b** CMC for increasing batch size over threshold that reduces performance

results. We adopted the same learning value, weight decay approach of the ConvNet model as in [7]. All training images from Camera A and B are merged for training purpose, randomly shuffled and given as input to both our modified version of Deep Metric Learning and to our Strict Pyramidal Deep Metric Learning (SP-DML).

The impact of two main factors in the model will be described: (1) batch size used for the training stage, through stochastic minibatch gradient descent technique; (2) the regularization parameter.

3.1 Batch Analysis

We adopted the stochastic gradient descent approach for updating learning parameters in our experiments, as being done in [7]. Figure 1a shows the increase in performance as we increase batch size and reduce the difference among negative and positive pairs. After a certain limit, the performance clearly decreases, as can be seen in Fig. 1b. This shows that the batch size should not be too large, as it becomes too unbalanced, i.e. negative pairs are a huge number compare to positive pairs. Therefore, the performance decreases. As a result, the best ratio between $n1$ and $n2$, should be $28/378$ setting $sizeBatch = 28$.

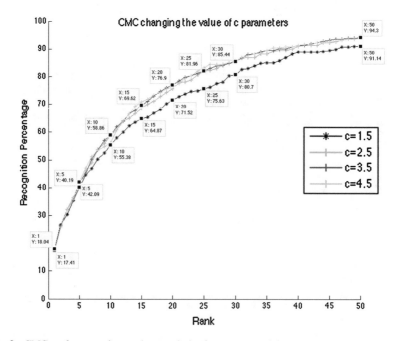

Fig. 2 CMC performance by varying regularization parameter 'c'

3.2 Regularization Term: Asymmetric Cost Analysis

After selecting the best batch size, we have improved performance with regularization parameters that gives different emphases to positive and negative pairs. By doing so, the dataset appears to be more well balanced in our experiments we tried to use it for different values of c i.e. ranging between 2.0 to 3.5 with an increase of 0.5 each time. We found that increasing this value better results are provided it can be judged from Fig. 2 that $c = 3.5$ gives optimal results for a greatest range rank values, as can be seen in Fig. 2.

4 Conclusions

We propose a strictly pyramid deep architecture model, ideal for PReID in real world scenarios. The model not only improved Rank1 performance by learning different view information with shared kernels. Furthermore, it shows efficient or similar results even with learning parameters less than other similar stat-of-art methodologies. Some effective training strategies are adopted to train network well for targeted application. We plan to extend our results also to other datasets to further check efficiency of the proposed model. However, for our experiments we do not utilize data augmentation, that in principle strongly enhances the results.

References

1. Baccouche, M., Mamalet, F., Wolf, C., Garica, C., Baskurt, A.: Sequential deep learning for human action recognition. Hum. Behav. Unterstanding Proc. **2** (2011)
2. Bedagkar-Gala, A., Shah, S.K.: A survey of approaches and trends in person re-identification. Image Vis. Comput. **32**(4), 270–286 (2014). http://www.sciencedirect.com/science/article/pii/S0262885614000262
3. Caianiello, E.-R., Petrosino, A.: Neural networks, fuzziness and image processing. In: Human and Machine Vision, pp. 355–370 (1994)
4. Cantoni, V., Petrosino, A.: Neural recognition in a pyramidal structure. IEEE Trans. Neural Netw. **13**(2), 472–480 (2002)
5. Christian Szegedy, A.T., Erhan, D.: Deep neural networks for object detection. Adv. Neural Inf. Process. Syst. **26**, 553–2561 (2013)
6. Iodice, S., Petrosino, A.: Salient feature based graph matching for person re-identi fi cation. Pattern Recognit. **48**(4), 1070–1081 (2014). http://dx.doi.org/10.1016/j.patcog.2014.09.011
7. Krizhevsky, A., Sutskever, I., Hinton, G.: Imagenet classification with deep convolutional neural networks. Adv. Neural Inf. Process. Syst. pp. 1097–1105 (2012)
8. LeCun, Y., Bottou, L., Bengio, Y., Haffner, P.: Gradient-based learning applied to document recognition. Proc. IEEE **11**(86), 2278–2324 (1998)
9. Maddalena, L., Petrosino, A., Ferone, A.: Object motion detection and tracking by an artificial intelligence approach. Int. J. Pattern. Recogn. Artif. Intell. **22**(5), 915–928 (2008)
10. Maddalena, L., Petrosino, A., Laccetti, G.: A fusion-based approach to digital movie restoration. Pattern. Recogn. **42**(7), 1485–1495 (2009)

11. Maddalena, L., Petrosino, A., Russo, F.: People counting by learning their appearance in a multi-view camera environment. Pattern. Recogn. Lett. **36**, 125–134 (2014)
12. Maresca, M.-E., Petrosino, A.: Clustering Local Motion Estimates for Robust and Efficient Object Tracking. In: Computer Vision-ECCV 2014 Workshops, pp. 244–253 (2014)
13. Melfi, R., Kondra, S., Petrosino, A.: Human activity modeling by spatio temporal textural appearance. Pattern. Recogn. Lett. **34**(15), 1990–1994 (2013)
14. Petrosino, A., Salvi, G.: A two-subcycle thinning algorithm and its parallel implementation on SIMD machines. IEEE. T. Image. Process. **9**(2), 277–283 (1999)
15. Phung, S.L., Bouzerdoum, A.: A pyramidal neural network for visual pattern recognition. IEEE Trans. Neural Netw. Publ. IEEE Neural Netw. Counc. **18**(2), 329–343 (2007)
16. Vezzani, R., Baltieri, D., Cucchiara, R.: People reidentification in surveillance and forensics. ACM Comput. Surv. **46**(2), 1–37 (2013)
17. Yi, D., Lei, Z., Li, S.Z.: Deep metric learning for practical person re-identification (2014). ArXiv e-prints
18. Zajdel, W., Zivkovic, Z., Kröse, B.J.A.: Keeping track of humans: have i seen this person before? In: Proceedings—IEEE International Conference on Robotics and Automation 2005 (April), pp. 2081–2086 (2005)

Part IV
Computational Intelligence Methods for Biomedical ICT in Neurological Diseases

A Pilot Study on the Decoding of Dynamic Emotional Expressions in Major Depressive Disorder

Anna Esposito, Filomena Scibelli and Alessandro Vinciarelli

Abstract Studies investigating on the ability of depressed patients to decode emotional expressions have mostly exploited static stimuli (i.e., static facial expressions of basic emotions) showing that (even though this was not always the case) depressed patients are less accurate (in literature this is reported as a bias) in decoding negative emotions (fear, sadness and anger). However, static stimuli may not reflect the everyday situations and therefore this pilot study proposes to exploit dynamic stimuli involving both visual and auditory channels. We recruited 16 outpatients with Recurrent Major Depressive Disorder (MDD) matched with 16 healthy controls (HC). Their competence to decode emotional expressions was assessed through an emotion recognition task that included short audio (without video), video (without audio) and audio/video tracks. The results show that depressed patients are less accurate than controls, even though with no statistical significant difference, in decoding fear and anger, but not sadness, happiness and surprise where differences are significant. This is independent of the communication mode (either visual, auditory, or both, even though MDDs perform more worse than HCs in audio/video) and the severity of depressive symptoms, suggesting that the MDDs poorer decoding accuracy towards negative emotions is latent and emerges only during and after stressful events. The poorer decoding accuracy of happiness and (positive) surprise can be due to anhedonia.

Keywords Major Depressive Disorder · Basic emotions · Bias · Communication modes · Multimodal dynamic stimuli · Disordered emotion perception

A. Esposito (✉)
Department of Psychology, Second University of Naples and IIASS, Caserta, Italy
e-mail: iiass.annaesp@tin.it

F. Scibelli
Università di Napoli Federico II, Naples, Italy
e-mail: filomena.scibelli@unina.it

F. Scibelli
Seconda Università di Napoli, Caserta, Italy

A. Vinciarelli
School of Computing Science, University of Glasgow, Glasgow, UK
e-mail: alessandro.vinciarelli@glasgow.ac.uk

© Springer International Publishing Switzerland 2016
S. Bassis et al. (eds.), *Advances in Neural Networks*, Smart Innovation,
Systems and Technologies 54, DOI 10.1007/978-3-319-33747-0_19

1   Introduction

The Major Depressive Disorder (MDD) is among the most frequent psychiatric illnesses and is revealed by several impairing symptoms at the cognitive, social, behavioral, and physical level [2]. A central disturbance generated by MDD is the impairment in social functioning. It has been suggested that a deficit in the ability to identify the emotional expressions of others may engender difficulties in social relations [36, 40, 42] given that emotional competence is crucial for successful interpersonal relationships [11, 13, 15]. This is because emotional expressions communicate information on the emotional state of others and enable appropriate social response to these signals [33]. Studies investigating the ability of depressed subjects to decode emotional expressions report contradictory findings. It was both found that depressed subjects are slow and less accurate [1, 3, 10, 12, 14, 20, 34, 41], as well as, fast and more accurate in decoding negative emotions [25, 26, 28, 39]. In general, it was found that depressed subjects may exhibit a global deficit in decoding emotions [4, 18, 31] and be less accurate than healthy subjects in decoding happiness [21, 23, 24, 27, 35, 37]. Since all the above results were obtained exploiting mostly static stimuli and/or only one communication mode (only audio or only video), the goal of the present study is to investigate on the ability of depressed patients to decode emotional multimodal dynamic stimuli.

2   Methods

2.1   Subjects

The study sample comprised 16 outpatients (10 males and 6 females; mean age = 53.3; s.d. = 9.8) with Recurrent Major Depressive Disorder (MDD) diagnosed according to DSM-IV criteria and 16 control subjects (6 males and 10 females; mean age = 52; s.d. = 13.3). All depressed patients (MDDs) received antidepressant medication, and were recruited at the Mental Health Service of Avellino, Italy. The control subjects (HCs) did not present psychiatric illness or a history of psychiatric illness and were recruited through telephone invitations. Table 1 shows clinical and demographic characteristics of two groups.

2.2   Stimuli and Procedure

The depressed patients participating to the experiments were diagnosed as major depressed at the Mental Health Service Center in Avellino (Italy) by a psychiatrist, who provided an anamnesis of the patients' clinical history (diagnosis, type of drugs, time of administration drugs).

Table 1 Demographic and clinical characteristics of two groups

	Groups				Statistics
	MDDs		HCs		
	N. = 16 (10M, 6F)		N. = 16 (6M, 10F)		
	Mean Scores	Standard Deviation	Mean Scores	Standard Deviation	
Age	53.3	9.8	52	13.3	$t_{27} = -0.39$
					p-value = 0.69
Years of education*	2.6	0.7	2.7	0.9	$H_1 = 0.002$
					p-value = 0.97
BDI-II**	37.4	10.3	2.6	3.3	$t_{30} = 12.9$
					p-value $\ll$ 0.001

*Scores coding: 1 = primary school; 2 = secondary school; 3 = high school; 4 = college
**Range of score: 0–9 = normal score; 10–18 = mild depression; 19–29 = moderate depression; 30–63 = severe depression

In addition, to assess the severity of the depressive symptoms, the MDD participants were administered with the Italian version of the Beck Depression Inventory-II Second Edition (BDI II) [8, 19]. To exclude the presence of depressive symptoms, the BDI-II was also administered to the HC participants.

The proposed emotion recognition task included three types of emotional stimuli: videotaped facial expressions (without audio); audiotaped (without video) vocal expressions; combined audio and video recordings. In these recordings, an Italian actor/actress expressed five basic emotions: happiness, fear, anger, surprise, sadness [16, 17]. The total number of stimuli used was 60 (20 for each communication mode) and their duration varied between 2 and 3 s.

The experiment was conducted in a quiet room. Informed consents were obtained from all the participants after the study had been described to them. Each participant first completed the BDI-II and then the emotion recognition task. The stimuli were presented one at a time through a personal computer. The subject was asked to select, from a set of options (happiness, fear, anger, surprise, sadness, a different emotion, no emotion), the basic emotion that best described the emotional state represented by the actor/actress in the played mute video, audio or audio/video recording.

2.3 Statistical Analysis

A set of statistical tests were performed to assess the collected data. The Students' t-test was performed to evaluate whether there were significant differences in the participants' age and BDI-II scores. The Krushal-Wallis test was performed to assess differences between the groups with respect to years of education.

A repeated measure analysis of variance (ANOVA), with emotions (the five basic emotions) and communication modes (only audio, mute video, and combined audio/video) as within factors and groups (MDDs and HCs) as between factor was performed on the number of correct responses to the emotional stimuli. Significance was established for $\alpha = 0.05$.

Confusion matrices were computed on the percentage of correct responses to the stimuli for each communication mode to assess possible misperceptions among categories.

Finally, the Pearson correlation coefficient was used to identify relationships between BDI-II scores and answers to the emotion recognition task.

3   Results

3.1   Clinical and Demographic Characteristics

According to the statistical analysis (see also Table 1), the two groups (MDDs and HCs) involved in the experiment did not show significant differences in age ($t = 0.39$; p-value > 0.05) and years of education ($H_1 = 0.002$; p-value > 0.05).

A significant difference was found (as expected) for the BDI-II scores ($t = 12.9$ p-value < 0.05): MDDs showed a severe degree of depression; HCs obtained scores indicating the absence of depression.

3.2   Beck Depression Inventory-II

For the MDD group, no correlation was found between the BDI-II scores and answers to the emotion recognition task (r Pearson = -0.34; p-value = 0.19) suggesting that the patients' performance to the emotion recognition task did not correlate with the severity of their symptoms.

3.3   The Emotion Recognition Task

Figure 1 shows the percentage of correct responses to each emotion and Fig. 2 the percentage of correct responses in each communication mode. Table 2 reports MDD and HC accuracy mean scores together with the F of Fisher and p-values for each emotion and Table 3 reports MDD and HC accuracy mean scores together with the F of Fisher and p-values for each communication mode. Tables 4 report confusion matrices for each emotion and each communication mode both for the MDDs and HCs respectively.

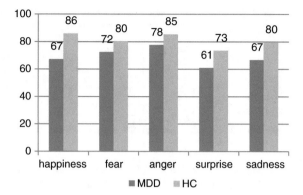

Fig. 1 Percentages of MDD (*blue bars*) and HC (*green bars*) correct responses to each emotion

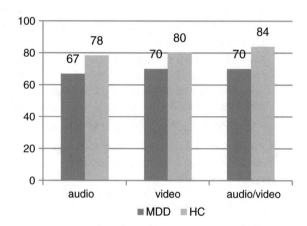

Fig. 2 Percentages of MDD (*blue bars*) and HC (*green bars*) correct responses to each emotion in each communication mode

Table 2 MDD and HC accuracy mean scores of each emotion

Emotions	MDDs		HCs		$F_{(1,30)}$ of Fisher	*p*-value
	Mean scores	Standard deviation	Mean scores	Standard deviation		
Happiness	2.688	0.209	3.438	0.209	6.45	0.017
Fear	2.896	0.155	3.208	0.155	2.04	0.163
Anger	3.104	0.168	3.417	0.168	1.72	0.199
Surprise	2.438	0.173	2.938	0.173	4.15	0.050
Sadness	2.667	0.165	3.188	0.193	4.97	0.033

F of Fisher and *p*-values are reported to assess significant differences

The ANOVA analysis shows as main effects:

(a) a significant difference among the groups ($F_{(1,30)} = 13.39$, *p*-value = 0.001): HCs were more accurate than MDDs (HCs accuracy mean score = 3.24, standard deviation (s.d.). = 0.09; MDDs accuracy mean score = 2.75, s.d. = 0.09);

Table 3 MDD and HC accuracy mean scores in each communication mode

Communication modes	MDDs		HCs		$F_{(1,30)}$ of Fisher	p-value
	Mean scores	Standard deviation	Mean scores	Standard deviation		
Audio	2.67	0.120	3.14	0.120	7.41	0.011
Mute video	2.80	0.120	3.21	0.110	5.90	0.021
Audio/Video	2.80	0.119	3.36	0.119	11.13	0.002

F of Fisher and p-values are reported to assess significant differences

(b) significant effects were found for emotions ($F_{(4,120)} = 3.21$, p-value = 0.034). Post hoc comparisons show a significant difference between surprise and anger (p-value = 0.016). In general, surprise is the less recognized and anger is the most recognized emotion by both groups. In addition, happiness is the most recognized emotion by HCs (see mean scores in Table 2);

(c) no significant effects were found for the communication modes ($F_{(2,60)} = 1.77$, p-value = 0.178) in each group. In general, both groups made more errors in the only audio and less errors in audio/video (see mean scores in Table 3);

(d) the ANOVA analysis also indicated significant interactions between emotions and communication modes ($F_{(8,240)} = 7.28$, p-value $\ll 0.001$) and emotions, communication modes and groups ($F_{(8, 240)} = 3.33$, p-value = 0.001).

With respect to emotions, post hoc comparisons show a significant difference between MDDs and HCs for happiness ($F_{(1,30)} = 6.45$ p-value = 0.017), sadness ($F_{(1,30)}=4.97$, p-value = 0.033) and to the limit of significance, surprise ($F_{(1,30)} = 4.158$ p-value = 0.05). No significant differences were found for fear ($F_{(1,30)} = 2.04$, p-value = 0.163) and anger ($F_{(1,30)} = 1.72$, p-value = 0.199), even though MDDs made more errors than HCs (see Fig. 1 and accuracy mean scores reported in Table 2).

With respect to communication modes, post hoc comparisons show that MDDs made significantly more errors than HCs in the only audio ($F_{(1,30)} = 7.41$, p-value = 0.011), the mute video ($F_{(1,30)} = 5.90$, p-value = 0.021) and the audio/video ($F_{(1,30)} = 11.14$, p-value = 0.002, see accuracy mean scores reported in Table 3 and displayed in Fig. 2).

Post hoc comparisons to explain the significant interaction between emotions and communication modes ($F_{(8,240)} = 7.288$; p-value $\ll 0.001$) indicate that differences in emotion recognition accuracy were significant in the audio, only for surprise ($F_{(1,30)} = 15.49$, p-value $\ll 0.001$), in the mute video, only for happiness ($F_{(1, 30)} = 11.68$, p-value = 0.002) and in the audio/video, only for sadness ($F_{(1,30)} = 7.55$, p-value = 0.010). In addition, surprise was the less recognized emotion for both MDDs and HCs in the mute video (Bonferroni multiple comparisons: p-value < 0.05).

Examining the confusion matrices (see Table 4), related to the percentage of correct recognition accuracy for each emotion and each group, it can be observed that:

Table 4 MDDs and HCs confusion matrices for each emotion in each communication mode (audio alone, mute video and audio/video)

Audio MDDs	Happiness	Fear	Anger	Surprise	Sadness	Different emotion	No emotion
Happiness	**69**	0	0	16	2	8	6
Fear	0	**77**	11	3	8	2	0
Anger	0	14	**67**	5	5	6	3
Surprise	20	5	2	**52**	0	17	5
Sadness	3	2	5	5	**70**	9	6
Audio HCs							
Happiness	**83**	0	0	11	0	3	3
Fear	0	**77**	2	8	3	11	0
Anger	0	6	**78**	2	5	6	3
Surprise	9	0	2	**84**	0	3	2
Sadness	2	3	3	0	**70**	22	3
Mute video MDDs							
Happiness	**66**	0	5	6	2	17	5
Fear	3	**80**	6	0	5	6	0
Anger	0	3	**83**	3	5	5	2
Surprise	5	2	13	**55**	3	14	9
Sadness	2	0	6	5	**67**	11	9
Mute video HCs							
Happiness	**92**	0	2	5	0	2	0
Fear	0	**89**	2	2	3	3	2
Anger	0	6	**88**	2	0	5	0
Surprise	8	2	20	**50**	3	14	3
Sadness	0	3	3	0	**83**	11	0
Audio/Video MDDs							
Happiness	**67**	0	2	9	2	20	0
Fear	0	**61**	6	5	11	14	3
Anger	0	3	**83**	0	5	8	2
Surprise	2	0	9	**77**	3	6	3
Sadness	0	6	8	5	**63**	17	2
Audio/Video HCs							
Happiness	**83**	0	0	9	0	8	0
Fear	0	**75**	0	6	5	13	2
Anger	0	5	**91**	0	0	5	0
Surprise	2	2	3	**86**	2	6	0
Sadness	0	5	2	0	**88**	6	0

The reported numbers are the percentages of correct responses to each emotion

When stimuli are portrayed through the audio, substantial differences (more than 10 %) in the decoding accuracy are seen for:

(a) happiness (69 % for MDDs against 83 % accuracy for HCs), which is mostly confused with surprise (16 %);
(b) surprise (52 % for MDDs against 84 % accuracy for HCs) which is mostly confused with happiness (20 %) or a different emotion (17 %);
(c) anger (67 % for MDDs against 78 % accuracy for HCs) which is mostly confused with fear (14 %).

When stimuli are portrayed through the mute video (see Table 4), substantial differences (more than 10 %) in the decoding accuracy are seen for:

(a) happiness (66 % for MDDs against 92 % accuracy for HCs), which is mostly confused with a different emotion (17 %);
(b) sadness (67 % for MDDs with respect to 83 % accuracy for HCs) which is mostly confused with no emotion (9 %) or a different emotion (11 %).

When stimuli are portrayed through the audio/video (see Table 4), substantial differences (more than 10 %) in the decoding accuracy are seen for:

(a) happiness (67 % for MDDs against 83 % accuracy for HCs), which is mostly confused with a different emotion (20 %) or surprise (9 %);
(b) sadness (63 % for MDDs with respect to 88 % accuracy for HCs) which is mostly confused with a different emotion (17 %) or anger (8 %);
(c) fear (61 % for MDDs with respect to 75 % accuracy for HCs) which is mostly confused with a different emotion (14 %) or sadness (11 %).

In addition, in the audio/video, decoding accuracy differences between MDDs and HCs were still high for surprise (for 9 %) and anger (8 %) respectively, indicating that this communication mode was the mostly impaired for MDD subjects.

4 Discussion

The results of this study show that outpatients with Recurrent Major Depressive Disorder exhibit a poorer decoding accuracy of happiness in the mute video, surprise in the audio and sadness in the audio/video. These results are not supported by the data reported in literature asserting a strong MDDs' bias toward negative emotions (either positively [25, 28] in the sense that MDDs are more accurate, or negatively [1, 3, 10, 12, 41], in the sense that they are less accurate than HCs). The present results partially support studies reporting a deficit towards negative emotions depending on the communication mode, since it has been found that MDDs are significantly less accurate than HCs in decoding sadness in the audio/video. Our results departure from these previous studies, as already discussed in the introduction, can be attributed to their use of static rather than dynamic visual stimuli. Dynamicity may account for performance's differences in both MDDs and HCs.

The presence of a bias for anger and fear cannot be completely excluded, since a general worst performance (even though, not significant) was exhibited by MDDs in decoding these emotions (see Table 4). Our data suggest that the negative bias for anger and fear may be latent. In particular, MDDs show poorer recognition accuracy in the audio for anger, and in the audio/video for sadness, fear and, to a certain extent, also anger. Therefore, as an alternative explanation, it can be suggested that the bias towards anger and fear may be concealed, pending on the occurrence of stressful events. During or after stressful event, it may become active and dominant, impairing MDDs daily ability to decode these emotions. This hypothesis is supported by Beck's theory [5–7, 9] according to which the "negative bias" is a depression vulnerability factor originating from dysfunctional cognitive schemes that arise during and/or right after stressful events.

What is clear from the reported data is that MDDs are really poor in decoding happiness (in the mute video this is significant while in the other modes there are substantial differences in the recognition accuracy percentages) and surprise (in the audio). The MDDs poorer decoding accuracy towards happiness and (positive) surprise may be due to anhedonia. Anhedonia is a deficit in the ability to experience pleasure in activities and situations considered rewarding [30]. Anhedonic symptoms are considered a trait of a pre-depressed personality. After stressful events, anhedonia can originate an overt depressive disorder [29], causing an impairment in the decoding of positive information. MDDs and HCs poorer decoding accuracy for surprise observed in the mute video can be attributed to the facial features associated with such an emotional stimulus that brings subjects to confuse it either with anger or a different emotion (see Table 4). As it has been reported by several authors, independently of the depressive disorder [16, 17], surprise is an emotion difficult to decode given that it can be associated either with positive or negative emotions.

In our study, it has been found that the severity of depressive symptoms (measured with BDI II) is not correlated with patients' performance to the emotion recognition task, proving that MDDs poorer emotional decoding accuracy is independent of their clinical state. This suggests that the poorer MDDs processing of emotional categories is a trait characteristic and can be a risk factor to depression relapse.

To our knowledge, this is the first study investigating the capability of depressed patients to decode emotional dynamic stimuli administered through three different communication modes: only visual, only auditory, and multimodal (visual and auditory). These stimuli better reflect the way people perceive and decode emotions in everyday situations with respect to static stimuli.

The three communication modes exploited in this study allowed investigating MDDs performance in the visual channel (much more investigated by previous studies through static stimuli) and matching it with performance in the auditory and combined visual and auditory channels. The results show that the poorer decoding accuracy towards basic emotions is present, in general, in all three communication modes, even though some differences emerge considering each mode in relation to each basic emotion. Particularly, depressed subjects have difficulty in decoding happiness in all three communication modes supporting the evidence of anhedonia as a symptom of depression. In addition, MDDs have difficulties in decoding emotional expressions of anger and surprise in the auditory channel; sadness and fear in the visual channel, and when the emotional signal is multimodal, exhibited a general worsening in decoding all the emotions, even though it was more accentuated for sadness.

The error increase in the decoding accuracy of emotional multimodal (audio/video) information suggests that multimodality does not always add to the amount of emotional information conveyed by the single channels. Rather it may mystify the decoding process of some emotions. Table 4 shows that depressed subjects often exploit the "different emotion" label, suggesting that they are either not able to decode the correct emotion or use the appropriate label for it.

The MDDs poorer decoding accuracy towards basic emotions found in this study is partly consistent with other studies that used different dynamic stimuli. Schneider et al. [38] used short video clips (combined audio and video information) and found that MDDs exhibited a general emotion processing deficit. Peron et al. [32] used emotionally prosodic produced pseudo-words (only audio) and found an impaired recognition of happiness, fear and sadness. Punkanen et al. [34] used musical examples (only audio mode) and found that depressed subjects mostly confused fear and sadness with anger and were less accurate in decoding happiness and tenderness.

Our results add to these previous studies the knowledge that depressed subjects, in their daily life, misinterpret the emotional expressions of others, independently of the communication mode, but are particularly in difficulty in the multimodal mode where they perform worse than HCs. This suggests that MDDs are generally poor in decoding realistic and dynamic emotional stimuli. In particular they show a general disturbance in decoding multimodal stimuli, making substantial errors (more than 10 %) for happiness, sadness and fear, but also misinterpreting surprise (8 %) and anger (9 %). Their misinterpretation of others' emotions not only contributes to their interpersonal difficulties, but also does not allow them to correctly decode their experienced emotions, further aggravating the depressive symptoms. This interpretative bias becomes more debilitating when depressed subjects have to face stressful and negative events.

Future works in this direction would require increasing the number of participants in order to have more precise assessments of depressive behaviors. In addition, since the recruited MDD patients were under drugs, it can be of interest to match this group with subjects that have never assumed drugs, since acute antidepressant administrations may have regularized the negative affective bias [22].

Acknowledgments The patients, the healthy controls and the doctors (psychiatrists) of the Mental Health Service Center in Avellino (Italy) are acknowledged for their participation and support to this study. This paper was inspired by the stimulating discussions raised during the participation of the first author in Dagstuhl Seminar 13451 "Computational Audio Analysis" held from Nov 3rd to 8th 2013 in Wadern, Germany.

References

1. Aldinger, M., Stopsack, M., et al.: The association between depressive symptoms and emotion recognition is moderated by emotion regulation. Psychiatry Res. **205**, 59–66 (2013)
2. American Psychiatric Association: Manuale Diagnostico e Statistico dei Disturbi Mentali (IV ed.). Washington, DC. Masson MI (2000)
3. Anderson, I.M., Shippen, C., Juhasz, G., et al.: State-dependent alteration in face emotion recognition in depression. Br. J. Psychiatry **198**, 302–308 (2011)
4. Asthana, H.S., Mandal, M.K., Khurana, H., Haque-Nizamie, S.: Visuospatial and affect recognition deficit in depression. J. Affect. Disord. **48**, 57–62 (1998)
5. Beck, A.T.: Thinking and depression: idiosyncratic content and cognitive distortions. Arch. Gen. Psychiatry **9**, 324–333 (1963)
6. Beck, A.T.: Thinking and depression: 2. Theory and therapy. Arch. Gen. Psychiatr. **10**, 561–571 (1964)
7. Beck, A.T., Rush, A.J., Shaw, B.F., Emery, G.: Cognitive Therapy of Depression. The Guildford Press, New York (1979)
8. Beck, A.T., Steer, R.A., Brown, G.K.: Comparison of the Beck depression inventories-IA and II in psychiatric outpatients. J. Pers. Assess. **67**, 588–597 (1996)
9. Beck, A.T.: Cognitive models of depression. In: Leahy, R.L., Dowd, E.T. (eds.) Clinical Advances in Cognitive Psychotherapy: Theory on Application, pp. 29–61. Springer Publishing Company, New York (2002)
10. Cerroni, G., Tempesta, D., et al.: Il riconoscimento emotivo delle espressioni facciali nella schizofrenia e nella depressione. Epidemiologia e psichiatria sociale **16**(2), 179–182 (2007)
11. Cooley, E.L., Nowicki, S.: Discrimination of facial expressions of emotion by depressed subjects. Genet. Soc. Gen. Psychol. Monogr. **115**(4), 449–465 (1989)
12. Csukly, G., Czobor, P., et al.: Facial expression recognition in depressed subjects. The impact of intensity level and arousal dimension. J. Nerv. Met. Dis. **197**, 98–103 (2009)
13. Darwin, C.: The Expression of the Emotions in Man and Animals. Murray, London (1872)
14. Douglas, K.M., Porters, R.J.: Recognition of disgusted facial expressions in severe depression. Br. J. of Psychiatry **197**, 156–157 (2010)
15. Ekman, P.: Emotion in the human face. Cambridge University Press, New York (1982)
16. Esposito, A., Riviello, M.T., Di Maio, G.: The Italian audio and video emotional database. In: Apolloni, B., Bassis, S., Morabito, C.F. (eds.) Neural Nets WIRN, vol. 204, pp. 51–61 (2009)
17. Esposito, A., Riviello, M.T.: The New Italian audio and video emotional database. In: COST 2102 International Training School 2009, LNCS 5967, pp. 406–422. Springer, Berlin, Heidelberg (2010)
18. Feinberg, T., Rifkin, A., Schaffer, C., Walker, E.: Facial discrimination and emotional recognition in schizophrenia and affective disorders. Arch. Gen. Psychiat. **43**, 276–279 (1986)
19. Ghisi, M., Flebus, G.B., Montano, A., Sanavio, E., Sica, C.: Beck Depression Inventory-II. Manuale italiano. Organizzazioni Speciali, Firenze (2006)
20. Gollan, J.K., Pane, H., McCloskey, M., Coccaro, E.F.: Identifying differences in biased affective information processing in major depression. Psychiatry Res. **159**(1–2), 18–24 (2008)
21. Gur, R.C., Erwin, R.J., Gur, R.E., et al.: Facial emotion discrimination: II. Behavioral findings in depression. Psychiatry Res. **2**, 241–251 (1992)
22. Harmer, C.J., O'Sullivan, U., et al.: Effect of acute antidepressant administration on negative affective bias in depressed patients. Am. J. Psychiatry **166**(10), 1178–1184 (2009)
23. Joormann, J., Gotlib, I.H.: Is this happiness I see? Biases in the identification of emotional facial expressions in depression and social phobia. J. Abnorm. Psychol. **115**(4), 705–714 (2006)
24. LeMoult, J., Joormann, J., Sherdell, L., Wright, Y., Gotlib, I.H.: Identification of emotional facial expressions following recovery from depression. J. Abnorm. Psychol. **118**(4), 828–833 (2009)

25. Liu, W., Huang, J., Wang, L., Gong, Q., Chan, R.: Facial perception bias in patients with major depression. Psychiatry Res. **97**(3), 217–220 (2012)
26. Mandal, M.K., Bhattacharya, B.B.: Recognition of facial affect in depression. Percept. Mot. Skills **61**, 13–14 (1985)
27. Mikhailova, E.S., Vladimirova, T.V., Iznak, A.F., et al.: Abnormal recognition of facial expression of emotions in depressed patients with major depression disorder and schizotypal personality disorder. Biol. Psychiatry **40**, 697–705 (1996)
28. Milders, M., Bell, S., Platt, J., Serrano, R., Runcie, O.: Stable expression recognition abnormalities in unipolar depression. Psychiatry Res. **179**, 38–42 (2010)
29. Myerson, A.: The constitutional anhedonia personality. Am. J. Psychiatry **22**, 774–779 (1946)
30. Pelizza, L., Pupo, S., Ferrari, A.: L'anedonia nella schizofrenia e nella depressione maggiore: stato o tratto? Rassegna della letteratura. J. Psychopathol. **18**, 1–12 (2012)
31. Persad, S.M., Polivy, J.: Differences between depressed and nondepressed individuals in the recognition of and response to facial emotional cues. J. Abnorm. Psychol. **102**, 358–368 (1993)
32. Peron, J., et al.: Major depressive disorder skews the recognition of emotional prosody. Neuro-Psychopharmacol. Biol. Psychiatry **35**, 987–996 (2011)
33. Phillips, M.L., Drevets, W.C., Rauch, S.L., Lane, R.: Neurobiology of emotion perception I: the neural basis of normal emotion perception. Biol. Psychiatry **54**, 504–514 (2003)
34. Punkanen, M., Eerola, T., Erkkilä, J.: Biased emotional recognition in depression: perception of emotions in music by depressed patients. J. Affect. Disord. **130**, 118–126 (2011)
35. Rubinow, D.R., Post, R.M.: Impaired recognition of affect in facial expression in depressed patients. Biol. Psychiatry **31**, 947–953 (1992)
36. Segrin, C.: Social skills deficit associated with depression. Clin. Psychol. Rev. **20**(3), 379–403 (2000)
37. Schlipt, S., Batra, A., Walter, G., et al.: Judgment of emotional information expressed by prosody and semantics in patients with unipolar depression. Front. Psychol. (2013). doi:10.3389/fpsyg.2013.00461
38. Schneider, D., Regenbogen, C., et al.: Empathic behavioral and physiological responses to dynamic stimuli in depression. Psychiatry Res. (2012). doi:10.1016/j.psychres.2012.03.054
39. Surguladze, S.A., Young, A.W., Senior, C., et al.: Recognition accuracy and response bias to happy and sad facial expressions in patients with major depression. Neuropsychology **18**(2), 212–218 (2004)
40. Teo, A., Choi, H., Valenstein, M.: Social relationships and depression: ten-year follow-up from a nationally representative study. PLoS ONE **8**(4), e62396 (2013). doi:10.1371/journal.pone.0062396
41. Wright, S.L., Langenecker, S.A., Deldin, P.J., Rapport, L.J., et al.: Gender-specific disruptions in emotion processing in younger adults with depression. Depression Anxiety **26**, 182–189 (2009)
42. Zlotnick, C., Khon, R., Keitner, G., Della Grotta, S.A.: The relationship between quality of interpersonal relationships and major depressive disorder: findings from the National Comorbidity Survey. J. Affect. Disord. **59**, 205–215 (2000)

Advances in a Multimodal Approach for Dysphagia Analysis Based on Automatic Voice Analysis

K. López-de-Ipiña, Antonio Satue-Villar, Marcos Faundez-Zanuy,
Viridiana Arreola, Omar Ortega, Pere Clavé,
M. Pilar Sanz-Cartagena, Jiri Mekyska and Pilar Calvo

Abstract Parkinson's disease (PD) is the second most frequent neurodegenerative disease with prevalence among general population reaching 0.1–1 %, and an annual incidence between 1.3–2.0/10,000 inhabitants. The most obvious symptoms are movement-related such as tremor, rigidity, slowness of movement and walking difficulties and frequently these are the symptoms that lead to the PD diagnoses but also they could have In this sense voice analysis is a safe, non-invasive, and reliable

K. López-de-Ipiña (✉) · P. Calvo
Universidad del País Vasco/Euskal Herriko Unibertsitatea, Europa Pz 1, 20008 Donostia, Spain
e-mail: karmele.ipina@ehu.es

A. Satue-Villar · M. Faundez-Zanuy
Fundació Tecnocampus, Avda. Ernest Lluch 32, 08302 Mataró, Spain
e-mail: satue@tecnocampus.cat

M. Faundez-Zanuy
e-mail: faundez@tecnocampus.cat

V. Arreola · O. Ortega · P. Clavé · M.P. Sanz-Cartagena
Hospital de Mataró, Consorci Sanitari del Maresme, Barcelona, Spain
e-mail: varreola@csdm.cat

O. Ortega
e-mail: oortega@csdm.cat

P. Clavé
e-mail: pere.clave@ciberehd.org

M.P. Sanz-Cartagena
e-mail: sanz@csdm.cat

V. Arreola · O. Ortega · P. Clavé · M.P. Sanz-Cartagena
Centro de Investigación Biomedica en Red de Enfermedades Hepaticas y Digestivas, Barcelona, Spain

J. Mekyska
Brno University of Technology, Brno, Czech Republic

© Springer International Publishing Switzerland 2016   201
S. Bassis et al. (eds.), *Advances in Neural Networks*, Smart Innovation,
Systems and Technologies 54, DOI 10.1007/978-3-319-33747-0_20

screening procedure for PD patients with dysphagia, which could detect patients at high risk of clinically significant aspiration. In this paper we will present a part of an ongoing project that will evaluate automatic speech analysis based on linear and non-linear features. These can be reliable predictors/indicators of swallowing and balance impairments in PD. An important advantage of voice analysis is its low intrusiveness and easy implementation in clinical practice. Thus, if a significant correlation between these simple analyses and the gold standard video-fluoroscopic analysis will imply simpler and less stressing diagnostic test for the patients as well as the use of cheaper analysis systems.

Keywords Speech analysis · Dysphagia · Parkinson disease · Database

1 Introduction

In the full study we will focus on three kinds of signals, but the first step will be focused on speech signals and dysphagia. It is based on a collaboration between an engineering faculties and a Hospital. A large number of group of people suffer dysphagia, as well as their effect such as: elderly people, stroke sufferers, multiple sclerosis, PD, Alzheimer's Disease, people with cancer of the throat and/or mouth, people with head and neck injuries.

The medical term for any difficulty or discomfort when swallowing is dysphagia. A normal swallow takes place in four stages, and involves 25 different muscles and five different nerves. Difficulties at different stages cause different problems and symptoms. The four stages of swallowing are the following ones:

1. The sight, smell, or taste of food and drink triggers the production of saliva, so that when you put food in your mouth (usually voluntarily) there is extra fluid to make the process of chewing easier.
2. When the food is chewed enough to make a soft bolus, your tongue flips it towards the back of the mouth to the top of the tube, which leads down to your stomach. This part of your throat is called the pharynx. This part of swallowing is also voluntary.
3. Once the bolus of food reaches your pharynx, the swallowing process becomes automatic. Your voice box (the larynx) closes to prevent any food or liquid getting into the upper airways and lungs, making the food bolus ready to pass down your throat (known as the oesophagus).
4. The oesophagus, which is a tube with muscular walls that contract automatically, then propels the food down to the stomach.
    Some signs of dysphagia are:

- Swallow repeatedly.
- Cough and splutter frequently.

- Voice is unusually husky and you often need to clear your throat.
- When you try to eat you dribble. Food and saliva escape from your mouth or even your nose.
- Find it easier to eat slowly.
- Quite often keep old food in your mouth, particularly when you have not had a chance to get rid of it unseen.
- Feel tired and lose weight.

1.1 Dysphagia and Voice Analysis

In the PD patient, dysphagia is usually accompanied by other oro-bucal symptoms such as hypokinetic dysarthria. Some studies have reported that the presence of both symptoms usually correlates and that voice disorders could be anticipatory of swallowing impairment [1]. Other studies concluded that a clear post-swallow voice quality provides reasonable evidence that penetration-aspiration and dysphagia are absent [2]. Voice analysis is a safe, non-invasive, and reliable screening procedure for patients with dysphagia which can detect patients at high risk of clinically significant aspiration [3]. The volume-viscosity swallow test (V-VST) was developed at the Hospital de Mataró to identify clinical signs of impaired efficacy (labial seal, oral and pharyngeal residue, and piecemeal deglutition) and impaired safety of swallow (voice changes, cough and decrease in oxygen saturation ≥ 3 %) [4–6]. The V-VST allows quick, safe and accurate screening for oropharyngeal dysphagia (OD) in hospitalized and independently living patients with multiple etiologies. The V-VST presents a sensitivity of 88.2 % and a specificity of 64.7 % to detect clinical signs of impaired safety of swallow (aspiration or penetration). The test takes 5–10 min to complete and is an excellent tool to screen patients for OD. It combines good psychometric properties, a detailed and easy protocol designed to protect safety of patients, and valid end points to evaluate safety and efficacy of swallowing and detect silent aspirations [3]. However, nowadays voice assessment is usually done by subjective parameters and a more exhaustive and objective evaluation is needed to understand its relationship with dysphagia and aspiration, as well as the potential relevance of voice disorders as a prognostic factor and disease severity marker. Hypokinetic dysarthria is a speech disorder usually seen in PD which affects mainly respiration, phonation, articulation and prosody. Festination is the tendency to speed up during repetitive movements. It appears with gait in order for sufferers to avoid falling down and also in handwriting and speech. Oral festination shares the same pathophysiology as gait disorders [7]. Voice analysis allows the assessment of all these parameters and has been used to evaluate the improvement of PD after treatment [8–13]. Voice impairments appear in early stages of the disease and may be a marker of OD even when swallow disorders are not clinically evident, which would allow to establish early measures to prevent aspiration and respiratory complications. Oropharyngeal dysphagia is a common condition in PD patients. In a recent meta-analysis, the prevalence of PD patients who perceive difficulty in swallowing was estimated at 35 % but when an objective swallowing assessment was performed, the estimated prevalence of OD reached 82 % [14]. This underreporting calls for a proactive clinical approach to dysphagia, particularly in light of the serious clinical consequences associated to OD in these patients. Dysphagia can produce two types of severe complications; (a) alterations in the efficacy of deglutition that may cause malnutrition or dehydration which may occur in up to 24 % of PD patients [15], and (b) impaired safety of swallow, which may lead to aspiration pneumonia with high mortality rates (up to 50 %) [16, 17]. Aspiration pneumonia remains the leading cause of death among PD patients.

2   Gold Standard Methods

Videofluoroscopy (VFS) is the gold standard to study the oral and pharyngeal mechanisms of dysphagia. VFS is a dynamic exploration that evaluates the safety and efficacy of deglutition, characterizes the alterations of deglutition in terms of videofluoroscopic signs, and helps to select and assess specific therapeutic strategies. Since the hypopharynx is full of contrast when the patient inhales after swallowing. Thereafter, VFS can determine whether aspiration is associated with impaired glossopalatal seal (predeglutitive aspiration), a delay in triggering the pharyngeal swallow or impaired deglutitive airway protection (laryngeal elevation, epiglottic descent, and closure of vocal folds during swallow response), or an ineffective pharyngeal clearance (post swallowing aspiration) [5]. In those cases where a possible dysphagia problem exists, a videofluoroscopic analysis is performed. This diagnose is more invasive as it implies radiation, but it is the procedure to have physical evidence of swal-lowing problems. The main goal of the first step of this study is to evaluate if an automatic tool based on speech analysis can be developed to support medical decision during the test depicted in Fig. 1. The current approach for dysphagia analysis has been developed by some of the medical authors of this paper, and can be summarized in Fig. 1. Process for dysphagia analysis based on three liquids of different viscosity and three different volumes per liquid. After swallowing each liquid and volume a word is pronounced by the patient and a speech therapist evaluates the voice quality in a subjective way (just listening to the speech signal).

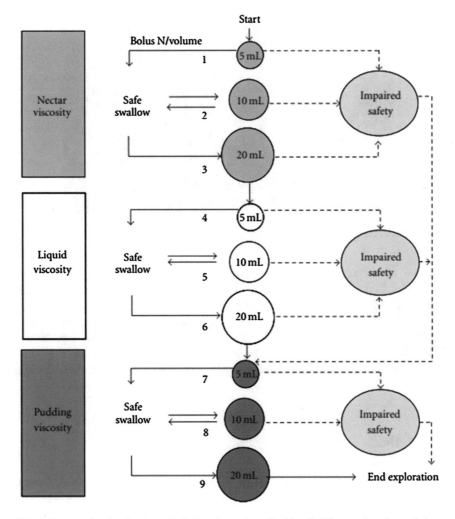

Fig. 1 Process for dysphagia analysis based on three liquids of different viscosity and three different volumes per liquid

3 Materials

At this moment a speech database is being acquired in the protocol depicted in Fig. 2, one sample after deglutition of each liquid and volume. Thus, a total of 9 realizations per patient are acquired. Figure 2 shows the acquisition scenario at Mataro's Hospital. The acquisition setup is based on a capacitor microphone Rode NT2000 (positioned at a distance of approximately 20 cm from the speaker's mouth) and external sampling card (M-AUDIO, FAST TRACK PRO Interface

Fig. 2 Acquisition scenario at Mataro's Hospital

audio 4 × 4) operating at 48 kHz sampling rate, 16 bit per sample, monophonic recording. Currently we are acquiring 3 patients per week. All the work was performed strictly following the ethical consideration of the organizations involved in the project. The initial database consist of 15 patients, 5 people with PD and 10 people in the control group. The uttered word during the test of liquids was "HOLA" a very natural way to obtain voice samples. Moreover this word has two vowels inside very useful to analysed pitch, harmonicity and another standard features.

4 Methods

This multimodal analysis has started on speech signals in the context of dysphagia test.

4.1 Feature Extraction

The voice analyzing approach, after database collection will be based on two kind of features linear and non-linear ones:

(a) Voiced/unvoiced classification and then to check the harmonic to noise ratio (HNR), pitch and its variations, jitter, shimmer, MFCCs, etc.
(b) Permutation entropy

4.2 Permutation Entropy

Permutation entropy directly accounts for the temporal information contained in the time series; furthermore, it has the quality of simplicity, robustness and very low computational cost [18–21]. Bandt and Pompe [22] introduce a simple and robust method based on the Shannon entropy measurement that takes into account time causality by comparing neighboring values in a time series. The appropriate symbol sequence arises naturally from the time series, with no prior knowledge assumed [1].

The Permutation entropy is calculated for a given time series $\{x_1, x_2, \ldots, x_n\}$ as a function of the scale factor s. In order to be able to compute the permutation of a new time vector X_j, $S_t = [X_t, X_{t+1}, \ldots, X_{t+m-1}]$ is generated with the embedding dimension m and then arranged in an increasing order: $[X_{t+j_1-1} \leq X_{t+j_2-1} \leq \cdots \leq X_{t+j_n-1}]$. Given m different values, there will be m! possible patterns π, also known as permutations. If $f(\pi)$ denotes its frequency in the time series, its relative frequency is $p(\pi) = f(\pi)/(L/s - m + 1)$. The permutation entropy is then defined as:

$$PE = - \sum_{i=1}^{m!} p(\pi_i) \ln p(\pi_i) \tag{1}$$

Summarising, Permutation entropy refers to the local order structure of the time series, which can give a quantitative measure of complexity for dynamical time series. This calculation depends on the selection of the m parameter, which is strictly related with the length n of the analysed signal. For example Bandt and Pompe [20] suggested the use of $m = 3, \ldots, 7$ following always this rule:

$$m! < n \tag{2}$$

If m is too small (smaller than 3) the algorithm will work wrongly because it will only have few distinct states for recording but it depends on the data. When using long signals, a large value of m is preferable but it would require a larger computational time.

5 Results and Conclusions

The speech data analyzed in this section are the materials described in Sect. 2. Figure 3 shows "HOLA" word uttered by a control individual for water and 20 ml and Fig. 4 "HOLA" word uttered by an individual with dysphagia also after water an 20 ml. About 40 linear features have been analyzed and Permutation Entropy for m = 6 and r = 6. Despite of the sample size is very small, some features appear with as discriminative for linear features. There variations not only in the Pitch (Fig. 4) but stronger variations for HNR and PE. Figure 5 shows these features for

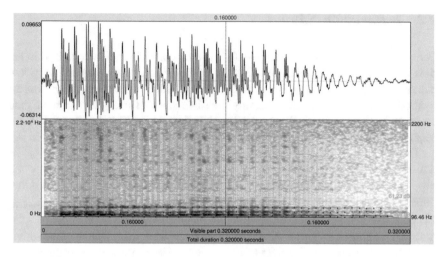

Fig. 3 "HOLA" word uttered by a control individual for Nectar and 20 ml

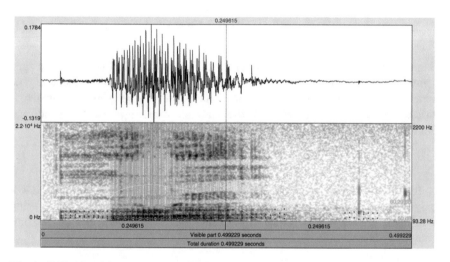

Fig. 4 "HOLA" word uttered by an individual with dysphagia for Nectar and 20 ml

Nectar and 5 and 20 ml. Moreover the work describes a part of an ongoing project that evaluates automatic speech analysis based on linear and non-linear features for the detection of dysphagia. These can be reliable predictors/indicators of swallowing and balance impairments in PD. An important advantage of voice analysis is its low intrusiveness and easy implementation in clinical practice. Thus, if a significant correlation between these simple analyses and the gold standard video-fluoroscopic analysis will imply simpler and less stressing diagnostic. Linear features such as HNR and PE could be powerful correlation elements to detect this

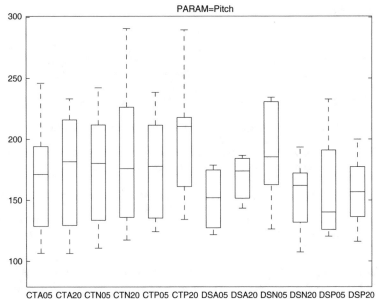

Fig. 5 Pitch after Nectar = A, Liquid = N and Pudding = P for 5 ml, and 20 ml

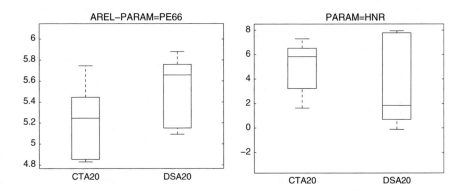

Fig. 6 Boxplot of PE m = 6, r = 6 (*left*) and HNR (*right*) for Nectar and 5 and 20 ml

impartment. In future works the database will be extended and new methodologies to detect suitable changes will be developed (Fig. 6).

Acknowledgments This work has been supported by FEDER and Ministerio de ciencia e Innovación, TEC2012-38630-C04-03 and the University of the Basque Country by EHU-14/58. The described research was performed in laboratories supported by the SIX project; the registration number CZ.1.05/2.1.00/03.0072, the operational program Research and Development for Innovation.

References

1. Perez-Lloreta, S., Negre-Pages, L., Ojero-Senarda, A., et al.: Oro-buccal symptoms (dysphagia, dysarthria, and sialorrhea) in patients with Parkinson's disease: preliminary analysis from the French COPARK cohort. Eur. J. Neurol. **19**, 28–37 (2012)
2. Waito, A., Bailey, G.L., Molfenter, S.M., et al.: Voice-quality abnormalities as a sign of dysphagia: validation against acoustic and videofluoroscopic data. Dysphagia **26**(2), 125–134 (2011)
3. Ryu, J.S., Park, S.R., Choi, K.H.: Prediction of laryngeal aspiration using voice analysis. Am. J. Phys. Med. Rehabil. **83**(10), 753–757 (2004)
4. Rofes, L., Arreola, V., Clavé, P.: The volume-viscosity swallow test for clinical screening of dysphagia and aspiration. Nestle Nutr Inst Workshop Ser. **72**, 33–42 (2012). doi:10.1159/000339979. Epub 2012 Sep 24
5. Rofes, L., Arreola, V., Almirall, J., Cabré, M., Campins, L., García-Peris, P., Speyer, R., Clavé, P.: Diagnosis and management of oropharyngeal Dysphagia and its nutritional and respiratory complications in the elderly. Gastroenterol. Res. Pract. **2011**, 818979 (2011). doi:10.1155/2011/818979. Epub 2010 Aug 3. PubMed PMID: 20811545; PubMed Central PMCID: PMC2929516
6. Clavé, P., Arreola, V., Romea, M., Medina, L., Palomera, E., Serra-Prat, M.: Accuracy of the volume-viscosity swallow test for clinical screening of oropharyngeal dysphagia and aspiration. Clin. Nutr. **27**(6), 806–815 (2008). doi:10.1016/j.clnu.2008.06.011. Epub 2008 Sep 11. PubMed PMID: 18789561
7. Moreau, C., Ozsancak, C., Blatt, J.-L., Derambure, P., Destee, A., Defebvre, L.: Oral festination in Parkinson's disease: biomechanical analysis and correlation with festination and freezing of gait. Mov. Disord. **22**(10), 1503–1506 (2007)
8. Nagulic, M., Davidovic, J., Nagulic, I.: Parkinsonian voice acoustic analysis in real-time after stereotactic thalamotomy. Stereotact. Funct. Neurosurg. **83**(2–3), 115–121 (2005)
9. Gobermana, A.M., Coelho, C.: Acoustic analysis of Parkinsonian speech I: speech characteristics and L-Dopa therapy. NeuroRehabilitation **17**, 237–246 (2002)
10. Gobermana, A.M., Coelho, C.: Acoustic analysis of Parkinsonian speech II: L-Dopa related fluctuations and methodological issues. NeuroRehabilitation **17**, 247–254 (2002)
11. Stewart, C., Winfield, L., Junt, A., Bressman, S.B., Fahn, S., Blitzer, A., Brin, M.F.: Speech dysfunction in early Parkinson's disease. Mov. Disord. **10**(5), 562–565 (1995)
12. Eliasova, I., Mekyska, J., Kostalova, M., Marecek, R., Smekal, Z., Rektorova, I.: Acoustic evaluation of short-term effects of repetitive transcranial magnetic stimulation on motor aspects of speech in Parkinson's disease. J. Neural Transm. **120**(4), 597–605 (2013)
13. Faundez-Zanuy, M., Satue-Villar, A., Mekyska, J., Arreol, V., Sanz, P., Paul, C., Guirao, L., Serra, M., Rofes, L., Clavé, P., Sesa-Nogueras, E., Roure, J.: A multimodal approach for parkinson disease analysis. Advances in Neural Networks: Computational and Theoretical Issues, vol. 37 of the Series Smart Innovation, Systems and Technologies pp 311–318, 2015; Springer
14. Kalf, J.G., et al.: Prevalence of oropharyngeal dysphagia in Parkinson's disease: a meta-anaysis. Parkinson Rel. Disord. **1**, 311–315 (2012)
15. Sheard, J.M., et al.: Prevalence of malnutrition in Parkinsom's disease: a systematic review. Nutr. Rev. **69**, 520–532 (2011)
16. Fernandez, H.H., Lapane, K.L.: Predictors of mortality among nursing home residents with a diagnosis of Parkinson's disease. Med. Sci. Monit. **8**, CR241–CR246 (2002)
17. Williams-Gray, C.H., Mason, S.L., Evans, J.R., et al.: The CamPaIGN study of Parkinson's disease: 10-year outlook in an incident population-based cohort. J. Neurol. Neurosurg. Psychiatry **84**, 1258–1264 (2013)
18. Zanin, M., Zunino, L., Rosso, O.A., Papo, D.: Permutation entropy and its main biomedical and econophysics applications: a review. Entropy **14**(8), 1553–1577

19. Morabito, F.C., Labate, D., La Foresta, F., Bramanti, A., Morabit, G., Palamara, I.: Multivariate multi-scale permutation entropy for complexity analysis of Alzheimer's disease EEG. Entropy **14**(7), 1186–1202 (2012)
20. Eguiraun, H., López-de-Ipiña, K., Martinez, I.: Application of entropy and fractal dimension analyses to the pattern recognition of contaminated fish responses in aquaculture. Entropy
21. Costa, M., Goldberger, A., Peng, C.-K.: Multiscale entropy analysis of biological signals. Phys. Rev. E **71**, 021906:1–021906:18 (2005). http://www.mdpi.com/1099-4300/16/11/6133/htm#sthash.ufFUh4mq.dpuf (2005)
22. Bandt, C., Pompe, B.: Permutation entropy: a natural complexity measure for time series. Phys. Rev. Lett. **88**, 174102:1–174102:4 (2002)

Improving Functional Link Neural Network Learning Scheme for Mammographic Classification

Yana Mazwin Mohmad Hassim and Rozaida Ghazali

Abstract Functional Link Neural Network (FLNN) has become as an important tool used in classification tasks due to its modest architecture. FLNN requires less tunable weights for training as compared to the standard multilayer feed forward network. Since FLNN uses Backpropagation algorithm as the standard learning scheme, the method however prone to get trapped in local minima which affect its classification performance. This paper proposed the implementation of modified Bee-Firefly algorithm as an alternative learning scheme for FLNN for the task of mammographic mass classification. The implementation of the proposed learning scheme demonstrated that the FLNN can successfully perform the classification task with better accuracy result on unseen data.

Keywords Classification · Functional link neural network · Learning scheme · Modified Bee-Firefly algorithm · Mammographic

1 Introduction

Functional Link Neural Network (FLNN) is a type of Higher Order Neural Networks (HONNs) introduced by Giles and Maxwell [1]. It was make known as an alternative approach to standard multilayer feed forward network in Artificial Neural Networks (ANNs) by Pao [2]. The FLNN is a flat network (without hidden layers) where it reduced the neural architectural complexity while at the same time possesses the ability to solve non-linear separable problems. The flat architecture of FLNN also make the learning algorithm in the network less complicated [3].

Y.M.M. Hassim (✉) · R. Ghazali
Faculty of Computer Science and Information Technology,
Universiti Tun Hussein Onn Malaysia (UTHM), 86400 Batu Pahat, Johor, Malaysia
e-mail: yana@uthm.edu.my

R. Ghazali
e-mail: rozaida@uthm.edu.my

FLNN network is usually trained by adjusting the weight of connection between neurons. The most common method for tuning the weight in FLNN is using the Backpropagation (BP) learning algorithm. However one of the crucial problems with the standard BP-learning algorithm is that it can easily get trapped in local minima especially for the non-linear problems [4], thus affect the performance of FLNN network. The modified Artificial Bee Colony (mABC) optimization algorithm was proposed as an alternative learning scheme to optimize FLNN weights [5] for classification tasks. In this study, we proposed the integration of mABC algorithm with Firefly algorithm (FFA) to improve mABC local search strategy and classification accuracy. Our experimental results showed that the FLNN performance trained with mABC-FFA learning scheme yield better accuracy result on classifying the out-of-sample or unseen data.

2   Related Works

In this section, the properties and learning scheme of FLNN as well as the proposed swarm optimizations method are briefly discussed.

2.1   Functional Link Neural Network Training

Most previous learning algorithm used in the training of FLNN, is the BP-learning algorithm [3, 4, 6–10]. BP-learning algorithm is based on gradient descent method used for tuning the weight parameters in FLNN in such a way that the activation functions gradually approximate the network output to the desired output signal in response to the input data. In this work we focused on FLNN with generic basis architecture. This model uses a tensor representation. In tensor model, each component of the input features multiplies the entire input features vector. This transformation enhanced the input features representation, instead of being described as a set of components $\{x_i\}$, it is described as $\{x_i, x_i x_j\}$, where $j \geq i$, or as $\{x_i, x_i x_j, x_i x_j x_k\}$, where $k \geq j \geq i$, up until it may reach the highest order terms. Let the enhanced input nodes x be represented as $x_t = x_1, x_2, \ldots x_n, x_1 x_2, x_1 x_3, \ldots x_{n-1} x_n$. The output value of the FLNN is obtained by:

$$\hat{y} = f(wx_t + b) \tag{1}$$

where $\hat{y}$ is the output while f denote the output node activation function and b is the bias. In Eq. (1), wx_t is the aggregate value which is the inner product of weight, w and x_t. The squared error E, between the target output and the actual output will be minimized as:

$$E = \frac{1}{2} \sum_{i=1}^{n} (y_i - \hat{y}_i)^2 \qquad (2)$$

where y_i is the target output and $\hat{y}_i$ is the actual output of the ith input training pattern, while n is the number of training pattern. During the training phase, the BP-learning algorithm will continue to update w and b until the maximum epoch or the convergent condition is reached.

Although BP-learning is the mostly used algorithm in the training of FLNN, the algorithm however has several limitations; It is tends to easily gets trapped in local minima especially for those non-linearly separable classification problems. The convergence speed of the BP learning also can gets is too slow even if the learning goal, a given termination error, can be achieved. Besides, the convergence behavior of the BP-learning algorithm depends on the choices of initial values of the network connection weights as well as the parameters in the algorithm such as the learning rate and momentum [4].

2.2 Modified Artificial Bee Colony

A modified Artificial Bee Colony (mABC) was proposed as a learning scheme to train the FLNN network for solving classification problems [5, 11]. The mABC is based on the original standard Artificial Bee colony Optimization algorithm (ABC) by Karaboga [12]. The original standard ABC simulates the intelligent foraging behavior of bees (employed, onlooker and scout bees) for solving multi-dimensional and multimodal optimization problem. The foraging behavior (search strategy) of the employed, onlooker from original ABC is presented as in (3) where the algorithm will randomly select only one weight to exploit.

$$v_{i,j} = x_{i,j} + \Phi_{i,j}(x_{i,j} - x_{k,j}) \qquad (3)$$

In mABC, the search strategy in (3) is modified by removing the random selection of single neighbor, j as presented in (4). This modification prompted the bees' to exploit all weights in the FLNN weights vector instead of single random weight.

$$v_i = x_i + \Phi_i(x_i - x_k) \qquad (4)$$

In the FLNN-mABC the weight, w and bias, b of the network are treated as optimization parameters with the aim to finding minimum Error, E as in (2). This optimization parameters are represented as D-dimensional solution vector, $x_{i,j}$ where $(i = 1, 2, \ldots, FS)$ and $(j = 1, 2, \ldots, D)$ and each vector i is exploited by only one employed and onlooker bee. In order to produce a candidate food source v_i from the old one x_i in memory, the mABC uses Eq. (4) where Φ_i is a random number in the range $[-1, 1]$. Since we have removed the single random weight selection of j, both employed and onlooker are forced to update all parameters in j where, $j = 1, 2, \ldots, D$.

2.3 Firefly Algorithm

Firefly algorithm (FA) was developed by Yang [13] which is inspired by the flashing behavior of fireflies. The purpose of flashing behavior is to act as signal system to attract other fireflies. In FA, the attractiveness of firefly is determined by its brightness or light intensity. The attractiveness, β of a firefly is defined in (5):

$$\beta = \beta_0 e^{-\gamma r^2} \tag{5}$$

where β_0 is the attractiveness at distance, $r = 0$. The movement of a firefly i attracted to brighter firefly j at x_i and x_j is determined by Eq. (6).

$$x_i = x_i + \beta_0 e^{-\gamma r^2}(x_j - x_i) + \alpha_1(rand - 0.5) \tag{6}$$

In (6), the second term is due to the attraction while the third term is randomization with α_1 being the randomization parameter and *rand* is random number generator uniformly distributed in $[0, 1]$. Since the scaling parameter of γ in FA shows the variation of attractiveness, the exploitation become maximal (fireflies swarm around other fireflies) as the parameter γ decrease toward zero which lead to greater local search ability [14, 15]. Hence, this work adopts the local search ability of FA to combine with mABC algorithm so that the benefit from the advantage of both methods can be utilized for training the FLNN network.

3 Modified Bee-Firefly Algorithm

The mABC was introduced as an alternative learning scheme for training the FLNN network to overcome drawbacks on classifying a multiclass classification data. Although it gives promising results, sometimes it may be leads to slow convergence as the proposed search strategy in (4) tend to encourage the bees more on exploration in the search space. In population-based optimization algorithms, both exploration and exploitation are necessary and need to be well balanced to avoid local minima trapping and thus achieved good optimization performance [16]. Since Eq. (4) is used by both employed and onlooker bee, with x_k is a random selected neighboring bees in the population, therefore, the solution search dominated by Eq. (4) is random enough for exploration but poor at exploitation.

```
 1) Cycle = 0
 2) Initialize FLNN optimization parameters, D
 3) Initialize population of scout bee with random solution xᵢ, i=1,2…FS
 4) Evaluate fitness of the population
 5) Cycle = 1:MCN
 6) Form new population (vᵢ) for employed bees
      ┌─────────────────────────────────────────────────────────────────┐
      │ a) select solution, k in the neighborhood of i, randomly          │
      │ b) For j = 1:D                                                     │
      │ c) Direct employed bee to exploit nectar value of j in population (vᵢ,ⱼ) │
      │    using Eq. (4) where j = 1,2,…,D is a dimension vector in i      │
      │ d) j= j+1;                                                         │
      │ e) exit loop when j = D;                                          │
      └─────────────────────────────────────────────────────────────────┘
 7) evaluate the new population (vᵢ)
 8) Apply greedy selection between vᵢ and xᵢ
┌────────────────────────────────────────────────────────────────────────┐
│ 9) calculate light intensity and attractiveness, I using Eq. (5)        │
│10) Produce the new solution (vᵢ)for onlooker fireflies                  │
│     a) For i=1:FS                                                        │
│        a) For j=1:FS                                                     │
│        b) If Iᵢ>Iⱼ ,Move firefly i towards j using Eq. (6)             │
│        c) Evaluate new solution (vᵢ) and update light intensity I       │
│        d) End for j                                                      │
│     b) End for i                                                         │
│11) Rank fireflies and find current best solution                        │
└────────────────────────────────────────────────────────────────────────┘
12) Determine the abandoned solution for the scout bees, if exists, and
    replace it with a new randomly produced solution xᵢ
13) Memorize the best solution
14) cycle=cycle+1
15) Stop when cycle = Maximum cycle number (MCN).
```

Fig. 1 Pseudo code of Modified Bee-Firefly (MBF) algorithm for training the FLNN network

To overcome this disadvantage, we substitute the onlooker bee search strategy with firefly algorithm. The proposed modified bee-firefly algorithm (MBF) consists of two phases. The first phase employs the exploratory search by employed bee while the other phase employs the onlooker firefly local search strategy to improve the solution quality. Figure 1 shows the pseudo code for modified Bee-Firefly (MBF) algorithm for training the FLNN network, where the box indicates the integration of Firefly algorithm made into the MABC.

4 Classification of Mammographic Masses

4.1 Mammographic Mass Data Set

Mammography is the screening process to detect breast cancer at early stage through detection of characteristic masses. Mammography is the most effective method for breast cancer screening available today [17]. However, the low positive predictive value of breast biopsy resulting from mammogram interpretation leads to approximately 70 % unnecessary biopsies with benign outcomes. Thus methods that can accurately predict breast cancer are greatly needed to reduce the high number of unnecessary breast biopsies. In this work, we use the mammographic mass data set obtained from the UCI machine learning repository [18]. The dataset consists of 6 attributes as follows:

1. BI-RADS assessment: 1 to 5 (ordinal)
2. Age: patient's age in years (integer)
3. Shape: mass shape: round = 1, oval = 2, lobular = 3, irregular = 4 (nominal)
4. Margin mass margin: circumscribed = 1, microlobulated = 2, obscured = 3, ill-defined = 4, speculated = 5 (nominal)
5. Density: mass density high = 1 iso = 2 low = 3 fat-containing = 4 (ordinal)
6. Severity: benign = 0 or malignant = 1 (binomial)

4.2 Experiment Setting

The Simulation experiments were performed on the training of FLNN with standard BP-learning (FLNN-BP), FLNN with original standard ABC (FLNN-ABC), FLNN with modified ABC (FLNN-mABC) and FLNN with our proposed modified bee-firefly algorithm (FLNN-MBF). Table 1 summarized parameters setting for the experiment.

The best accuracy is recorded from these simulations. The activation function used for the both MLP and FLNN network output is Logistic sigmoid function. The mammographic mass datasets is divided into tenfolds where, ninefolds were used as the training set and the remaining fold was used as the test set. For the sake of convenience we set our FLNN input enhancement up to second order.

4.3 Simulation Results

Ten trials were performed on each simulation where 10 independent run were performed for every single fold. The average value of tenfold cross validation of each data set are used for comparison with other learning scheme. With each learning scheme, we have observed Accuracy, Precision, Sensitivity and Specificity which can be defined as follows:

Table 1 The parameter setting for the experiment

Parameters	MLP-BP	FLNN-BP	FLNN-ABC	FLNN-mABC	FLNN-MBF
Learning rate	0.1–0.5	0.1–0.5	–	–	–
Momentum	0.1–0.9	0.1–0.9	–	–	–
Colony size	–	–	50	50	50
Maximum epoch/cycle	1000	1000	1000	1000	1000
Minimum error	0.001	0.001	0.001	0.001	0.001

Accuracy: The proportion of true results among the total number of cases examined.

$$Accurracy = \frac{True\ Positives + True\ Negatives}{True\ Positives + False\ Positives + True\ Negatives + false\ Negatives}$$

Sensitivity: True positive rate the proportion of True positive results that are correctly identified

$$Sensitivity = \frac{True\ Positives}{True\ Positives + False\ Negatives}$$

Precision: Proportion of the true positives against all the positive results (both true positives and false positives).

$$Precision = \frac{True\ Positives}{True\ Positives + False\ Positives}$$

Specificity: True negative rate that is the proportion of True negative that are correctly identified

$$Specificity = \frac{True\ Negatives}{True\ Negatives + False\ Positives}$$

The classification accuracy is presented as in Table 2 below.

The classification accuracy results presented in Table 2 shows that training the FLNN network with MBF gives better accuracy result both on training set and test set on mammographic mass dataset. The FLNN-MBF outperformed the rest of FLNN-MABC, FLNN-ABC, FLNN-BP and MLP-BP models with significant result based on accuracy rate on unseen data (test set).

Table 2 Simulation result on classification accuracy

Mammographic mass datasets	Classification accuracy (%)				
	MLP-BP	FLNN-BP	FLNN-ABC	FLNN-MABC	FLNN-MBF
Training set	50.75	59.16	82.53	83.30	**84.12**
Test set	69.82	58.50	81.46	82.68	**83.45**

Table 3 Performance of network models on unseen data

	Sensitivity (%)	Precision (%)	Specificity (%)
MLP-BP	35.42	–	65.32
FLNN-BP	72.85	49.71	45.74
FLNN-ABC	79.91	82.01	83.1
FLNN-MABC	81.02	83.41	84.41
FLNN-MBF	**82.06**	**83.93**	**84.97**

Results in Table 3 shows the performance of each network models based on sensitivity, precision and specificity measurement. From Table 3, the high value of sensitivity on FLNN-MBF indicates that FLNN-MBF identifies high true positive rate. The specificity of FLNN-MBF is also higher than Accuracy, Precision and Sensitivity indicates that true negative rate is high. Hence it is clearly indicated that the proposed training scheme could facilitated better learning for FLNN network as compared to the standard ABC and standard BP-learning algorithm.

5 Conclusion

In this work, we implemented the FLNN-MBF model for the task of classification on mammographic mass data. The experiment has demonstrated that FLNN-MBF performs the classification task quite well. The simulation result also shows that the proposed modified Bee-Firefly algorithm (MBF) can successfully train the FLNN for mammographic mass classification task with better accuracy percentage on unseen data. Thus MBF can be considered as an alternative learning scheme for training the FLNN network that can give a promising result.

References

1. Giles, C.L., Maxwell, T.: Learning, invariance, and generalization in high-order neural networks. Appl. Opt. **26**(23), 4972–4978 (1987)
2. Pao, Y.H., Takefuji, Y.: Functional-link net computing: theory, system architecture, and functionalities. Computer **25**(5), 76–79 (1992)
3. Misra, B.B., Dehuri, S.: Functional link artificial neural network for classification task in data mining. J. Comput. Sci. **3**(12), 948–955 (2007)
4. Dehuri, S., Cho, S.-B.: A comprehensive survey on functional link neural networks and an adaptive PSO–BP learning for CFLNN. Neural Comput. Appl. **19**(2), 187–205 (2010)
5. Hassim, Y.M.M., Ghazali, R.: A modified artificial bee colony optimization for functional link neural network training. In: Proceedings of the First International Conference on Advanced Data and Information Engineering (DaEng-2013), pp. 69–78. Springer, Singapore (2014)
6. Haring, S., Kok, J.: Finding functional links for neural networks by evolutionary computation. In: Van de Merckt T et al. (ed.) Proceedings of the Fifth Belgian–Dutch Conference on Machine Learning, BENELEARN1995, pp. 71–78. Brussels, Belgium (1995)
7. Haring, S., Kok, J., Van Wesel, M.: Feature selection for neural networks through functional links found by evolutionary computation. In: ILiu X et al. (ed.) Advances in Intelligent Data Analysis (IDA-97), LNCS 1280, pp. 199–210 (1997)
8. Abu-Mahfouz, I.-A.: A comparative study of three artificial neural networks for the detection and classification of gear faults. Int. J. Gen. Syst. **34**(3), 261–277 (2005)
9. Sierra, A., Macias, J.A., Corbacho, F.: Evolution of functional link networks. IEEE Trans. Evol. Comput. **5**(1), 54–65 (2001)
10. Dehuri, S., Mishra, B.B., Cho, S.-B.: Genetic feature selection for optimal functional link artificial neural network in classification. In: Proceedings of the 9th International Conference on Intelligent Data Engineering and Automated Learning. pp. 156–163. Springer, Daejeon, South Korea (2008)

11. Hassim, Y.M.M., Ghazali, R.: Optimizing functional link neural network learning using modified bee colony on multi-class classifications. In: Advanced in Computer Science and its Applications, pp. 153–159. Springer Berlin Heidelberg. (2014)

12. Karaboga, D.: An Idea Based on Honey Bee Swarm for Numerical Optimization, Erciyes University, Engineering Faculty, Computer Science Department, Kayseri/Turkiye (2005)

13. Yang, X.S.: Firefly algorithm. Eng. Optim. 221–230 (2010)

14. Guo, L., Wang, G.-G., Wang, H., Wang, D.: An effective hybrid firefly algorithm with harmony search for global numerical optimization. Sci. World J. **2013**, 9 (2013)

15. Bacanin, N., Tuba, M.: Firefly algorithm for cardinality constrained mean-variance portfolio optimization problem with entropy diversity constraint. Sci. World J. **2014**, 721521 (2014)

16. Trelea, I.C.: The particle swarm optimization algorithm: convergence analysis and parameter selection. Inf. Process. Lett. **85**(6), 317–325 (2003)

17. Chou, S.-M., Lee, T.-S., Shao, Y.E., Chen, I.F.: Mining the breast cancer pattern using artificial neural networks and multivariate adaptive regression splines. Expert Syst. Appl. **27** (1), 133–142 (2004)

18. Frank, A., Asuncion, A.: UCI Machine Learning Repository. http://archive.ics.uci.edu/ml. Irvine, CA, University of California, School of Information and Computer Science (2010)

Quantifying the Complexity of Epileptic EEG

Nadia Mammone, Jonas Duun-Henriksen, Troels Wesenberg Kjaer,
Maurizio Campolo, Fabio La Foresta and Francesco C. Morabito

Abstract In this paper, the issue of automatic epileptic seizure detection is
addressed, emphasizing how the huge amount of Electroencephalographic (EEG)
data from epileptic patients can slow down the diagnostic procedure and cause mis-
takes. The EEG of an epileptic patient can last from minutes to many hours and the
goal here is to automatically detect the seizures that occurr during the EEG recording. In other words, the goal is to automatically discriminate between the interictal
and ictal states of the brain so that the neurologist can immediately focus on the
ictal states with no need of detecting such events manually. In particular, the attention is focused on absence seizures. The goal is to develop a system that is able
to extract meaningful features from the EEG and to learn how to classify the brain
states accordingly. The complexity of the EEG is considered a key feature when dealing with an epileptic brain and two measures of complexity are here estimated and
compared in the task of interictal-ictal states discrimination: Approximate Entropy

N. Mammone (✉)
IRCCS Centro Neurolesi Bonino-Pulejo, Via Palermo c/da Casazza,
SS. 113, Messina, Italy
e-mail: nadiamammone@tiscali.it; nadia.mammone@irccsme.it

J. Duun-Henriksen
HypoSafe A/S, Diplomvej 381, 2800 Kongens Lyngby, Denmark
e-mail: jh@hyposafe.com

T.W. Kjaer
Department of Neurology, Neurophysiology Center, Roskilde University Hospital,
Koegevej 7-13, 4000 Roskilde, Denmark
e-mail: neurology@dadlnet.dk

M. Campolo · F. La Foresta · F.C. Morabito
DICEAM, Mediterranea University of Reggio Calabria, Reggio Calabria, Italy
e-mail: campolo@unirc.it

F. La Foresta
e-mail: fabio.laforesta@unirc.it

F.C. Morabito
e-mail: morabito@unirc.it

© Springer International Publishing Switzerland 2016
S. Bassis et al. (eds.), *Advances in Neural Networks*, Smart Innovation,
Systems and Technologies 54, DOI 10.1007/978-3-319-33747-0_22

(ApEn) and Permutation Entropy (PE). A Learning Vector Quantization network is then fed with ApEn and PE and trained. The ApEn+LVQ learning system provided a better sensitivity compared to the PE+LVQ one, nevertheless, it showed a smaller specificity.

Keywords Permutation entropy · Approximate entropy · Learning vector quantization · Electroencephalogram · Childhood absence epilepsy · Seizure detection

1 Introduction

Electroencephalography (EEG) is the recording of electrical activity along the scalp. EEG measures voltage fluctuations resulting from ionic current flows within the neurons of the brain. EEG is most often used to diagnose epilepsy, which causes abnormalities in the electroencephalogram. It is also used to diagnose sleep disorders, coma, encephalopathies, and brain death. Once the EEG has been recorded (which can last for minutes to days) a careful review of the entire recording is needed, in order to detect the presence of critical events and to come up with a diagnosis. Epileptic seizures are an example of such critical events that are worth to be marked and inspected for diagnostic purpose. In this paper, we will focus in particular on Childhood Absence Epilepsy (CAE) which is a common idiopathic generalized epilepsy syndrome [1, 2]. CAE affects the characteristics of the EEG and in particular its randomness. This is the reason why entropy was introduced by many researchers as a possible mathematical tool to process the EEG in order to extract diagnostic information.

Permutation Entropy (PE) attracted the attention of many researchers in this field. Introduced by Bandt and Pompe in 2002 [3] to detect dynamic complexity changes in real world time series, was proposed by Cao et al. [4] to detect dynamical changes on EEG data from epileptic patients. Li et al. [5] carried out a predictability analysis of absence seizures with PE. They tested the quality of PE as a tool to predict absence seizures in Genetic Absence Epilepsy Rats from Strasbourg (GAERS). Bruzzo et al. [6] applied PE to detect vigilance changes and preictal states from scalp EEG in epileptic patients. They evaluated the separability of amplitude distributions of PE resulting from preictal and interictal phases. Zanin et al. [7] reviewed the theoretical foundations of the PE, as well as the main applications to the understanding of biomedical systems. Nicolau et al. [8] investigated the use of PE as a feature for automated epileptic seizure detection. Ouyang et al. [9] investigated by multiscale PE the transition of brain activities towards an absence seizure. Zhu et al. [10], showed that PE is able to identify epileptic seizures in EEG and intracranial EEG (iEEG). The same authors proposed an unsupervised Multi-Scale K-means (MSK-means) algorithm to distinguish epileptic EEG signals and identify epileptic zones [11]. Mateos et al. [12] developed a method based on PE to characterize EEG from different stages in the treatment of a chronic epileptic patient. Li et al. [13] carried out statistical experiments to explore the utility of using relevance feedback on EEG data to distinguish between different activity states in human absence epilepsy. Yang et al. [14] analysed the EEG of children with electrical status epilepticus during slow-wave sleep (ESES) syndrome and of control subjects. They could observe that the entropy measures of EEG were significantly different between the ESES patients and normal control subjects.

Approximate Entropy is another powerful measure of complexity that attracted the attention of many researchers. It was introduced by Pincus [15]. Sakkalis et al. investigated three measures capable of detecting absence seizures with increased sensitivity based on different underlying assumptions. Namely, an information-based method known as Approximate Entropy (ApEn), a nonlinear alternative (Order Index), and a linear variance analysis approach. The authors highlight how ApEn parameters affect the method's behaviour, suggesting that a more detailed study and a consistent methodology of their determination should be established [16, 17].

Guo et al. proposed a methodology based on approximate entropy features derived from multiwavelet transform and combines with an artificial neural network to classify ictal states [18].

Burioka et al. evaluated the ApEn of the EEG in 8 healthy volunteers and in 10 patients with absence epilepsy. Nonlinearity was clearly detected in EEG signals from epileptic patients during seizures but not during seizure-free intervals or in EEG signals from healthy subjects [19].

In the recent years, the authors proposed a spatial-temporal analysis of epileptic EEG recordings based on measures of complexity of the EEG [20–28]. The EEG of patients affected by absence epilepsy showed a particular PE topography over the scalp, especially during seizures. This result was inferred revising movies that showed the evolution of PE mapping [21]. This result was inferred revising movies that showed the evolution of PE mapping [21]. During seizures, electrodes tended to cluster together in a particular way, according to their PE levels. The EEG of healthy subjects did not show any similar behavior nor exhibited any recurrent portrait in PE topography. In the same paper, the electrodes were clustered according to their PE levels but no link with the brain state (ictal or interictal) was investigated.

In this paper, the goal is to develop a system that is able to extract meaningful features from the EEG and to learn how to classify the brain states accordingly. The complexity of the EEG is considered a key feature and two measure of complexity are estimated and compared in the task of interictal-ictal states discrimination: Approximate Entropy and Permutation Entropy. In particular, the present work aims to investigate how to associate the particular ictal entropy (both PE and ApEn) topography with cerebral state. In this way, we will find out how the spatial distribution of the entropy extracted from the EEG matches the brain state. As this is a problem of supervised classification based on preliminar clustering, our attention was focused on Learning Vector Quantization. In particular, a LVQ network will be fed with entropy data and it will be trained to discriminate between interictal and ictal states. The paper is organized as follows: Sect. 2 will introduce the methodology, providing details of EEG recording, PE and ApEn estimation and LVQ training, Sect. 3 will illustrate the results and Sect. 4 will address the Conclusions.

2 Methodology

The procedure can be itemized as follows: (1) the n-channels EEG is recorded and stored; (2) the EEG is partitioned into m 3 s overlapping windows (with a 2 s overlap) and processed window by window; (3) Given a window under analysis, PE and ApEn are estimated channel by channel and the n values are arranged in a $nx1$ PE/ApEn vector (once the EEG is fully processed, we will have a nxm PE matrix and a nxm ApEn matrix; (4) The PE/ApEn vectors are presented as input to the LVQ network that is the trained to discriminate between *ictal* PE/ApEn vectors and *interictal* (seizure-free) PE/ApEn vectors; (5) projecting the classification of the PE/ApEn vectors back to the corresponding EEG windows, we can infer which windows were classified as *ictal* and which ones as *interictal*. In the end of this process, the EEG segments likely to be ictal will be automatically marked.

2.1 EEG Recording

Standard EEG recording from a patient, diagnosed with childhood absence epilepsy, was used for training and testing. The patient was a young girl and her age was 9 years. The EEG was acquired according to the international 10/20 system (Fig. 1) with Cadwell Easy II (Cadwell Laboratories, Inc., Kennewick, WA). All channels were filtered with a pass band of 0.53–25 Hz, and digitized at a rate of 200 Hz. In total, 8 paroxysms longer than 2 s were identified by a board-certified clinical neurophysiologist. In total, 15.8 min of electroencephalogram data were recorded. All data analysis was performed using MATLAB (The MathWorks, Inc., Natick, MA).

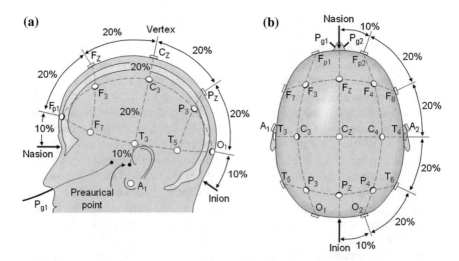

Fig. 1 The International 10–20 system seen from (A) left and (B) above the head. A = ear lobe, C = central, Pg = nasopharyngeal, P = parietal, F = frontal, Fp = frontopolar, O = occipital

2.2 Permutation Entropy and Approximate Entropy

In the estimation of PE, the EEG multi-dimensional time series is turned into a multi-dimensional symbolic time series by fragmenting the continuous signal into a sequence of symbols (motifs). By counting the number of the different types of motifs in a time-interval of fixed length being shifted along the time series, the probability of occurrence of each motif in the signal is estimated. ApEn is derived from Kolmogorov entropy. It can be used to analyze a finite length signal and describe its unpredictability or randomness. Its computation involves embedding the signal into the phase space and estimating the rate of increment in the number of phase space patterns within a predefined value r, when the embedding dimension of phase space increases from m to $m + 1$.

The n-channels EEG recording was processed by means of sliding temporal windows. The EEG samples were buffered in 3 s overlapping windows with a 2 s overlap (since the sampling rate is 200 Hz, a window includes $N = 600$ EEG samples), then PE and ApEn estimated, channel by channel and window by window. Each EEG window includes n time series, where n is the number of EEG channels. This window is stored in the computer as a nxN matrix. Within each window, a sample of PE/ApEn per channel is computed and the n values are arranged in a $nx1$ PE/ApEn vector, therefore, a nxN EEG matrix corresponds to a $nx1$ PE/ApEn vector. In this way, a compressed temporal representation of the original time series is produced.

2.3 PE and ApEn Parameter Optimization

First of all, our aim was to find out which is the optimal m-L configuration for PE and ApEn, in order to optimize their ability to discriminate between interictal EEG and ictal EEG.

Several trials were carried out under different parameter settings:

- The embedding dimension m ranged from 3 to 7;
- The time lag L ranged from 1 to 10;

Therefore, there are 50 possible parameter settings for either PE and ApEn.

Once the EEG recording is converted into PE and ApEn matrices (see Sect. 2.2), they are partitioned into interictal and ictal sub-matrices. The two sub-matrices are then averaged with respect to the rows (channels) and to the columns (Entropy sample vectors) so that an average interictal Entropy value ($avgPE(interictal)$ and $avgApEn(interictal)$) and an average ictal Entropy value ($avgPE(ictal)$ and $avgApEn(ictal)$) are estimated for the recording under analysis, in each specific parameter configuration. In order to compare the behaviour of PE and ApEn during the interictal and the ictal states, we computed two indices:

- the difference $D_{PE} = avgPE(interictal) - avgPE(ictal)$;
- the ratio $R_{PE} = avgPE(interictal)/avgPE(ictal)$.
- the difference $D_{ApEn} = avgApEn(interictal) - avgApEn(ictal)$;
- the ratio $R_{ApEn} = avgApEn(interictal)/avgApEn(ictal)$.

2.4 LVQ Classification of PE/ApEn Vectors

As we have already claimed in the Introduction, since the aim was to cluster the electrodes with respect to their PE levels and to classify the cerebral state accordingly, it was a matter of unsupervised clustering preliminar to supervised classification. Self-organizing networks can learn to detect regularities and correlations in their input and adapt their future responses to that input accordingly and LVQ is a method for training competitive layers in a supervised manner. A competitive layer automatically learns to classify input vectors according to the distance between input vectors [25]. In our experiment, once the PE/ApEn profiles were estimated from the EEG (a $nx1$ PE vector contains the n PE values extracted from the n EEG channels within the specific EEG window under analysis), such PE/ApEn vectors were presented as input to the network. Given a PE/ApEn vector, the LVQ network can be trained to cluster the electrodes in a competitive way (so that the electrodes showing relatively similar PE/ApEn values are clustered together) and then to associate the vector to a specific target (the cerebral state: ictal or interictal). We have trained a LVQ network with 3 hidden neurons because, looking at the topographical distribution of PE values [21], 3 clusters of activity could be identified: high-, average- and low-PE. The learning function adopted was *lvq1*.

3 Results

The EEG duration 15.5 min and the width of the overlapping windows was 3 s, with a 2 s overlap, therefore a $nx1$ PE vector was estimated every 1 s and, globally, 942 PE/ApEn vectors were computed.

According to the procedure described in Sect. 2.3, PE and PEr are estimated in every possible parameter configuration, each ith parameter configuration provides a pair (Di_{PE}, Ri_{PE}) and (Di_{ApEn}, Ri_{ApEn}) that can be represented as a point (see Fig. 2, (Di_{PE}, Ri_{PE}) are represented as blue circles, and (Di_{ApEn}, Ri_{ApEn}) are represented as green circles). First of all, we can observe that ApEn (green circles) provided better results because green circles lie in the right-upper part of the 2D plot whereas blue points (PE's points) lie in the left-down part, thus showing smaller D and R. In order to identify the optimal parameter settings for PE and for ApEn, the parameters that ensured the highest D and R were selected. As regards PE, the optimal configuration was $m = 4$, $L = 8$ (bold blue circle in Fig. 2). As regards ApEn, the configuration that provided the highest D was: $m = 3$, $L = 2$ (bold green circle in Fig. 2).

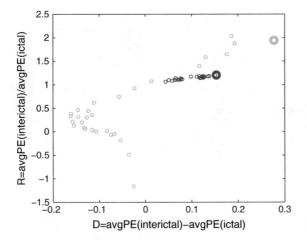

Fig. 2 D versus R in every possible parameter setting for either PE (*blue circles*) and ApEn (*green circles*). The parameter setting that ensured the highest *Di* and *Ri* for PE was $m = 4$, $L = 8$ (bold *blue circle*) whereas the parameters that ensured the best performance for ApEn with respect to either D and R was $m = 3$, $L = 2$

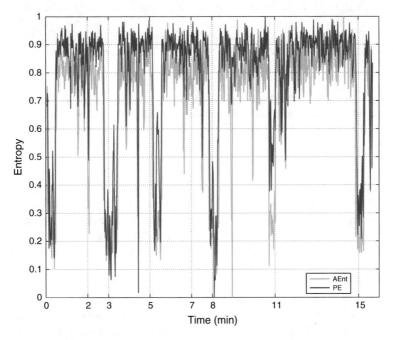

Fig. 3 PE and ApEn profiles estimated from Fp1 channel. X-axis represents time (in minutes) The seizure onset is marked with a *vertical dashed line*. PE is depicted in *blue* and ApEn is depicted in *green*

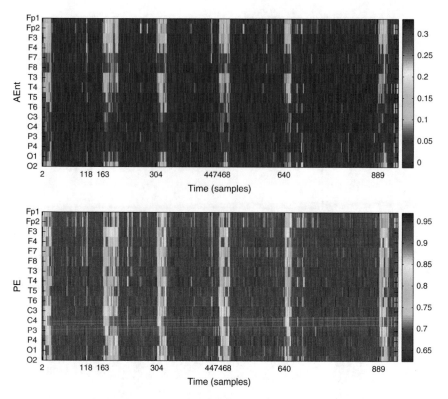

Fig. 4 PE and ApEn profiles of the entire EEG recording. X-axis represents time (in samples) and seizure onset are marked with *vertical dashed lines* whereas Y-axis represent the electrodes (16 channels). The intensity is represented with a coloration going from *red* (high values) to *blue* (low values)

To visually evaluate the behaviour of PE and ApEn, (Fig. 3) shows the PE and ApEn profiles of a single channel EEG. PE and ApEn clearly drop during seizures, as endorsing that EEG becomes more regular and its complexity appears reduced.

In order to give an overall view of the PE and ApEn profiles of the entire EEG recording, estimated for each EEG channel, Fig. 4 shows PE an ApEn trend as a 2D image.

The EEG recording included 8 seizures and, in particular, 165 EEG windows out of 942 were ictal whereas the remaining were interictal. 75 % of ictal entropy vectors and 75 % interictal entropy vectors were used for training and the remaining 25 % was used for testing. The corresponding target was the cerebral state *ictal* (=1) *interictal*(=0). The LVQ was trained with PE vectors and with ApEn vectors separately.

As regards the simulations carried out on PE profiles, PE+LVQ algorithm detected 6 out of 8 seizures, with no false prediction. Therefore a 75 % Sensitivity together with a 0 False Positive Rate (FPR) were achieved. As regards ApEn profiles, ApEn+LVQ algorithm detected 7 out of 8 seizures, with 1 false prediction. Thus a 87.5 % Sensitivity together with a 0.02 FPR were achieved. The six main paroxysms were successfully detected by both methodologies. Two short paroxysms were missed by PE+LVQ and one of them was detected by ApEn+LVQ. This means that for this patient, given a EEG window under analysis, the corresponding PE/ApEn spatial topography over the scalp can properly be described grouping the electrodes in three clusters: high-PE, average-PE, low-PE, because a good classification of the brain states was achieved starting from three clusters.

4   Conclusions

In this paper, a system that extracts measures of complexity from the EEG and learns how to classify the brain states accordingly, was developed. To this purpose, two powerful measures of complexity, Approximate Entropy (ApEn) and Permutation Entropy (PE), were compared with respect to their ability to discriminate between interictal (seizure free) and ictal (seizure) states. First of all, their parameter setting was optimized in order to extend the gap between the mean interictal entropy levels and the mean ictal entropy levels. Once the optimal parameter setting is identified, the PE and ApEn profiles are extracted from the EEG accordingly, buffering the EEG into 3 s overlapping windows and processing the EEG window by window. Each EEG window has n rows, where n is the number of channels, therefore n entropy values ($nx1$ entropy vector) is extracted from a given EEG window. A Learning Vector Quantization network is trained to associate each entropy vector to the corresponding brain state. The ApEn+LVQ learning system provided a better sensitivity compared to the PE+LVQ one, nevertheless, it showed a smaller specificity. In the future, this study will be extended to a large number of patients as well as to a large number of healthy subjects in order to assess the ability of such a system to properly classify the cerebral states and to provide information about the spatial-temporal distribution of entropy the scalp over time.

Acknowledgments This work was cofunded by the Italian Ministry of Health, project code: GR-2011-02351397.

References

1. Duun-Henriksen, J., Kjaer, T., Madsen, R., Remvig, L., Thomsen, C., Sorensen, H.: Channel selection for automatic seizure detection. Clin. Neurophysiol. **123**(1), 84–92 (2012)
2. Duun-Henriksen, J., Madsen, R., Remvig, L., Thomsen, C., Sorensen, H., Kjaer, T.: Automatic detection of childhood absence epilepsy seizures: toward a monitoring device. Pediatr. Neurol. **46**(5), 287–292 (2012)

3. Bandt, C., Pompe, B.: Permutation entropy: a natural complexity measure for time series. Phys. Rev. Lett. **88**(17) (2002)
4. Cao, Y., Tung, W.W., Gao, J.B., Protopopescu, V.A., Hively, L.M.: Detecting dynamical changes in time series using the permutation entropy. Phys. Rev. E **70**(046217), 1–7 (2004)
5. Li, X., Ouyangb, G., Richards, D.A.: Predictability analysis of absence seizures with permutation entropy. Epilepsy Res. **77**, 70–74 (2007)
6. Bruzzo, A.A., Gesierich, B., Santi, M., Tassinari, C.A., Birbaumer, N., Rubboli, G.: Permutation entropy to detect vigilance changes and preictal states from scalp EEG in epileptic patients. A preliminary study. Neurol. Sci. **29**, 3–9 (2008)
7. Zanin, M., Zunino, L., Rosso, O., Papo, D.: Permutation entropy and its main biomedical and econophysics applications: a review. Entropy **14**(8), 1553–1577 (2012)
8. Nicolaou, N., Georgiou, J.: Detection of epileptic electroencephalogram based on permutation entropy and support vector machines. Expert Syst. Appl. **39**(1), 202–209 (2012)
9. Ouyang, G., Li, J., Liu, X., Li, X.: Dynamic characteristics of absence EEG recordings with multiscale permutation entropy analysis. Epilepsy Res. **104**(3), 246–252 (2013)
10. Zhu, G., Li, Y., Wen, P., Wang, S., Xi, M.: Epileptogenic focus detection in intracranial EEG based on delay permutation entropy. **1559**, 31–36 (2013)
11. Zhu, G., Li, Y., Wen, P., Wang, S.: Classifying epileptic EEG signals with delay permutation entropy and multi-scale K-means. Adv. Exp. Med. Biol. **823**, 143–157 (2015)
12. Mateos, D., Diaz, J., Lamberti, P.: Permutation entropy applied to the characterization of the clinical evolution of epileptic patients under pharmacological treatment. Entropy **16**(11), 5668–5676 (2014)
13. Li, J., Liu, X., Ouyang, G.: Using relevance feedback to distinguish the changes in EEG during different absence seizure phases
14. Yang, Z., Wang, Y., Ouyang, G.: Adaptive neuro-fuzzy inference system for classification of background EEG signals from ESES patients and controls
15. Pincus, S.M.: Entropy as a measure of system complexity. In: Proceedings of the National Academy of Sciences of the USA, vol. 88, pp. 2297–2301 (1991)
16. Giannakakis, G., Sakkalis, V., Pediaditis, M., Farmaki, C., Vorgia, P., Tsiknakis, M.: An approach to absence epileptic seizures detection using approximate entropy. In: Conference on Proceedings of IEEE Engineering in Medicine and Biology Society, pp. 413–416. IEEE (2013)
17. Sakkalis, V., Giannakakis, G., Farmaki, C., Mousas, A., Pediaditis, M., Vorgia, P., Tsiknakis, M.: Absence seizure epilepsy detection using linear and nonlinear EEG analysis methods. In: Conference on Proceedings of IEEE Engineering in Medicine and Biology Society, pp. 6333–6336. IEEE (2013)
18. Guo, L., Rivero, D., Pazos, A.: Epileptic seizure detection using multiwavelet transform based approximate entropy and artificial neural networks
19. Burioka, N., Cornlissen, G., Maegaki, Y., Halberg, F., Kaplan, D., Miyata, M., Fukuoka, Y., Endo, M., Suyama, H., Tomita, Y., Shimizu, E.: Approximate entropy of the electroencephalogram in healthy awake subjects and absence epilepsy patients
20. Ferlazzo, E., Mammone, N., Cianci, V., Gasparini, S., Gambardella, A., Labate, A., Latella, M., Sofia, V., Elia, M., Morabito, F., Aguglia, U.: Permutation entropy of scalp EEG: a tool to investigate epilepsies: suggestions from absence epilepsies. Clin. Neurophysiol. **125**(1), 13–20 (2014)
21. Mammone, N., Labate, D., Lay-Ekuakille, A., Morabito, F.C.: Analysis of absence seizure generation using EEG spatial-temporal regularity measures. Int. J. Neural Syst. **22**(6) (2012)
22. Mammone, N., Morabito, F.C., Principe, J.C.: Visualization of the short term maximum lyapunov exponent topography in the epileptic brain. In: Proceedings of 28th IEEE EMBS Annual International Conference (EMBC 2006), pp. 4257–4260. New York City, USA (2006)
23. Mammone, N., Morabito, F.: Analysis of absence seizure EEG via permutation entropy spatio-temporal clustering. In: Proceedings of International Joint Conference on Neural Networks (IJCNN), pp. 1417–1422 (2011)

24. Mammone, N., Principe, J., Morabito, F., Shiau, D., Sackellares, J.C.: Visualization and modelling of STLmax topographic brain activity maps. J. Neurosci. Methods **189**(2), 281–294 (2010)
25. Kohonen, T.: Learning vector quantization. In: The Handbook of Brain Theory and Neural Networks, pp. 537–540. MIT Press, Cambridge, MA (1995)
26. Mammone N., Morabito F. C.: Independent Component Analysis and High-Order Statistics for Automatic Artifact Rejection. In: Proceedings of the 2005 International Joint Conference on Neural Networks. Vol. 4, pp. 2447–2452 (2005)
27. La Foresta F., Inuso G., Mammone N., Morabito F. C.: PCA-ICA for automatic identification of critical events in continuous coma-EEG monitoring. BIOMEDICAL SIGNAL PROCESSING AND CONTROL, Vol. 4, pp. 229–235 (2009)
28. Mammone N., Morabito F. C.: Analysis of absence seizure EEG via Permutation Entropy spatio-temporal clustering. In: Proceedings of the 2011 International Joint Conference on Neural Networks, pp. 1417–1422 (2011)

What Relatives and Caregivers of Children with Type 1 Diabetes Talk About: Preliminary Results from a Computerized Text Analysis of Messages Posted on the Italian Facebook Diabetes Group

Alda Troncone, Crescenzo Cascella, Antonietta Chianese
and Dario Iafusco

Abstract Although Facebook groups dedicated to diabetes are widely available on the Internet, little is known about the conversation in them. This study aims to evaluate the content of communication in an Italian Facebook group dedicated to type 1 diabetes "Mamme e diabete". Messages posted by participants from June to September 2014 were collected and assessed using computerized text analysis. Textual analysis revealed 5 dominant thematic clusters: "*food and correction*" (33.6 %), "*diabetes and life*" (18.47 %), "*hi group*" (16.87 %), "*bureaucracy*" (15.82 %) and "*needle*" (15.32 %). Findings suggested that the focus of conversations is on daily management of diabetes and all the topics related. Mothers use online discussion boards also as a place to compare and share experiences regarding diabetes and related impact on their lives and seek and give encouragement by means of the group to better face the burden associated with the illness, its duration and effects on the child and family.

Keywords Type 1 diabetes · Facebook group · Text analysis · Disease management · Social networks

A. Troncone (✉) · C. Cascella · A. Chianese
Department of Psychology, Second University of Naples, Caserta, Italy
e-mail: alda.troncone@unina2.it

C. Cascella
e-mail: crescenzo.cascella@gmail.com

A. Chianese
e-mail: antonietta.chianese1@gmail.com

D. Iafusco
Department of the Woman, of the Child and of the General and Specialized Surgery,
Second University of Naples, Naples, Italy
e-mail: dario.iafusco@unina2.it

© Springer International Publishing Switzerland 2016
S. Bassis et al. (eds.), *Advances in Neural Networks*, Smart Innovation,
Systems and Technologies 54, DOI 10.1007/978-3-319-33747-0_23

235

1   Introduction

For people suffering from chronic diseases, social networking websites are becoming a way to exchange information and provide to members a virtual place to receive support and freely talk about concerning issues [1]. On Facebook several groups originated for health reasons and there participants meet each other in order to share their experiences [2, 3].

Many individuals with type 1 diabetes and their relatives frequently use the Internet when "googling" for diabetes-related information and social networking [4] or chatting [5, 6]. Most evidences showed that the Internet use is improving the ability of patients in the process of self-care and is positively affecting the management of their disease [7, 8].

Although online chronic disease groups are widely available, to date, few research works investigated on the information that patients and/or their relatives request and share, and little is known about the nature of the virtual communities that congregate on Facebook. Studies on online chronic disease groups (e.g. Internet cancer support group, HIV/AIDS support group) have used content analysis in order to understand the typology of responses posted by participants directly affected by the illness [9, 10]. The few studies examining the messages posted by individuals suffering from type 1 diabetes at Facebook and Twitter groups [11, 12] and web-based forum [4] identified several topics and different purposes and use of virtual community.

The present study was designed to evaluate a sample of discussion on a specific Italian Facebook group dedicated to type 1 diabetes. "Mamme e diabete" (moms and diabetes) is a Facebook group where caregivers of children with type 1 diabetes (in the majority mothers) exchange messages. In order to better understand the nature of interactions taking place among participants and to quantitatively describe what Facebook users share, messages posted by users were assessed using a computerized text analysis.

This preliminary study seeks to address the following questions:

1. Which are the concepts associated to the word diabetes?
2. What are the most frequent topics posted?

2   Methods and Data

The text corpus that served as input to the computerized text analysis is a text sample of messages exchanged on the web site "Mamme e Diabete" from June to September 2014.

The participants' writings were copied and corrected for grammatical errors. Names and surnames of the writers, dates and times of the posts were deleted. In all approximately 40355 posts were copied and pasted into a text file.

Textual analysis of the participants' writing was performed by means of a content analysis software package called the T-Lab 5.1 [13]. The T-Lab operates through a distinction between context units (CU's—fixed chunks of text, which serve to divide the total body of input text, e.g. paragraphs) and lexical units (LU's —the different words themselves or categorizations of words, e.g. lemma's, semantic classes or dictionary categorizations). Through this fundamental distinction, an occurrence matrix (LUm × CUn) and a co-occurrence matrix (LUn × LUn) can be constructed, which together serve as the basis for all the T-Lab's operations. This gives each paragraph text (CU) and each word (LU) a particular *profile*. Each CU is characterized according to the profile of occurrences of each LU within it. Each LU is characterized both according to the profile of its occurrences in every CU and according to its profile of co-occurrences with all other LU's. Theories based on linguistics and statistics, which regulate the T-Lab's operations, translate in more complex transformation rules and organize the relationships between 'data' and their representation [14]. T-Lab is able to identify the semantic contents active in the text and categorize it through them. The procedure is grounded on the assumption that the semantic content of a text can be depicted in terms of how the words tend to associate with each other. The meaning of a word consists of the associations it has with the other words in the text [15]. According to the above assumption, an active semantic content corresponds to a pattern of words' co-occurrences (i.e., a set of words that tend to be present together across the text).

This research exploits two T-Lab's functions: (1) the co-occurrence analysis (that computes word's associations) and (2) the thematic analysis of the context units (thematic analyses of elementary contexts). The computation of word's associations is used for producing an association chart on each single word that gives an impression of the contextual use of that word in terms of which other words the term under consideration often co-occurs. To analyze the contextual meaning of a word, the software reviews the co-occurrence profile of such word with respect to other sample words. The relative closeness in the chart of one word to another is devised according to a parameter called the cosine coefficient [16]. Thematic analyses of elementary contexts (ECs are a type of context unit) provide a representation of the corpus' content through a small number of significant thematic clusters. The clusters divide the media discourse into a number of 'themes' that represent the content of the media discourse. Thematic clusters are internally homogeneous and externally heterogeneous.

For further information see Lancia [13] and http://www.t-lab.it.

3   Results and Discussion

An association chart of the word "diabetes" has been constructed by T-Lab in order to find out which are the associated concepts in the entire text.

Within the first 12 relevant words surrounding the word "diabetes", the word "son" (occurrence = 1018; cosine coeff. = 0.454) is the most important one

(Fig. 1). The word "years" (occurrence = 901; cosine coeff. = 0.436), "child" (occurrence = 675; cosine coeff. = 0.377), "ours" (occurrence = 508; cosine coeff. = 0.377), "understanding" (occurrence = 565; cosine coeff. = 0.334), "seeing" (occurrence = 564; cosine coeff. = 0.325), "diabetic" (occurrence = 433; cosine coeff. = 0.322), "life" (occurrence = 347; cosine coeff. = 0.313), "feeling" (occurrence = 442; cosine coeff. = 0.307), "thinking" (occurrence = 459; cosine coeff. = 0.296), "month" (occurrence = 406; cosine coeff. = 0.289) and "illness" (occurrence = 243; cosine coeff. = 0.285) are closely related. These simple associations showed that diabetes-related words, from the mothers' perspective, were mainly focused on domains regarding the burden associated with the illness and its duration, the effects of illness on the child and family, the understanding of the disease mechanisms, and its acceptance of the illness.

The output of ECs thematic analyses converged on a structure of five thematic clusters. Each cluster received a label based on the qualitative interpretation carried out by analyzing the ECs grouped in each theme and the words connected to each cluster. The five clusters were interpreted as follows: "food and correction" is characterized by issues concerning correction of insulin dose, carbohydrate counting level, food, and eating; "bureaucracy" contains issues regarding entitlement to compensation, bureaucracy procedures, and practices; "diabetes and life" focuses on emotional correlates connected to diabetes and its management; "needle" is a cluster characterized by words closely related to all the procedures associated with the insulin administration, such as injections, pricks, and insulin pumps;

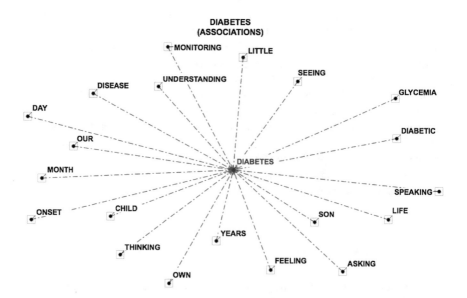

Fig. 1 "Diabetes" association chart. The words in the graph are closely related to the word "diabetes" in the text corpus. The higher a word's cosine coefficient with respect to the word "diabetes", the closer that word is positioned to the center

"hi group!" refers to contents related to the group's participation, encouragement, and emotional support. Considering the relative weight of each thematic cluster, the largest cluster was "food and correction" (33.6 %), followed by "diabetes and life" (18.47 %), "hi group" (16.87 %), "bureaucracy" (15.82 %) and "needle" (15.32 %) (Fig. 2). In Table 1, the five thematic clusters are presented listing the chi-square test values of the most important LU's in each cluster.

Our results suggest that mothers of children with type 1 diabetes use online discussion boards as a place to seek and provide mainly information about food and insulin corrections, compare and share experiences regarding diabetes and related impact on their lives, seek and give encouragement and share coping strategies to better face the bureaucratic tasks and duties, and better manage the disease's daily demands, especially on tasks linked to insulin administration (multiple daily insulin injections, blood glucose monitoring by finger pricking, subcutaneous administration of insulin with portable infusion pump, etc.).

These findings provide a useful contribution to enhance knowledge about the typology and nature of responses posted by participants in a social networking website. This knowledge is critical for the implications it has on improving the clinical management of diabetes:

(1) The knowledge and nature of responses posted by participants can help health care providers to be cognizant of the main issues that cause concerns to caregivers. This knowledge is essential for improving the doctor-patient communications.

(2) Within the larger effort to use computer technology to address problems and improve the quality of end-user life, the comprehension of needs of social networking users can help to better configure Web context (e.g. creation of

Fig. 2 Percentage of thematic clusters in the corpus

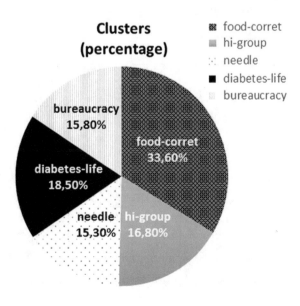

Table 1 Five clusters obtained from the thematic clustering of elementary contexts from the corpus. A high chi-square value means the corresponding lexical unit is central to a cluster

Cluster n. 1 Word	Food-correction χ^2	Cluster n. 2 Word	Diabetes-life χ^2	Cluster no. 3 Word	Hi-group χ^2	Cluster no. 4 Word	Bureaucracy χ^2	Cluster no. 5 Word	Needle χ^2
Glycaemia	1309.614	Life	918.595	Strength	3462.731	Benefit	1027.592	Insulin pump	1755.574
Carbohydrates	1187.025	Diabetes	496.939	Hug	2999.255	Allowance	1000.225	Using	1488.553
Correction	1123.757	Son	461.941	Giada	2552.92	Month	678.803	Needle	1272.155
Eating	1091.214	Our	427.762	Praying	1757.411	Inps*	655.866	Pen (injection)	1105.69
Hypoglycemia	916.296	Living	333.729	Strong	1711.66	Paying	609.71	Chemist	597.854
High	714.212	Friend	332.713	Welcome	1225.376	Law	534.378	Putting	579.181
Insulin	688.914	Thinking	229.167	Good	1005.919	Disability	495.038	Refrigerator	381.553
Glucose	618.755	Feeling	219.828	Hello	867.462	Request	481.578	Test strips	373.385
Dose	594.613	Understanding	215.519	Courage	794.185	Diabetic	423.386	I-port	353.797
Lunch	560.286	Anger	191.823	Mariagrazia	727.5	Exemption	388.467	Pricking	346.37
Night	467.425	Crying	187.985	Big kisses	682.22	Municipality	362.756	Belly	332.194
Dinner	461.825	Talking	186.773	You	615.922	Welfare ag.	352.351	Injection	321.918
Monitoring	400.864	Parent	178.19	Closeness	485.764	Years	344.696	Sensor	307.482
Values	396.984	People	164.315	God	458.526	Certificate	342.931	Finger	272.703
Pizza	357.377	Being aware	160.367	Mommy	441.454	Medical ex	342.917	Use	261.046
Milk	351.689	Illness	157.318	News	433.669	Breastfeed	337.479	Band-aid	255.399
Biscuit	345.141	Child	155.1	Heart	363.917	Commission	334.575	To change	228.647
Lantus	338.567	Worse	151.948	Giada	327.39	Carer	310.034	To buy	193.729
Lowering	319.926	Reasoning	151.228	Big	310.252	Benefit from	309.405	Diabetologist	190.217
Hyperglycemia	319.517	Moment	139.223	Compliments	289.166	Vaccine	298.031	Cartridge**	183.775

*Inps = Social security service

**Preparations of insulin coming with a pen that has a built-in cartridge made of glass

other virtual communities through online groups) in order to assist people to achieve improved health and well-being.

As emphasized by Barrera et al. [17], with additional outcome research, online groups could be items in a menu of validated resources that can be offered to the public in order to make patients with type 1 diabetes able to manage their health and cope and self manage their chronic disease.

It will be necessary to collect further data to deepen the understanding of what ingredients allow configuring different Web contexts in order to reach different aims (e.g. to diffuse information or to provide peer support) and different targets (e.g. patients, caregivers, and the general public) [18]. To this aim, future works will be devoted to analyze a larger amount of text collected from the same Facebook group. Thematic analyses of elementary contexts will be re-conducted in order to plot the evolution of themes in terms of their relative weights in each month during a 12-month's period. The thematic cluster "diabetes and life" will be used as a new corpus and re-analyzed in order to identify more specific sub themes.

References

1. Shaw, B.R., McTavish, F., Hawkins, R., Gustafson, D.H., Pingree, S.: Experiences of women with breast cancer: exchanging social support over the CHESS computer network. J. Health Commun. **5**, 135–139 (2000)
2. Farmer, A.D., Bruckner Holt, C.E., Cook, M., Hearing, S.D.: Social networking sites: a novel portal form communication. Postgrad. Med. J. **85**(1007), 455–459 (2009)
3. Hawn, C.: Take two aspirin and tweet me in the morning: how Twitter, Facebook, and other social media are reshaping health care. Health Aff. (Milwood) **28**, 361–368 (2009)
4. Ravert, R.D., Hancock, M.D., Ingersoll, G.M.: Online forum messages posted by adolescents with type 1 diabetes. Diabetes Educ. **30**(5), 827–834 (2003)
5. Iafusco, D., Ingenito, N., Prisco, F.: The chat line as a communication and educational tool in adolescents with insulin-dependent diabetes: preliminary observations. Diabetes Care **12**, 1853 (2000)
6. Iafusco, D., Galderisi, A., Nocerino, I., Cocca, A., Zuccotti, G., Prisco, F., Scaramuzza, A.: Chat line for adolescents with type 1 diabetes: a useful tool to improve coping with diabetes: a 2-year follow-up study. Diabetes Technol. Ther. J. **13**(5), 551–555 (2011)
7. De Boer, M.J., Versteegen, G.J., Van Wijhe, M.: Patients' use of the internet for pain-related medical information. Patient Educ. Couns. **68**(1), 87–97 (2007)
8. Sayers, S.L., et al.: Social support and self-care of patients with heart failure. Ann. Behav. Med. **35**(1), 70–79 (2008)
9. Klemm, P., Hurst, M., Dearholt, S.L., Trone, S.R.: Gender differences on internet cancer support groups. Comput. Nurs. **17**(2), 65–72 (1998)
10. Mo, P.K., Coulson, N.S.: Exploring the communication of social support within virtual communities: a content analysis of messages posted to an online HIV/AIDS support group. Cyberpsychology Behav. **11**(3), 371–374 (2008)
11. De la Torre-Díez, I., Díaz-Pernas, F.J., Antón-Rodríguez, M.: A content analysis of chronic diseases social groups on Facebook and Twitter. Telemedicine e-Health **18**(6), 404–408 (2012)

12. Greene, J.A., Choudhry, N.K., Kilabuk, E., Shrank, W.H.: Online social networking by patients with diabetes: a qualitative evaluation of communication with Facebook. J. Gen. Intern. Med. **26**(3), 287–292 (2011)
13. Lancia, F.: Strumenti per l'Analisi dei Testi [Methods for Texts Analysis]. Franco Angeli. Milan, Italy (2004)
14. Sengers, F., Raven, R.P., Van Venrooij, A.H.T.M.: From riches to rags: biofuels, media discourses, and resistance to sustainable energy technologies. Energy Policy **38**(9), 5013–5027 (2010)
15. Salvatore, S., Gelo, O., Gennaro, A., Manzo, S., Al Radaideh, A.: Looking at the psychotherapy process as an intersubjective dynamic of meaning-making: a case study with discourse flow analysis. J. Constructivist Psychol. **23**(3), 195–230 (2010)
16. Salton, G.: Automatic Text Processing: The Transformation, Analysis and Retrieval of Information by Computer. Addison-Wesley, Reading, Massachusetts (1989)
17. Barrera, Jr., M., et al.: Do internet-based support interventions change perceptions of social support?: an experimental trial of approaches for supporting diabetes self-management. Am. J. Community Psychol. **30**(5), 637–654 (2002)
18. Libreri, C., Graffigna, G:. Mapping online peer exchanges on diabetes. Neuropsychological Trends 125–134 (2012)

A New ICT Based Model for Wellness and Health Care

Domenico Mirarchi, Patrizia Vizza, Eugenio Vocaturo,
Pietro Hiram Guzzi and Pierangelo Veltri

Abstract Health promotion represents the principal process to empowerment the citizens. The health models focus on helping people to prevent illnesses through their behavior, and on looking at ways in which a person can pursue better health. Information and Communications Technologies (ICTs) may play an important role in the definition and use of these health models; ICT solutions allow exchange of information between health professionals, communities, producers of health research and other actors health. Starting from the current models, we propose a new model that identifies the principle actions to improve behavior and health of the citizens, highlighting specific aspects of daily life and proponing ICT solutions to allow the implementation of these actions.

Keywords ICT solutions · Health care model

1 Introduction

The main goal of the development of a health model is to improve the quality of community life members by empowering the citizens with the right knowledges and skills to take responsibility for their wellness, starting by their voluntary actions

D. Mirarchi (✉) · P. Vizza · P.H. Guzzi · P. Veltri
Department of Surgical and Medical Science, University Magna
Graecia of Catanzaro, Catanzaro, Italy
e-mail: d.mirarchi@unicz.it

P. Vizza
e-mail: vizzap@unicz.it

P.H. Guzzi
e-mail: hguzzi@unicz.it

P. Veltri
e-mail: veltri@unicz.it

E. Vocaturo
Department of Computer Science, Modelling, Electronics and Systems
Engineering (DIMES), University of Calabria, Rende, Italy
e-mail: vocaturo@deis.unical.it

© Springer International Publishing Switzerland 2016
S. Bassis et al. (eds.), *Advances in Neural Networks*, Smart Innovation,
Systems and Technologies 54, DOI 10.1007/978-3-319-33747-0_24

and behavior [1]. An important concept in the health model is the empowerment of citizen for enhances factors socially, as a social cohesion.

The model for health prevention can be divided in three general typologies of models [2]: (i) Model to give information (allows the introduction and the spread of medical definition in the field of health); (ii) Model of the *self−empowerment* (considers the individual experiences important to understand the health processes); (iii) Model of community development (involves people in the development, planning and structuring of health education). The most important models at the base of healthcare are following reported: **the Health Belief Model (HBM)** motivates people to take positive for health with the desire to avoid negative health consequences [3, 4]; **the Protection Motivation Theory (PMT)** is an education and motivational model that explains why people engage in unhealthy practices and offers suggestions for changing those behaviors [5]; **the Theory of Planned Behavior (TPB)** is based on the consideration that the control of the behavior is difficult and, to solve this problem, the perception of the control on the behavior has been introduced in the model [6]; **the Locus of Control and the Self-Efficacy** inferred in [7] states that the control is the central structure of almost all the aspects of the behavior and the human experiences; lastly, **the Social Influence Model** provides a better understanding of all the factors that most influence the health behavior in different areas of society.

The Italian National Health System has been based on the theories of Beveridge models that focusing on the social right of the health for the citizen [8].

In the Italian Health Model, the State has the power to set the essential levels of care (called *LivelliEssenziali diAssistenza*, LEAs), which must be available to all residents throughout the country, while the regions have virtually exclusive responsibility for the organization and administration of publicly financed health care, according to the study done by France et al. in [9].

The solution at the perennial crisis of the health system in according with the research done in [10] is representing by the technological innovations that in the National Health System are still in an early stage of development. The use of Information and Communication Technology (ICT) represents the basic way to support the promotion of healthy lifestyles.

We propose a new health education model that has been developed for four main areas in the daily life of a citizen. The model has been proposed introducing and developing new ICT solutions. An example of an ICT solution proposed in this paper is *Forum−Health*: forum or chat are dedicated to health problem, as alcohol, drugs and smoke; the user can discuss, participate and/or describe their experiences in an anonymous way.

The example is illustrated in Fig. 1. The proposed model is described in detail; in particular, in a first part, we describe the actions of citizens empowerment in environments of work, school and in interaction with the natural environment; in the second part, instead, the integrated ICT solutions will be presented.

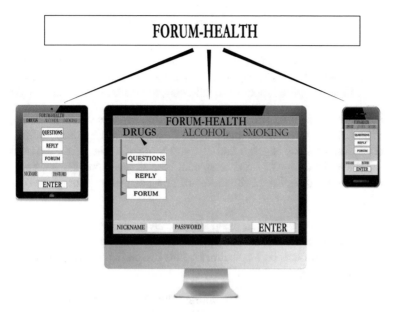

Fig. 1 Forum-Health: an example of ICT solution for citizen empowerment

2 The Model for Health Education

The main goal of the proposed model is to guarantee the promotion, maintenance and recovery of health, improving the satisfaction of life expectancy of the population. The proposed model has been developed in the fields of: (i) citizen empowerment, (ii) natural environment, (iii) workplaces and (iv) school. The actions suggested for each field have been distinguished in two actions: (i) Recommend actions (represent the indispensable and necessary solutions) and (ii) Desired actions (represent the extra solutions).

2.1 Actions for Citizen Empowerment

Through the empowerment, the citizen acquires the necessary knowledge for the control of his health and he will have a social improvement. The identification of strategies needed for a policy of the citizen empowerment is one of the principal aspect for the implementation of this model. The planned actions can be synthesized as following:

- Recommend action: Personal Wellness: the attention and the management of personal wellness is the principal element to develop a health education in the everyday life;
- Desired action: Community Wellness: people must share the information and work together to realize a collaboration system for an improvement of the health level.

2.2 Actions for Natural Environment Control

The surrounding environment is one of the main elements that affects the health of people. Sport-wellness-nature is the basic concept for the grow of environmental awareness. In the proposed model, the following actions have been introduced:

- Recommend actions: (i) simplify the access to the spaces dedicated to physical activity; (ii) build pedestrian and bicycle paths; (iii) realize decentralize set of specific structures for physical activity; (iv) reorganize the urban area with the planning and the design of residential, commercial, green and game areas.
- Desired actions: (i) optimize the means of transport; (ii) encourage walking and cycling.

Moreover, in the list of recommend actions, the environment of factors control, as air, water, land and food, must be considered; the polluting substances released in these factors cause effects on the health. For this reason in the model are introduced specific approaches:

- Air: in outdoor environment, the polluting substances are produced by human and nature actions. In indoor environment, instead, the air contamination is related to the building criteria that must encourage the air exchange.
- Water: water quality control consists of periodic verifies of the water in according to specific guidelines.
- Land: to avoid the land contamination, rules of control and correct behavior have been defined.
- Food: the control activity is made through the definition of correct recommendation about the preservation, treatment and cooking of the food.

2.3 Actions for Work Environment Control

Health promotion in workplace is an essential activity designed to support healthy behavior in the workplace and to improve health outcomes. In this direction, an appropriate system of control and prevention in workplace must be aimed to improve the knowledge of risks. Today, there are a lot of control systems but they are confined mainly to the prevention of injuries, giving a passive role to the worker. On the contrary, the actions of health promotion should aim to propose new ideas that introduce the worker as an active actor. Examples of these actions could be:

- Recommend actions: (i) the creation of courses in healthy eating, exercises and personal training; (ii) the production of an informative card containing the proposed courses that must be repeated periodically to assess the performed activity; (iii) the control of fatigue level of the worker.
- Desired action: the programming of community activities.

2.4 Actions for School Environment Control

The school is a particularly rich soil for the spread of health and wellness concept because it breeds the new citizens of tomorrow. The planning of strategies for health promotion and education in the schools aims to inform and educate the kids and youth about the risk factors, the behaviors and the more correct attitude. The principal figure for health education is the health personnel but, in school environment, he has not a good pedagogical approach and uses a specific medical language that could be not comprehensible for the youth. To overcome this problem, health personnel must be supported by teachers to give the basic wellness information in a more simple language. In this way, the student could acquire the appropriate knowledge and apply them in their life. In the proposed model, the suggested idea are described following:

- Recommend action: the introduction of a new health figure in the schools to program promotion activities and to give specific information, for example, to prevent posture, smoke and teeth problems. This new figure must be endowed with medical and pedagogical knowledge to better communicate with students.
- Desired action: the proposal of education games to promote health and wellness giving correct information in appropriate times and ways.

2.5 ICT Solutions for Citizen Empowerment

The role of ICT solutions for the development of this model is to support and facilitate the proposed actions, involving more people in every activity. For the wellness of community and citizen, the desirable ICT solutions are:

- Forum-Health: specific online sessions are created to discuss of health social problem. The users access to the forum making a registration with own birth data, opportunely privacy protected, and discuss, participate and/or recount their experiences in an anonymous way.
- Sport-Together: a platform is developed to quickly and easily organize sports team, identifying the people who want to participate and sending them the notification of date and schedule.
- Who do you care: a platform is defined to help citizen empowerment by indicate possible health facilities or professional giving information about their offered services, starting from citizen position. Moreover, an approval rating or notes by other users (citizens) can be associated with each structure.

The Sport-Together solution has been illustrated in Fig. 2.

Fig. 2 Sport-Together: an ICT solution for citizen empowerment

2.6 ICT Solutions for Natural Environment Control

For the natural environment control, the proposed ICT solutions are:

- Space-Health: the application is designed to indicate the areas in which physical activities can be performed, providing information on how to reach the nearest locations. In addition, for private spaces, entrance fees and activities permitted are listed. The user makes a registration to identify his geographical position.
- Physical-Activity: starting from existing system, the solution consists to monitor physical activity for each user, giving information about his exercises and results. Moreover, the proposed system sends an alert notification to user after an inactivity time:
- Clean-Environment: a platform is created to select the interest environment (indoor or outdoor) and analyze the environment factors (air, water, land and food) through specific sensors. These sensors measure the contamination levels and send a message with the results by acquired data.

In Fig. 3, the example of Space-Health solution has been reported.

SPACE - HEALTH

Fig. 3 Space-Health: a ICT solution for natural environment control

2.7 ICT Solutions for Work Environment Control

In work environment control, the ICT solution presented in this model concerns the development of a platform to monitor stress levels of workers, both mental and physical. The solution Health-and-Work uses wearable devices that, during few hours job, measures and monitors the relative activity, as shown in Fig. 4. In particular, the platform sends to the worker some questions for understand the level of his health in general.

For each question, the worker must answers with a score in the range of 0–10 points. In this way, the total points will be compared with a threshold reference level: if they are above threshold, an alert is sent and solutions to restore the correct conditions will be suggested. The collection of these answers is also useful to organize focused activities to protect health in the workplace.

2.8 ICT Solutions for School Environment Control

For the prevention in school age, the *Health−Game* ICT solution in Fig. 5 is a platform with playful interface that checks the health knowledge acquired in the school so that the students could compare ideas and opinions. The questions must be prepared by specialist figures with a suitable pedagogical preparation.

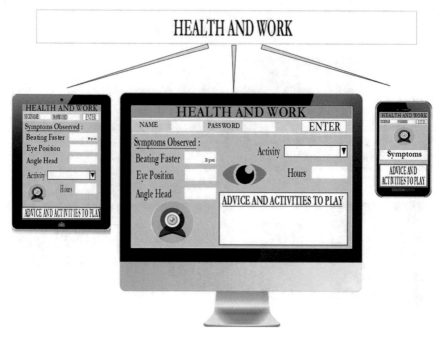

Fig. 4 Health-and-Work: the ICT solution for work environment control

Fig. 5 Health-Game: the ICT solution for school environment control

3 Discussion

For the realization of the model, some problems (cultural, technical and structural) could be represent an obstacle. One of the most significant problems is the lack or the inefficiency of specific structures; for this reason, the creation of adequate and suited infrastructures represents the basic step for the design of the proposed model. Another important aspect that must be considered in the developed of this type of model regards the empowerment of the citizen; the ability of citizens to acquire specific skills in the field of wellness and health is linked to their social and economic conditions. To overcome this problem, a solution could be to provide for a supervision of authoritative sites accessible by citizens; these sites have a list of specific instructions for safe web surfing.

Another idea regards the integration of a specialized figure that can support citizen in the use of new technologies for clarification and support.

4 Conclusions

A new model for the development of wellness and health concepts has been proposed reporting the actions that must be performed and, consequently, the proposed ICT solutions. The proposed project develops these actions in the principal areas of citizen empowerment, natural environment, workplace and schools, which are the areas where is essential that the health concepts are correctly acquired and applied. We believe that the acquisition and the correct application of the basic empowerment concepts reduce the workload and even waste of the National Health System. The use of ICT enables citizens to learn the proper rules of health in a smoother and simpler way. This makes citizens more prepared and careful to the choices that affect their future care.

References

1. Green, E.C., Murphy, E.: Health belief model, The Wiley Blackwell Encyclopedia of Health. Behavior, and Society, Illness (2002)
2. Taylor, S.E.: Health Psychology, McGraw-Hill (1999)
3. Rosenstock, I.M.: Historical origins of the health belief model, Health Education Monographs, pp. 328–335 (1974)
4. Becker, M.H., Maiman, L.A.: Sociobehavioral determinants of compliance with health and medical care recommendations. Med. Care 13(1), 10–24 (1975)
5. Rogers, R.V., Ronald, W., Cacioppo, J.T., Petty, R.: Cognitive and physiological processes in fear appeals and attitude change: A revised theory of protection motivation, Guil for Press, Social Psychophysiology: A sourcebook, pp. 153–177 (1983)
6. Ajzen, I.: Attitudes, Personalty and Behavior. Dorsey Press, Chicago, IL (1988)
7. Skinner, E.A.: Perceived Control, Motivation, & Coping. Sage Publications, vol. 8, (1995)

8. France, G., Taroni, F.: The evolution of health policy making in Italy. J. Health Polit. Policy Law **30**(1) (2005)
9. France, G.: I livelli essenziali di assistenza: un caso italiano di policy innovation. In I Servizi Sanitari in Italia, Il Mulino, Bologna (2003)
10. Santarelli, C., Di Carlo, E.: E-health in Italia: un modello di valutazione. Mondo Digitale AICA, Milano (2013)

The Implicit and Explicit Attitudes Predict Electoral Decision-Making in 2008 Italian Political Campaign

Antonio Pace, Angiola Di Conza, Antonietta Nocerino,
Augusto Gnisci, Anna M. Raucci and Ida Sergi

Abstract This contribution applies the social judgment based on warmth and competence in the Italian political context (2008 electoral campaign; N = 625), by integrating insights coming from the literature on the Big Two and on the dual cognition theories. It aims at identifying how the evaluations toward leaders and parties (in terms of assigned warmth and competence, and ingroup/outgroup perception) predict the electoral decision-making. Furthermore, we test the hypothesis that the evaluations of the opposite party (group) and leader (group member) differ. The results are discussed considering the relationship between the Big Two and the following social evaluations and behaviors, in terms of intergroup competition and dual cognitive processes, the peculiar Italian political context and its increasing level of political personalization.

Keywords Warmth and competence · Implicit attitude · Decision-Making · Political elections

A. Pace (✉) · A. Di Conza · A. Nocerino · A. Gnisci · A.M. Raucci · I. Sergi
Department of Psychology, Second University of Naples, Caserta, Italy
e-mail: antoniopace84@gmail.com

A. Di Conza
e-mail: angiola.diconza@gmail.com

A. Nocerino
e-mail: a.nocerino_psyche@libero.it

A. Gnisci
e-mail: augusto.gnisci@unina2.it

A.M. Raucci
e-mail: anna.r@hotmail.it

I. Sergi
e-mail: ida.sergi@unina2.it

© Springer International Publishing Switzerland 2016   253
S. Bassis et al. (eds.), *Advances in Neural Networks*, Smart Innovation,
Systems and Technologies 54, DOI 10.1007/978-3-319-33747-0_25

1 Introduction

Recent developments in the social psychology research have focused on the relevance of two basic dimensions in the formulation of the social judgment regarding the individuals and the social group. These two dimensions have been labeled as Warmth and Competence [1]. The former generally refers to friendliness, helpfulness, sincerity, trustworthiness, and good intention, whereas the latter concerns intelligence, skill, creativity, efficacy, and confidence. This dichotomy has been applied to the studies about attitudes, stereotypes, social judgment and decision-making processes, and proved useful both for intergroup and interpersonal evaluations [2].

Cuddy and colleagues [2] maintained that Warmth and Competence are central in social judgments; they both depend on social structural variables, namely perceived competition and status, respectively; and they have emotional and behavioral consequences, as pointed out by the Stereotype Content Model [3] and by the BIAS map [2]. According to the Stereotype Content Model, combining the evaluations on the two dimensions, it is possible to identify the evaluated group or individual as an ingroup (part of the group the individual identifies himself with and/or he feels to belong to) or as an outgroup (part of the group the individual feels not to fit in). The evaluation of an individual's ingroup is characterized by high levels on both the dimensions, whereas the evaluation of an outgroup can be univalent (with low levels for both the dimensions) or ambivalent (when it is characterized by a high level on a single dimension).

Other authors stated the warmth primacy over competence [4], indicating that the warmth dimension account for more variance in trait ratings [5], is evaluated more quickly than the competence one [6, 7], and constrains judgment on the other dimension [8]. In fact, according to some authors [1], meeting anyone for the first time, immediately an individual evaluates if his/her intentions (warmth) are positive or negative and, then, if he/she is ready to act according to them (competence) [9]. According to Wojciszke and Abele [10], however, the competence dimension can assume a greater relevance when the personal wellbeing is perceived as depending on the abilities held by the evaluated target.

The analysis of the two fundamental dimensions has also been applied to the study of political evaluations and decisions [11, 12]. Although politicians are normatively evaluated according to their perceived competence [12], warmth still have its relevance, being indicative of the perceived ability of the politician to apply politics favoring its electorate.

A politician's competence is often inferred starting from facial appearance and this inference can influence the electoral choice [13]. The rapid automatic inferences can influence the subsequent processing of new information about the candidates that, in turn, dilute the effect of initial impressions. This is predicted by the dual cognition systems [14].

Dual cognition systems account for different kinds of information processing, resting on the observation that people can process information in at least two different ways. On one hand, they can rely on a quickly, not accurate and non-effort requiring approach (automatic or impulsive) and, on the other hand, they can rely on a more accurate, time- and effort-based processing (deliberative or reflective). These two processing ways can lead to qualitatively distinct outcomes, differently accurate, but equally effective [14, 15]. A similar distinction between "automatic" and "delibera-tive" has been applied to the constructs of attitude, evaluation and judgment, under the labels "implicit" and "explicit" [15, 16]. Although both implicit and explicit attitudes regard the association of an object with its evaluation, each one has its own peculiar characteristics: explicit attitudes concern evaluations of which individuals are aware, while implicit attitudes refer to evaluative elements whose origin, activation and/or influence on behavior are unknown to the individual [17, 18].

The research on implicit and explicit attitudes was also applied to political evaluations in order to investigate the relations between them [19, 20], that proved extremely variable ($0.37 < r < 0.78$) [21], and to point out if implicit (and explicit) attitudes could predict political decision-making [21, 22].

Integrating insights coming from the literature on the Big Two, the Stereotype Content Model, intergroup evaluations, and implicit attitudes, our study represents an application of these models and concepts in a real evaluative context. The study was conducted during the 2008 Italian political campaign, in a specific national scenario, characterized by the fall down of the previous government led by the wide center-left coalition, in office only for two years, and by several transformations of the political parties' alliances for both the left and right wings and by the change of the left-wing main representative [23]. Silvio Berlusconi was the leader of the new center-right main party, Popolo della Libertà (PdL), formed mainly by the union of two former main right parties, Forza Italia and Alleanza Nazionale; Walter Veltroni was the leader of the new center-left main party, Partito Democratico (PD), resulting mainly by the union of the former main two left parties, Democratici di Sinistra and Margherita.

1. According to ingroup-outgroup theory [3, 24], we hypothesize that, during the 2008 Italian pre-electoral campaign, the electoral choice expresses one's own perception of the leaders and of the political parties he/she sympathizes with as an ingroup and therefore: (a) the center-right electors will judge both Silvio Berlusconi and Popolo della Libertà (PdL) warm and competent; (b) the center-left electors both Walter Veltroni and Partito Democratico (PD) warm and competent.

2. For similar reasons, we hypothesize that the electoral choice expresses one's own perception of the opposite leaders and parties he/she does not sympathize with as an outgroup, which can be evaluated in a univalent or an ambivalent way. In particular, we argue that the opposite party is judged as univalent (namely low warm and low competent), whereas being the leader an individual, the most important to-be-evaluated element in the political arena (according to a growing level of personalization in politics) [12], he/she is characterized as competent but not warm. A voter can consider a leader able to act according to his/her own intention (competent), but against the voters' interests (low warm, since he/she belongs to the opposite political wing). For this we hypothesize

that: (a) center-right electors will judge PD low warm and low competent (univalent), whereas Veltroni low warm but high competent (ambivalent); (b) center-left evaluate PdL low warm and low competent (univalent), whereas Berlusconi low warm but high competent (ambivalent).

3. Moreover, starting from the idea that the sharing of political issues, interests and ideals leads the individual to favor one wing over another, and considering warmth linked to the intentions attributed to a group or a person to act in favor or against voters' interests [2, 5, 25], we suggest that the level of granted warmth is, any case, lower than competence. Applying this to the 2008 pre-electoral campaign, we hypothesize that: (a) center-right electors will assign Veltroni and PD warmth levels lower than competence levels; (b) center-left electors assign Berlusconi and PdL warmth levels lower than competence ones.

4. Finally, in line with the abovementioned studies about the relation between explicit and implicit attitudes in the political field, and according to the statement that past negative or positive evaluations of leaders and political parties can be stored in the long term memory as automatic affective reactions [26, 27], and considering the evaluation of warmth as much characterized by an affective connotation and by an immediate reaction [4], we suggest that, in general, the relation between warmth and implicit attitude is higher than the relation between competence and implicit attitude referred to leaders and parties as evaluative objects. More specifically, we hypothesize that this difference can depend on the object, namely it will be less relevant for the leaders, whose competence is a paramount evaluative dimension, which can be inferred after 1 s exposure to the politician's face [14].

2 Method

2.1 Participants

One thousand thirty-seven electors (M = 456, F = 581) from Campania (Italy), aged 18–72, took part in the present study (M = 26.00, SD = 9.00). In our analyses, we refer exclusively to the evaluations of PdL and PD electors, that were at times the main center-right (N = 225) and center-left parties (N = 400). Therefore, the final sample consists of 625 Italian citizens (M = 249, F = 376) aged from 18 to 69 (M = 25.00, SD = 8.00).

2.2 Measures

The two dimensions of explicit evaluations toward leaders (Berlusconi and Veltroni) and their political parties (PdL and PD) were assessed through the Warmth and Competence Rating Scale (WCS) [3, 24]. This scale consists of twelve items: six referred to competence (competent, confident, capable, efficient, intelligent, skillful) and six referred to warmth (friendly, well-intentioned, trustworthy, warm, good-natured, sincere). The electors rated each item on a 7-point scale (from $1 = $ "not at all" to $7 = $ "extremely"). The use of the WCS allows detecting if and how much center-left and center-right electors perceive as warm (W) and competent (C): the leaders, Veltroni ($\alpha_C = 0.91$, $\alpha_W = 0.92$) and Berlusconi ($\alpha_C = 0.86$, $\alpha_W = 0.91$); the parties, PD ($\alpha_C = 0.93$, $\alpha_W = 0.94$) and PdL ($\alpha_C = 0.90$, $\alpha_W = 0.94$). The level of warmth and competence assigned to each object was obtained averaging the answers given to each pool of items [3, 24].

Implicit attitudes toward leaders ($\alpha_L = 0.80$) and parties ($\alpha_P = 0.78$) were assessed by two IAT procedures (Implicit Association Test) [28]. The test required a rapid choice of photos or words, representing two conceptual categories (five pictures for each leader and each political party) and two attribute categories (five positive words for the pleasant category: joy, love, pleasure, happiness, smile; and five negative words for the unpleasant category: terrible, horrible, unpleasant, bad, awful). This procedure provided two "D" indexes [28] of the relative preference for Berlusconi with respect to Veltroni, and for PdL with respect to PD.

2.3 Procedure

Implicit and explicit attitudes procedures were administered by a computer during the last month preceding the elections. After the Election Day, participants were rung up and invited to indicate their actually expressed political choice. The participation to the study was voluntary and anonymity was guaranteed.

Three types of counterbalancing were performed, concerning the administration of explicit and implicit attitude measures, the implicit measures of attitudes toward leaders and parties, and the concept-attribute association within each IAT (in order to contrast the prior task effect) [28].

2.4 Data Analysis

First of all, we averaged the warmth and competence perceived by PdL electors and by PD electors to each considered political object (Berlusconi, Veltroni, PdL and PD).

To test the hypotheses 1 and 2, following the procedure of Cuddy, Fiske and Glick [29], the means were compared with the middle point of the scale ($4 = $ "moderately"), considering it as the touchstone to establish if each level of explicit attitude was high or low (higher or lower than the middle point of the scale). Each mean was compared with the middle point of the scale by a one-sample t-test. Moreover, the evaluations associated to each object by the two electorates (center-right and center-left electors) were compared via eight Analyses of Variance

(ANOVAs), having the voting behavior as independent variable and the evaluations on the warmth and competence dimensions attributed by the participants to each political object (Berlusconi, Veltroni, PdL and PD) as dependent variables.

To test the hypothesis 3, four ANOVAs with repeated measures (GLM method) were conducted, each one having as independent variable the vote, indicative of a preference toward a political party and its leader, and as dependent variable the warmth and competence scores associated to each evaluated political object (e.g., Veltroni's warmth and competence; or PD's warmth and competence, etc.).

To test the hypothesis 4, we analyzed the relationship between explicit and implicit attitudes toward political leaders and parties, carrying out two correlation analyses using Pearson's r coefficient. The results were subjected to a z-test to verify if the differences were significant.

3   Results

3.1   Ingroup Evaluation

Referring to the ingroup evaluation (hypothesis 1), from the results, shown in Table 1, we noted that the means of warmth and competence (a) attributed to Berlusconi and to PdL by the center-right electors, and (b) the ones attributed to Veltroni and to PD by the center-left electors were all significantly higher than the middle point of the scale (for all t-tests, $p < 0.01$ or $p < 0.001$).

3.2   Outgroup Evaluation: Leaders and Parties

As shown in Table 2, referring to the outgroup evaluation (hypothesis 2), we obtained that (a) the center-right electors judged Veltroni competent, but low warm, and PD low competent and low warm, whereas (b) the center-left voters judged Berlusconi competent, but low warm, and PdL low competent and low warm. Moreover, as displayed in Table 3, the conducted ANOVAs resulted significant,

Table 1 Means of competence and warmth toward their own party and leader for center-right (N = 225) and center-left (N = 400) electors (cut-off point = 4)

		Competence		Warmth	
		M	SD	M	SD
Center-right electors	Berlusconi	5.9	1.0	4.8	1.2
	PdL	5.2	1.2	4.8	1.2
Center-left electors	Veltroni	5.3	1.0	5.0	1.1
	PD	4.9	1.1	4.8	1.1

Table 2 Means of competence and warmth toward the opposite leader and party for center-right (N = 225) and center-left (N = 400) electors (cut-off point = 4)

		Competence		Warmth	
		M	SD	M	SD
Center-right electors	Veltroni	4.2	1.2	3.7	1.2
	PD	3.7	1.2	3.4	1.1
Center-left electors	Berlusconi	4.4	1.3	2.6	1.2
	PdL	3.7	1.4	2.7	1.3

Table 3 ANOVAs about levels of competence and warmth toward each leader and each party (N = 625)

	Competence			Warmth		
	F	p	η^2	F	p	η^2
Berlusconi	233.94	<0.001	0.27	464.79	<0.001	0.43
Veltroni	138.80	<0.001	0.18	182.88	<0.001	0.23
PdL	186.44	<0.001	0.23	393.08	<0.001	0.39
PD	160.34	<0.001	0.20	196.86	<0.001	0.24

Note For all the analyses: gdl = 1, 623

indicating that the levels of warmth and competence attributed to the opposite leader differ from the ones attributed to his party.

3.3 Outgroup Evaluation: Assigned Warmth Is Lower Than Assigned Competence

As expected, for each political object judged as outgroup by the electors, the levels of warmth are lower than the levels of competence (hypothesis 3): (a) the center-right electors judged Veltroni significantly less warm than competent ($F(1, 224) = 89.54$, $p < 0.001$, $\eta^2 = 0.29$), and (b) the center-left electors evaluated Berlusconi significantly less warm than competent ($F(1, 399) = 977.65, p < 0.001$, $\eta^2 = 0.71$). Finally, warmth is lower than competence also when referred to (a) the evaluation of PD ($F(1, 224) = 46.08$, $p < 0.001$, $\eta^2 = 0.17$), for the center-right electors, and (b) of PdL ($F(1, 399) = 382.18$, $p < 0.001$, $\eta^2 = 0.49$), for the center-left electors.

3.4 Correlation Between Implicit Attitudes and the Big Two

Analyzing the relations between the implicit attitude and the two explicit dimensions of warmth and competence toward leaders and political parties (hypothesis 4), our findings show that, for the leaders' evaluation, the correlation between warmth and implicit attitude is $r = 0.48$ ($p < 0.001$), and the correlation between competence and implicit attitude is $r = 0.43$ ($p < 0.001$). Referring to parties' evaluations, the correlation between warmth and implicit attitude is $r = 0.58$ ($p < 0.001$), and the correlation between competence and implicit attitude is $r = 0.51$ ($p < 0.001$).

Using the z-test to check the links concerning leaders' and parties' evaluations, the results show that the correlations between warmth and implicit attitudes are significantly higher than the correlations between competence and implicit attitudes for parties' evaluations ($z = 1.73$, $p < 0.04$), but not for leaders' ones ($z = 1.18$, $p = 0.12$).

4 Discussion and Conclusions

The two basic dimensions of social judgment, warmth and competence, are relevant for the decision process mechanisms leading to the political choice to vote for one leader/party rather than another. The present study shows that the electors differently rely on these two dimensions, when asked to judge different political targets (the leaders and the parties), belonging to the voted political wing or to the opposite one.

The results indicate that, as expected for hypothesis 1, during the 2008 pre-electoral campaign, both the electors of our sample voted for the leaders and the parties they perceived as ingroup (evaluating them as warm and competent). In fact, both the center-right and the center-left electors evaluated the respective political ingroups (the party and its leader) warm and competent. In the political field, a positive evaluation on both the dimensions can be predictive of a tendency to support the actions of members in the group an individual feels to belong, and this disposition can induce co-operational behaviors, expressed by the assignment of one's own vote.

As expected for hypothesis 2, the evaluation as low warm and competent exclusively characterized the opposite political party (PD and PdL), whereas the leaders (Veltroni and Berlusconi) were ambivalently judged as high competent and low warm. Therefore, in the present study, the electors seem to consider the main opposition party not only incompetent, and then unable to rule, but also lacking in the "affective" characteristics, necessary for a completely positive judgment. Conversely, the electors evaluate the leaders of opposite party as competent. Indeed, they represent their parties nationally, so that some level of abilities is recognized to each one also by the opposite electorate. On the other side, the opposite leaders are judged as low warm, because their intentions are recognized as negative, and not matching the exigencies of the opposite electorate.

Referring to the hypothesis 3, assuming that perception of warmth and competence have different intensity degrees depending on the evaluated political object (leader/party), it resulted that in every case the evaluation of competence is higher than warmth. This mixed evaluation shows that the elector perceives a threat toward him/herself or his/her group [25], being the warmth predictive of positive or negative intentions by the evaluated object toward the evaluator.

Finally, the hypothesis 4, suggesting a higher correlation between implicit attitudes and perceived warmth compared with the correlation between implicit attitudes and perceived competence, is confirmed only for parties' evaluations. As abovementioned, previous studies showed that competence is the most relevant dimension for politicians' evaluations [12] and it is inferred after 1 s exposure to the politician's image [13]. Moreover, Wojciszke and Abele [10] proposed that, even if warmth is generally the most important and primary evaluated dimension, competence becomes more relevant when the abilities held by the to-be-evaluated object have a relevant impact on the evaluator's life. We argue that warmth and competence are used by the electors at a first stage of the process of leader's evaluation and, in line with models of cognitive emotions, as the dual cognition theories and the hot cognition theory, both the dimensions tend to be consolidated and automated, characterizing the automatic affective reactions (corresponding to the implicit attitude) toward political representatives [26]. For the evaluation of the party, instead, the attribution of competence proves less linked to the development of an early impression. Some authors [4, 10] hypothesized that the evaluation of the warmth dimension precedes the evaluation of the competence; the presented results confirm this hypothesis for parties evaluations, which probably follow a more complex process based on reasoning and information processing.

We can conclude that warmth and competence are based on more affective and more cognitive evaluations, respectively. However, the evaluative process through which they are attributed to a social object can be either automatic or reflective, depending on the characteristics of the object. Even if these achievements are limited for a single campaign in a specific country, they highlight the importance of considering all the aspects that influence the final decision of electorate, and suggesting the need to propose a psychological model for best predicting the final behavior (i.e., the electoral choice).

Future research should deepen these characteristics, starting from the relevance held by each dimension for the to-be-evaluated object, in order to provide a deeper understanding of the joint role of automatic and reflective processing and how this integrates with the universal recourse to the Big Two to produce political evaluations and to influence the electoral behavior. It would be also interesting comparing the results of this study with the emerging tri-factorial model of psychosocial evaluations (i.e., morality, sociability, and competence) [30].

References

1. Abele, A.E., Cuddy, A.J.C., Judd, C.M., Yzerbyt, V.Y.: Fundamental dimensions of social judgment: a view from different perspectives. Eur. J. Soc. Psychol. **38**, 1063–1065 (2008)
2. Cuddy, A.J.C., Fiske, S.T., Glick, P.: Warmth and competence as universal dimensions of social perception: the stereotype content model and the BIAS map. Adv. Exp. Soc. Psychol. **40**, 62–149 (2008)

3. Fiske, S.T., Cuddy, A.J.C., Glick, P., Xu, J.: A Model of (often mixed) stereotype content: competence and warmth respectively follow from perceived status and competition. J. Pers. Soc. Psychol. **82**, 878–902 (2002)
4. Abele, A.E., Bruckmüller, S.: The bigger one of the "big two"? Preferential processing of communal information. J. Exp. Soc. Psychol. **47**, 935–948 (2011)
5. Abele, A.E., Wojciszke, B.: Agency and communion from the perspective of self versus others. J. Pers. Soc. Psychol. **85**, 768–776 (2007)
6. Willis, J., Todorov, A.: First impressions: making up your mind after a 100-ms exposure to a face. Psychol. Sci. **17**, 592–598 (2006)
7. Wojciszke, B.: Affective concomitants of information on morality and competence. Eur. Psychol. **10**, 60–70 (2005)
8. Yzerbyt, V., Kervyn, N.O., Judd, C.M.: Compensation versus halo effect: the unique relations between the fundamental dimensions of social judgment. Pers. Soc. Psychol. Bull. **34**, 1110–1123 (2008)
9. Brambilla, M., Rusconi, P., Sacchi, S., Cherubini, P.: Looking for honesty: the primary role of morality (vs. sociability and competence) in information gathering. Eur. J. Soc. Psychol. **41**, 135–143 (2011)
10. Wojciszke, B., Abele, A.E.: The primacy of communion over agency and its reversals in evaluations. Eur. J. Soc. Psychol. **38**, 1139–1147 (2008)
11. Castelli, L., Carraro, L., Ghitti, C., Pastore, M.: The effects of perceived competence and sociability on electoral outcomes. J. Exp. Soc. Psychol. **45**, 1152–1155 (2009)
12. Funk, C.: The impact of scandal on candidate evaluations: an experimental test of the role of candidate traits. Polit. Behav. **18**, 1–24 (1996)
13. Todorov, A., Mandisodza, A.N., Goren, A., Hall, C.C.: Inferences of competence from faces predict elections outcomes. Science **308**, 1623–1626 (2005)
14. Chaiken, S., Trope, Y.: Dual-Process Theories in Social Psychology. Guilford Press, New York (1999)
15. Strack, F., Deutsch, R.: Reflective and impulsive determinants of social behaviour. Pers. Soc. Psychol. Rev. **8**, 220–247 (2004)
16. Greenwald, A.G., Banaji, M.R.: Implicit social cognition: attitudes, self-esteem, and stereotypes. Psychol. Rev. **102**, 4–27 (1995)
17. Banaji, M.R.: Implicit attitudes can be measured. In: Roediger III, H.L., Nairne, J.S., Neath, I., Surprenant, A. (eds.) The Nature of Remembering: Essays in Remembering Robert G. Crowder, pp. 117–150. American Psychological Association, Washington (2001)
18. Gawronski, B., Hofmann, W., Wilbur, C.J.: Are implicit attitudes unconscious? Conscious. Cogn. **15**, 485–499 (2006)
19. De Houwer, J., De Bruycker, E.: The implicit association test outperforms the extrinsic affective Simon task as an implicit measure of inter-individual differences in attitudes. Br. J. Soc. Psychol. **46**, 401–421 (2007)
20. Gnisci, A., Di Conza, A., Senese, V.P., Perugini, M.: Negativismo Politico, Voto e Atteggiamento. Uno Studio su un Campione di Studenti Universitari alla Vigilia delle Elezioni Europee del 12–13 Giugno 2004. Rassegna di Psicologia **26**, 57–82 (2009)
21. Di Conza, A., Gnisci, A., Perugini, M., Senese, V.P.: Atteggiamento Implicito ed Esplicito e Comportamenti di Voto. Le Europee del 2004 in Italia e le Politiche del 2005 in Inghilterra. Psicologia Sociale **2**, 301–329 (2010)
22. Friese, M., Bluemke, M., Wänke, M.: Predicting voting behavior with implicit attitude measures: the 2002 German parliamentary election. Exp. Psychol. **54**, 247–255 (2007)
23. Ignazi, P.: Italy. Eur. J. Polit. Res. **47**, 1025–1027 (2008)
24. Fiske, S.T., Cuddy, A.J.C., Glick, P.: Universal dimensions of social perception: warmth and competence. Trends Cogn. Sci. **11**, 77–83 (2007)
25. Abdollahi, A., Fiske, S.: Social judgment: warmth and competence are universal dimensions. Mind Mag. **8** (2008)
26. Abelson, R.: Computer Simulation of 'Hot' Cognition. In: Tomkins, S.S., Messick, S. (eds.) Computer Simulation of Personality, pp. 277–298. Wiley, New York (1963)

27. Taber, C., Lodge, M., Glathar, J.: The Motivated Construction of Political Judgments. In: Kuklinski, J.H. (ed.) Citizens and Politics: Perspectives from Political Psychology, pp. 197–226. Cambridge University Press, London (2001)

28. Greenwald, A.G., Nosek, B.A., Banaji, M.R.: Understanding and using the implicit association test: I. An improved scoring algorithm. J. Pers. Soc. Psychol. **85**, 197–216 (2003)

29. Cuddy, A.J.C., Fiske, S.T., Glick, P.: The BIAS map: behaviors from intergroup affect and stereotypes. J. Pers. Soc. Psychol. **92**, 631–648 (2007)

30. Leach, C.W., Ellemers, N., Barreto, M.: Group virtue: the importance of morality (vs. competence and sociability) in the positive evaluation of in-groups. J. Pers. Soc. Psychol. **93**, 234–249 (2007)

Internet Dependence in High School Student

Ida Sergi, Antonio Pace, Augusto Gnisci, Mariella Sarno
and Anna M. Raucci

Abstract With increased accessibility on Internet, Internet addiction is becoming a serious problem, especially among adolescents. They usually do these activities for entertainment, excitement, challenge seeking, or emotional coping. However, both excessive play of games and frequent use of SNSs may lead to certain negative outcomes. Factors that may influence the tendency for adolescents to become dependent are various. Most existing studies on Internet addiction focus primarily on internal and individual factors that may predispose individuals to problematic Internet use, as personality traits. The main aim of this study was to assessing the risk of Internet addiction in a sample of Italian young adults of High School by looking background factors, attitudes, behavioral habits and personality traits. The results indicate that different dimensions of internet addiction can be predicted by a combination of different users' characteristics. In particular, Agreeableness-Conscientiousness and Extroversion were inversely related to Internet addiction. The results were discussed in terms of prevention, the identification of specific variables associated with Internet addiction allows for targeting individuals who appear to be at risk for developing Internet addiction.

Keywords Internet dependence · High school · Personality · IAT · HEXACO

I. Sergi (✉) · A. Pace · A. Gnisci · M. Sarno · A.M. Raucci
Department of Psychology, Second University of Naples, Caserta, Italy
e-mail: ida.sergi@unina2.it

A. Pace
e-mail: antoniopace84@gmail.com

A. Gnisci
e-mail: augusto.gnisci@unina2.it

M. Sarno
e-mail: mama2428@hotmail.it

A.M. Raucci
e-mail: anna.r@hotmail.it

© Springer International Publishing Switzerland 2016
S. Bassis et al. (eds.), *Advances in Neural Networks*, Smart Innovation, Systems and Technologies 54, DOI 10.1007/978-3-319-33747-0_26

1   Introduction

The Internet has great value for modern society. It has changed the way we work and the way we play, creating new possibilities for self-expression and communication. However, it also enables, and possibly encourages, compulsive behavior. This tendency has been labeled in several ways, including Internet addiction, Internet addiction disorder, Internet dependence and problematic, pathological, excessive or compulsive Internet use [1, 2]. Although there are different meanings of "Internet Addiction", the terms refer to the concept that a person cannot control his/her use of Internet, leading noticeable sorrow and functional impairment [3].

Several measures have been proposed to assess Internet dependence. Goldberg [4] was the first to empirically focus on the addictive of Internet use. He constructed a rating scale, the Internet Addictive Disorder (IAD). Young [5] introduced the Internet Addiction Test (IAT) [5, 6] Del Miglio, Gamba and Cantelmi [7] proposed a scale to assess Internet-correlated diseases called the Use, Abuse and Dependence on Internet (UADI) inventory. These are only the most popular but there are many others [8–10].

With increased accessibility on Internet, Internet addiction is becoming a serious problem, especially among adolescents. Generally, playing video, using Social Networking Sites (e.g., Facebook, Twitter) or Internet games are popular online activities among young. They usually do these activities for entertainment, excitement, challenge seeking, or emotional coping. However, both excessive play of games and frequent use of SNSs may lead to certain negative outcomes. Block [11] identified three subtypes of Internet addiction: excessive gaming, sexual preoccupations and email/text messaging. Only after 4 years, a review indicates that social networking is another type of online activity that may generate addiction among young users [12].

Factors that may influence the tendency for adolescents to become dependent are various. Most existing studies on Internet addiction focus primarily on internal and individual factors that may predispose individuals to problematic Internet use, as personality traits. Personality traits appear particularly important given the theory that addiction shapes personality, leading to an addictive personality. Individuals who exhibit certain personality traits and/or do not show other personality traits may be more likely to develop an addiction to different forms of technology [13]. A few studies on adolescents [14–18] college students have suggested a correlation of neuroticism or psychoticism with Internet addiction. For instance, a study suggested that introverted persons were more susceptible to Internet addiction [17]. Another study suggested that extroversion was a significant predictor [19]. Additional research supports the theoretical link between Internet addiction and personality traits. A study of Dutch adolescents found that emotional stability, agreeableness, and conscientiousness as measured via the Big Five Scale (QBF) decreased the risk of Internet addiction operationalized with the Compulsive Internet Use Scale (CIUS) [20]. In a sample of individuals with Internet addiction disorder from universities and high schools, it was found that Internet addicts had lower extraversion and higher psychoticism scores [21] as measured via Eysenck's Personality Questionnaire Revised (EPQ-R) [22]. In a sample of adolescents was found that the likelihood of developing Internet addiction as assessed via the CIUS [20] was associated increased with low agreeableness and emotional stability, and high introversion [16]. Additionally, in a sample of university students, Internet addicts as classified via Young's Internet Addiction Test [5] scored more highly on the EPQ's neuroticism and psychoticism subscales than non-addicts.

This study aimed at assessing the presence of Internet addiction in a large sample of young adults, and at investigating the interactions between personality traits and the use of Internet as risk factors for Internet addiction.

To identify the potential predictors related to personal characteristics we chose a personality questionnaire based on six factors: the HEXACO model [23]. This model incorporates a sixth factor in addition to the factors included in the Big Five model: Honesty-Humility factor. This additional trait has been shown to be related to Agreeableness [23]. Research has shown that this model, with the inclusion of the Honesty-Humility trait, explains incremental variance beyond the classical five-factor approach [24]. Integrating all the previous studies, the main aims of this study was exploring the attitude toward Internet use and the HEXACO personality traits, in increasing the risk of developing Internet addiction. The hypotheses of the current study were: (1) student's Internet usage would be risk factors; and (2) some HEX-ACO personality traits would increase the risk for being addicted to the Internet.

2 Method

2.1 Participants

Participants were 310 Italian students attending High School in Milan (67 males and 243 females) aged from 14 to 21 years ($M = 16.23$, $SD = 1.72$). Data were collected in five weeks.

2.2 Measures

Measures of Internet dependence. Internet dependence was measured through the following scales: Internet Addiction Test (IAT) [5]; Addictive Internet Behaviour (AIB) [25]; Internet Addiction Diagnostic Questionnaire (IADQ) [25]; Internet Dependence Checklist (IDC) [26].

The Internet Addiction Test (IAT) [5]; see also [1] is a 20-item questionnaire. The items are rated on a 5-point Likert scale (1 = Never; 5 = Always).

The Addictive Internet Behaviour scale (AIB) [25] is a unidimensional scale composed of 9 "yes-no" items, adapted from the DSM-IV criteria for psychoactive substance addiction [26].

The Internet Addiction Diagnostic Questionnaire (IADQ) [25] is a unidimensional scale composed of 8 "yes-no" items, adapted from the DSM-IV criteria for psychoactive substance addiction [27].

The Internet Dependence Checklist (IDC) [26] is a unidimensional scale composed of 11 "yes-no" items developed to parallel the symptoms of drug abuse and addiction.

Psychological measures. The attitude toward Internet use was measured through three ad hoc psychological scales: Negative Attitude toward Internet; Internet Importance; Fear to Use Internet.

The Negative Attitude toward Internet is a unidimensional 8-item scale. The items are rated on a 7-point semantic differential.

The Internet Importance is a unidimensional 11-item scale. The items are rated on a 7-point Likert scale (1/7 = Much less/more important for me).

The Fear to Use Internet is a unidimensional 10-item scale. The items are rated on a 7-point Likert scale (1/7 = Much less/more probable).

Personality measures. Personality traits were measured using the HEXACO-60 [23]. The items are rated on a 5-point Likert scale (1/5 = Strongly disagree/agree) and are thought to measure six dimensions: Honesty-Humility; Emotionality; eXtraversion; Agreeableness; Conscientiousness; Openness to experience.

Behavioral checklist. In addition to the previous scales, a behavioral checklist was submitted, asking for the following variables: Internet connection; Time spent daily using the Internet; Time spent daily playing videogames; Time spent daily using the computer; Using the Internet by morning, afternoon or night.

2.3   Design and Procedure

The questionnaires were administered during the school day in the classroom. After providing demographic information, participants completed a research booklet that included the self-reported questionnaires detailed in the previous sections. The assessments typically lasted 30–40 min. All data were collected and analyzed anonymously.

2.4   Data Analysis

First, Principal Component (ACP) and Cronbach's α analyses were executed on all the scales, obtaining a factor score for each dimension: (a) IAT ($\alpha = 0.87$, *explained variance* = 30.31 %); (b) Negative Attitude toward Internet ($\alpha = 0.67$, *expl. var.* = 38.02 %); (c) Internet Importance ($\alpha = 0.71$, *expl. var.* = 26.39 %); (d) Fear to Use Internet ($\alpha = 0.87$, *expl. var.* = 46.81 %); (e) HEXACO-60 (honesty-humility, $\alpha = 0.65.$, *expl. var.* = 5.03 %; emotionality, $\alpha = 0.63$, *expl. var.* = 7.06 %; extraversion, $\alpha = 0.68$,   *expl.*   *var.* = 8.23 %;   agreeableness-conscientiousness, $\alpha = 0.63$, *expl. var.* $= 8.96$ %; openness to experience, $\alpha = 0.66$, *expl. var.* $= 6.49$ %).

To investigate if Internet dependence (measured by IAT) was predicted by demographic variables (age, sex, time spent daily using the Internet), psychological measures (negative attitude toward Internet, Internet importance, fear to use Internet) and personality factors (honesty-humility, emotionality, extraversion, agreeableness-conscientiousness, openness to experience), we executed a three-step linear regression. In the first step, we inserted demographic variables (only age and time spent daily using the Internet, resulted significant in a preliminary regression with only these predictor variables); in the second, we introduced psychological measures (each one resulted significant in the preliminary regression); in the final step, we inserted the personality factors (only extraversion and agreeableness-conscientiousness, resulted significant in the preliminary regression).

3 Results

3.1 Descriptive Data

Most of subjects have an Internet connection (Yes $= 96.8$ %, No $= 3.2$ %). They spend a lot of hours using the Internet (Nothing $= 0.3$ %, 0–2 h $= 33.2$ %, 2–4 h $= 35.2$ %, 4–6 h $= 16.3$ %, 6–8 h $= 8.1$ %, 8 + h $= 6.8$ %), playing videogames (Nothing $= 39.4$ %, 0–2 h $= 46.8$ %, 2–4 h $= 6.4$ %, 4–6 h $= 3.6$ %, 6 + h $= 3.9$ %), and using the computer (Nothing $= 2.3$ %, 0–2 h $= 38.7$ %, 2–4 h $= 35.8$ %, 4–6 h $= 14.2$ %, 6–8 h $= 4.9$ %, 8 + h $= 4.2$ %). Moreover, they prefer to use Internet during the afternoon (Nothing $= 0.3$ %, Morning $= 4.8$ %, Early afternoon $= 11.6$ %, Late afternoon $= 43.9$ %, Evening $= 37.7$ %, Night $= 1.6$ %).

Most of subjects are not addicted to Internet (i.e., they do not to use Internet in a pathological way), obtaining low scores at the measures of Internet dependence: (a) IAT (1–5 scale: 1–2 $= 44.5$ %, 2–3 $= 45.5$ %, 3–4 $= 9.7$ %, 4–5 $= 0.3$ %); (b) AIB (0–9 scale: 0 $= 27.4$ %, 1 $= 20.6$ %, 2 $= 12.3$ %, 3 $= 13.5$ %, 4 $= 11.0$ %, 5 $= 8.1$ %, 6 $= 4.2$ %, 7 $= 2.9$ %); (c) IADQ (0–8 scale: 0 $= 29.4$ %, 1 $= 22.9$ %, 2 $= 20.3$ %, 3 $= 11.9$ %, 4 $= 8.4$ %, 5 $= 5.8$ %, 6 $= 0.6$ %, 7 $= 0.6$ %); (d) IDC (0–11 scale: 0 $= 29.7$ %, 1 $= 21.0$ %, 2 $= 13.9$ %, 3 $= 8.4$ %, 4 $= 10.6$ %, 5 $= 5.8$ %, 6 $= 5.5$ %, 7 $= 2.9$ %, 8 $= 1.9$ %, 10 $= 0.3$ %).

3.2 Predictor Variables of Internet Dependence

As expected, the IAT correlated with all the other measures of Internet Dependence (for AIB $r = 0.65$, for IADQ $r = 0.63$ and for IDC $r = 0.61$; for all $p < 0.01$). Therefore, for the sake of brevity, we used only IAT as target variable in the

Table 1 Linear regression, with IAT as dependent variable, and demographic variables (step 1), psychological measures (step 2), and personality factors (step 3) as predictor variables

Steps	Measures	B	SE	Beta	F	df	R^2
1	Age	−0.10	0.03	−0.16[a]	40.54[b]	2	0.21
	Time spent daily using the Internet	0.42	0.05	0.42[b]			
2	Age	−0.10	0.03	−0.17[b]	27.46[b]	5	0.31
	Time spent daily using the Internet	0.37	0.05	0.37[b]			
	Negative attitude	−0.13	0.05	−0.13[a]			
	Internet importance	0.20	0.05	0.20[b]			
	Fear to use internet	0.18	0.05	0.18[b]			
3	Age	−0.08	0.03	−0.13[a]	23.85[b]	7	0.36
	Time spent daily using the Internet	0.34	0.05	0.34[b]			
	Negative attitude	−0.14	0.05	−0.14[a]			
	Internet importance	0.20	0.05	0.20[b]			
	Fear to use internet	0.16	0.05	0.16[a]			
	Extraversion	−0.15	0.05	−0.15[a]			
	Agreeableness-conscientiousness	−0.15	0.05	−0.15[a]			

[a] $p < 0.01$
[b] $p < 0.001$

regression. Furthermore, it is the most widespread measure of Internet dependence and this allows comparisons with preceding literature.

The linear regression on IAT showed significant effects among the predictor variables, with a final index of $R^2 = 0.36$ (see Table 1). Results showed that Internet dependence was predicted by demographic variables (namely, age and time spent daily using the Internet), psychological measures (negative attitude toward Internet, Internet importance, fear to use Internet), and personality factors (namely, extraversion and agreeableness-conscientiousness).

In particular, Internet dependence was predicted by: (a) being younger; (b) spending more time using the Internet; (c) having a positive attitude toward Internet; (d) evaluating as important the use of Internet; (e) being afraid to use Internet; (f) being less extroverted; (g) being less agreeable-conscientious.

4 Discussion and Conclusions

The main aim of this study was to assessing the risk of Internet addiction in a sample of Italian young adults of High School by looking background factors, attitudes, behavioral habits and personality traits.

The results indicate that different dimensions of internet addiction can be predicted by a combination of different users' characteristics. Age and time spent daily using the Internet, negative attitude toward Internet, attributing importance to

Internet, fear to use Internet, and extraversion and agreeableness-conscientiousness were significant predictors of internet addiction dimensions.

More in details, about the prevalence of Internet addiction in our sample the results from the IAT questionnaire underlined that none of the students scored high levels of Internet addiction, although a moderate trend regarding partially addicted students was found.

The results show that personality traits were important predictors. The regression analysis produced some interesting suggestions, showing that Agreeableness-Conscientiousness and Extroversion were inversely related to Internet addiction. This result could reflect the behavior of students who have some problems in establishing real interpersonal relationships, or they could be less inclined to share teamwork experiences with other colleagues, and thus could spend their available time on the Internet. Subjects with a low level of Agreeableness-Conscientiousness may be more oriented towards developing problematic Internet use as a means of satisfying their personal needs. Low level in Conscientiousness may indicate that an unstructured environment, such as the Internet, could be experienced as more interesting to explore than the real social context. Less conscientious adolescents would chose using the Internet over other, less pleasurable activities, and they may be at increased risk of using Internet excessively. Our results showed that also Extraversion was another personality trait with a negative correlation in predicting Internet addiction. Indeed, low extroversion individuals tend to spend their time interacting with Internet to engage in activities that cover their lack of social and interpersonal relationships [28, 29]. In other words, younger individuals that spending more time using internet, with a personality less extroverted and less agreeableness-conscientious might not being able to control and moderate their Internet-use behaviors.

Several limitations should be considered in this study: first, the diagnosis of Internet addiction needs to be refined to improve its reliability and validity. Second, given that our sample comes from the north of Italy, next step should be collecting data in the Center and South.

The present research has a number of implications for prevention. The expectation of this study is to encourage internal debate on Internet addiction for monitoring high school student behavior. In terms of prevention, the identification of specific variables associated with Internet addiction allows for targeting individuals who appear to be at risk for developing Internet addiction. These adolescents and their parents may be approached and instructed about the problems their Internet usage may cause.

References

1. Chang, M.K., Law, S.P.M.: Factor structure for young's internet addiction test: a confirmatory study. Comput. Hum. Behav. **24**, 2597–2619 (2008)
2. Widyanto, L., Griffiths, M.D.: Internet addiction: A critical review. Int. J. Ment. Health Addict. **4**, 31–51 (2006)

3. Research, The, Addiction, Controversy Surrounding Internet: 21.27. Young, K. Cyberpsychol. Behav. **2**, 381–383 (1999)
4. Goldberg, I.: Internet Addiction Disorder (1996). www.urz.uniheidelberg.de/Netzdienste/anleitung/wwwtips/8/addict.html
5. Young, K.: Caught in the Net: How to Recognize the Signs of Internet Addiction and a Winning Strategy for Recovery. Wiley, New York (1998)
6. Beard, K., Wolf, E.: Modification in the proposed diagnostic criteria for internet addiction. Cyberpsychol. Behav. **4**, 377–383 (2001)
7. Del Miglio, C., Gamba, A., Cantelmi, T.: Costruzione e Validazione Preliminare di Uno Strumento UADI per la Rilevazione delle Variabili Psicologiche e Psicopatologiche Correlate all'Uso di Internet. Giornale Italiano di Psicopatologia **7**, 293–306 (2001)
8. Anderson, K.J.: Internet use among college students: an exploratory study. J. Am. Coll. Health **50**, 21–26 (2001)
9. Kubey, R.W., Lavin, M.J., Barrows, J.R.: Internet use and collegiate academic performance decrements: early findings. J. Commun. **51**, 366–382 (2001)
10. Pratarelli, M., Browne, B., Johnson, K.: The bits and bytes of computer/internet addiction: a factor analytic approach. Behav. Res. Methods Instrum. Comput. **31**, 305–314 (1999)
11. Block, J.J.: Issues for DSM-V: internet addiction. Am. J. Psychiatry **165**, 306–307 (2008)
12. Kuss, D.J., Griffiths, M.D.: Online social networking and addiction—a review of the psychological literature. Int. J. Environ. Res. Public Health **8**, 3528–3552 (2011)
13. Takao, M.,Takahash,S., Kitamura,M.: Addictive personality and problematic mobile phone use. Cyberpsychol. Behav. **12**(5): 501–507 (2009)
14. Ko, C.H., Yen, J.Y., Chen, C.C., Chen, S.H., Wu, K., Yen, C.F.: Tridimensional personality of adolescents with internet addiction and substance use experience. Can. J. Psychiatry **51**, 887–894 (2006)
15. Ko, C.H., Hsiao, S., Liu, G.C., Yen, J.Y., Yang, M.J., Yen, C.F.: The characteristics of decision making, potential to take risks, and personality of college students with internet addiction. Psychiatry Res. **175**, 121–125 (2010)
16. Van der Aa, N., Overbeek, G., Engels, R.C., Scholte, R.H., Meerkerk, G.J., Van den Eijnden, R.J.: Daily and compulsive internet use and well-being in adolescence: a diathesis-stress model based on big five personality traits. J. Youth Adolesc. **38**, 765–776 (2009)
17. Yan, W., Li, Y., Sui, N.: The relationship between recent stressful life events, personality traits, perceived family functioning and internet addiction among college students. Stress Health **30**, 3–11 (2014)
18. Yao, M.Z., He, J., Ko, D.M., Pang, K.: The influence of personality, parental behaviors, and self-esteem on internet addiction: a study of Chinese college students. Cyberpsychol. Behav. Soc. Netw. **17**, 104–110 (2014)
19. Zamani, B.E., Abedini, Y., Kheradmand, A.: Internet addiction based on personality characteristics of high school students in Kerman. Iran. Addict. Health **3**, 85–91 (2011)
20. Meerkerk, G.J, Van Den Eijnden, R.J, Vermulst, A.A., Garretsen, H.F.: The Compulsive Internet Use Scale (CIUS): some psychometric properties. Cyberpsychol. Behav. **12**(1):1–6 (2009)
21. Xiuqin, H., Huimin, Z., Mengchen, L., Jinan, W., Ying, Z., Ran, T.: Mental health, personality, and parental rearing styles of adolescents with Internet addiction disorder. Cyberpsychol. Behav. Soc. Netw. **13**(4):401–6 (2010)
22. Eysenck, H.J., Eysenck, S.B.G.: Manual of the Eysenck Personality Questionnaire (Junior and Adult). Hodder & Stoughton, Kent, UK (1975)
23. Ashton, M.C., Lee, K.: The HEXACO–60: a short measure of the major dimensions of personality. J. Pers. Assess. **91**, 340–345 (2009)
24. Hilbig, B.E., Heydasch, T., Zettler, I.: To boast or not to boast: Testing the humility aspect of the Honesty-Humility factor. Pers. Individ. Differ. **69**, 12–16 (2014)
25. Young, K.: Internet Addiction: The Emergence of a New Clinical Disorder. Cyberpsychol. Behav. **3**, 237–244 (1996)

26. Scherer, K.: College life online: Healthy and unhealthy Internet use. In: Paper presented at the 104th Annual Meeting of the American Psychological Association, Chicago (1997)
27. American Psychiatric Association: Diagnostic and Statistical Manual of Mental Disorders IV. American Psychiatric Association, Washington (1994)
28. Correa, T., Hinsley, A.W., de Zuniga, H.G.: Who interacts on the web?: the intersection of users' personality and social media use. Comput. Hum. Behav. **26**, 247–253 (2010)
29. Rice, L., Markey, P.M.: The role of extraversion and neuroticism in influencing anxiety following computer-mediated interactions. Pers. Individ. Differ. **46**, 35–39 (2009)

Universal Matched-Filter Template Versus Individualized Template for Single Trial Detection of Movement Intentions of Different Tasks

Muhammad Akmal, Mads Jochumsen, Muhammad Samran Navid, Muhammad Shafique, Syed Muhammad Tahir Zaidi, Denise Taylor and Imran Khan Niazi

Abstract Brain-computer interfaces (BCIs) have been proposed for neurorehabilitation after stroke by inducing cortical plasticity. To transfer the technology from the controlled settings in the lab to the clinic several issues must be addressed. In this study, it was investigated how the performance was affected by using a universal task template to detect movement intentions associated with movements performed with two different levels of force and speed. The performance of the universal template was compared to an individualized template constructed for each

M. Akmal · M. Shafique
Faculty of Engineering, Riphah International University, Islamabad, Pakistan
e-mail: Muhammad.akmal@riphah.edu.pk

M. Shafique
e-mail: muhammad.shafique@riphah.edu.pk

S.M.T. Zaidi
National University of Sciences and Technology, Islamabad, Pakistan
e-mail: tahirzaidi@ceme.nust.edu.pk

M. Jochumsen · M.S. Navid · I.K. Niazi (✉)
Department of Health Science and Technology, Aalborg University,
Aalborg, Denmark
e-mail: imrankn@hst.aau.dk; imran.niazi@nzchiro.co.nz; iniazi@aut.ac.nz

M. Jochumsen
e-mail: mj@hst.aau.dk

M.S. Navid
e-mail: samran@hst.aau.dk

I.K. Niazi
Center for Chiropractic Research, New Zealand College of Chiropractic,
Auckland, New Zealand

D. Taylor · I.K. Niazi
Health & Rehabilitation Research Institute, Auckland University of Technology,
Auckland, New Zealand
e-mail: Denise.taylor@aut.ac.nz

© Springer International Publishing Switzerland 2016   275
S. Bassis et al. (eds.), *Advances in Neural Networks*, Smart Innovation,
Systems and Technologies 54, DOI 10.1007/978-3-319-33747-0_27

task. Twelve healthy subjects performed four types of dorsi-flexions while continuous electroencephalography (EEG) was recorded from ten channels. The movement intentions were detected ($\sim$200–300 ms before the movement onset) from the continuous EEG using a matched-filter approach. The true positive rate was significantly higher (P = 0.001) when using the individualized template where 68–76 % of the movements were correctly detected on the contrary to 65–70 % when using the universal template. The number of false positive detections per 5 min was lower (P = 0.036) when using the universal template ($\sim$13) compared to the individualized template ($\sim$14). Despite the lower performance when using the universal detection template, the performance of the detector is in the range of what has been reported previously for inducing cortical plasticity.

1 Introduction

Over the past years, brain-computer interface (BCI) technology has been a means for controlling external devices by paralyzed patients suffering from e.g. spinal cord injury. A BCI can be used to detect the intention to perform a task using only the brain activity and send a control command to e.g. a wheel chair (left/right/forward/backward) [1, 2]. Another application of BCIs is neurorehabilitation [3, 4]. It has been shown that BCIs can be used to induce cortical plasticity which has been linked to motor learning which is essential in e.g. motor recovery after stroke [5, 6].

Recently, a protocol was proposed where Hebbian-associated plasticity was induced by using a BCI [7]. The intention to move was paired with electrical stimulation of the nerve innervating the muscle involved in the imagined movement. The sensory feedback from the electrical stimulation must reach the cortical level of the brain during the movement intention for inducing plastic changes. Since the sensory feedback takes time ($\sim$30–50 ms depending on the site of stimulation) to propagate from the point of stimulation to the cortical level of the brain the intention to move must be predicted.

The control signal that was used was the movement-related cortical potential (MRCP). The MRCP is observed as an increase in negativity in the electroencephalogram (EEG) up to 2 s prior a voluntary movement (executed or imagined) which makes it possible to predict when a person intends to move [8, 9]. The initial negative phase of the MRCP (movement intention) has been detected on a single trial basis in the continuous EEG with detection latencies in the range of 50–300 ms prior the onset of the movement [9–12].

Clinical studies on the use of BCIs in rehabilitation have started to emerge [13, 14], but before the BCI technology can be transferred from the lab to the clinic several issues must be addressed such as ease of setup and automation of subject training [15]. In the clinic, the BCI shall likely be operated by clinic personnel with

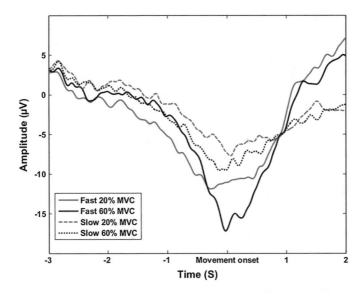

Fig. 1 Average of 50 movements for each task for a representative subject

limited time to setup the BCI which can be a time consuming task. Therefore, it has been investigated to use only a single channel of EEG [12]; dry electrodes could potentially be used as well. Another approach to promote the transfer of technology to the clinic was proposed by removing the cumbersome training of the subject by using a global (universal) approach of subject-independent data instead of subject-specific data [16, 17]. In the study by Niazi et al., [17], MRCPs associated with uniformly performed movements were detected. However, MRCPs performed with variation in force and speed differ in morphology [11, 12, 18] (see Fig. 1); therefore the performance of the detector may differ when movements are performed with different levels of force and speed which will likely be the case when stroke patients are attempting to perform the movements. In this study, the aim was to evaluate a matched-filter approach to detect movement intentions with task-specific and task-independent detection templates for the matched-filter.

2   Methods

2.1   Subjects

Twelve healthy subjects participated in the study (four females and eight males: 27 ± 6 years old). All procedures were approved by the local ethical committee (N-20100067), and the subjects gave their informed consent before participation.

2.2 Experimental Protocol

The subjects were seated in a comfortable chair with their right foot fixed to a pedal with an attached force transducer. At the beginning of the experiment the maximum voluntary contraction (MVC) was determined when the subjects performed a dorsi-flexion of the ankle joint. Based on the MVC, the subjects performed four visually cued tasks: (I) 0.5 s to reach 20 % MVC, (II) 0.5 s to reach 60 % MVC, (III) 3 s to reach 20 % MVC and (IV) 3 s to reach 60 % MVC. The subjects were constrained to spend the given time to reach the desired level of MVC. In the following, movement type I and II will be called 'Fast' and type III and IV will be called 'Slow' followed by the level of MVC. The subjects were visually cued by a custom made program (Knud Larsen, SMI, Aalborg University) where force was used as input. The order of the tasks was randomized in blocks, and ~5 min of practice were spent by the subjects to familiarize with the tasks.

2.3 Signal Acquisition

EEG: Ten channels of EEG (EEG amplifiers, Nuamps Express, Neuroscan) were recorded from self-adhesive scalp electrodes with a sampling rate of 500 Hz. The electrodes were placed at FP1, F3, Fz, F4, C3, Cz, C4, P3, Pz and P4 according to the international 10–20 system. The electrodes were grounded at nasion and referenced to the right earlobe.

Force and MVC: The force from the force transducer on the pedal was sampled with 2000 Hz and recorded using the Mr. Kick software (Knud Larsen, SMI, Aalborg University). The onset of the force for each contraction was determined when all values of the force signal in a 200 ms window exceeded the baseline. The window was shifted by one sample. The baseline was determined by the mean value of the force in the interval 4-1 s before the cued onset of the task [11]. The MVC was determined as the highest value out of three attempted MVCs. Each repetition was separated by 1 min of rest.

2.4 Movement Detection

The proposed method for detecting the movements has been described previously [9, 11], but will be summarized in the following. Initially, the signals were bandpass filtered from 0.05–10 Hz with a 2nd order Butterworth filter in the forward and reverse direction and down-sampled to 20 Hz. Afterwards, a surface Laplacian filter was applied to extract a surrogate channel (SC) of the Cz-electrode with the eight surrounding electrodes. FP1 was not used to calculate the surrogate channel, but to register electrooculography (EOG). The following formula was used to calculate the SC:

$$SC = Cz - \frac{1}{8}(F3 + Fz + F4 + C3 + C4 + P3 + Pz + P4) \tag{1}$$

After the SC was calculated, two different approaches for detecting the movements were performed: an individualized and a universal approach.

In the individualized approach, a template was constructed from half of the data (from a single task) and the other half was used for testing (randomly selected). The template was the average of the training data from the onset of each movement and 2 s prior this point (see Fig. 1). Afterwards the other half was used for testing and training, respectively (2-fold cross-validation).

For the universal approach, the data for each of the four tasks was randomly divided into two sets as described for the individualized approach. The detection template was calculated in the same way as described for the individualized template, but the average was calculated across the training data of each of the four tasks.

The true positive rate (TPR) was calculated based on the average number of detected movements in the two testing sets.

From the training data a receiver operating characteristics (ROC) curve was obtained through 3-fold cross-validation on the training data. A trade-off between the TPR and number of false positive (FP) detections was obtained by selecting the detection threshold associated with the turning point of the ROC curve. Detections occurred based on the likelihood ratio (Neyman Pearson lemma) computed between the SC in the testing data and the template. The window was shifted 200 ms each time; in this way an online system was simulated since no information about the time index of the movements was used for the detection. Detection occurred when two out of three windows exceeded the detection threshold and was below the EOG threshold in FP1 (125 µV).

Besides the TPR, the number of FPs/5 min and the detection latency with respect to the onset of the movement were used as performance metrics of the detector.

2.5 Statistics

Three Wilcoxon signed ranked tests were performed to determine if the performance of the universal and individualized detector differed significantly. The tests were performed for the three detection performance metrics: TPR, FPs and detection latencies.

3 Results

The results of the detection performance when using the universal and individualized templates are summarized in Table 1. It was found that higher TPRs were obtained when using the individualized templates ($\sim$73 %) compared to the

Table 1 All values are presented as mean ± standard deviation

	TPR (%)	FPs/5 min	Detection latency (ms)
	Universal template		
Fast 20 % MVC	65 ± 10	14 ± 3	−251 ± 188
Fast 60 % MVC	68 ± 8	12 ± 5	−204 ± 395
Slow 20 % MVC	69 ± 7	13 ± 4	−292 ± 198
Slow 60 % MVC	70 ± 19	12 ± 4	−350 ± 247
	Individualized template		
Fast 20 % MVC	68 ± 13	14 ± 4	−310 ± 179
Fast 60 % MVC	74 ± 12	13 ± 4	−296 ± 166
Slow 20 % MVC	76 ± 18	14 ± 3	−294 ± 231
Slow 60 % MVC	73 ± 19	15 ± 4	−382 ± 262

The duration of each session was ~10 min. The detection latencies are reported with respect to the onset of the movement (0 ms)

universal template (~68 %). The Wilcoxon signed ranked test revealed that the TPRs for the individualized approach were significantly ($P = 0.001$) higher than the universal approach. The TPRs were quite similar across the tasks, but slightly higher TPRs were obtained for the slow movements.

The number of FPs/5 min was a bit higher for the individualized approach (~14) compared to the universal approach (~13); this difference was significant ($P = 0.036$). Lastly, the detection latencies were determined, and they were similar for the two approaches. The difference between the latencies for the two approaches was not significant ($P = 0.19$).

4 Discussion

In this study, it was shown that movement intentions can be detected 200–300 ms before the movement onset when using task universal and task individualized detection templates. The TPR was significantly higher for the individualized templates, but on the expense on more FPs/5 min.

The differences in the performance of the two templates may be explained by two factors; the detection threshold and signal morphology. The detection threshold was selected based on the ROC curve; this value was manually selected (at the turning point of the curve). Since the detection threshold was calculated based on a trade-off between the TPR and number of FPs, by selecting different thresholds there will be a difference in performance. This is consistent with the finding that the TPR was lower for the universal template, but the number of FPs was also lower indicating a higher detection threshold on the contrary to a higher TPR and more FPs (lower detection threshold). The other factor, signal morphology, can explain why greater TPRs were obtained when using the individualized templates. It can be seen from Fig. 1 that a more accurate output of the matched-filter can be obtained when using the individualized templates. Using a universal template, where an average of the four takes is calculated, discriminative details of the signal morphology may be lost leading to a reduction in the detector performance.

The performance that was obtained in the current study is in the range of what has been reported previously [9, 11, 17, 19]. However, the findings regarding the detection performance for fast versus slow movements are a bit contradicting to what has been reported previously [11] and according to Fig. 1 where greater signal-to-noise ratios were obtained for fast movements; this was expected to be reflected in the detector performance. An explanation for the discrepancy may be the cross-validation procedure that was utilized in this study compared to the findings in Jochumsen et al. [11].

The lower limit for the detection performance needed to induce plasticity is not known [4], but it has been reported that a TPR of ~60 % is sufficient to induce these changes [20]. However, it was also shown that the effect of inducing plasticity increases with improved performance of the detector [20]. Therefore, it will be a trade-off of how much time that must be spent on calibrating the detector to improve the performance.

The findings suggest that universal templates can be created to promote the translation of BCI technology from the lab to the clinic with detector performance in the range of clinical applicability. It should be noted that the analysis was performed offline and that it should be tested on the intended users; stroke patients. It is expected that it will work as an online system with a similar due to the simplicity of the detector. The detector has been tested in a stroke population and implemented as an online system [9, 20].

References

1. Wolpaw, J.R., Birbaumer, N., McFarland, D.J., Pfurtscheller, G., Vaughan, T.M.: Brain-computer interfaces for communication and control. Clin. Neurophysiol. **113**, 767–791 (2002)
2. Millán, J.R., Rupp, R., Müller-Putz, G.R., Murray-Smith, R., Giugliemma, C., Tangermann, M., Vidaurre, C., Cincotti, F., Kübler, A., Leeb, R., Neuper, C., Müller, K., Mattia, D.: Combining brain–computer interfaces and assistive technologies: state-of-the-art and challenges. Front. Neurosc. **1** (2010)
3. Daly, J.J., Wolpaw, J.R.: Brain-computer interfaces in neurological rehabilitation. Lancet Neurol. **7**, 1032–1043 (2008)
4. Grosse-Wentrup, M., Mattia, D., Oweiss, K.: Using brain-computer interfaces to induce neural plasticity and restore function. J. Neural Eng. **8**, 025004 (2011)
5. Mrachacz-Kersting, N., Niazi, I.K., Jiang, N., Pavlovic, A., Radovanović, S., Kostic, V., Popovic, D., Dremstrup, K. Farina, D.: A novel brain-computer interface for chronic stroke patients. In: Converging Clinical and Engineering Research on Neurorehabilitation Anonymous, pp. 837–841. Springer (2013)
6. Pascual-Leone, A., Dang, N., Cohen, L.G., Brasil-Neto, J.P., Cammarota, A., Hallett, M.: Modulation of muscle responses evoked by transcranial magnetic stimulation during the acquisition of new fine motor skills. J. Neurophysiol. **74**, 1037–1045 (1995)

7. Mrachacz-Kersting, N., Kristensen, S.R., Niazi, I.K., Farina, D.: Precise temporal association between cortical potentials evoked by motor imagination and afference induces cortical plasticity. J. Physiol. (Lond.), **590**, 1669–1682 (2012)
8. Shibasaki, H., Hallett, M.: What is the Bereitschaftspotential? Clin. Neurophysiol. **117**, 2341–2356 (2006)
9. Niazi, I.K., Jiang, N., Tiberghien, O., Nielsen, J.F., Dremstrup, K., Farina, D.: Detection of movement intention from single-trial movement-related cortical potentials. J. Neural Eng. **8**, 066009 (2011)
10. Lew, E., Chavarriaga, R., Silvoni, S., Millán, J.R.: Detection of self-paced reaching movement intention from EEG signals. Front. Neuroeng. **5**, 13 (2012)
11. Jochumsen, M., Niazi, I.K., Mrachacz-Kersting, N., Farina, D., Dremstrup, K.: Detection and classification of movement-related cortical potentials associated with task force and speed. J. Neural Eng. **10**, 056015 (2013)
12. Jochumsen, M., Niazi, I.K., Rovsing, H., Rovsing, C., Nielsen, G.A., Andersen, T.K., Dong, N.P., Sørensen, M.E., Mrachacz-Kersting, N., Jiang, N., Farina, D., Dremstrup, K.: Detection of movement intentions through a single channel of electroencephalography. In Replace, Repair, Restore, Relieve–Bridging Clinical and Engineering Solutions in Neurorehabilitation Anonymous, pp. 465-472. Springer (2014)
13. Ang, K.K., Chua, K.S.G., Phua, K.S., Wang, C., Chin, Z.Y., Kuah, C.W.K., Low, W., Guan, C.: A randomized controlled trial of EEG-based motor imagery brain-computer interface robotic rehabilitation for stroke. Clin. EEG Neurosci. (2014)
14. Ang, K.K., Guan, C.: Brain-Computer Interface in Stroke Rehabilitation. J. Comput. Sci.Eng. **7**, 139–146 (2013)
15. Popescu, F., Blankertz, B., Müller, K.: Computational challenges for noninvasive brain computer interfaces. IEEE Intell. Syst. **23**, 78–79 (2008)
16. Blankertz, B., Dornhege, G., Krauledat, M., Müller, K., Kunzmann, V., Losch, F., Curio, G.: The Berlin Brain-Computer Interface: EEG-based communication without subject training. IEEE Trans. Neural Syst. Rehabil. Eng. **14**, 147–152 (2006)
17. Niazi, I.K., Jiang, N., Jochumsen, M., Nielsen, J.F., Dremstrup, K., Farina, D.: Detection of movement-related cortical potentials based on subject-independent training. Med. Biol. Eng. Comput. **51**, 507–512 (2013)
18. Nascimento, O.F., Nielsen, K.D., Voigt, M.: Movement-related parameters modulate cortical activity during imaginary isometric plantar-flexions. Exp. Brain Res. **171**, 78–90 (2006)
19. Xu, R., Jiang, N., Lin, C., Mrachacz-Kersting, N., Dremstrup, K., Farina, D.: Enhanced Low-latency Detection of Motor Intention from EEG for Closed-loop Brain-Computer Interface Applications. IEEE Trans. Biomed. Eng. **61**, 288–296 (2013)
20. Niazi, I.K., Kersting, N.M., Jiang, N., Dremstrup, K., Farina, D.: Peripheral Electrical Stimulation Triggered by Self-Paced Detection of Motor Intention Enhances Motor Evoked Potentials. IEEE Trans. Neural Syst. Rehabil. Eng. **20**, 595–604 (2012)

Sparse fNIRS Feature Estimation via Unsupervised Learning for Mental Workload Classification

Thao Thanh Pham, Thang Duc Nguyen and Toi Van Vo

Abstract Recent studies have demonstrated that functional near-infrared spectroscopy (fNIRS) is a potential non-invasive system for human mental workload (MWL) evaluation in both off-line and on-line manners. While most of the studies have been based on supervised classification of different MWL levels, which requires much effort to collect labeled training data, investigation on unlabeled data seems to be more promising. In this paper, we developed unsupervised learning and classification techniques of fNIRS parameters to support human workload classification. In the experimental setup, five subjects engaged in ten-loop memorizing tasks that were devised into two MWL levels while fNIRS signals were being monitored over their frontal lobes. Independent component analysis (ICA) was applied on a set of unlabeled random fNIRS data to extract the basis and sparse functions. Then two-dimensional convolutional matrices, which were constructed as sets of convolutional coefficients of fNIRS signal with learned basis functions, were implemented as the inputs for MWL classification using convolutional neural network classifier. Study of generalized linear model demonstrated that basis functions extracted using ICA is more effective when illustrating the activation regions over measuring cortex than using the modeled hemodynamic response functions. Besides, ICA basis function demonstrates the sparseness so that it is superior to basis functions learned by the conventional method of principle component analysis (PCA) in mental classification and shows its potential for further study of fNIRS signals based on their hidden basis functions.

Keywords Functional near-infrared spectroscopy (fNIRS) · Mental work-load (MWL) · Unsupervised learning · Spare features

T.T. Pham · T.D. Nguyen · T. Van Vo (✉)
Biomedical Engineering Department, International University of Vietnam National Universities, Ho Chi Minh City, Vietnam
e-mail: vvtoi@hcmiu.edu.vn

© Springer International Publishing Switzerland 2016
S. Bassis et al. (eds.), *Advances in Neural Networks*, Smart Innovation, Systems and Technologies 54, DOI 10.1007/978-3-319-33747-0_28

283

1 Introduction

To date, neuroimaging techniques like electroencephalogram (EEG) and functional magnetic resonance imaging (fMRI) have achieved several successes in the analysis of neural activities based on the assumption of available constructed canonical basic functions. In details, fMRI analysis draws statistical information from a priority assumption of temporal shapes of the hemodynamic response, such as the canonical temporal basis set as combination of one or more Γ-variant or Gaussian functions to model the response [1, 2]. On the other hand, EEG studies focus on event-related potentials (ERPs) as time-frequency perturbations of underlying field potential processes [3]. In other words, a common assumption is made in these studies that the evoked response may be modeled as a linear combination of an available sets of basis functions.

Recently, analysis of basis functions of optical measurements using fNIRS has generally been confined under traditional time-frequency methods such as Fourier and wavelet transforms [4]. Some studies demonstrated that the available modeled basis functions like canonical hemodynamic response (HRF), which have been widely applied in fMRI analysis, offer several advantages for the analysis of fNIRS data [5]. While these approaches can offer vast improvements to the detection of evoked response and greatly simplify statistical testing as well as enhance study of hemodynamic responses using fNIRS, there exist several issues to consider, such as their appropriateness and adaptation when applying on divergent subjects and measuring conditions [6]. Moreover, unlike fMRI which typically reports a single type of functional contrast, fNIRS measures two distinct parameters including oxyHemoglobin (HbO) and deoxyHemoglobin (HbR), which expected to have a slightly different in phase and amplitude [7]. Thus, the use of a single common available set of canonical basis functions to model both of the hemoglobin responses will give rise to non-uniform detection deficiencies of these HbO and HbR changes. In other words, when using a canonical basis set to test timing-based hypotheses, one must choose the basis functions that can adapt appropriately to the analyzed signals to achieve uniform sensitivity to different dynamics of the evoked response. In this approach, an adaptive basis sets derived from unsupervised learning on a large random and unlabeled set of fNIRS data, which is varied in timing and frequency, could become well-suited to be applied in the time-series analysis of fNIRS signals.

Unsupervised feature learning techniques have been widely applied in basis feature extraction in both one–and two-dimensional data to simulate nervous system in mammals [8–10]. Previous studies [6, 11] revealed that basis functions of audiotory signals closely correspond to neural responses in early audio processing in mammals. For example, when sparse coding models were applied to natural sounds or speech, the learned basis vectors showed a striking resemblance to the cochlear filters in the auditory cortex. Similarly, the basis features of images revealed the edge-like bases similar to the early processing of visual cortex [6]. In this study, we proposed unsupervised learning using independent component analysis (ICA) to extract basis functions of fNIRS data and then applied them to reveal the hidden hemodynamic activities in fNIRS measurements. ICA has been successfully applied to extract independent physiological sources from fNIRS signals for artifact removal and spatial representation [12, 13]. Besides, we also implemented principle component analysis (PCA), which is the one of the most fundamental feature extraction methods, for comparison. Finally, the extracted basis features from fNIRS dataset using PCA and ICA were convolved with fNIRS signal to present the hemodynamic activities in a temporal two-dimensional representation and applied this representation as inputs for classification of two discrete human mental workload levels.

2 Unsupervised Feature Learning of fNIRS Signals

For the application of unsupervised feature learning with PCA and ICA to fNIRS data, we first trained the proposed PCA and ICA learning frameworks using a large random fNIRS dataset to extract the basis functions. This dataset was gathered from fNIRS measurement database and is varied in timing, sampling frequencies, subjects, experimental procedures and measurement conditions. We expected that the use of this dataset would give abundant information to support the learning of adaptive basis functions that contain useful information of human hemodynamic responses. This dataset of raw fNIRS records was segmented into 17,421 frames of 1000 data points (corresponding to approximately 56 seconds with sampling rate of 18 Hz). The training frames were pre-processed with DC removal and whitening. Basis functions learned from the ICA and from PCA are denoted as A_{ica} and A_{pca} respectively. They were sorted according to the increase in oscillation term of the functions. The test fNIRS signals were then convolved with learned basis functions to obtain convolutional matrices C_{ica} and C_{pca} (Fig. 1).

In this study, only the feature learning and representations of HbO were investigated.

2.1 Feature Learning using Principle Component Analysis

The idea behind PCA [9] is to represent the data with respect to a new basis, where the vector that form this new basis are the eigenvectors of the data covariance matrix. Since the data covariance matrix is symmetric, its eigenvectors are orthogonal. As a feature detection mechanism, it is often the case that the eigenvectors give interesting underlying feature of the data.

Let $\mathbf{x}^{(t)} = [x_1^{(t)}, \ldots, x_n^{(t)}]$ be a data item, and $\mathbf{e}_i = [e_1, \ldots, e_n]^T$ be the ith eigenvector (eigenvectors are ordered in decreasing eigenvalues). The projection of $\mathbf{x}^{(t)}$ onto $\mathbf{e}_i$ is denoted as $y_i^{(t)}$, and is given by the inner product of these two vectors

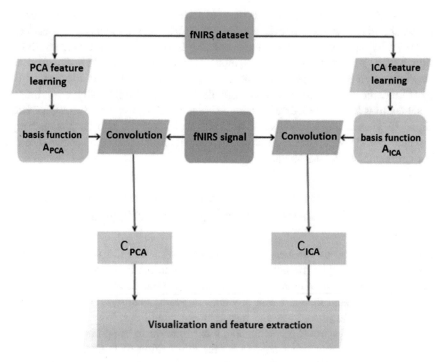

Fig. 1 Overall process of feature learning on fNIRS signals based on unsupervised learning

$$y_i^{(t)} = [\mathbf{x}^{(t)}]^T \mathbf{e}_i = \mathbf{e}_i^T \mathbf{x}(t). \tag{1}$$

That is, $\mathbf{y}^{(t)} = [y_1^{(t)}, \ldots, y_n^{(t)}]^T$ is the representation of $\mathbf{x}^{(t)}$ with respect to the new basis. We considered the set $A_{pca} = [e_1, \ldots, e_n]$ as the matrix whose ith column is the ith eigenvector and as the set of basis functions of PCA.

2.2 Feature Learning using Independent Component Analysis

The generative model in ICA [9] is defined as a linear transformation of the latent independent components. Consider dataset $\mathbf{I} \in \mathrm{R}^{nxm}$ with n samples and m dimensions. In ICA, $\mathbf{I}$ is generated as a linear superposition of some features $A_i \in R^{nxm}$

$$\mathbf{I} = \sum_{i=1}^{n} A_i s_i \tag{2}$$

where s_i are coefficients that are different from sample to sample. As typical in linear models, the estimation of A_i is equivalent to determining the values of W_i which gives s_i as outputs of linear feature detectors with some weights W_i

$$s_i = \sum W_i \mathbf{I}. \tag{3}$$

The features A_i are the same for all samples and defined as the basis functions of $\mathbf{I}$. The set of basis functions A is called a dictionary of $\mathbf{I}$. ICA attempts to learn a maximally non-redundant code. For this reason, latent coefficients s_i are assumed to be statistically independent i.e.,

$$p(s) = \prod_{i=1}^{n} p(s_i). \tag{4}$$

The marginal probability distributions $p(s_i)$ are usually assumed to be sparse (i.e., having high kurtosis). In this present work, logistic distribution was used [9]. The basis functions were learned by maximizing the log-likelihood of the model via gradient ascent to finally obtain $A_{ica} = A$.

2.3 Convolutional Matrix

With regards to an input signal $S(t)$, the signal is assumed to be a combination of basis functions. Therefore, by applying the linear filter as each basis function functions A_i ($0 < i \leq K$ where K is the number of basis functions) from the set basis functions. As trained from an large unlabeled dataset, we obtain the output C_i that reveals the correlation of the basis function and signal. In this study, we hypothesized that this output would give a new approach to illustrate the signal by the activities of hidden basis components. According to the filter definition [9], the filtering operation using the impulse response (i.e., the basis functions) and the input signal $S(t)$ is simply the convolution operation.

$$C_i(t) = S(t) * A_i. \tag{5}$$

In this present study, we defined the convolutional matrix $\mathbf{C}$ as a two-dimensional set of convolutional outputs of signal $S(t)$ with each feature basis A_i from a set of basis functions A, $\mathbf{C} = [C_1 \quad C_2 \quad \cdots \quad C_N]$. The convolutional matrix was then presented in two-dimensional visualization to demonstrate the signal. Moreover, the mental workload classification was investigated using these convolutional matrices from unsupervised learning.

3   Experimental Setup and Results

3.1   Experiment

Five adult volunteers (2 females, mean age of 20.6 ± 0.5 years) participated in this study. None of them reported neurological and vision abnormalities. The local review board has approved the studies and written informed consents were obtained from all subjects before experiments. The tenets of Declaration of Helsinki were followed.

fNIRS measured the blood oxygenation over prefrontal cortex while subjects were operating memorizing tasks. Levels of oxygenated and deoxygenated hemoglobin concentration changes could be estimated by applying modified Beer-Lambert law [14]. FOIRE—3000 (Shimadzu, Japan) instrument was used in this experiment to obtain fNIRS signals. A 2-by-3 array of light sources and sensors with 3-cm source-detector separation were attached to the subject forehead, corresponding to totally seven channels with acquisition sampling rate of 18 Hz.

In this experiment, a graphical rotating three-dimensional cube was presented on computer screen in front of the subject [15]. During each task, the cube rotated and showed five sides (one top and four lateral), each of which consisted of four small squares. Each square could have a color in such a way that the whole cube had totally two or four colors (randomly red, green, blue and yellow) corresponding to two MWL of easy and difficult levels (denoted as MWL2 and MWL4). Subject was asked to count and memorize total number of squares and their colors, which was pseudo-randomized to MWL2 or MWL4. Then subject relaxed for additional 10 s to shift fNIRS signals to baseline. One complete task took 45 s. Each experiment protocol last 15 min and each subject was required to perform the entire experiment protocols separately ten times.

3.2   fNIRS Basis Functions

PCA and ICA basis functions were learned from large random unlabeled fNIRS dataset and were hypothesized to demonstrate hemodynamic responses of fNIRS signals. This dataset was gathered from fNIRS measurement database with varied timing, sampling frequencies, subjects, experimental procedures and measurement conditions. This dataset of 777 raw fNIRS records was segmented into 17,421 frames of 1000 points (corresponding to approximately 56 seconds with sampling rate of 18 Hz). Fifty basis functions were learned by ICA (denoted as ICs) and sorted according to the increase in oscillation term of the functions. ICA basis functions illustrated the strong sparseness, which indicated that the responses are only active corresponding to certain specific stimulation.

Visualization of the bases shows that the basis functions of fNIRS signals are a set of activities in different oscillations (i.e., frequencies). Specifically, in Fig. 2, the

(a)   **(b)**

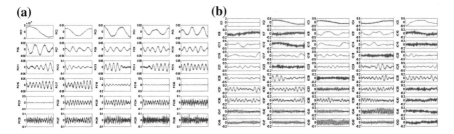

Fig. 2 Basis functions of HbO signals learned by **a** PCA and **b** ICA respectively

PCA basis functions demonstrate the discrete Fourier transforms arranged in order of increasing frequencies. Meanwhile, unlike PCA basis functions, the activities of ICA bases illustrate the strong sparseness, which indicates that the responses are only active at certain time during stimulation.

Generalized linear model was employed to evaluate the performance of PCA and ICA basis functions in comparison with HRF. In this study, t-test over seven channels was used to illustrate the activation maps over prefrontal cortex and to evaluate the performance of basis functions when applying to generalized model in comparison with delta function and HRF. As indicated by Fig. 3, only designed matrices from 33 PCA basis functions are statistically similar to the fNIRS signal than delta functions. Meanwhile, there are only 2 PCA basis functions (1st and 3rd basis functions) have designed matrix more statistically similar to fNIRS signal than HRF. In addition, activation maps of these two basis functions are more localized than that of HRF. All designed matrices of ICA basis functions are more statistically similar to fNIRS signal than that of event-related response or delta function.

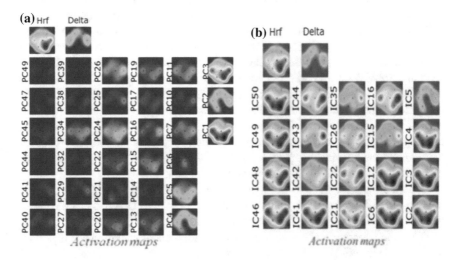

Fig. 3 Activation maps generated from generalized linear models of HRF, Delta, PCA and ICA basis functions. **a** PCA, **b** ICA

In particular, twenty ICA basis functions have t-values higher than that of HRF (as shown in Fig. 3). Moreover, activation maps of most of these basis functions (the 2nd, 3rd, 4th, 6th, 12th, 16th, 21st, 22nd, 41st, 44th, 46th, 48th, 49th and 50th) are more localized than that of HRF.

3.3 Application to Mental Workload Classification

In this section, we examined the utilization of proposed framework for unsupervised feature learning in the process of mental workload classification. The proposed sets of PCA and ICA functions were convolved with fNIRS signals recorded in mental workload experiment. The derived two-dimensional convolutional matrices of mental workload data were used for classification on labeled training/test data to discriminate two mental workload levels MWL2 and MWL4.

In this study, we employed convolutional neural network (CNN) as the supervised classifier. The classification was processed separately among subjects and between different hemoglobin parameters. CNNs are artificial neural networks with one or more convolutional layers [16, 17]. The network structure contain local connections and tied weights followed by some form of pooling which results in translation invariant features. The advantage of CNNs is that they are easier to train and have many fewer parameters than fully connected networks with the same number of hidden units.

For each subject, each of the two-dimensional convolutional matrices was segmented into a set of 65-second segments (10 s of pre-rest—45 s of task—10 s of post-rest).

We processed and classified each fNIRS channel individually and followed the practice of supervised classification to partition each set into training and test sets with 10-fold random sampling process, which means that the dataset is randomly divided into half for training and half for test. The final results were the averages of 10-fold random sampling. The label of each segment was either 0 or 1, corresponding to MWL2 and MWL4. In this study, we employed a CNN with one convolutional layer and one pooling. Classification model was trained for 10 epochs using stochastic gradient descent with exponential decay of the learning rate after each epoch and momentum. The classification model was selected with the initial learning rate of 0.1, momentum of 0.95 and the exponential learning rate decay to haft of previous learning rate per epoch. The size of convolutional layer was chosen as 9 with 20 filters and the size of pooling layer is 4.

The classification showed computationally acceptable results (Table 1) when applying both PCA and ICA convolutional matrices of HbO signals to the classification of MWL. Although the classification performances changed with different subjects, the experimental results suggested that the classification using ICA basis features is more effective than those from PCA.

Table 1 Averaged accuracies of mental workload classification on basis features from PCA and ICA obtained from fNIRS signals of five subjects (Mean ± S.D. %)

Subject	PCA	ICA
1	62.63 ± 5.08	70.16 ± 2.35
2	70.05 ± 4.99	73.07 ± 7.48
3	74.60 ± 1.37	73.06 ± 2.02
4	72.10 ± 7.58	78.48 ± 3.13
5	63.04 ± 10.43	66.40 ± 5.69

4 Conclusion

This study proposes a novel method for analyzing fNIRS signals. It has been shown that the convolutional matrices of basis functions and signal demonstrated the detailed oscillations and responses of basis features in raw fNIRS signal. ICA basis functions demonstrate the sparseness and they are superior to basis functions learned by the conventional principle component analysis (PCA) in mental workload classification.

Our study, therefore, suggests an idea of applying more advanced unsupervised feature learning, such as deep neural network, in future for robust analysis of basis features from fNIRS data, as well as to provide user a tool to investigate deep information of fNIRS signals.

Practical classification tasks based on fNIRS signals are much more complex than the two-class classification proposed in this study. While the challenge of larger datasets often lies in the considering harder tasks, our objective by using the convolutional neural networks combined with two-dimensional convolutional representations of fNIRS signals was to enhance more effective classification that can be learned with large number of features. Though the classification results were still in the acceptable range, in which we hypothesized that this is due to the limit in the amount of data available for unsupervised learning, we believe that more data for classification will enhance the results of MWL discrimination.

Acknowledgments This work has been supported by grants No. 106.99-2010.11 of Vietnam National Foundation for Science and Technology Development (NAFOSTED) and No. B2011-28-01 of Vietnam National Universities—HCMC. We thank the European Union's Seventh Framework Programme for Research, Technological Development and Demonstration for grant No. 611014 CONNECT2SEA project.

References

1. Chen, H., Yao, D., Liu, Z.: A comparison of Gamma and Gaussian dynamic convolution models of the fMRI BOLD response. Magn. Reson. Imaging **23**, 83–88 (2005)
2. Penny, W., Fladin, G., Trujillo-Barreto, N.: Chapter 25-Spatio-temporal models for fMRI. In: Statistical Parametric Mapping. Academic Press, London (2007)

3. John, R.H., Wilson, W.P.: EEG and Evoked Potentials in Psychiatry and Behavior Neurology. Butterworth-Heinemann (1999)
4. Gregory, B., Qianqian, F., Stefan, A.C., Eric, L.M., Dana, H.B., Juliette, S., et al.: Spatio-temporal imaging of the hemoglobin in the compressed breast with diffuse optical tomography. Phys. Med. Biol. **52**, 3619–3641 (2007)
5. Ferrari, M., Quaresima, V.: A brief review on the history of human functional near-infrared spectroscopy (fNIRS) development and fields of application. NeuroImage. **63**, 921–935 (2012)
6. Huppert, T.J., Hoge, R.D., Diamond, S.G., Franceschini, M.A., Boas, D.A.: A temporal comparison of BOLD, ASL, and NIRS hemodynamic responses to motor stimuli in adult humans. NeuroImage. **19**, 368–382 (2006)
7. Yuan, Z.: Spatiotemporal and time-frequency analysis of functional near infrared spectroscopy brain signals using independent component analysis. J. Biomed. Opt. **18**, 106011 (2013)
8. Cohen, L.: Convolution, filtering, linear systems, the Wiener-Khinchin theorem: generalizations. In: Proceedings of SPIE 1770, Advanced Signal Processing Algorithms, Architectures, and Implementations III, 378–393 (1992)
9. Hyvrinen, A., Hurri, J., Hoyer, P.O.: Natural Image Statistics: A Probabilistic Approach to Early Computational Vision. Springer Publishing Company, Incorporated, London (2009)
10. Friston, K.J., Fletcher, P., Josephs, O., Holmes, A., Rugg, M.D., Turner, R.: Event-Related fMRI: Characterizing Differential Responses. NeuroImage. **7**, 30–40 (1998)
11. Smith, E.C., Lewicki, M.S.: Efficient auditory coding. Nature **439**, 978–982 (2006)
12. Olshausen, B.A., Field, D.J.: Emergence of simple-cell receptive field properties by learning a sparse code for natural images. Nature **381**, 607–609 (1996)
13. Huppert, T.J., Diamond, S.G., Franceschini, M.A., Boas, D.A.: HomER: a review of time-series analysis methods for near-infrared spectroscopy of the brain. Appl. Opt. **48**, D280–D298 (2009)
14. Calvert, G.A., Campbell, R., Brammer, M.J.: Evidence from functional magnetic resonance imaging of crossmodal binding in the human heteromodal cortex. Curr. Biol. **10**, 649–657 (2000)
15. Sassaroli, A., Zheng, F., Hirshfield, M., Girouard, A., Solovey, E., Jacob, R., Fantini, S.: Discrimination of mental workload levels in human subjects with functional near-infrared spectroscopy. J. Innov. Opt. Health Sci. **1**, 227–237 (2008)
16. Cecotti, H.: A time-frequency convolutional neural network for the offline classification of steady-state visual evoked potential responses. Pattern Recognit. Lett. **32**, 1145–1153 (2011)
17. Sainath, T.N., Kingsbury, B., Saon, G., Soltau, H., Mohamed, A., Dahl, G., et al.: Deep Convolutional Neural Networks for Large-scale Speech Tasks. Neural Networks, **64**, 39–48 (2015)

Analyzing Correlations Between Personality Disorders and Frontal Functions: A Pilot Study

Raffaele Sperandeo, Anna Esposito, Mauro Maldonato and Silvia Dell'Orco

Abstract This paper reports the results of an open cross-sectional study in which socio-demographic, clinical, psychopathological and neuropsychological features of outpatients—who consecutively enter for treatments in a private service of psychiatry and psychotherapy—are examined. The involved participants (63 subjects, 24 males and 39 females aged from 18 to 66, mean age 34 years) were assessed for personality disorders (PDs) by administering the Structured Clinical Interview for Diagnosis of axis II disorders (SCID-II) and for frontal lobe hypo-functioning activities by administering the Frontal Assessment Battery (FAB). 21 subjects reached a FAB total score less than 13.5, indicating the presence of frontal function deficits. 21 subjects had a diagnosis of Personality Disorder (PD). The remaining 42 subjects did not meet sufficient criteria for any diagnosis of PD even though they showed one or more typical PD symptoms. It was found that the PD syndromic diagnosis did not significantly correlate with frontal functions deficits, while some PD symptoms (such as abnormal behavior, emotional experiences and pathological cognitive processes) correlated negatively with the FAB total scores and therefore with a frontal lobe hypo-functioning. These significant

R. Sperandeo (✉)
Department of Human Sciences DISU, University of Basilicata, Potenza, Italy
e-mail: raffaele.sperandeo@unibas.it; raffaele.sperandeo@gmail.com

A. Esposito
Department of Psychology, Seconda Università di Napoli, and IIASS, Caserta, Italy
e-mail: iiass.annaesp@tin.it

M. Maldonato · S. Dell'Orco
Department of European and Mediterranean Cultures, Università della Basilicata, Potenza, Italy
e-mail: m.maldonato@unibas.it

S. Dell'Orco
e-mail: silviadellorco@gmail.com

© Springer International Publishing Switzerland 2016
S. Bassis et al. (eds.), *Advances in Neural Networks*, Smart Innovation,
Systems and Technologies 54, DOI 10.1007/978-3-319-33747-0_29

negative correlations suggest that PD symptoms have more ecological value than the DSM V diagnostic categories confirming their artificial nature and uselessness for clinical practices.

Keywords Personality disorders · Frontal lobe hypo-functioning · Executive functions · Personality dimensional model

1   Introduction

Ever since the publication of the fourth version of the *Diagnostic and Statistical Manual of Mental Disorders* (DSM IV) a large number of studies have challenged the nosographic value of Personality Disorders (PD) classified in this manual among the Axis II disorders. The DSM V [1] has partially acknowledged the criticisms made in recent years to the model of PD syndromes and proposed, an alternative description of personality disorders based on a mixed, dimensional and prototype model. However, while waiting that this new and more complex diagnostic approach is shared by all, the official PDs classification has remained unchanged. The DSM V recognizes ten Personality Disorders or syndromes and groups them into the three following clusters:

1. Cluster A: Schizoid Personality Disorder (SPD), Schizotypal Personality Disorder (STPD), Paranoid Personality Disorder (PPD);
2. Cluster B: Histrionic Personality Disorder (HPD), Narcissistic Personality Disorder (NPD), Borderline Personality Disorder (BPD), Antisocial Personality Disorder (ASPD).
3. Cluster C: Avoidant Personality Disorder (APD), Dependent Personality Disorder (DPD), Obsessive-Compulsive Personality Disorder (OCPD)

Patients affected by one of these 10 PD disorders or syndromes, are described by a list of psychopathological traits and abnormal behavior typical of the specific disorder. The psychopathological and behavioral characteristics of each disturbance are not hierarchically organized and the DSM V give to all of them the same weight. For example, the borderline personality disorder (BPD) is described by a list of nine symptoms.[1] According to the DSM V, a diagnosis of BPD can be made if the patient shows at the least 5 of the nine listed symptoms [1]. However, several psychiatric patients do not present a sufficient number of symptoms to obtain a diagnosis of a specific personality disorder, even though they show some of the psychopathological traits typical of these disorders. Therefore, on a clinical level they are bearers of a strong suffering and cannot be officially cured.

---

[1]The symptoms are: avoidance of abandonment, identity disturbance, inappropriate anger, affective instability, suicidal or self-mutilative behavior, impulsivity in two areas, unstable relationships, chronic feeling of emptiness and stress-related paranoid ideation.

There are many reasons to consider unsatisfying the current PD classification among which: the excessive diagnostic heterogeneity within each disorder; the poor reliability and validity of the categories; the high rates of diagnostic overlap among various disorders of axis II; and the scarce relationship with the level of the patient's clinical impairments [2]. Many clinicians and researchers have agreed in considering "syndromes" artificial constructs and believe that the constituent elements of the syndromic picture can be conceptualized as psychopathological dimensions expressed along a continuum of intensity, frequency and degree of clinical severity [3, 4].

It seems indeed that the current syndromes of axis II combine in artificial constructs the character traits, psychopathological characteristics, clinical symptoms and symptomatic behaviors that underlie different and independent pathogenic processes. Numerous studies, dating back to the last years of the past century, have examined the internal consistency of each nosographic category of PD and showed that many DSM criteria to define a specific diagnosis, are useless and redundant [5]. Empirical studies indicate that the dimensional approaches have greater validity than the DSM diagnostic categories [6–8]. Skodol et al. [9] found that the DSM IV diagnostic categories for PDs are, over time, less stable than the symptoms used for their definitions [10] to the extent of which it becomes more useful to account for the symptoms rather than the associated diagnosis. Although the nosographic criterion for the identification of PD symptoms is categorical (present or absent), the majority of diagnostic tools offer a dimensional description where symptoms can be marked as absent, subthreshold, or present. The subthreshold tagging requires that symptoms can be displayed by patients, even though to an extreme that would determine its clinical significance.

In 2001, an expert committee of the *American Psychiatric Association* and of the *National Institute of Mental Health* stressed the need to develop a dimensional model of PDs in order to achieve more effective diagnostic tools [11], and to improve the quality of therapeutic programs [12].

In an attempt to discriminate effectively the symptoms of different origin present in the so called "personality syndromes", exploratory and confirmatory factor analyses studies have been conducted to identify subtended components of the personality traits [3–5]. However, only the research on the borderline personality disorder produced results of some use showing how this disorder consists of three factors unrelated among each other [6].

Alternatively the symptoms of the DSM V diagnostic categories have been validated exploiting their relationships with psychopathological external elements developed as clinical constructs bearing information on PDs regardless of the PDs DSM V defined syndromes.

Three psychopathological constructs external to personality disorders are currently considered related to them: (a) the dimensional models of personality and temperament, such as the Five Factor Model (FFM) and/or the Temperament and Character Inventory (TCI) [8]; (b) disorders of the self-experience as the dissociative phenomena [13]; (c) dysfunctions of the prefrontal cortex as t dis-executive syndromes [14].

Several anatomical, functional and neuropsychological studies have shown a significant relationship between prefrontal cortex functions (orbitofrontal and dorsolateral) and PDs [14, 15]. Executive functions are complex cognitive operations and affect the ability to formulate and monitor action plans to achieve desired goals [16]. They include: executive attention, working memory, the filter of the relevant data, inhibition to irrelevant stimuli, and programming and verification processes [17]. Some PD symptoms seem to overlap with dis-executive phenomena as for example, the fragmentation of the temporal experience (in borderline personalities) and the working memory's ability to select and store in the "here and now" the current perceptual information [18].

In this paper we report a pilot study carried out on a sample of 63 outpatients, with or without a PD diagnosis, consecutively admitted to a program of psychotherapeutic and psychopharmacological treatment. A large number of psychiatric outpatients, even without a diagnosis of PDs, show psychopathological traits typical of PD syndromes. The study aims to assess, if any, relationships between PD symptoms and executive function hypo-functioning in order to validate the PD diagnostic criteria exploiting external psychopathological constructs.

The patients were assessed for PDs through the SCID II [19] and for frontal lobe hypo-functioning activities through the FAB [20].

To evaluate the relationship between PDs or PD symptoms and frontal function hypo-functioning, the scores obtained from the two tests were subjected to the Spearman correlation test.

The correlations emerged are described and discussed with respect to their psychopathological and clinical meanings in the following sections.

2 Methods

2.1 Ethics

The participants were informed that the questionnaires they were requested to fill in, would be exploited both for their clinical assessment and (in respect of anonymity and confidentiality) for research protocols. To preserve anonymity, alphanumeric codes were attributed to all participants. In addition, participants were informed that the diagnostic evaluations would not involve any risk for their physical or mental integrity and would not affect the treatment required for their clinical condition. The study was planned in accordance with advices given by the "CONSORT statement" [21].

2.2 Procedure and Experimental Set-up

The involved participants were evaluated by clinicians with at least 5 years of experience to identify the exclusion criteria. In this phase, an anamnestic evaluation and a clinical assessment of their mental state are made. Subsequently, participants were assessed for personality disorders by administering the SCID-II [19] and for frontal lobe hypo-functioning activities by administering the FAB. [20].

The SCID-II is a semi-structured interview for the diagnostic evaluation of the ten PDs reported by the DSM V. As mentioned above these PDs are grouped into clusters A, B, and C described above.

The SCID-II is applicable in clinical research, where it is used to outline PD profiles as well as compare other questionnaires for assessing PDs. It allows to assess the PD symptoms as dimensional constructs. The evaluator can indeed code a PD symptom as absent, present or at a subthreshold level. In the last case, the PD symptom is present but does not affect the patient's daily life.

The FAB consists of six subtests related cognitive activities such as: conceptualization, mental flexibility, motion programming, sensitivity to interference, inhibitory control, and environmental autonomy. It has good inter-raters reliability ($\kappa = 0.87$, $p < 0.001$), internal consistency (Cronbach's alpha = 0.78) and discriminant validity of 89.1 %. It significantly correlates with the Wisconsin Card Sorting Test [20] and can be performed in 10 min.

The FAB was designed as a fast and efficient test to detect frontal lobe dysfunctions in a variety of patients and has been shown to be easy to administer and not frustrating for patients. The battery allows to evaluate severe disfunctions of the executive processes monitored by both the orbitofrontal cortex and dorsolateral cortex [22].

2.3 Participants

The patients were involved in the present study after giving their informed consent. The exclusion criteria were applied for subjects with mild cognitive deficits (both congenital and/or acquired), acute psychotic symptoms, acute manic symptoms, severe or moderate acute depressive symptoms and acute anxiety symptoms of severe degrees. At present the pilot project included 63 subjects, 24 males and 39 females aged from 18 to 66 years, mean age 34 years (SD = 11.75 years). Among these patients, only 21 reached a FAB total score less than 13.5, indicating the presence of frontal function deficits (The FAB total score is 14.28 (SD = 2,64). The FAB subtest more affected by the examined patients is "conceptualization": 42 subjects (66,7 %) showed a deficit in this area.

Table 1 reports the percentage of subjects that reached a FAB score denoting a fontal lobe hypo-functioning. In addition it reports, for each of the six FAB cognitive subtests, the percentages of subjects whose deficit falls into each of them.

Table 1 Percentage of subjects that reached a FAB score denoting a fontal tr lobe hypo-functioning for each of the six FAB cognitive subtests

	No (%) of subjects with deficit
FAB total score <13.5	21 (33.3 %)
Conceptualization	42 (66.7 %)
Mental flexibility	22 (34.9 %)
Motion programming	19 (30.2 %)
Sensitivity to interference	15 (23.8 %)
Inhibitory control	23 (36.5 %)
Environmental autonomy	1 (1.6 %)

3   Results

Table 2 reports the participants' socio-demographic variables. As it can be read from it, of the 63 subjects considered, 43 were employed, 19 unemployed and 1 was retired. In addition, 37 subjects were single, 22 married, 3 divorced, and 1 was a widower. On the education side, 15 subjects were graduated, 38 had a high school diploma, 8 a secondary school diploma and 1 subject completed only the primary school.

There was no evidence of correlation between the socio-demographic variables presented in Table 2, the PD diagnosis obtained with SCID II and the FAB scores.

Among the participants, 21 were diagnosed with one or more PDs through the SCID II. The remaining 42 subjects did not meet sufficient criteria for any PD diagnosis, even though they showed one or more typical PD symptoms. The PD trait more represented in the sample was the narcissistic one that was evidenced in the 65 % (41) of patients.

Table 2 Participants' socio-demographic variables

	No (%) of subjects
Males	24 (38 %)
Females	39 (61 %)
Employed	43 (68.3 %)
Unemployed	19 (30.25)
Pensioners	1 (1.6 %)
Bachelor degree	15 (23.8 %)
High school	38 (60.3 %)
Secondary school	8 (12.7 %)
Primary school	1 (1.6 %)
Single	37 (58.7 %)
Married	22 (34.9 %)
Divorced	3 (4.8 %)
Widower/widow	1 (1.6 %)

Table 3 List of Personality Disorders (PD) found in 21 subjects participating to the experiment

	No (%) of subject with a PD diagnosis	No (%) of subjects with at least one PD Trait
Avoidant PD	6 (9.5 %)	14 (22.2 %)
Dependent PD	0 (0 %)	30 (47.6 %)
Obsessive compulsive PD	5 (7.9 %)	31 (49.2 %)
Paranoid PD	1 (1.6 %)	20 (31.7 %)
Schizotypal PD	1 (1.6 %)	15 (23.8 %)
Schizoid PD	1 (1.6 %)	5 (7.9 %)
Histrionic PD	0 (0 %)	10 (15.9 %)
Narcissistic PD	5 (7.9 %)	41 (65 %)
Borderline PD	9 (14.3 %)	32 (50.8 %)
1 diagnosis of PD	17 (27 %)	–
2 diagnoses	3 (4.8 %)	–
5 diagnoses	1 (1.6 %)	–

The PD diagnoses are reported in Table 3 and grouped in clusters A, B and C as defined in the DSM IV.

It can be seen from Table 3 that 11 subjects group into the cluster C, with 6 subjects diagnosed as having an Avoidant Personality (APD), and 5 having an Obsessive-Compulsive Personality Disorder (OCPD). 3 subjects grouped into the cluster C with 1 subject having a Paranoid Personality Disorder (PPD), 1 a Schizoid Personality Disorder (SPD) and 1 a Schizotypal Personality Disorder (STPD).

14 subjects grouped into the cluster B, with 5 as having a Narcissistic Personality Disorder (NPD), and 9 a Borderline Personality Disorder (BPD).

Table 4 reports the results of the Spearman correlation test for the FAB total scores, and FAB components vs PD symptoms.

Diagnoses originated exploiting the DSM V syndromic criterion show no correlation with FAB total scores, while some PD symptoms, taken singularly, are negatively correlated with frontal function deficits. These PD symptoms are the following: anxiety of abandonment, magical thinking, unusual perceptions, odd thinking and speech, impulsiveness, identity alteration, tendency to develop psychotic phenomena under stress, chronic feeling of emptiness and boredom, inappropriate and intense anger, as it can be seen in Table 4. These correlations suggest a possible link between a specific PD symptom and a determinate executive function hypo-functioning. However, more data are needed to obtain a statistical consistency.

Table 4 FAB total score, FAB components vs PD symptoms: Spearman correlation test

FAB	Personality disorder symptoms	Rho	p
Total score	Anxiety of abandonment	−0.291	<0.05
	Magical thinking	−0.283	<0.05
	Unusual perceptions	−0.355	<0.01
	Odd thinking and speech	−0.269	<0.05
	Impulsiveness	−0.443	<0.01
	Psychotic phenomena under stress	−0.397	<0.01
	Alteration of the identity	−0.266	<0.05
	Chronic feeling of emptiness and boredom	−0.289	<0.05
	Inappropriate and intense anger	−0.264	<0.05
Inhibitory control	Unusual perceptions	−0.296	<0.05
	Odd thinking and speech	−0.301	<0.05
Planning	Excessive relational intimacy	−0.314	<0.05
Conceptualization	Difficulty expressing disagreement	−0.374	<0.01
	Psychotic phenomena under stress	−0,359	<0.05
	Anxiety of abandonment	−0.424	<0.01
	Unusual perceptions	−0.329	<0.01
	Chronic feeling of emptiness and boredom	−0.266	<0.05
Mental flexibility	Excessive conscientiousness	−0.314	<0.05
	Greediness	−0.298	<0.05

4 Discussion

The diagnostic manual of the American Psychiatric Association describes personality disorders on the basis of syndromic criteria. According to this model, each personality disorder is defined by a list of symptoms hierarchically equivalent. To make a diagnosis of PD is necessary to identify a fixed number of symptoms listed by the syndromic criteria.

However, the listed symptoms (or pathological personality traits) arise from the classical descriptions of psychopathology; they correlate with neurobiological phenomena, are normally distributed in the population and have a stable genetic structure [6, 7]. The diagnoses derived from the DSM V diagnostic criteria (e.g. Avoidant Personality Disorders APD) are however, artificial categories often disconnected from data exploited in practices.

Many of the subjects involved in the present experiment did not meet sufficient criteria for a PD diagnosis even though they suffer for one or more typical PD symptoms and express a great clinical burden. In addition 30 % of them exhibit major frontal function impairments. In particular, they were mostly impaired in the ability to conceptualize.

In our analysis no correlation was found among PD diagnoses derived from DSM V syndromic criteria and FAB total scores. However a significant negative correlation was found between PD symptoms and frontal function deficits, supporting the debate that PD syndromic constructs proposed by DSM V are basically artificial categories of little relevance to clinical work.

The negative correlation between PD pathological traits and executive function deficits open to a PD psychopathological analysis based on the validation of the syndromic constructs.

The above findings suggest interesting new clinical relapses relating PD symptoms and frontal lobe impairments. However, the relatively small number of involved patients does not allow to make solid and consistent conclusions. To this aim more data and more proper experimental designs are required.

References

1. American Psychiatric Association: Diagnostic and Statistical Manual of Mental Disorders, 5th Edition (DSM-IV TR). American Psychiatric Press, Washington DC (2013)
2. Krueger, R.F., Skodol, A.E., Livesley, W.J., Shrout, P.E., Huang, Y.: Synthesizing dimensional and categorical approaches to personality disorders: refining the research agenda for DSM-V Axis II. Int. J. Methods Psychiatr. Res. 16(S1), S65–S73 (2007)
3. Livesley, W.J.: A framework for integrating dimensional and categorical classifications of personality disorder. J. Pers. Disord. 21(2), 199–224 (2007)
4. Paris, J.: Neurobiological dimensional models of personality: a review of three models. In Widiger, T.A., Simonsen, E., Sirovatka, P.J., Regier, D.A. (eds.): Dimensional Models of Personality Disorders. Refining the Research Agenda for DSM-V, pp. 61–72. American Psychiatric Association, Washington, DC
5. Grilo, C.M., McGlashan, T. H., Morey, L.C., Gunderson, J.G., Skodol, A.E., Shea, et al.: Internal consistency, intercriterion overlap and diagnostic efficiency of criteria sets for DSM-IV schizotypal, borderline, avoidant and obsessive-compulsive personality disorders. Acta Psychiatr. Scand. 104(4), 264–72 (2001)
6. Livesley, W.J.: Diagnostic dilemmas in classifying personality disorder In: Phillips K.A., First, M.B., Pincus, H.A (eds.), Advancing DSM: Dilemmas in psychiatric diagnosis, pp. 153–190. American Psychiatric Association, Washington, DC (2003)
7. Trull, T.J., Durrett, C.A.: Categorical and dimensional models of personality disorder. Annu. Rev. Clin. Psychol. 1, 355–380 (2005)
8. Widiger, T.A., Simonsen, E., Krueger, R., Livesley, W.J., Verheul, R.: Personality disorder research agenda for the DSM-V. J. Pers. Disord. 19(3), 315–338 (2005)
9. Skodol, A.E., Pagano, M.E., Bender, D.S., et al.: Stability of functional impairment in patients with schizotypal, borderline, avoidant, or obsessive-compulsive personality disorder over two years. Psychol. Med. 35, 443–451 (2005)
10. Cantone, D., Sperandeo, R., Maldonato, M.: A dimensional approach to personality disorders in a sample of juvenile offenders. Rev. Latinoam. Psicopat. Fund. 15(1), 42–57 (2012)
11. Rounsaville, B.J., Alarcon, R.D., Andrews, G., Jackson, J.S., Kendell, R.E., Kendler, K.: Basic nomenclature issues for DSM-V. In: Kupfer, D.J., First, M.B., Regier, D.E. (eds.) A research agenda for DSM-V, pp. 1–29. American Psychiatric Association, Washington, DC (2002)
12. Verheul, R.: Clinical utility of dimensional models for personality pathology. J. Pers. Disord. 19(3), 283–302 (2005)
13. Meares, R., Gerull, F., Stevenson, J., Korner, A.: Is Self Disturbance the Core of Borderline Personality Disorder? An Outcome Study of Borderline Personality Factors. Aust. N. Z. J. Psychiatry 45, 214–222 (2011)

14. Berlin, H.A., Rolls, E.T., Iversen, S.D.: Borderline Personality Disorder, Impulsivity, and the Orbitofrontal Cortex. Am. J. Psychiatry 162, 2360–2373 (2005)
15. Soloff, P.H., Meltzer, C.C., Becker, C., Greer, P.J., Kelly, T.M., Constantine, D.: Impulsivity and prefrontal hypo-metabolism in borderline personality disorder. Psychiatry Res. 123(3), 153–163 (2003)
16. Fabbro, F.: Manuale di neuropsichiatria infantile. Roma, Carocci (2012)
17. Gazzaniga, M.S., Ivry, R.B., Mangun, G.R.: Neuroscienze cognitive. Bologna, Zanichelli (2005)
18. Hart., S.: Brain, Attachment, Personality. An introduction to neuroaffective development. London, Karnac Book (2008)
19. First, M.B., Gibbon, M., Spitzer, R.L., Williams, J.B.W., Benjamin, L.: User's guide for the Structured Clinical Interview for DSM-IV Axis II Personality Disorders (SCID-II). Biometrics Research Department, New York State Psychiatric Institute, New York (1996)
20. Dubois, B., Slachevsky, A., Litvan, I., Pillon, B.: A frontal assessment battery at bedside. Neurology 55(11), 1621–1626 (2000)
21. Schulz, K.F., Altman, D.G., Moher, D.: CONSORT Statement 2010: linee guida aggiornate per il reporting di trial randomizzati a gruppi paralleli. Evidence 4(7), e1000024 (2012)
22. Chapados, C., Petrides, M.: Impairment only on the fluency subtest of the Frontal Assessment Battery after prefrontal lesions. Brain 136(10), 2966–2978 (2013)
23. Diamond, A., Lee, K.: Interventions shown to aid executive function development in children 4 to 12 years old. Science 333(6045), 959–964 (2011)

Processing Bio-medical Data
with Class-Dependent Feature Selection

Nina Zhou and Lipo Wang

Abstract In this paper, we show how to select different feature subsets for different classes, i.e., class-dependent feature subsets, for biomedical data. A feature importance ranking measure, i.e., class separability measure, is used to rank features for each class and obtain class-dependent feature importance ranking. Then several feature subsets for each class are formed and an "optimal" one for each class is determined through a classifier, e.g., the support vector machine (SVM). Our method of class-dependent feature selection is applied on several biomedical data sets and compared with class-independent feature selection. The experimental result shows that our approach to class-dependent feature selection is efficient in reducing feature dimension and producing satisfactory classification accuracy.

Keywords Support vector machines · Class-dependent feature selection · Class-independent feature selection · Class separability measure

1 Introduction

When dealing with biomedical data, the data dimensionality, i.e., the number of input features, can be quite large. To reduce computational burden and noise, it is often desirable to reduce the data dimensionality. An effective approach is feature extraction, for example, principal component analysis (PCA) [10, 25] and singular vector decomposition [22]. The resultant features via feature extraction are obtained by certain tranforms from the original features and are therefore different from the original

N. Zhou
Institute for Infocomm Research, 21-01 Connexis (South Tower),
1 Fusionopolis Way 138632, Singapore

L. Wang (✉)
School of Electrical and Electronic Engineering, Nanyang Technological University,
Block S1, 50 Nanyang Avenue, Central Area 639798, Singapore
e-mail: elpwang@ntu.edu.sg

© Springer International Publishing Switzerland 2016   303
S. Bassis et al. (eds.), *Advances in Neural Networks*, Smart Innovation,
Systems and Technologies 54, DOI 10.1007/978-3-319-33747-0_30

features. In this paper, we focus on another approach, i.e., feature selection, which chooses a subset of input features from the entire input feature set.

Based on the different measures used to find the best feature subset, feature selection methods are divided into the following two categories: filter approaches [19] and wrapper approaches [18]. The classical RELIEF algorithm [19] and its extended version RELIEFF [21] are examples of filter approaches. They assign a weight to each feature and then update the weight according to training instances. These weights represent relevance of features, therefore, all features are ranked according to their weights and those features with weights above a predefined threshold are selected. Wrapper approaches "wrap" feature selection around a classifier. Most wrapper approaches also utilize heuristic search techniques, such as sequential forward and backward search [35], hill climbing [2], and best-first search [18], to first search for possible feature subsets, then evaluate those feature subsets through the classifier, and finally determine an optimal subset in terms of classification accuracy. In this paper, we adopt a wrapper approach to select features for better accuracy.

Considering the possibility that different groups of features may have different abilities in distinguishing different classes [1, 7, 8, 13, 14, 23, 24, 26, 28–30, 34, 37, 38], we will choose different feature subsets for different classes, which is called class-dependent feature selection [29], as opposed to the usual class-independent feature selection [4–6, 11, 19, 36], which is to select a common feature subset for all the classes in a given classification problem. We note that class-independent feature selection is in fact a special case of class-dependent feature selection, that is, if all feature subsets in class-dependent feature selection for all classes happen to be the same, one obtains class-independent feature selection. The filter and wrapper approaches mentioned in [19, 21] belong to class-independent feature selection. For class-dependent feature selection, Baggenstoss [1] provided related theoretical analysis and utilized it on some artificial data sets. Oh et al. [28, 29] proposed a filter approach to selecting class-dependent feature subsets for the CENPARMI handwritten numerical database. The experimental results [28, 29] showed that classification accuracies of class-dependent feature selection were better compared to those of class-independent feature selection. In this paper, we will demonstrate a wrapper approach to selecting class-dependent features for biomedical data using the support vector machine (SVM) [31, 32] as the classifier.

This paper is organized as follows. In Sect. 2, we review the class separability measure (CSM), and introduce our approach to class-dependent feature selection. In Sect. 3, we provide experiment results of our method on four biomedical data sets from the UCI machine learning repository databases [27], and compare the results with those of class-independent feature selection. In the end, we present conclusions about the present work.

2 Methodology

2.1 The Class Separability Measure

Class separability measure (CSM) has been used by many researchers with different versions. The class separability proposed by Oh et al. [28, 29] is represented by $S(c_i, c_j, x)$, where c_i and c_j represent class i and class j of the data set, respectively, and x is a training sample. Each feature's class separability is calculated individually, e.g., $S(c_i, c_j, x_p)$ for feature p, and features are ranked according to their class separation values. Fu and Wang [11] defined another class separability measure to rank each feature's classification capability. This CSM includes two distance elements: the within-class distance (distance between patterns within each class) and the between-class distance (the distance between patterns among different classes), which are described in Eqs. 1 and 2. According to [11], we will adopt the ratio of the within-class distance to the between-class distance to measure each feature's classification capability. For the whole training data, the within-class distance S_w [11] is calculated as:

$$S_w = \sum_{c=1}^{C} P_c \sum_{j=1}^{n_c} (x_{cj} - m_c)(x_{cj} - m_c)^T \tag{1}$$

and the between-class distance S_b [11] is calculated as:

$$S_b = \sum_{c=1}^{C} P_c (m_c - m)(m_c - m)^T \tag{2}$$

Here C denotes the number of classes and P_c denotes the probability of class c. n_c refers to the number of samples in class c and x_{cj} refers to sample j in class c. m_c refers to the mean vector of class c and m refers to the mean vector of all the training samples. As mentioned above, the smaller the ratio S_w/S_b, the better the separability. When evaluating one feature's classification capability, we calculate the ratio (S_w/S_b) with the current feature removed, i.e., denoted as S_w'/S_b'. The greater S_w'/S_b', the more important the removed feature is. Hence, we may evaluate the importance level of the features according to the ratio [11] with an attribute deleted each time in turn.

2.2 Our Approach to Class-Dependent Feature Selection

We describe our class-dependent feature selection method in three steps. In step one, we convert a C-class classification problem to C 2-class classification problems. Each problem only has two classes: the current class and the other one including all the other classes. In step two, for each 2-class problem, we adopt the ranking measure CSM to evaluate the importance of each feature. In step three, based on

each class-dependent feature importance ranking list, we form different feature subsets for each class by sequentially adding one feature into the previous subset. Each
feature subset is evaluated through an SVM and the feature subset corresponding to
the highest classification accuracy will be our choice for this class.

During the process, a feature mask is introduced to describe features' states, i.e.,
kept or removed. The feature mask is a vector, each element of which has only two
values '0' and '1', in which '0' represents the absence of a particular feature and
'1' represents the presence of a feature. For example, considering a data set with 5
features $\{x_1, x_2, x_3, x_4, x_5\}$, if the optimal feature subset obtained is $\{x_1, x_3, x_5\}$, the
corresponding feature mask should be the vector $\{1, 0, 1, 0, 1\}$.

2.3   SVM with Class-Dependent Features

We first build several SVM models and then combine them together for accommodat-
ing class-dependent feature subsets. Each model is a binary classifier and is specific
for one class. In the following, we will introduce the construction process.

1. The training process: In this process, we construct C SVM models by training
   patterns, i.e., each class has its own model according to its specific feature subset. For example, the model i is trained with all the training examples in class i
   having positive labels and all the examples in other classes having negative labels. Specifically, all the training examples need to be filtered by a feature mask
   of class i before they are input for training. For instance, if the feature mask of
   class i has $n^{(i)}$ '1', all the training examples to form class i will have $n^{(i)}$ features
   as the input and those features corresponding to '0' in the feature mask are removed. The output can be either '+1' or '−1'. If the input pattern x_j belongs to
   class i, we consider it as a positive sample ('+1'). Or we consider it as a negative
   sample ('−1'). The ith SVM model solves the following problem [16]:

$$\min_{\omega^i, b^i, \xi^i} \frac{1}{2} \omega^{i^T} \omega^i + \varsigma^i \sum_{j=1}^{l} \xi_j^i$$
$$\omega^{i^T} \phi(x_j) + b^i \geq 1 - \xi_j^i, \text{ if } y_j = i,$$
$$\omega^{i^T} \phi(x_j) + b^i \leq -1 \xi_j^i, \text{ if } y_j \neq i, \tag{3}$$
$$\xi_j^i \geq 0, \quad j = 1, \dots, l$$

where ϕ is the mapping function. ς^i is the penalty parameter for class i, and ξ_j^i are
"slack variables" for class i. x_j corresponds to sample j in l samples. Minimizing
$\frac{1}{2}(\omega^i)^T \omega^i$ means maximizing the margin between two groups of data. $\varsigma^i \sum_{j=1}^{l} \xi_j^i$ is
a penalty term used to reduce the number of training errors in case of nonlinear
separable data.

2. The testing process: After the class-dependent models are constructed, we will use them to test unlabeled patterns. Same as the training process, each testing pattern is filtered with one class's feature mask before input into the corresponding SVM model, i.e., the original attributes corresponding to '0' in the feature mask are removed. Among the C outputs, the testing pattern x_j, belongs to the class with the largest output value:

$$\text{Class of } x_j \equiv \text{argmax}_{i=1,2,...,C}(\omega^{i^T}\phi(x_j) + b^i) \tag{4}$$

3 Experiments and Discussions

In order to demonstrate whether class-dependent feature selection is more efficient than class-independent feature selection, we conduct the experiment on two biomedical data sets from the UCI machine learning repository databases [27]. Two terms used for comparison between the two methods are the number of features deleted and the classification accuracy.

3.1 Experimental Data

The first data set is the Ecoli data set. It has 7 attributes (localization sites of the protein) and 8 classes. The number of instances is 336. The second data set is the processed Cleveland data set. It mainly concerns heart disease diagnosis and is collected from the Cleveland Clinic Foundation. There are originally 303 samples, 13 features and 5 classes. Because there are 6 samples with unknown feature values, we remove the 6 samples from our experiment.

3.2 Experiment and Results

From various kinds of SVM software packages, LIBSVM 3.1 [3] with the RBF kernel was chosen in our experiment. 10-fold corss validation method is used to calculate the accuracy. In Table 1, the results for the Ecoli data set show us that Ecoli data set [15] has very different numbers of features deleted for different classes with class-dependent feature selection. The result on the Cleveland Heart Disease data (Table 2) [9] also show that different classes have very different feature subsets. Class 1 has few features removed, i.e., on average 1.9 (Table 3). While for class 2, 3, 4 and 5, the number of features deleted in the 10 simulations are on average within the range of [9, 12].

Table 1 Feature selection results for the Ecoli data set

Feature selection approach	Classes	The number of features deleted in each of the 10 simulations	Average number of features deleted
Class-independent	All classes	0 0 0 0 1 0 0 1 0 0	0.2
Class-dependent	Class 1	1 0 0 0 0 1 1 0 0 1 0	0.3
	Class 2	0 0 0 0 0 0 0 0 0 0	0
	Class 3	0 0 1 1 1 1 0 1 1 1	0.7
	Class 4	2 4 1 4 2 4 4 5 4 4	3.4
	Class 5	2 1 3 0 0 3 1 2 2 3	1.7
	Class 6	4 4 4 4 4 2 4 4 4 4	3.8
	Class 7	6 6 6 6 6 4 6 6 6 6	5.8
	Class 8	6 6 6 6 6 6 6 6 6 6	6.0

Table 2 Feature selection results for the Cleveland heart disease data set

Feature selection approach	Classes	The number of features deleted in each of the 10 simulations	Average number of features deleted
Class-independent	All classes	1 3 7 1 0 7 4 7 3 2	3.5
Class-dependent	Class 1	1 7 1 4 1 1 1 1 1 1	1.9
	Class 2	12 12 12 12 12 2 12 12 12 12	11
	Class 3	12 12 9 4 12 12 0 12 12 12	9.7
	Class 4	8 5 1 12 5 12 12 12 12 11	9.0
	Class 5	12 12 12 12 12 12 12 12 12 12	12

Table 3 Classification accuracy comparisons among different feature selection methods for the two biomedical data sets

Methods and data	Ecoli data set (%)	Cleveland data set (%)
Without feature selection	86.81	55.56
Class-independent feature selection	86.61	56.23
Class-dependent feature selection	86.91	58.61

In Table 3, we present classification accuracies for 3 different conditions, i.e., without feature selection and with class-dependent and class-independent feature selection. The obvious improvement on the classification accuracy is for the Cleveland data set. Compared with the accuracy on the data without feature selection, our method has the accuracy increased by about 3 %. Compared with that of the class-independent method, our method has increased the accuracy from 56.23 to 58.61 %.

4 Conclusions

In this paper, we demonstrated an approach to class-dependent feature selection. We adopted class separability measure [11] to evaluate feature importance, based on which an optimal feature subset was determined for each class through the SVM. The experimental results for two biomedical data sets [27] show that each class has a different feature subset which includes representative features for classifying the current class from the other classes, and the corresponding classifier can improve or at least maintain the classification accuracy using those class-dependent feature subsets.

References

1. Baggenstoss, P.M.: Class-specific features in classification. IEEE Trans. Signal Process. pp. 3428–3432 (1999)
2. Caurana, R.A., Freitag, D.: Greedy attribute selection. In: Proceedings of the Eleventh International Conference on Machine Learning, pp. 28–36. Morgan Kaufmann Publishers, NEW Brunswick, NJ (1994)
3. Chang, C.-C., Lin, C.-J.: LIBSVM: a library for support vector machines. http://www.csie.ntu.edu.tw/~cjlin/libsvm (2001)
4. Chu, F., Wang, L.P.: Gene expression data analysis using support vector machines. In: Proceedings of the International Joint Conference on Neural Networks 2003, vols. 1–4, pp. 2268–2271 (2003)
5. Chu, F., Xie, W., Wang, L.P.: Gene selection and cancer classification using a fuzzy neural network. In: Proceedings of the North-American Fuzzy Information Processing Conference (NAFIPS 2004), vol. 2, pp. 555–559 (2004)
6. Chu, F., Wang, L.P.: Applications of support vector machines to cancer classification with microarray data. Int. J. Neural Syst. 15(6), 475–484 (2005)
7. Crawford, M.M., Kumar, S., Ricard, M.R., Gibeaut, J.C., Neuenschwander, A.: Fusion of airborne polarimetric and interferometric SAR for classification of coastal environments. IEEE Trans. Geosci. Remote Sens. 37, 1306–1315 (1999)
8. Desai, M., Shazeer, D.J.: Acoustic transient analysis using wavelet decomposition. In: IEEE Conference on Neural Networks for Ocean Engineering, pp.29–40 (1991)
9. Detrano, R.: The Cleveland Heart Disease Data Set. V.A. Medical Center, Long Beach and Cleveland Clinic Foundation (1988)
10. La Foresta, F., Morabito, F.C., Azzerboni, B., Ipsale, M.: PCA and ICA for the extraction of EEG components in cerebral death assessment. In: IJCNN 05. Proceedings of 2005 IEEE International Joint Conference on Neural Networks, vol. 4, pp. 2532–2537 (2005)
11. Fu, X.J., Wang, L.P.: Data dimensionality reduction with application to simplifying RBF network structure and improving classification performance. IEEE Trans. Syst. Man Cybern. B Cybern. 33(3), 399–400 (2003)
12. Fu, X.J., Wang, L.P.: A GA-based novel RBF classifier with class-dependent features. In: Proceedings of 2002 Congress on Evolutionary Computation, no. 2, pp. 1890–1894 (2002)
13. Fu, X.J., Wang, L.P.: Rule extraction from an RBF classifier based on class-dependent features. In: CEC2002: Proceedings of the 2002 Congress on Evolutionary Computation, vols. 1 and 2, pp. 1916–1921 (2002)
14. Fu, X.J., Wang, L.P.: A rule extraction system with class-dependent features. In: Ghosh, A., Jain, L.C. (eds.) Evolutionary Computing in Data Mining, pp. 79–99. Springer, Berlin (2005)

15. Horton, P., Nakai, K.: A probablistic classification system for predicting the cellular localiza-
    tion sites of proteins. In: Intelligent Systems in Molecular Biology, pp.109–115 (1996)
16. Hsu, C.-W., Lin, C.-J.: A Comparison of methods for multi-class support vector machines.
    IEEE Trans. Neural Netw. **13**(2), 415–425 (2002)
17. Hsu, C.W., Chang, C.C., Lin, C.J.: A practical guide to support vector classification. National
    Taiwan University, Department of Computer Science and Information Engineering, Taipei,
    Taiwan (2003)
18. John, G.H., Kohavi, R., Pfleger, K.: Irrelevant features and the subset selection problem. In:
    Proceedings of the Eleventh International Conference on Machine Learning, pp. 367–370.
    AAAI Press, Portland (1994)
19. Kira, K., Rendell, L.A.: The feature selection problem: traditional methods and a new algo-
    rithm. In: Proceedings of 10th National Conference on Artificial Intelligence, pp. 129–134.
    AAAI Press/MIT press, Park, CA (1992)
20. Koller, D., Sahami, M.: Toward Optimal Feature Selection. In: Proceedings of the 13th Inter-
    national Conference on Machine Learning (ML), pp. 284–292, Bari, Italy (1996)
21. Kononenko, I.: Estimating attributes: analysis and extensions of RELIEF. In: Proceeding of the
    European Conference on Machine Learning (ECML94), pp. 171–182. Springer-Verlag, Berlin,
    Heidelberg (1994)
22. Liu, B., Wan, C.R., Wang, L.P.: An efficient semi-unsupervised gene selection method via
    spectral biclustering. IEEE Trans. Nano Biosci. **5**(2), 110–114 (2006)
23. Marchiori, E.: Class dependent feature weighting and K-nearest Neighbor classification. Patt.
    Recogn. Bioinform. LNCS **7986**, 69–78 (2013)
24. Mohammadi, M., Raahemi, B., Akbari, A., Nassersharif, B.: New class-dependent feature
    transformation for intrusion detection systems. Secur. Commun. Netw. **5**, 1296–1311 (2012)
25. Morabito, C.F.: Independent component analysis and feature extraction techniques for NDT
    data. Mater. Eval. **58**(1), 85–92 (2000)
26. Musselman, M., Djurdjanovic, D.: Time-frequency distributions in the classification of
    epilepsy from EEG signals. Expert Syst. Appl. **39**, 11413–11422 (2012)
27. Newman, D.J., Hettich, S., Blake, C.L., Merz, C.J.: UCI repository of machine learning data-
    bases. University of California, Department of Information and Computer Science, Irvine, CA
    (1998). http://www.ics.uci.edu/~mlearn/MLRepository.html
28. Oh, I.S., Lee, J.S., Suen, C.Y.: Using class separation for feature analysis and combination
    of class-dependent features. In: Fourteenth International Conference on Pattern Recognition,
    no.1, pp. 453–455 (1998)
29. Oh, I.S., Lee, J.S., Suen, C.Y.: Analysis of class separation and combination of class-dependent
    features for handwriting recognition. IEEE Trans. Patt. Anal. Mach. Intell. no.21, pp. 1089–
    1094 (1999)
30. Tian, J., Li, M., Chen, F., Feng, N.: Learning subspace-based rbfnn using coevolutionary al-
    gorithm for complex classification tasks. IEEE Trans. Neural Netw. Learn. Sys. (2015)
31. Vapnik, V.: Statistical Learning Theory. Wiley, New York (1998)
32. Wang, L.P. (ed.): Support Vector Machines: Theory and Applications. Springer, New York
    (2005)
33. Wang, L.P., Chu, F., Xie, W.: Accurate cancer classification using expressions of very few
    genes. IEEE/ACM Trans. Bioinf. Comput. Biol. **4**(1), 40–53 (2007)
34. Wang, L.P., Zhou, N., Chu, F.: A general wrapper approach to selection of class-dependent
    features. IEEE Trans. Neural Netw. **19**(7), 1267–1278 (2008)
35. Wang, L.P., Fu, X.J.: Data Mining with Computational Intelligence. Springer, Berlin (2005)
36. Zhou, N., Wang, L.P.: Effective selection of informative SNPs and classification on the
    HapMap genotype data. BMC Bioinf. **8**, 484 (2007)
37. Zhou, N., Wang, L.P.: Class-dependent feature selection for face recognition. In: Advances in
    Neuro-Information Processing, Part II, vol. 5507, pp. 551–558 (2009). Proceedings of 15th
    International Conference on Neural Information Processing, ICONIP 2008, Auckland, New
    Zealand, 2008
38. Zhou, W.G., Dickson, J.: A novel class dependent feature selection method for cancer bio-
    marker discovery. Comput. Biol. Med. **47**, 66–75 (2014)

LQR Based Training of Adaptive Neuro-Fuzzy Controller

Usman Rashid, Mohsin Jamil, Syed Omer Gilani and Imran Khan Niazi

Abstract The focus of this paper is the design and implementation of adaptive network based fuzzy inference (ANFIS) controller by using the training data obtained from a system controlled by Linear Quadratic Regulator (LQR). This work is motivated by the need to remove stochastic observer required for LQR in noisy environments while at the same time to have optimal performance. This theory is validated by taking the well investigated case of Active Suspension System of a Quarter Vehicle. The performance of the obtained ANFIS controller is tested using *Quanser*$^{©}$ Active Suspension Plant. It is observed that the ANFIS controller gives good close loop performance, while removing the requirement of a stochastic observer.

Keywords Active suspension system · Quarter vehicle model · LQR · Fuzzy inference system · Adaptive neuro fuzzy inference system · Hardware-in-loop

U. Rashid (✉) · M. Jamil · S.O. Gilani
Department of Robotics and Artificial Intelligence, National University of Sciences and Technology, Islamabad, Pakistan
e-mail: usman@smme.edu.pk

M. Jamil
e-mail: mohsin@smme.nust.edu.pk

S.O. Gilani
e-mail: omer@smme.nust.edu.pk

I.K. Niazi
Department of Health Science and Technology, Aalborg University, Aalborg, Denmark
e-mail: imrankn@hst.aau.dk; imran.niazi@nzchiro.co.nz; iniazi@aut.ac.nz

I.K. Niazi
Center for Chiropractic Research, New Zealand College of Chiropractic, Auckland, New Zealand

I.K. Niazi
Health & Rehabilitation Research Institute, Auckland University of Technology, Auckland, New Zealand

© Springer International Publishing Switzerland 2016
S. Bassis et al. (eds.), *Advances in Neural Networks*, Smart Innovation, Systems and Technologies 54, DOI 10.1007/978-3-319-33747-0_31

1 Introduction

Linear quadratic regulator (LQR) is a widely used technique in optimal control applications [1, 2]. It provides control regulation while satisfying a quadratic cost function. As is evident from its name, it is a linear controller and, thus, suitable for systems which exhibit linear response characteristics about an equilibrium point. LQR requires state feedback for its implementation. This requirement can be removed by using an observer. However, in noisy environment (such as output from an accelerometer), the design and tuning of the observer becomes a cumbersome task.

On the other hand, fuzzy inference controllers are universal nonlinear systems. Their ability to model and control nonlinear systems has been established by extensive research [3, 4]. A fuzzy inference controller is obtained by translating the knowledge of an expert into fuzzy sets, membership functions, if-then rules and inference mechanism [5]. However, expert knowledge becomes limited in case of more complex systems. This limitation can be overcome by using Adaptive Network based Fuzzy Inference System [6]. This technique models the fuzzy system as a neural network. Algorithms are available which can be used to train this fuzzy neural network by using input/output data from the plant to be controlled. Inverse model learning from open-loop data along with hybrid training algorithm is used to achieve this task [7]. Another technique is that of acquiring data from the closed loop controlled system [8].

The aim of this work is to bring together the advantages of both these control strategies by training the ANFIS controller using the data obtained from a system controlled by LQR. Plant output along with LQR control law are recorded. This data is then used in off-line training of various ANFIS controllers having different number and type of input/output membership functions. Their learning accuracy is then checked against test data obtained from the system. A comparison of each controller against the performance of LQR controller is also presented.

In order to implement and validate this idea, the active suspension system of a quarter car model is taken as a test case. Scaled down active suspension plant made by *Quanser*© is used [9]. It is well suited for this study because it uses an accelerometer as in commercial suspension systems to measure the vertical acceleration of vehicle body. This accelerometer has noisy output and, thus, requires a stochastic observer such as Kalman filter to produce system states. Also, this test-bench has sensors to measure all its physical states.

Section 2 briefly introduces the active suspension system and the Quanser testbench used for this study. A mathematical model of this test-bench is presented in this section. This model is used to the obtain LQR controller. This model also lists the states that need to be measured/observed when a LQR controller is used. The design requirements and implementation limits are also discussed. In Sect. 3, an introduction to ANFIS is presented. The methodology proposed in this work to train an ANFIS controller for the active suspension system and implementation is also provided. A comparison of results from open-loop, LQR controlled and ANFIS controlled system are given in Sect. 4. Section 5 concludes the paper.

2   Active Suspension System

Suspension system is an indispensable part of modern vehicles. Suspension systems can be classified into three categories with respect to external input [10]. Passive suspensions are made up of shock absorbing dampers and non-controlled springs. Semi-active suspensions are same as passive systems with the addition of electronically selectable damping rates. In such systems, a switch is usually used to select between soft and stiff dampers. Active suspensions on the other hand have actuators that provide continuous control over damping rates. The control force of these actuators is determined by feedback control using sensors attached to the vehicle. Accelerometers, gyros, deflection sensors are usually used. Whereas, electromagnetic and hydraulic actuators are used to provide damping forces.

A suspension system serves following purposes in vehicles [10].

- Isolating the body from road disturbances.
- Isolating the body from inertial disturbances caused by breaking and turning. These disturbances result in body roll and pitch.
- Minimizing vertical wheel travel thus increasing handling and adhesion.
- Maintaining optimum tire-to-road contact on all four wheels.

However, it is not possible to achieve optimum performance from a suspension system for each of theses goals at the same time. Inherently, these design objectives contradict each other. A detailed discussion of this phenomenon can be found in [10].

This complexity of design objectives can be reduced by studying quarter vehicle model (QVM). This model considers the suspension system of just one tire. This simplification eliminates the second and the fourth design objective.

2.1   Quarter Vehicle Model

A QVM reduces the full vehicle model to just one tire and other parts as required. Figure 1 shows a QVM usually used to study the suspension system. It consists of following components:

- M_s: Sprung mass; represents vehicle's body.
- M_{us}: Unsprung mass; represents the tire.
- K_s: Represents the spring connecting body and the tire.
- B_s: Represents the damper connecting body and the tire.
- K_{us}, B_{us}: Model the contact between tire and the road.
- F_c: Represents the active damper.

Fig. 1 Quarter vehicle
model

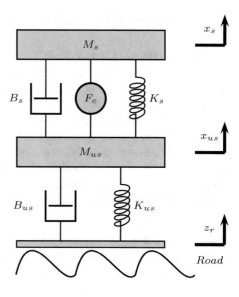

The state space representation of this QVM is given by following equations.

$$\dot{x}(t) = Ax(t) + Bu(t) \tag{1}$$

$$y(t) = Cx(t) + Du(t) \tag{2}$$

$$x(t) = \begin{bmatrix} x_s(t) - x_{us}(t) \\ \frac{d}{dt}x_s(t) \\ x_{us}(t) - z_r(t) \\ \frac{d}{dt}x_{us}(t) \end{bmatrix} = \begin{bmatrix} \text{\textit{Suspension travel}} \\ \text{\textit{Body vertical velocity}} \\ \text{\textit{Tire travel}} \\ \text{\textit{Tire vertical velocity}} \end{bmatrix} \tag{3}$$

$$u(t) = \begin{bmatrix} \frac{d}{dt}z_r(t) \\ F_c \end{bmatrix} = \begin{bmatrix} \text{\textit{Road velocity}} \\ \text{\textit{Control force}} \end{bmatrix} \tag{4}$$

$$y(t) = \begin{bmatrix} x_s(t) - x_{us}(t) \\ \frac{d^2}{dt^2}x_s(t) \end{bmatrix} = \begin{bmatrix} \text{\textit{Suspension travel}} \\ \text{\textit{Body acceleration}} \end{bmatrix} \tag{5}$$

The objectives discussed in Sect. 2 can be translated in terms of plant's states. The first objective, i.e. isolating the body from road disturbances, is achieved by minimizing "body vertical velocity". The third objective is achieved when "tire travel" is minimized.

Fig. 2 Quanser's Active Suspension Plant used for this study

2.2 Quanser Active Suspension Plant

Figure 2 shows a Quanser Active Suspension System Plant. This plant is a scaled down test-bench based on quarter vehicle model. An ultra-fast response electric motor connected with a capstan mechanism is used for active damping. Along with an accelerometer, this plant has sensors which measure all the states used in state space model as mentioned by Eq. 3. However, it should be noted that in an actual suspension system, the only outputs which can be measured affordably are suspension travel and body acceleration. This plant is connected with a PC and a software interface is available in SIMULINK/MATLAB. A LQR controller implementation is provided with the this plant by *Quanser*© [9]. The same implementation for LQR controller is used in this work, and its details are not included to save space.

3 ANFIS

A conventional type-1 and type-2 fuzzy inference system is usually employed to create a controller using expert knowledge about the plant. The resulting fuzzy controller is a nonlinear controller which provides better performance as compared to linear controllers. On the other hand, the adaptive network based fuzzy inference system [6] models the type-3 Takagi-Sugeno fuzzy inference system as a neural network. The advantage of network based approach is that the expert knowledge is no longer required. Instead, input/output training data from the plant is used to create and tune the fuzzy controller.

Figure 3 shows a two input one output ANFIS controller used in this work. It is a type-3 Takagi-Sugeno fuzzy inference system. This system has two membership functions corresponding to each input. Similarly, it has two linear output membership functions. Its two inputs, i.e. y_1 and y_2, refer to the two outputs of suspension system given by Eq. 5. Its output, i.e. F_c refers to the input of the suspension system.

Fig. 3 Adaptive network based fuzzy inference system with two inputs and one output

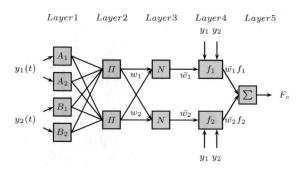

This network can be extended to include more input, output membership functions. Hybrid algorithm [6] is used to train this neural network as a controller.

3.1 Acquiring Training Data

Training data is acquired from active suspension test-bench as shown in Fig. 4. The LQR is implemented as full state feedback controller. Plant output and control force produced by LQR are recorded for known road profile. Figure 5 shows road velocity for the applied road profile. The data is saved for 3 s. It has velocity impulses in both directions. The control force applied by LQR to damp system oscillations produced as a result of applied road profile can be seen in Fig. 6. The recorded output of the system can be seen in Figs. 7 and 8. It is worth mentioning that Fig. 7 shows suspension travel, which is the first output as mentioned in state-space model in Eq. 5. Figure 8 shows body acceleration which is the second output.

3.2 Training and Implementation

The data acquired from the plant is used to obtain six ANFIS controllers. A list of these controllers can be seen in Table 1. All of these controllers are type-3 Takagi-

Fig. 4 Data is acquired from active suspension plant controlled by LQR. Signal labels refer to Eqs. 1, 2 and 4

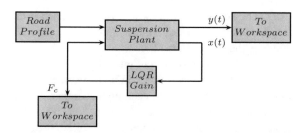

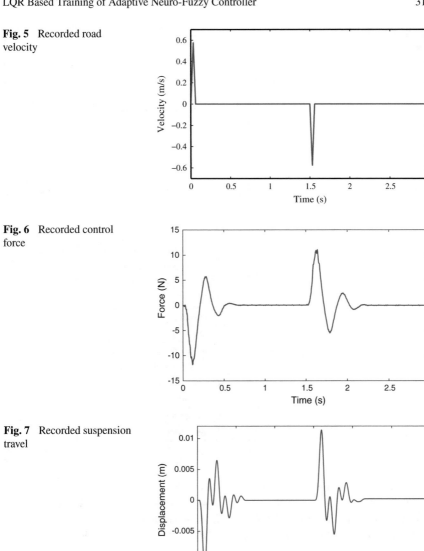

Fig. 5 Recorded road velocity

Fig. 6 Recorded control force

Fig. 7 Recorded suspension travel

Sugeno inference systems. These controllers only differ in number and type of input/output membership functions. It can be seen in Table 1 that two type of input member ship functions are used, i.e. triangular and gaussian. Similarly, there are two type of output membership functions, i.e. constant and linear. The number of membership functions is shown in braces ("{ }"). Following points describe the training/testing methodology for each controller.

Fig. 8 Recorded body
acceleration

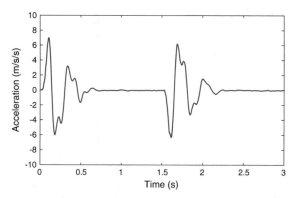

Table 1 Test errors for
different ANFIS controllers

S. no.	ANFIS	Test error
1	Triangular input mfs {3 3}	2.05638
	Constant output mfs{9}	
2	Gaussian input mfs{3 3}	2.10172
	Constant output mfs{9}	
3	Gaussian input mfs{3 3}	1.90727
	Linear output mfs{25}	
4	Gaussian input mfs{5 5}	1.96889
	Constant output mfs{25}	
5	Gaussian input mfs{7 7}	1.76433
	Constant output mfs{49}	
6	Gaussian input mfs{7 7}	1.54358
	Linear output mfs{49}	

- MATLAB ANFIS toolbox is used for training and testing purposes.
- Data shown in Figs. 7 and 8 is used as input data for the anfis network.
- Data shown in Fig. 6 is used as output data for the anfis network.
- Half of the data from previous steps is used as training data, while rest is used as test data.
- Hybrid learning algorithm [6] is used for training of anfis network.
- Root Mean Squared value is used as error measure for training and testing purposes.
- Each anfis network/controller is generated using the toolbox with "grid-partition" option.
- Each network is trained for 25 epochs.

It is evident from the table that the test error decreases nominally with the increase in number of membership functions. Figure 9 shows the testing of controller no. 2 and controller no. 5. The testing results of controller 1, 3 and 4 are same as that of controller 2. While testing results of controller 6 are similar to controller 5. The comparison of these test plots show that as the number of input/output membership

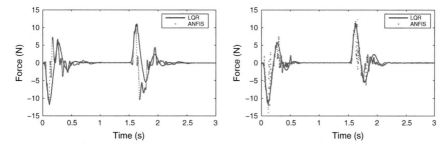

Fig. 9 Testing of ANFIS controllers and their comparison with LQR controller data. On left is controller no. 2 and on right is controller no. 5. Controller numbers refer to S. no. column in Table 1

Fig. 10 Control force of ANFIS controller no. 5—when implemented on actual system

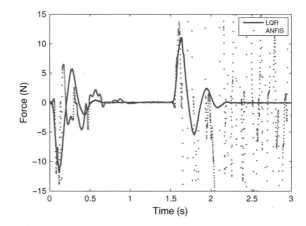

functions increase, the output of anfis network becomes more and more non linear. All these controllers are validated by implementing them in SIMULINK/MATLAB with the active suspension plant in loop. The performance of these controllers is compared with the LQR controller. Figure 10 shows that controller 5 becomes unstable when implemented in close loop with the plant. Controller 6 also becomes unstable in close loop.

4 Results

The discussion in previous section shows that only controller 1 to 4 are stable when implemented in close loop with the plant. Controller 5 and 6 become unstable.

The comparison of testing error from Table 1 hints that the close loop performance of controller 4 should be better than controller 1. However, there is not much difference in terms of settling time, rise time, and peak to peak magnitude of generated control force. From this comparison controller 2 stands out in terms of design simplicity, close loop performance and robustness.

Fig. 11 Suspension travel
for ANFIS controller no. 2

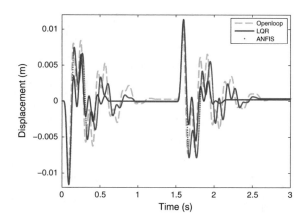

Fig. 12 Body acceleration
for ANFIS controller no. 2

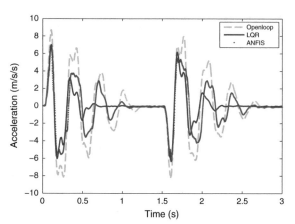

The performance of controller 2 is compared with lqr controller and open loop
response in Figs. 11, 12, 13 and 14. These figures compare the four states of the
plant. It is evident that the performance of ANFIS controller is better than openloop
response however it is poor as compared to lqr controller.

5   Conclusion

In this work, six ANFIS controllers were obtained and trained for Active suspension
system by using the close loop data from the LQR controlled system. The motive
was to obtain a controller that can give optimum performance like LQR without the
need of a stochastic observer. A number of different ANFIS systems having various
types and number of input/output membership functions were considered as candidate for such a controller. Results showed that as the number of membership functions increase, the training/testing error decrease. However, the performance of the

Fig. 13 Body velocity for
ANFIS controller no. 2

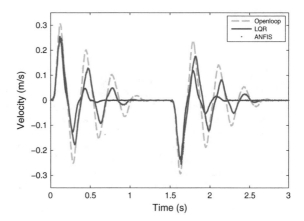

Fig. 14 Tire travel for
ANFIS controller no. 2

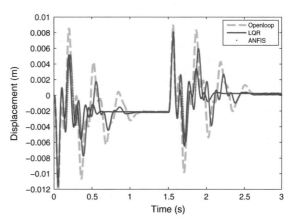

controller on the actual system did not improve substantially. It was also observed
that ANFIS systems with linear output membership functions became unstable when
they were implemented on actual system. Based on this comparison a particular
ANFIS controller was chosen to replace the LQR controller.

On the other hand, the comparison of the chosen ANFIS controller with the open
loop system showed that the ANFIS system is stable and adequately damps oscillations. However, its performance in comparison with LQR is poor. LQR controller's
performance is a limit on ANFIS's performance as the latter is trying to imitate the
former. Nonetheless, the ANFIS controller is a good alternative of LQR controller
in noisy environments where a stochastic observer is required.

Thus, this work is concluded by observing that anfis controller when trained by
close loop lqr data cannot match the performance of lqr controller. However, it performance is adequate enough to replace lqr controller in noisy environment which
requires a stochastic observer.

References

1. Liu, H., Lu, G., Zhong, Y.: Robust LQR attitude control of a 3-DOF laboratory helicopter for aggressive maneuvers. IEEE Trans. Ind. Electron. **60**(10), 4627–4636 (2013)
2. Olalla, C., Leyva, R., El Aroudi, A., Queinnec, I.: Robust LQR control for PWM converters: an LMI approach. IEEE Trans. Ind. Electron. **56**(7), 2548–2558 (2009)
3. Hussein, B.N., Sulaiman, N., Raja, R.K., Marhaban, M.H.: Design of PSO-based fuzzy logic controller for single axis magnetic levitation system. IEEJ Trans. Electr. Electron. Eng. **6**(6), 577–584 (2011)
4. Li, H., Yu, J., Hilton, C., Liu, H.: Adaptive sliding-mode control for nonlinear active suspension vehicle systems using TS fuzzy approach. IEEE Trans. Ind. Electron. **60**(8), 3328–3338 (2013)
5. Passino, K.M., Yurkovich, S.: Fuzzy Control, 1st edn., pp. 26. Addison-Wesley (1997)
6. Jang, J.S.R.: ANFIS: adaptive-network-based fuzzy inference systems. IEEE Trans. Syst. Man Cybern. **23**(3), 665–685 (1993)
7. Toha, S.F., Tokhi, M.O.: Dynamic nonlinear inverse-model based control of a twin rotor system using adaptive neuro-fuzzy inference system. In: Third UKSim European Symposium on Computer Modeling and Simulation, pp. 107–111. Athens (2009)
8. Mahmoud, T.S., Mohammad, H.M., Tang, S.H.: ANFIS: self tuning fuzzy PD controller for twin rotor mimo system. IEEJ Trans. Electri. Electron. Eng. **5**(3), 369–371 (2010)
9. Quanser. Active Suspension User's Manual. Quanser Consulting Inc. Ontario, Canada (2009)
10. Appleyard, M., Wellstead, P.E.: Active suspensions : some background. IEE Proc. Control Theory Appl. **142**(2), 123–128 (1995)

Structured Versus Shuffled Connectivity in Cortical Dynamics

Ilenia Apicella, Silvia Scarpetta and Antonio de Candia

Abstract Previous papers have studied a leaky Integrate-and-Fire (IF) model whose connectivity was designed in such a manner as to favor the spontaneous emergence of collective oscillatory spatiotemporal patterns of spikes. In this model the alternation of up and down states does not depend on a kind of neuron bistability, nor on synaptic depression, but is rather a network effect. In order to check if the transition region with bimodal distribution of the firing rate survive even after changes in the topology of the network, in this work we shuffle all the connections. For each chosen connection we change the presynaptic neuron, choosing as the new presynaptic one a random neuron of the network. After shuffling the connections, not only the number of excitatory and inhibitory connections is the same as before, but also the strengths of the connections are the same, and only the topology is changed. We observe that shuffling the connections changes the features of the dynamics dramatically. Before shuffling the system has a transition from a regime of Poissonian noisy activity to a regime of spontaneous persistent collective replay, and at the transition point the network dynamics shows an intermittent reactivation of the stored patterns, with alternation of up and down state, and bimodal distribution of spiking rate. Shuffling all the connections we observe that the transition region with bimodal distribution disappears, and the dynamics is Poissonian with unimodal rate distribution for all the investigated parameters. These results show the role of topology in dictating the emerging collective dynamics of neural circuits.

Keywords Topology · Phase transition · Spontaneous cortical activity

I. Apicella (✉) · S. Scarpetta
Department of Physics "E.R.Caianiello", INFN, University of Salerno,
84084 Fisciano, Salerno, Italy
e-mail: ilenia.apicella@hotmail.it

S. Scarpetta
e-mail: sscarpetta@unisa.it

A. de Candia
Department of Physics, INFN sezione di Napoli,
University of Napoli "Federico II", Naples, Italy

© Springer International Publishing Switzerland 2016
S. Bassis et al. (eds.), *Advances in Neural Networks*, Smart Innovation,
Systems and Technologies 54, DOI 10.1007/978-3-319-33747-0_32

1 Introduction

Many experiments have showed the presence of alternation between "up states" with high firing rate activity and "down states" of network quiescence in cortical spontaneous activity. Although the precise mechanism by which this spontaneous transition occurs is still unclear, it has been observed in different systems and conditions, for example during slow-wave sleep, anesthesia and quiet waking, and it is similar to up states transition produced by sensory stimulation. Some results have suggested that these up states transitions occur with repeating stereotyped spatiotemporal patterns of cellular activity, so that it has been conjectured that up states are attractors in a landscape in which the brain can move [1].

Many of most recent works [2–6] have confirmed the idea that brain operates near the critical point of a phase transition, and this seems to have some advantages in terms of optimization of dynamical range, information transmission and capacity.

Several models have been made to explain power law distributions of the avalanches that emerges in spontaneous cortical activity. In previous works [7, 8] we have proposed a model of network composed by N leaky Integrate-and-Fire (IF) neurons, whose connections have synaptic strengths designed in order to store in the network a set of spatiotemporal patterns, showing a transition between two regime of quiescence (down states) and high rate coherent activity (up states). This transition happens even in absence of short term depression or of any kind of single neuron bistability. It is related to network connectivity. Indeed, the mean time the network spends in the down or in the up state depends on noise intensity and connection strength. However it is not clear how much influence have the connections efficacies and how much it is important the topology. In order to study the role of topology of the network, in this work we shuffle all the connections of the network, i.e. for each connection we change the presynaptic neuron, choosing a new presynaptic neuron randomly among the $N-1$ neurons of the network. We indicate with cs the ratio between the number of times we make this procedure and the number of connections. In this paper we study the extreme cases, i.e. $cs = 0$, that indicates the situation of fixed structured connectivity [7], in which no connection has been shuffled, and the case $cs = 1$, that indicates a completely shuffled connectivity. The set of connections strengths (i.e. values of J_{ij}) are exactly the same in $cs = 0$ and $cs = 1$ case, but the units connected by these connections are randomized in the case $cs = 1$, while they are those dictated by learning in $cs = 0$ case.

In the previous work we have studied the case of $cs = 0$ and we have shown two different regimes of high and low activity changing the connection strength and noise intensity. In fact, we have a low activity regime characterized by a nearly exponential distribution of firing rates with a maximum at rate zero, and a high activity regime, characterized by a nearly Gaussian distribution peaked at a high rate. A bimodal distribution of firing rates with alternation of up and down state emerge between this two region, showing a transition region in which one observes neuronal avalanches exhibiting power laws in size and duration. These three different behaviors disappear when we shuffle the pre-synaptic units ($cs = 1$).

2   The Model

We consider the model described in [7] and we reshuffle a fraction cs of the connections. We briefly resume the model here.

We consider a network of $N = 3000$ spiking neurons, represented by the Spike Response Model of Gerstner [9] with a Poissonian noise distribution. If we label with index i the postsynaptic neuron and with index j the presynaptic one, when the neuron i does not fire, the postsynaptic membrane potential is given by the equation

$$u_i(t) = \sum_j \sum_{t_i < t_j < t} J_{ij} \left(e^{-(t-t_j)/\tau_m} - e^{-(t-t_j)/\tau_s} \right)$$
$$+ \sum_{t_i < \hat{t}_i < t} \hat{J}_i \left(e^{-(t-\hat{t}_i)/\tau_m} - e^{-(t-\hat{t}_i)/\tau_s} \right) \tag{1}$$

where τ_m is the characteristic time of membrane, τ_s is characteristic time of synapse (in this paper $\tau_m = 10$ ms and $\tau_s = 5$ ms), t_j are the spike times of neuron j, $\hat{t}_i$ are the times of noise events releasing a random charge at some point of the membrane of neuron i. J_{ij} is connection strength between pre- and postsynaptic neurons and $\hat{j}_i$ is related to the noise and it is extracted from a Gaussian distribution with mean 0 and standard deviation $\sigma = \sqrt{\frac{\alpha}{\rho} \sum_j J_{ij}^2}$, where α represents the "noise level" of the network and $\rho = 1$ ms^{-1}.

When the potential $u_i(t)$ reaches the threshold (that we set the same θ for all neurons) the neuron i emits a spike and its potential is reset to the resting value $u_i = 0$.

At the beginning of simulation we set synapse strengths J_{ij} with a "learning procedure", inspired by Spike Timing Dependent Plasticity (STDP) [10, 11] by which we store P patterns in the network connections. After this procedure there is a "pruning procedure", by which only a fraction of connections J_{ij} survives. A third procedure occurs, that does not concern the connection strengths, but their topology; in this procedure we create a randomized structure of connections.

A phase-coded pattern $\mu = 1, \ldots, P$ is a periodic ordered train of spikes $\{t_i^\mu\}$ with period T^μ and with one spike per neuron and per cycle. In this work, as in [7], we use P $= 2$ patterns and $T^\mu = 333$ ms, i.e. a frequency $\nu^\mu = 1$ Hz. After the "learning procedure" we get

$$\delta J_{ij} = H_i \sum_{n=-\infty}^{\infty} A\left(t_i^\mu - t_j^\mu + nT^\mu \right) \tag{2}$$

where t_j^μ and t_i^μ are the pre- and postsynaptic spikes in pattern μ and the function $A(\tau)$ is the "learning window" inspired to STDP and described in [7, 8].

In the Eq. (2) H_i is a constant that sets the strength of the connections, and depends on the postsynaptic neuron. We use two values of H_i, H_0 and H_1. The first one is for "normal" neurons, the second one is $H_1 = 3H_0$ for "leader" neurons, i.e.

neurons that with higher incoming connection strengths amplify activity initiated by noise and give rise to a cue able to initiate the short collective replay. The leader neurons are chosen as a fraction of 3 % of neurons that have consecutive phases, for each pattern μ.

With the pruning procedure we delete a fraction f^+_{prune} of positive (excitatory) connections with the lowest value and a fraction $f^{-,i}_{prune}$ of negative (inhibitory) connections with the lowest absolute value. With a good choice of $f^{-,i}_{prune}$, that can depend on neuron i, at the end we have the sum of incoming connections to neuron i very close to zero. In this work we use $f^+_{prune} = 70\%$. As a consequence, after the pruning procedure, about 12 % of the $N(N - 1)$ connections survives as positive connections, and 27 % as negative ones, with statistical fluctuations of order $1/\sqrt{N}$.

The last procedure that concerns the connections is the "shuffling procedure". The strength remains the same of those we have considered above, but now we change the topology of connections, because we want investigate the effect of the structure of network on the dynamics. In order to study this problem, we shuffle a fraction cs of connections. We pick up a randomly chosen connection J_{ij} and we use the strength J_{ij} to connect the postsynaptic connection i with another randomly chosen presynaptic neuron k. Then we put the connection J_{ij} between i and j to zero $(J_{ij} = J_{ik}, J_{ij} = 0)$. We repeat this shuffling procedure until a fraction cs of connections has been shuffled. Note that this procedure does not change the strength of connections and it does not change the ratio between inhibitory and excitatory connections, but it changes only the topology of the network. The postsynaptic neuron i remains fixed. In this way we also conserve the presence of leaders, i.e. the fraction of neurons that has stronger input (and noise input) than others.

After this three procedures, we study the dynamics of network, with fixed connections and without short term depression, comparing the structured connections $(cs = 0)$ with the shuffled connections $(cs = 1)$.

3   Results

Because we want to understand the effects of the connections topology on the network, we study its dynamics in function of two parameters: the strength of connections H_0, expressed in units of the threshold θ of the neurons, and the noise level α, that has dimensions of ms^{-1}, changing the structure of connections. With the fixed value of the fraction of shuffled connections, cs (0 or 1) we change the parameters H_0 and α, measuring the firing rate of network. Then we compare this two cases.

We indicate with letter A the results when $cs = 0$ (structured connections) and with letter B when $cs = 1$ (shuffled connections). In previous work [7], we have shown a transition from a Poissonian regime with a nearly exponential distribution to a nearly Gaussian distribution, changing the values of H_0/θ and α. Between this two regime a bimodal distribution exists.

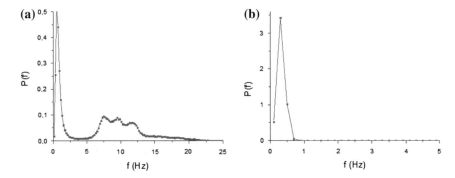

Fig. 1 Firing rate for $H_0/\theta = 0.221$ and $\alpha = 0.06\text{ms}^{-1}$. We observe in the case of model A ($cs = 0$) a bimodal distribution, that disappears when we consider the model B ($cs = 1$)

In Fig. 1 the distribution of firing rate is shown for a particular value of H_0/θ and α. While in the model A we observe a bimodal distribution showing both picks, one at low rates and one at high rates, in model B we always observe a unimodal distribution dominated by noise, even for the other values of H_0/θ and α.

The replay of patterns in a raster plot confirms the result that the transition disappears in the case of $cs = 1$ and we have a quiescence state for all investigated parameters.

In the model A we have a collective coherent replay of one of two stored patterns during the state of higher activity; so we can see (Fig. 2a) both patterns initiated intermittently. Conversely, in Fig. 2b we see a low rate activity.

At the same time we evaluate the average rates in Hz per neuron, shown in Fig. 3.

In Fig. 3a we can see that the average rate increases continuously as either of the parameters is increased and we can distinguish three different regions, corresponding to three different dynamical behaviors. Yellow points indicate a low average rate. In this region the distribution is nearly exponential and the dynamics is dominated by noise, so that the potential of neuron, although with some probability crosses the threshold and emits a spike, is not able to generate a spreading activity in the network. Red points indicate a high average rate. In this region, in which the distribution is nearly Gaussian, the noise triggers the replay of one of the patterns encoded in the network, and once the replay of pattern has started, it is permanent because the noise is not able to stop it. In the intermediate region (green points), with a bimodal distribution of firing rate, we have an alternation of states with a high rate ("up" states) and states with a low rate ("down" states). In this case the noise is able to start the replay of a pattern, but also to stop it. This region separates the region of quiescence with low activity and the region of permanent replay of a spatiotemporal pattern. In the case $cs = 1$ the average spiking rate is always unimodal, completely composed by yellow points, with low firing rate dominated by noise (Fig. 3b).

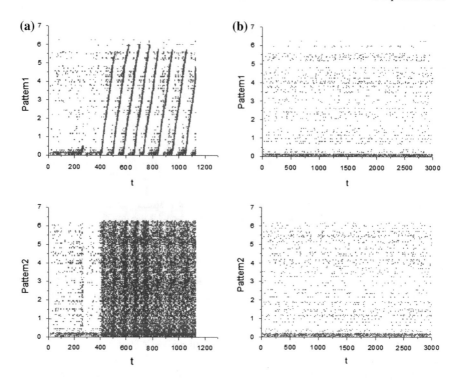

Fig. 2 Raster plot for the same parameters of Fig. 1, for **a** cs = 0 and **b** cs = 1

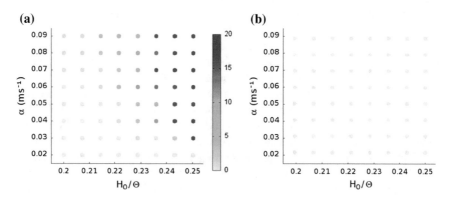

Fig. 3 Average spiking rate

So far we have studied the dynamics of shuffled network in a small range of parameters H_0/θ and α, the same investigated in [7] and we can conclude that for these values of parameters there is no transition. For a larger range of H_0/θ and α there is a region with high rate, but we can't call it an up state. Indeed, if we analyzed the behavior of variance and normalized variance, we can see a different behavior in the case of structured and reshuffled network. We remember that the normalized variance is defined as follow:

$$\hat{\sigma}^2 = N\Delta \frac{\langle r^2 \rangle - \langle r \rangle^2}{\langle r \rangle} = \frac{\langle n_{tot}^2 \rangle - \langle n_{tot} \rangle^2}{\langle n_{tot} \rangle} \tag{3}$$

where Δ is the time interval, $r = \frac{n_{tot}}{N\Delta}$ is the rate, n_{tot} is the total number of spikes in a time interval Δ. For a Poissonian distribution the normalized variance is 1, because the variance is equal to the average number of events, i.e.

$$\langle n_{tot}^2 \rangle - \langle n_{tot} \rangle^2 = \langle n_{tot} \rangle.$$

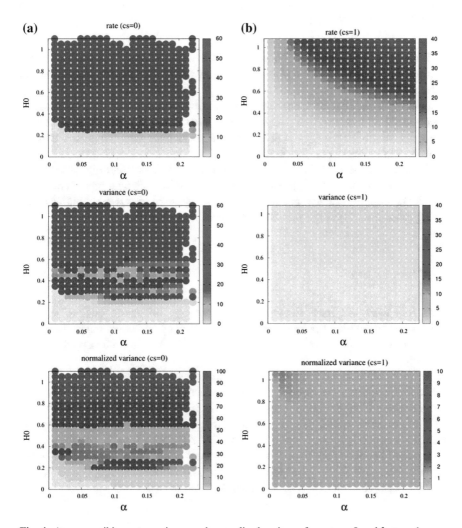

Fig. 4 Average spiking rate, variance and normalized variance for **a** cs = 0 and **b** cs = 1

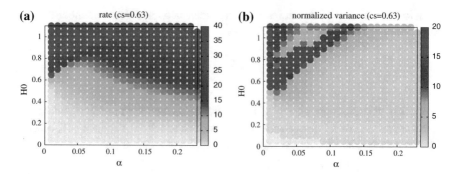

Fig. 5 Average spiking rate (**a**) and normalized variance (**b**) for a shuffle fraction cs = 0.63. We recognize two regions with high rate, one area with correlated activity on the *left*, and one with Poissonian activity on the *right*

We compare the behavior of firing rate, variance and normalized variance of the two cases of network we have studied, structured and reshuffled, for a large range of parameters H_0/θ and α.

Looking at the Fig. 4 we observe the normalized variance changing in the colors correspondent to values from 0 to 100 in the case of $cs = 0$, different from the case of $cs = 1$ where we have, even at high rate, a normalized variance always close to 1, feature of a Poissonian distribution.

Interestingly, as shown by the rate and normalized variance in Fig. 5, when we shuffle only a fraction $cs = 0, 63$ of connections we have both a regime of correlated activity (at high H0 and low noise α) and also a regime of uncorrelated high activity (at high noise α, with high rate but low normalized variance).

4 Conclusion

With this work we have shown that the topology influences qualitatively the dynamics of network, because the transition between up and down states disappears when connections are completely shuffled, while keeping conserved all the strengths.

In many neural models the dynamics is studied with random connectivity. Here we have shown that the topology of the connections is highly relevant, since the two networks have the same sparse connections (12 % of N(N − 1) are positive, 27 % are negative, and all others are zero) with exactly the same strengths, only one reshuffled with respect to the other, and we have got very different dynamics.

It would be interesting to understand which are the features of the topology that characterizes $cs = 0$, and therefore which topological feature allows a such rich dynamic behavior not observed in the randomized $cs = 1$ case.

Here we have focused mainly on the extreme cases $cs = 0$ and $cs = 1$; in the future we will investigates intermediate cases, changing gradually the connectivity. Preliminary results are very intriguing, we see indeed that the region of phase-space with correlated activity gradually reduces as we increases cs, and then disappears at $cs = 1$.

References

1. Cossart, R., Aronov, D., Yuste, R.: Attractor dynamics of network UP states in the neocortex. Nature **423**, 283–288 (2003). doi:10.1038/nature01614
2. Petermann, T., Thiagarajan, T.C., Lebedev, M.A., Nicolelis, M.A., Chialvo, D.R., Plenz, D.: Spontaneous cortical activity in awake monkeys composed of neuronal avalanches. Proc. Natl. Acad. Sci. U.S.A. **106**, 15921–15926 (2009). doi:10.1073/pnas.0904089106
3. Ribeiro, T.L., Copelli, M., Caixeta, F., Belchior, H., Chialvo, D.R., Nicolelis, M., et al.: Spike avalanches exhibit universal dynamics across the sleep-wake cycle. PLoS ONE **5**, e14129 (2010). doi:10.1371/journal.pone.0014129
4. Plenz, D.: Neuronal avalanches and coherence potentials. Eur. Phys. J. Spec. Top. **205**, 259–301 (2012). doi:10.1140/epjst/e2012-01575-5
5. Haimovici, A., Tagliazucchi, E., Balenzuela, P., Chialvo, D.R.: Brain organization into resting state networks emerges at criticality on a model of the human connectome. Phys. Rev. Lett. **110**, 178101 (2013). doi:10.1103/PhysRevLett.110.178101
6. Chialvo D. : Emergent complex neural dynamics. Nat. Phys. **6**, 744–750 (2010). doi:10.1038/nphys1803
7. Scarpetta S., de Candia A.: Alternation of up and down states at a dynamical phase-transition of a neural network with spatiotemporal attractors. Front. Syst. Neurosci. (2014)
8. Scarpetta, S., de Candia, A.: Neural avalanches at the critical point between replay and non-replay of spatiotemporal patterns Plos One (2013)
9. Gernster W.: Time structure of the activity in neural network models. Phys. Rev. **51**(1) (1995)
10. Bi, P.: Precise spike timing determines the direction and extent of synaptic modifications in cultured hippocampal neurons. J. Neurosci. **18**, 10464–10472 (1998)
11. Markram, H., Lubke,J., Frotsher,M., Sakmann,B: Regulation of synaptic efficacy by coincidence of postsynaptic Aps and EPSPs. Science **275**, 213–215 (1997). doi:10.1126/science.275.5297.213

Part V
Neural Networks-Based Approaches to Industrial Processes

Machine Learning Techniques for the Estimation of Particle Size Distribution in Industrial Plants

Damiano Rossetti, Stefano Squartini and Stefano Collura

Abstract This paper aims to evaluate the effectiveness of different Machine Learning algorithms for the estimation of Particle Size Distribution (PSD) of powder by means of Acoustic Emissions (AE). In industrial plants it is very useful to use non-invasive and adaptable systems for monitoring the particle size, for this reason the AE represents an important mean for detecting the particle size. To create a model that relates the AE with the powder size, Machine Learning is a viable approach to model a complex system without knowing all the variables in details. The test results show a good estimation accuracy for the various Machine Learning algorithms employed in this study.

Keywords Machine learning algorithms · Particle size distribution · Acoustic emission

1 Introduction

The particle size of the powder is an important parameter in many industrial processes, as it affects to the physical and chemical properties of materials. In most cases the powder particles have irregular shapes and speed, and travelling within structures that change its characteristics over time due to wear. Furthermore, it is not interesting to describe the size of single particle but the size of an ensemble of particles, so it is necessary to use cumulative parameters to describe it, such as the

D. Rossetti · S. Squartini (✉)
Department of Information Engineering,
Università Politecnica delle Marche, Ancona, Italy
e-mail: s.squartini@univpm.it

D. Rossetti
e-mail: d.rossetti@pm.univpm.it

S. Collura
Loccioni Group, Angeli di Rosora (AN), Italy
e-mail: s.collura@loccioni.com

© Springer International Publishing Switzerland 2016
S. Bassis et al. (eds.), *Advances in Neural Networks*, Smart Innovation,
Systems and Technologies 54, DOI 10.1007/978-3-319-33747-0_33

Particle Size Distribution (PSD). The PSD is a list of values that denote the amount of particles, usually a percentage, whose size lies within predefined ranges, sorted in ascending or descending order. In industrial plants, the evaluation of the PSD is typically performed by introducing a probe into the duct and collecting a sample of the powder and analyzing it in laboratory. This method produces an accurate estimation of the PSD for a given time instant, but it is time consuming and difficult to use for continuous monitoring. To have a continuous monitoring of powder size, it is necessary to employ a system that carries out the estimation in a non invasive way, for the all time horizon of interest.

The main challenge is to find physical model able to describe a very complex system, with unknown and uncontrolled variables, and then to select a set of physical quantities related to the particles size.

In this work, Acoustic Emission (AE) signals produced by a powder impacting on a metallic surface have been identified as meaningful quantities in order to obtain PSD measure. Leach et al. [6, 7] were the first to use AE signals for particle sizing. They collected AE spectra in the range 50–200 kHz, from the impact among particles. By measuring the beat frequencies from different resonance frequencies of particles with varying diameters, they could determine their average diameter and size range. The method gave satisfactory results just for regularly shaped particles, i.e. spheres and cylinders. Unfortunately, this explicit method is impractical for most of industrial applications where are involved fluxes of irregular shaped of particles. However it was demonstrated, at least from a theoretical point of view, that a particle impinging on a metallic surface generate an AE signal containing the information about its size.

Another issue in an industrial plant is that there are several sources of AE, so it is also not easy to distinguish between the AE produced by powder and the other sources.

With these premises, suitable techniques have to be developed in order to create a mathematical model without knowing all the variables and all noise sources that influence the process. In this context, Machine Learning algorithms represent an interesting approach that allows to overcome all the raised issues.

2   Methodological Approach

To evaluate the accuracy of the models created by means of the Machine Learning algorithms, the results obtained with an industrial system used in plants for the production of energy, named POWdER [2], have been taken as useful reference. This system continuously monitors the PSD of coal powder conveyed in ducts from grinding mills up to the boiler. Monitoring is done using sensors installed on the outer surface of the ducts which detect the acoustic emissions produced by the powder. The sensors are installed near a duct curve because in this point there is the highest probability that the particles hits the inner surface of the duct and so generate the

AE. For the training of their regression model, the POWdER uses a dataset derived from measurement campaigns carried out at different working conditions.

Each dataset contains some observations, each one made by 64 features and 3 targets. From the raw signals collected with the sensors, are extracted the 64 features that characterize the energy of the acoustic emission signal. A Principal Component Analysis was performed in order to reduce the features number, but the results obtained with the reduction were worst than those obtained with all features, for this in all tests are used all features to train the models.

The targets represent the values of the PSD associated with the AE. For this system, the reference measurement of PSD is carried out through the sampling and sieving method. A sampling of powder is performed at the plant by introducing a probe into the duct and collecting a sample of the powder. After the sampling, the sample is sent to the laboratory where it is sieved and classified through a nested column of three sieves of decreasing screen openings, respectively, corresponding to MESH 50 (300 μm), MESH 100 (150 μm) and MESH 200 (75 μm). The outcomes are the three numerical values, corresponding to the percentage of coal particles in the initial sample whose dimensions are, respectively, lower than 300, 150 and 75 μm. For the simulations in this work, were used 3 datasets related to different ducts of a plant for the production of energy. Hence forward the datasets will be named Duct 1, Duct 2 and Duct 3, in reference to the duct from which are measured the data.

There are many different machine learning algorithms in the literature, in this preliminary work the most common algorithms used for regression problems have been considered: the Artificial Neural Network (ANN), the Support Vector Machine (SVM) and Extreme Learning Machine (ELM).

Each algorithm has been used with a procedure of Cross Validation (CV), in the form K-Folds with folds containing 15 % of the entire dataset available. At each cycle of the CV is selected the 15 % of the total dataset that is then used as Testing set, while the remaining 85 % is used as Training set. In total, 6 different pairs of Training and Testing are involved, making sure that each Testing set had different observations than the others. In order to perform the optimization of the parameters required for the training of the models, a further K-Folds CV has been performed in which, at each cycle, 15 % of the Training set is used as a Validation Set. Also in this case 6 combinations of the Training and Validation set have been used. The parameters yielding the lower estimation error in the Validation set has been selected as the optimal parameters.

In all tests, features and targets have been normalized in the range [0:1].

2.1 Artificial Neural Networks

Generally, a Neural Network (NN) [3] is composed of neurons organized in layers, denoted as input, output, and hidden layers. At the start of training process, the weights of the neurons are initialized with random numbers and, during the training process, these weights are update with the Backpropagation [8] algorithm. Different activation functions can be chosen for the neurons, in the experiments, the unipolar sigmoid has been used. The standard structure selected for the test, using one input layer, two hidden layer and one output layer. The number of hidden layers nodes has been varied, from 20 to 50, in order to identify the configuration that minimizes the estimation error. Above the 50 neurons, the performances of regression models decrease so it has not been performed an exhaustive investigation.

2.2 Support Vector Machine for Regression

According to Support Vector Machine for Regression (or Support Vector Regression) theory [9], the predicted output is computed as:

$$\mathbf{y}(\mathbf{x}) = \sum_{n=1}^{N} (a_n - \hat{a}_n) k(\mathbf{x}, \mathbf{x}_n) + b , \tag{1}$$

where $\sum_{n=1}^{n} (a_n - \hat{a}_n) = 0$ with $a_n, \hat{a}_n \in [0, C]$, and k denotes the kernel function. The support vector $\mathbf{x}_n$ and the bias b, together with a_n and $\hat{a}_n$ are computed during the training phase. All the experiments have been performed using LIBSVM [1] (a library for Support Vector Machines) and the ϵ-SVR mode. The Radial Basis Function (RBF) has been chosen as kernel function and the search grid approach has been performed assuming the following values for C and γ:

$$C = \left\{ 2^0, 2^1, \dots, 2^8 \right\}, \ \gamma = \left\{ 2^{-5}, 2^{-4}, \dots, 2^3 \right\}.$$

The parameter ϵ, which regulates the permissible error for the objective function, has been set to 0.01.

2.3 Extreme Learning Machine

Extreme Learning Machine has been presented by Huang et al. [5] as a fast learning algorithm for single hidden layer feed forward neural networks (SLFNs). Differently from standard ANNs approaches, the input weights are randomly generated and the output ones are tuned by a least-square method. In later work [4], an unified solution for regression, binary, and multi-class classification has been presented. The random generation of the input weights leads however to a certain variability of the results. This variability was partly offset by training 200 models on the same Training set and then applied to the Testing set, mediating then the results in order to have a more stable and accurate estimation which have less variability. To identify the best configuration, the number of hidden layer nodes has been varied from 80 to 110. The test showed that this range of neurons provide the best results.

3   Computer Simulation and Results

The performance evaluation of various models is performed by taking into account the values of the mean and variance of the Root Mean Square Error (RMSE) on the Testing set obtained with the CV. In Table 1 is reported the partition in Training, Validation and Testing for each employed dataset.

All simulations have been performed in Matlab 2013a® running on a Windows 7® OS.

In Tables 2, 3 and 4 the error values obtained from tests on selected datasets have been reported. Each table refers to one different duct and reports the results for three Mesh values. The first column shows the values of the system reference POWdER and subsequent columns the error results obtained with the three algorithms of machine learning studied.

In the tables, the better results for the Testing set are reported in bold.

Table 1   Datasets partition

Ducts	Observations number			
	Total	Train	Valid	Test
Duct 1	272	188	38	46
Duct 2	251	174	35	42
Duct 3	286	198	40	48

Table 2   Algorithms performance on DUCT 1 dataset

Rmse	Duct 1			
	POWdER	ANN	ELM	SVR
Mesh 50				
Avg train	0.0750	0.0780	0.0763	0.0813
Var train		1.83e-4	2.20e-5	5.06e-5
Avg test	0.1390	0.1386	0.1094	**0.1043**
Var test		7.34e-4	2.94e-4	3.29e-4
Mesh 100				
Avg train	0.7600	0.8533	0.8874	0.8461
Var train		3.30e-3	1.00e-3	9.41e-2
Avg test	1.4000	1.6845	**1.3494**	1.4007
Var test		1.90e-2	1.62e-2	2.01e-1
Mesh 200				
Avg train	1.4200	1.2191	1.7546	0.6798
Var train		1.25e-2	3.70e-3	2.02e-2
Avg test	3.2200	2.8612	2.4950	**2.3023**
Var test		1.27e-1	2.45e-2	6.83e-2

Table 3 Algorithms performance on DUCT 2 dataset

Rmse	Duct 2			
	POWdER	ANN	ELM	SVR
Mesh 50				
Avg train	0.0600	0.0968	0.0842	0.0408
Var train		3.44e-4	8.63e-6	7.55e-6
Avg test	0.1400	0.1453	0.1359	**0.1134**
Var test		8.43e-4	1.07e-4	4.08e-4
Mesh 100				
Avg train	0.5700	1.3245	0.6519	0.4421
Var train		4.08e-2	2.20e-3	3.80e-3
Avg test	1.1100	2.5003	1.0583	**1.0383**
Var test		2.32e-1	1.20e-2	5.50e-2
Mesh 200				
Avg train	0.9200	1.3331	0.7825	0.6591
Var train		2.50e-2	7.50e-3	2.33e-2
Avg test	1.9600	2.7403	1.6679	**1.6618**
Var test		2.24e-1	5.41e-2	8.72e-2

Table 4 Algorithms performance on DUCT 3 dataset

Rmse	Duct 3			
	POWdER	ANN	ELM	SVR
Mesh 50				
Avg train	0.0710	0.0935	0.1006	0.0443
Var train		2.56e-4	1.10e-4	3.15e-5
Avg test	0.1400	0.1776	0.1463	**0.1091**
Var test		1.16e-4	2.49e-4	1.41e-4
Mesh 100				
Avg train	0.6200	0.8489	0.8095	0.4370
Var train		2.42e-2	1.60e-3	1.50e-3
Avg test	1.0300	1.5623	1.1413	**0.9898**
Var test		1.33e-2	2.03e-2	6.40e-3
Mesh 200				
Avg train	1.0900	1.2111	1.5067	0.5963
Var train		3.88e-2	1.10e-3	1.50e-3
Avg test	1.6400	2.5308	2.1146	**1.7313**
Var test		6.17e-1	4.15e-2	2.31e-2

By analysing the average values of the error on the Testing data for the firsts two Ducts, it can be observed that the SVR and ELM provide the lower estimation error, while the ANN has the worse results. This trend is also confirmed for the Training data, where the SVR and ELM outperforms the ANN. The SVR and ELM perform also better than POWdER in all cases. Looking at the variance values, it can be seen that the SVR and the ELM to have a very similar variability, lower than ANN.

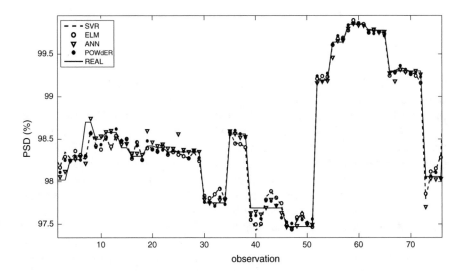

Fig. 1 Comparison of estimated output values with real target ones on the training set

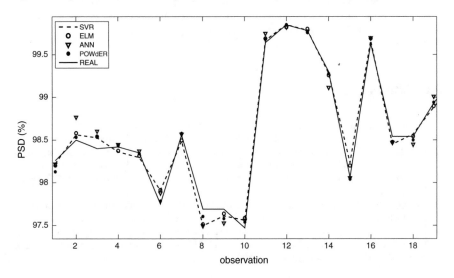

Fig. 2 Comparison of estimated output values with real target ones on the testing set

For the third Duct, only the algorithm SVR perform better than the system POWdER for he first two meshes, instead, for the Mesh 200, the system POWdER provides better result than all machine learning algorithms.

In the Figs. 1 and 2 the estimation results obtained with the various algorithms are plotted in comparison with the corresponding real targets and the targets estimated by POWdER. Both training and testing data related to Mesh 50 of Duct 1 have been used on purpose. In x axis is shown the number of observations and in y axis are shown the values of PSD in terms of percentage, the real and estimated values.

4 Conclusion

This work was aimed to demonstrate the possibility to using Machine Learning algorithms to estimate the PSD of powder measured by means of AE. The data used are those used by the commercial system POWdER. The tests made on the datasets corresponding to different ducts, show how the Machine Learning algorithms produce a good estimate of the values of the PSD, in most cases better than the POWdER's results that is been used as reference. In particular, the algorithm SVR returns the best results in terms of average error, algorithms ANN and ELM have worse outcomes than SVR but in line with the reference values. In the light of this, we can conclude that we demonstrated the possibility to use machine learning techniques for the estimation of the particle size of a powder using acoustic emissions, in non-intrusive systems.

Future developments of the study will aim to assess the applicability of machine learning techniques for PSD distribution estimation in different contexts (food powders, pharmaceutical, etc.), to expand the fields of application. Others machine learning algorithms, in combination with alternative feature sets, will be also tested to verify the possibility of further reducing the estimation error.

References

1. Chang, C.C., Lin, C.J.: LIBSVM: a library for support vector machines. ACM Trans. Intell. Syst. Technol. **2**, 27:1–27:27 (2011). http://www.csie.ntu.edu.tw/~cjlin/libsvm
2. Collura, S., Possanzini, D., Gualerci, M., Bonelli, L., Pestonesi, D.: Coal mill performances optimization through non-invasive online coal fineness monitoring. Powergen, Wien (2013)
3. Haykin, S.: Neural Networks: A Comprehensive Foundation, 2nd edn. Prentice Hall PTR, Upper Saddle River, NJ, USA (1998)
4. Huang, G.B., Zhou, H., Ding, X., Zhang, R.: Extreme learning machine for regression and multiclass classification. IEEE Trans. Syst. Man Cybern. Part B Cybern. **42**(2), 513–529 (2012)
5. Huang, G.B., Zhu, Q.Y., Siew, C.K.: Extreme learning machine: a new learning scheme of feedforward neural networks. In: Proceedings. 2004 IEEE International Joint Conference on Neural Networks, 2004, vol. 2, pp. 985–990 (2004)

6. Leach, M., Rubin, G., Williams, J.: Particle size determination from acoustic emissions. Powder Technol. **16**(2), 153–158 (1977). http://www.sciencedirect.com/science/article/pii/0032591077870010
7. Leach, M., Rubin, G., Williams, J.: Particle size distribution characterization from acoustic emissions. Powder Technol. **19**(2), 157–167 (1978). http://www.sciencedirect.com/science/article/pii/0032591078800242
8. Rumelhart, D.E., Hinton, G.E., Williams, R.J.: Learning representations by back-propagating errors. In: Neurocomputing: Foundations of Research, pp. 696–699. MIT Press, Cambridge, MA, USA (1988). http://dl.acm.org/citation.cfm?id=65669.104451
9. Vapnik, V.N.: The Nature of Statistical Learning Theory. Springer, New York, NY, USA (1995)

Combining Multiple Neural Networks to Predict Bronze Alloy Elemental Composition

Eleonora D'Andrea, Beatrice Lazzerini and Vincenzo Palleschi

Abstract The problem of predicting the composition of alloys measured by means of Laser-Induced Breakdown Spectroscopy (LIBS) analysis is frequently tackled in the literature. In this paper we propose the use of an ensemble of neural networks to model the functional relationship between LIBS spectra and the corresponding composition of bronze alloys, expressed in terms of concentrations of the constituting elements. The networks are trained independently and their inputs are determined by different feature selection processes. Their outputs are then combined by applying an averaging function. The results achieved allow to correctly predict the composition of *unknown* bronze alloy samples.

Keywords RBF · Feature selection · Laser-Induced breakdown spectroscopy

1 Introduction

Laser-Induced Breakdown Spectroscopy (LIBS) analysis is a well-known spectroscopic technique for the identification of material and chemical composition of several kinds of samples (e.g., soils, rocks, metal alloys). The LIBS technique allows obtaining qualitative and quantitative information about the composition of

E. D'Andrea
Research Center "E. Piaggio", University of Pisa, Largo Lucio Lazzarino,
56122 Pisa, Italy
e-mail: eleonora.dandrea@for.unipi.it

B. Lazzerini (✉)
Department of Information Engineering, University of Pisa,
Largo Lucio Lazzarino, 56122 Pisa, Italy
e-mail: b.lazzerini@iet.unipi.it

V. Palleschi
Applied Laser Spectroscopy Laboratory, Institute of Chemistry of Organometallic
Compounds, Research Area of CNR, Via G. Moruzzi 1, 56124 Pisa, Italy
e-mail: vincenzo.palleschi@cnr.it

© Springer International Publishing Switzerland 2016   345
S. Bassis et al. (eds.), *Advances in Neural Networks*, Smart Innovation,
Systems and Technologies 54, DOI 10.1007/978-3-319-33747-0_34

samples, since it leads to absolute concentration values for each chemical element [1]. During the LIBS analysis process, a high power laser beam is focused on the surface of the sample, causing the ablation of a very small amount of mass, due to a rapid rise in the temperature of the locally heated region. The ablated mass, by interacting with the laser pulse, causes the formation of a high-temperature plasma on the surface of the sample. Subsequently, when the laser pulse ends, the plasma expands and cools, by allowing the observation of the characteristic atomic emission lines of the elements. More in detail, the plasma emits light with discrete spectral peaks which are collected by means of a spectrograph, and analyzed to extract the chemical composition of the sample, as each chemical element in the periodic table is associated with unique LIBS spectral peaks. The main problems related to LIBS analysis are limited sensitivity and poor precision, which also affect the global accuracy of the results [2]. The usual approach to LIBS quantitative analysis makes use of calibration curves [3], built using appropriate reference standards. The calibration curves (emission line intensity vs. concentration of the corresponding element) are built using a few samples with *known* composition, and then are used to determine the composition of *unknown* samples. The main drawbacks of this approach are the need for calibration samples similar to the unknown ones, and constant experimental conditions. An alternative method, which overcomes the first problem, is called Calibration-Free LIBS (CF-LIBS) [4]. In this case, the plasma temperature and the electron number density are considered, and strict experimental condition assumptions are made. Drawbacks are long time analysis and need to detect one line of each element in the plasma with known atomic data.

More recently, statistical methods such as Partial Least Squares (PLS) [5] have been introduced. The negative side of these algorithms is the need of careful validation of the results; in fact, as in all the methods based on calibration curves or reference samples, the capability of the statistical algorithm to reproduce well the composition of the calibration set does not automatically imply that the composition of *unknown* samples would be predicted with the same accuracy [6]. To overcome such problems, neural networks (NNs) have been successfully employed in [7–9]. Of course, NNs are typically employed as individual models for approximating the input-output relationship.

On the other hand, in the literature, ensembles of learning machines are often proposed as an efficient way to solve classification and regression problems by combining the responses of independent learning machines, i.e., different experts performing the same task, so as to achieve a better response with respect to that of the individual experts [10–13]. The effectiveness of ensemble models depends on: (i) differences among the experts in terms of, e.g., training sets, representation of the inputs, training algorithms, (ii) model architecture, e.g., serial, parallel, hierarchical, (iii) combination criterion, e.g. combiner-oriented, selection-oriented. In fact, combining a set of imperfect experts allows managing and compensating the limitations of individual experts [12] and allows achieving more accurate results with respect to simply choosing the best expert. In particular, regarding the combination of the responses from the experts, several operators can be used, e.g., average, maximum, minimum, median, weighted average, e.g., based on the experts' performances.

Another problem related to LIBS analysis is the high-dimensionality of data, i.e., spectral lines. LIBS spectra are typically expressed in terms of N spectral lines (with N of the order of thousands). Of course using such a high number of features generally decreases, rather than increasing, the accuracy; so generally, a preliminary step of dimension reduction is adopted in order to extract important information initially unknown, although already contained within the spectrum, and to achieve the desired accuracy while keeping the complexity of the approximating model as low as possible. To this aim, a feature selection (FS) process can be used to reduce the number of inputs to the model to the minimum number n (with $n << N$) of significant variables (in our case the intensities of spectral lines). Well-known methods for FS are *forward feature selection* (FFS) [10] and *genetic algorithms* (GA) [14].

The objective of this work is to detect the elemental composition of new, previously *unknown*, bronze alloy samples, based only on the characteristics of a few *known* samples. By *unknown* sample we mean a physical sample not shown to the model during the training phase. To this aim, we propose to use an ensemble of NNs (i.e., independent experts), with different architecture and inputs, and the average as combination operator. In particular, we adopt Radial Basis Function (RBF) NNs [10]. Finally, a different FS is performed for each expert: the intensities of the selected spectral lines will be the input variables for the corresponding NN.

The paper has the following structure. Sections 2 and 3 describe, respectively, the bronze alloy data used in the paper, and the proposed approach to predict the composition of *unknown* alloy samples. Sections 4 and 5 provide, respectively, the experimental results obtained, and the conclusions.

2   Bronze Alloys Dataset

The data used in the paper were collected by exploiting 5 different physical samples of modern bronze alloys with known concentrations, of four chemical elements: Copper (Cu), Zinc (Zn), Tin (Sn), and Lead (Pb), shown in Table 1. We will use the word "sample" to refer to a specific modern bronze alloy, i.e., a given composition of four chemical elements. On the other hand, we will refer to a specific instance of a sample as a *spectrum*. This means that a sample corresponds to a set of spectra. A total of 750 spectra (150 spectra for each sample) were acquired and processed. The LIBS spectra of the samples represent the intensity of the LIBS signal at 3606 spectral lines, thus each spectrum is represented in $\Re^{3606}$. As a preprocessing phase, the intensity values were normalized in the interval $[0 \div 1]$ according to the max-min formula in order to have all the emission lines on the same intensity scale.

As we aim to predict the elemental composition of the *unknown* bronze alloys samples, thus performing the so called "out-of-sample" test, we will use only a subset of the available samples to train the model, and we will test it on the

Table 1 Elemental composition of 5 samples of modern bronze alloys (highlighted samples are used as out-of-sample data)

Sample Id	Composition (% in weight)			
	Copper (Cu)	Zinc (Zn)	Tin (Sn)	Lead (Pb)
S161	87.1	5	5.5	2.4
S162	90.9	0.3	7.8	1
S163	84.1	1.4	10.5	4
S164	85.7	9.5	3.6	1.2
S165	87.3	3.7	6.6	2.4

remaining samples. For each element we identify the sample containing the maximum value and the sample containing the minimum value for the concentrations of that element as done in [8]. Each sample containing the minimum or maximum value of the concentration of an element is included in the training set. The complete training set is composed by the samples {S162, S163, S164}, while {S161, S165} represents the "out-of-sample" test set.

3 Proposed Approach

The aim of this paper is twofold. On the one hand, we aim to build an ensemble of NNs able to reproduce the mapping from spectra to elemental composition of *unknown* bronze alloy samples, based only on the characteristics of a few *known* samples. On the other hand, we aim to solve a high-dimensional FS problem, by selecting the most meaningful spectral lines of the LIBS spectrum in order to reduce the dimensionality of the problem, maintaining high accuracy and small size of the approximating model.

We propose an ensemble of experts consisting of two RBF NNs. The inputs to each expert are different, as they are chosen according to different FS processes. The outputs from the experts are combined into a single response by averaging the experts' responses. We developed the ensemble within the Matlab® environment.

3.1 Feature Selection Step

The inputs to each expert are selected according to two different FS processes, namely, FFS and GA-based FS. In the latter case, due to computational reasons, the FS process was applied on a reduced set of features, where the relevant information typically lies. In the former case, we search for relevant information in the original set of features.

More precisely, the first expert is an RBF NN having as inputs the features selected by means of an FFS process applied to the 3606 features. At each step of the process, the FFS selects the best subset of features with the aim of optimizing the error made by the *criterion function*, which is an RBF NN having as inputs, at each step, the selected features, and as outputs the 4 concentrations. In addition we fixed a maximum number of hidden neurons equal to 20 and a *spread* value of 1 (the default value proposed in Matlab®). The resulting set of features from the FFS is the set of $n_1 = 3$ LIBS intensities {914, 977, 1884} corresponding to the wavelengths (in nm) {326.15, 334.45, 472.05}. The feature corresponding to the wavelength 334.45 nm, associated to a strong emission line of Zn I, resulted to be the most discriminant feature. Moreover, we noticed that, after adding the 4th feature, the MSE error does not decrease in a meaningful way, thus we considered only 3 input features.

The second expert is another RBF NN having as inputs a different set of spectral lines selected by means of a GA-based FS process applied on a reduced set of 480 features selected among the 3606 by a human expert and previously employed in [8, 15]. Thus, by representing each spectrum in $\mathfrak{R}^{480}$, we perform a GA-based FS process. Due to computational costs reasons, we made in this case another simplification and we fixed the maximum number of features to 10, by customizing the mutation function. The chromosome is represented as a binary-coded vector, where each gene is related to a feature and represents the presence or absence of that feature in the final set. We used stochastic uniform selection, single-point crossover with probability 0.1, and Gaussian mutation. The population consisted of 30 chromosomes, each including 480 genes, and the maximum number of generations was 200. Here too we defined as *fitness function* of the GA an RBF NN-based criterion function. The GA-based FS found the set of $n_2 = 6$ LIBS intensities {715, 912, 977, 1155, 2560, 2621} corresponding to the set of wavelengths (in nm) {299.56, 325.88, 334.45, 357.60, 657.01, 673.12}.

3.2 Development of the Ensemble

After determining the optimal sets of features, each RBF of the ensemble is trained on the samples {S162, S163, S164} and is tested in out-of-sample mode on the remaining two, i.e., the *unknown* samples {S161, S165}.

The RBF architecture employed in both models has one hidden layer, with a maximum number of hidden neurons equal to 20, n_i input variables (the spectral lines selected through the FS processes, with $n_1 = 3$ for the first expert, and $n_2 = 6$ for the second expert) and 4 output variables (the concentrations of the four chemical elements). The transfer functions for the hidden neurons and the output neurons are, respectively, Gaussian, and linear. Regarding the spread, we performed an experimental analysis for each expert model, in order to find its optimal value. We tried spread values ranging from 0.1 to 1 with step 0.1. For each value, the RBF was trained 10 times. In each trial, we generated different training and test sets (80 % and 20 %, respectively, of the total data). We chose as final best RBF NN the one corresponding to the minimum average MSE on the 10 different datasets. In both RBF NNs the optimal value for the spread resulted to be 0.1. Further, the number of hidden neurons resulted to be, respectively, 18 and 16 for the two chosen RBF NNs.

Finally, as combining strategy, we employed the most popular one, i.e., the ensemble averaging method [11].

4 Experimental Results

This section presents the experimental results. We used only three of the available samples to train the ensemble, and we tested it on the remaining samples, with the aim of determining the composition of the two *unknown* bronze samples ("out-of-sample" test). Each experiment was repeated 30 times, as suggested by [16] to be a typical value used in simulations. The error of the model is computed as the mean MSE over the 30 executions.

Table 2 shows the obtained results, achieved on the out-of-sample test set (samples S161 and S165) over the 30 trials. More in detail, the table shows the MSE achieved by the single RBF experts and the improvement (of about one order of magnitude) obtained by the ensemble. The obtained results outperform (of about one order of magnitude) previous results in [8] on the same out-of-sample data but using a single Multi-Layer Perceptron.

Table 3 highlights the obtained results, in terms of mean (target and estimated) concentrations, by chemical element, on sample S165 and provides a comparison with [15]. In [15], the authors apply the One-Point Calibration (OPC) CF-LIBS method, on the same samples (sample S161 cannot be compared because it was used as a reference in [15]). The OPC CF-LIBS method gives very similar (although slightly more accurate) results with respect to our ensemble model. By computing

Table 2 Mean out-of-sample MSEs obtained by the ensemble model

Sample id	MSE expert 1 (st. dev.)	MSE expert 2 (st. dev.)	MSE ensemble model (st. dev.)
{S161, S165}	1.165 $(7.05 \cdot 10^{-1})$	1.526 $(5.27 \cdot 10^{-1})$	$8.834 \cdot 10^{-1}$ $(2.60 \cdot 10^{-1})$

Table 3 Target and estimated mean concentrations on sample S165 by element

Element	Target conc. (%)	Ensemble model		OPC CF-LIBS [15]	
		Estimated conc. (%)	Error (%)	Estimated conc. (%)	Error (%)
Copper	**87.3**	86.07	−1.23	86.6	−0.7
Zinc	**3.7**	5.14	+1.44	4.0	+0.3
Tin	**6.6**	6.67	−0.07	7.6	+1
Lead	**2.4**	2.11	−0.29	1.9	−0.5

the Mean Absolute Error (MAE) over the four mean concentrations in Table 3, we achieve that the MAE made by our ensemble model is 0.757, while the MAE made by OPC is 0.625. However, this slight difference is obtained thanks to the adoption of complex algorithms which are not easy to apply to the analysis of a large number of spectra. The use of NNs, on the contrary, allows performing in a very short time the analysis of a large number of spectra in a completely automatic way.

Finally, we observe that the adoption of multiple NNs has allowed us to improve the results achieved on the same sample in [8], where the MAE results to be equal to 2.52.

5 Concluding Remarks

In this paper we have presented an ensemble model of multiple RBF NNs for predicting the elemental composition of *unknown* bronze alloys measured by means of the LIBS technique. After an FS process for input dimension reduction, we performed "out-of-sample" experiments by training the ensemble on a subset of samples, and successfully testing it on the *unknown* samples.

References

1. Melessanaki, K., Mateo, M., Ferrence, S.C., Betancourt, P.P., Anglos, D.: The application of LIBS for the analysis of archaeological ceramic and metal artifacts. Appl. Surf. Sci. **197–198**, 156–163 (2002)
2. Winefordner, J.D., Gornushkin, I.B., Correll, T., Gibb, E., Smith, B.W., Omenetto, N.: Comparing several atomic spectrometric methods to the super stars: special emphasis on laser induced breakdown spectrometry, LIBS, a future super star. J. Anal. At. Spectrom. **19**(9), 1061–1083 (2004)
3. Yoon, Y., Kim, T., Yang, M., Lee, K., Lee, G.: Quantitative analysis of pottery glaze by laser induced breakdown spectroscopy. Microchem. J. **68**, 251–256 (2001)
4. Ciucci, A., Palleschi, V., Rastelli, S., Salvetti, A., Tognoni, E.: New procedure for quantitative elemental analysis by laser-induced plasma spectroscopy. Appl. Spectrosc. **53**(8), 960–964 (1999)
5. Andrade, J.M., Cristoforetti, G., Legnaioli, S., Lorenzetti, G., Palleschi, V., Shaltout, A.A.: Classical univariate calibration and partial least squares for quantitative analysis of brass samples by laser-induced breakdown spectroscopy. Spectrochim. Acta B **65**(8), 658–663 (2010)
6. Palleschi, V.: Comment on a multivariate model based on dominant factor for laser-induced breakdown spectroscopy measurements. J. Anal. At. Spectrom, Wang, Z., Feng, J. Li, L., Ni, W., Li, Z. (eds.) **26**(11), 2300–2301 (2011)
7. D'Andrea, E., Pagnotta, S., Grifoni, E., Legnaioli, S., Lorenzetti, G., Palleschi, V., Lazzerini, B.: A hybrid calibration-free/artificial neural networks approach to the quantitative analysis of LIBS spectra. Appl. Phys. B **118**(3), 353–360 (2015)
8. D'Andrea, E., Pagnotta, S., Grifoni, E., Lorenzetti, G., Legnaioli, S., Palleschi, V., Lazzerini, B.: An artificial neural network approach to laser-induced breakdown spectroscopy quantitative analysis. Spectrochim. Acta B **99**, 52–58 (2014)

9. El Haddad, J., Bruyère, D., Ismaël, A., Gallou, G., Laperche, V., Michel, K., Canioni, L., Bousquet, B.: Application of a series of artificial neural networks to on-site quantitative analysis of lead into real soil samples by laser induced breakdown spectroscopy. Spectrochim. Acta B **97**(1), 57–64 (2014)
10. Haykin, S.S.: Neural Networks. A Comprehensive Foundation, 2nd ed. Prentice-Hall, New Jersey (1999)
11. Naftaly, U., Intrator, N., Horn, D.: Optimal ensemble averaging of neural networks. Netw. Comput. Neural Syst. **8**, 283–296 (1997)
12. Sharkey, A.J.C.: On combining artificial neural nets. Connect. Sci. **8**, 299–314 (1996)
13. Lazzerini, B., Volpi, S.L.: Classifier ensembles to improve the robustness to noise of bearing fault diagnosis **16**, 235–251 (2011)
14. Holland, J.H.: Adaptation in Natural and Artificial Systems. University of Michigan Press (1975)
15. Cavalcanti, G.H., Teixeira, D.V., Legnaioli, S., Lorenzetti, G., Pardini, L., Palleschi, V.: One-point calibration for calibration-free laser-induced breakdown spectroscopy quantitative analysis. Spectrochim. Acta B **87**, 51–56 (2013)
16. Kuncheva, L.I.: Combining Pattern Classifiers: Methods and Algorithms. Wiley Interscience, New Jersey (2004)

Neuro-Fuzzy Techniques and Industry: Acceptability, Advantages and Perspectives

Valentina Colla, Marco Vannucci and Leonardo M. Reyneri

Abstract The paper analyses the issues related to the use of neuro-fuzzy techniques in the industrial field focusing on the characteristics that influence the acceptance of the various paradigms. The advantages provided by these techniques and the limits that prevent their wide acceptance in the industrial framework are depicted. Exemplar case study are presented and future perspective and guidelines for the successful integration of soft computing techniques within industry are outlined.

1 Introduction

Neuro-Fuzzy (NF) methods include a wide set of computing techniques which hail from the seminal works of McCulloch and Pitts, Zadeh and Grossberg on Neural Networks (NN) and Fuzzy Systems (FS) dating back to the mid 20th century. The elements of novelty and potentiality of such approaches pushed the scientific community to define new theories, work up new methods and analyse limitations of newborn NF techniques. Driven by the enthusiasm of the first period, this effort led to the development of a plethora of new techniques of which only a small fraction survived. Nowadays these approaches are commonly referred as *soft computing*. This group includes, among the others, NN, fuzzy logic, genetic algorithms (GA), etc.

NF techniques have been widely employed through the years to face a multitude of different problems coming from diverse domains, from games to simulation of complex industrial systems.

V. Colla · M. Vannucci (✉)
TeCIP Institute, Scuola Superiore Sant'Anna, Via G. Moruzzi, 1, 56124 Pisa, Italy
e-mail: mvannucci@sssup.it

V. Colla
e-mail: colla@sssup.it

L.M. Reyneri
Politecnico di Torino, Corso Duca degli Abruzzi, 24, 10129 Torino, Italy
e-mail: leonardo.reyneri@polito.it

© Springer International Publishing Switzerland 2016
S. Bassis et al. (eds.), *Advances in Neural Networks*, Smart Innovation,
Systems and Technologies 54, DOI 10.1007/978-3-319-33747-0_35

In the last years NN and FS seem to have reached the mature stage in which their potentials and limitations are clearer. The enthusiasm of the first days slightly faded and NF techniques are applied with a higher level of awareness. Further, the proof of the substantial identity of mayor branches of NF approaches that has been presented in [12] limited the number of independent NF paradigms and directed the efforts of experts and researchers toward a sensibly smaller number of topics.

However, NN and FS suffer from a problem of acceptability and usability in the industrial environment (where the term *industrial* refers to a wide range of real world applications) where NF techniques are not used as widely as expected and, in many case, regarded in a suspicious light. This scepticism is sometimes due to scarce knowledge of the NF technologies or the lack of resources to learn, set-up and test these methods. However sometimes it is justified by some objective limits that NF approaches put into evidence if compared to commonly used (and accepted) standard techniques.

In this paper the problems of acceptability and usability of NF techniques in the industrial field is handled by putting into evidence the theoretical and practical advantages of this approach and the limits that restrained its proper use in such framework. This paper is organised as follows: in Sect. 2 the acceptability issues of NF techniques are discussed, putting into evidence criticisms and advantages of such approaches within the industrial field. In Sect. 3 some exemplar case studies are presented. Finally, in Sect. 4 some conclusive remarks and an outline of the aspects to focus on for improving NF techniques acceptability in the industrial field are provided.

2   Industrial Acceptance of Neuro-Fuzzy Techniques

Since the first approaches the connection between NF systems and industry has been controversial: on one hand affected by scepticism and on the other one characterized by the interest and attraction toward these techniques from industrial personnel when put into contact with the achievements and capabilities of these *novel* approaches [2].

Despite the evidence of the potentials of NF techniques when coping with industrial applications, these techniques are often neglected by industrial decision makers, engineers and technicians, who prefer to use standard methods which are commonly acknowledged. NF methods are usually introduced to the industrial environment by academics when collaborations are pursued. This latter aspect contributes to the conception of NF methods as a *niche* approach. On the other hand, the examples of industrial applications which took advantage from the use of NF techniques are numerous and contribute to increasing their acceptability. In the following the main criticisms and benefits related to the use of NN and FS for industrial applications are depicted.

2.1 Criticisms to the Acceptability of Neuro-Fuzzy Approaches in the Industrial Field

One of the main criticisms traditionally ascribed to NF methods is *cripticity*, specially when talking about neural networks. The term *black-box* is traditionally associated to NN since a trained network is perceived as a tool that can efficiently performs its task without giving any possibility to the human user to understand *why* it works so well. In facts, inner workings which lead to such good results are distributed among all the parameters of the network and the input/output relationships are practically hidden. These characteristics contribute to the generation of scepticism in not expert users within industry, who are more familiar with standard methods. In the applications based on analytical or empirical models, the role and interactions among the involved variables is clear, the number of models parameters limited and their effect interpretable, easily quantifiable and, sometimes, controllable. NF models are normally more complex, involve a much higher number of parameters, thus the meaning and control of each of them is nearly impossible. In this case the background and expertise of engineers working in the industry must be taken into consideration; their background rarely includes NN or FS, thus this personnel is much more inclined to the use of known standard techniques instead of soft computing. Further, the training of personnel inside the companies on these topics would represent a cost and requires time and this fact often makes the application of these techniques unaffordable.

Cripticity is far lower in the case of fuzzy systems which are based on fuzzy variables whose meaning can be easily interpreted together with the relationships among them. Traditional models, human expertise and knowledge expressed as fuzzy rules can be converted into a fuzzy system so as to restrict the cripticity of the models to the values of its parameters which can be set manually or through a training process similar to the one adopted for NNs. On the basis of this consideration, it is worth to note that the above.mentioned unification theorem [12] allows to *translate* NN models into fuzzy models, allowing their interpretation.

NFs techniques are limited within industry also by the absence of certification which makes impossible their practical use in certain fields such as automotive and transportation. Moreover, most of the industrial applications where NFs are employed are subordinated to Non disclosure Agreement (NdA) and by issues related to intellectual property that restrain the diffusion of these approaches.

Another fundamental aspect to consider in the acceptability of NF techniques is the availability and quality of *data* for the NF model training. Unfortunately suitable data to be used for models training are not always available within industry, since data gathering and maintenance presume the existence of an infrastructure devoted to it, which costs money and time and whose returns are not certain, immediate nor quantifiable. Although the quantity of data necessary to train a NF model varies depending on its type (i.e. amount of data required for NN training is much higher with respect to NF systems) the lack of suitable and reliable data definitely prevents the use of NF techniques and represents an obstacle to the diffusion of such techniques in the industrial field. One of the most appreciated characteristics of NF systems is their generalization capabilities: these approaches in facts, if efficiently trained, are able to successfully handle new patterns, unused for the tuning of the NF model. Generalization capabilities are influenced by the distribution of training data: if such data are *well* distributed over the input domain, generalization capabilities can be satisfactory; on the contrary, if data used for model training are too similar, the inductive process of the NF model can be compromised due to the lack of diversity within data that prevents the efficient extraction of general enough information.

Training data *quality* is important for the successful use of NF techniques since these approaches are *data-driven* and their performance is extremely related to the reliability of data employed for their training. Criticalities in the training dataset can dramatically compromise the performance of a NF model. Typical issues related to industrially gathered data are linked to the presence of noise, outliers, sensors failures, communication problems and compromise data reliability. The use of unreliable data for NF training leads to the modelling of mistaken input/output relationships: for this reason in literature a lot of works aiming at filtering training data [3] can be found.

Another issue concerning training data is related to the *variables* which should be fed to the NF models. This is not a simple issue since on one hand not expert users are tempted to employ numerous variables under the naive belief that the NF model can do miracles, by autonomously selecting those variables which actually provide the necessary information for the successful design of the model. Unfortunately, even if NF techniques can perform a sort of internal selection of variables by limiting the weights related to non-influential ones, the use of unnecessary variables introduces noise which decreases models performance. On the other hand, by using a too small set of input variables, the risk is to exclude features that affect the handled phenomenon, even in a way unknown to human experts. This exclusion could be detrimental as well for the NF model, that otherwise could exploit the informative content of discarded variables. For this reason, also variables selection is a topic that has been deeply investigated by the scientific community leading to numerous publication which can be found in literature [4].

2.2 Relevant Advantages of the Neuro-Fuzzy from an Industrial Point of View

If on the one hand there are objective critical issues which inhibit the spread of NF techniques in the industrial framework, on the other hand some features of this paradigms favour such cohesion. These characteristics are mainly related to the achievements of NF techniques, to their flexibility and usability.

In many cases, when suitably applied to industrial tasks, NF techniques achieve amazing results, exploiting with success their intrinsic computational capabilities. Although in literature there are thousands of papers describing the advantages of NF approaches, most of them concern *toy problems*, very different from the real ones. On the other hand, few paper put forth dramatical increase of performance associated to the use of NF techniques [15]. Unfortunately in many papers the comparison between the achievements of NF and standard methods is not present or *unfair* (i.e. comparing the best NF technique with a suboptimal traditional technique). A fair comparison, clearly putting into evidence the actual advantages of the NF approaches, would improve the acceptability of these methods. Moreover NF approaches have been shown to be able to solve critical industrial problems unsuccessfully handled otherwise [14].

Flexibility is another relevant characteristic strongly appreciated in the industrial field. NF systems can be employed to face virtually any kind of task and application: they have been in facts proven to be universal approximators [8]. Finally, the design time necessary to put-up a NF model is lower than for a standard model; further the required knowledge on the modelled phenomenon is minimal if compared to the one employed for designing a traditional model and allows the achievement of suboptimal but satisfactory solutions also from users not expert of NFs. These two latter characteristics are particularly taken into consideration in the industrial field since they allow to rapidly test the models and achieve results.

3 Case Studies

Here some examples of use of NF techniques on real world industrial problems are presented. The aim of this section is to put into evidence the aspects and issues related to NF methods which influence industrial managers and engineers whether to employ or not these techniques.

3.1 Prediction of the End-Point in a Converter

In the integrated steel-making cycle a step which takes place in the Blast Oxygen Furnace (BOF) is devoted to the burning off of carbon by oxygen blowing. This process is controlled on the basis of the measurements of a sublance which periodically takes information on various chemicals concentrations, steel temperatures and other parameters. The analysis of the sublance normally ends some minutes before the end of the blowing, thus there is no information on the final carbon content [C] which is a fundamental process outcome. Numerous works have been carried out for its estimation by exploiting available information. Usually this task is performed by using the following empirical formula:

$$Log\left(\frac{P_c}{[C]^n[O]}\right) = \frac{A}{T} + B \tag{1}$$

where P_c is a constant pressure value, and A, B are two empirical constants [O] is the final oxygen content and [T] is the steel temperature.

Unfortunately the performance of this empirical model is not satisfactory from the industrial point of view, since the phenomenon to reproduce is extremely complex due to the strong and unknown interaction among the involved variables. In [13] a 2 layers feed-forward NN was employed, which tries to reproduce the relationships between model inputs and output that are not adequately represented by the empirical model. The performance obtained by the NN is extremely satisfactory: the prediction error is reduced of 64 % with respect to equation Eq. 1, there is no problem on data availability and the training time is negligible. Nevertheless the model was not positively perceived among industrial personnel, who found difficult to *read* any physical meaning within the NN structure or parameters.

3.2 Train Speed Prediction Through NF Techniques

An Automatic Train Protection (ATP) system is formed by two communicating subsystems: the ground and the on board ones. The on board subsystem estimates the actual train position and speed, the ground one communicates to the train several information including the maximum allowed speed in the successive parts of the railway. According to these figures the train has to control its speed thus, a correct estimation of the actual speed is crucial.

In [1, 7, 11] the following different standard and NF methods for train speed estimation are attempted and compared:

1. a set of formulae provided by experts, which implement traditional *crisp* reasoning. These formulas are applied alternatively on the basis of measured variables describing the train state. These variables include wheels angular acceleration and speed, and axles adhesion;
2. a fuzzy inference system (FIS) with two inputs (difference between the velocity on the two controlled axles, axle adhesion). The parameters of the designed FIS have been determined by means of a training procedure exploiting a large amount of collected data;
3. a more sophisticated version of the basic set of formulae, where some numerical parameters are set by means of a GA based optimization exploiting the same set of data used for the training of the FIS and aiming at minimizing the difference between actual and predicted train speed;
4. a two-layers feed-forward NN which takes as inputs angular speeds and accelerations and adhesion conditions.

According to the results of the tests, the best performing model was the one based on pure NN, but excellent performance are obtained also by the FIS. However, the following criticisms were raised about these approaches:

- it is hard to interpret the NN and some of the fuzzy rules generated by the FIS training process are hard to justify;

- it is difficult to implement the specification requirements on these approaches and especially on the NN based one;
- when tested on faulty conditions, the crisp method achieves the best results, as it explicitly puts into practice the specifications provided by the expert personnel.

On the basis of the results and of the above mentioned issues, the GA-tuned crisp model was preferred. The main motivation lies in its reliability: the rules driving it have been designed by the personnel thus the rationale is easily interpretable and the values of the model internal parameters have clear meaning. Moreover this model works well also in case of malfunctions. Finally a satisfactory cohesion with training data is guaranteed by the GA tuning. Going back to the concept of acceptability of NF systems, in this case the GA tuned crisp rules model is the best trade-off between accuracy and comprehensibility.

3.3 Prediction of Jominy Profiles of Steels

The Jominy profile is obtained by measuring the hardness values h_i at increasing distances d_i from the quenched end of a previously heated steel specimen. This test provides useful information on several mechanical characteristics of the steel but is time consuming and destructive, thus several studies aimed at building a model able to reproduce such profile on the basis of the steel chemical composition, such as the work presented in [9] where a linear empirical model is employed.

However, the results achieved by traditional empirical models are not satisfactory due to the complex relationships between steel chemistry and microstructure which influence hardness. For this reason different NN-based approaches have been attempted. In [6] the Jominy profile is approximated with a parametric curve, and then the shape of each profile is predicted through a neural network which links the steel chemistry to the curve parameters. This approach is satisfactory when the *shape* of the profile is *standard* (i.e. limited number of steel grades are considered) but fails when many different types of steel are considered. A different approach based on a neural sequential predictor is proposed in [5]: here each point of the curve (apart the first two ones) is singularly predicted by a NN having as inputs the contents of some chemical elements and some of the previously predicted hardness values. This sequential NN model achieves very satisfactory results, even when facing different and varying steel grades.

In this case the NN model was appreciated by the industrial personnel normally responsible for the Jominy test. The reasons that determined the acceptance of this model are related to the clear improvement it brings, to the interpretability of the result it provides (i.e. an usual Jominy curve) and to the user-friendly software that manages the model.

4 Conclusions

On the basis of the concepts expressed about acceptability and on the field experience shortly depicted in the presented case studies, it stands out that the use of NF techniques in the industrial field can be extremely fruitful, in terms of accuracy, design time, knowledge and experience required for models development and flexibility.

Despite these irrefutable advantages, industrial personnel reluctantly accepts these approaches, although the industries are nowadays ready from the technical point of view (i.e. data availability, computational instruments) to adopt these tools. The motivations lie mainly in the well known *black-box* nature of NF methods which inhibit the explicit comprehension of the models inner workings and their direct control from the human users. In this context, fuzzy systems seem to be a promising paradigm that allows to merge human knowledge to the capabilities of an efficient training based on data.

There are several applications where the use of NF systems seems to be unavoidable to achieve satisfactory results. It is the case, for instance, of data mining, processes that embed randomness, complex systems: in these applications the knowledge to exploit is hidden in a wide amount of data, often noisy and imprecise or containing randomness, and traditional methods cannot be successful [10].

The acceptability of NF system in industry can be improved by means of a closer collaboration with NF experts, mostly academics, who have to outline *if* and under what *conditions* these techniques are more convenient than traditional ones.

References

1. Allotta, B., Colla, V., Malvezzi, M.: Train position and speed estimation using wheel velocity measurements. Proc. Inst. Mech. Eng. Part F J. Rail Rapid Transit **216**(3), 207–225 (2002)
2. Bonissone, P., Chen, Y.T., Goebel, K., Khedkar, P.: Hybrid soft computing systems: industrial and commercial applications. Proc. IEEE **87**(9), 1641–1667 (1999)
3. Cateni, S., Colla, V., Vannucci, M.: A fuzzy logic-based method for outliers detection. In: Proceedings of the IASTED International Conference on Artificial Intelligence and Applications, AIA 2007, pp. 561–566 (2007)
4. Cateni, S., Colla, V., Vannucci, M.: Variable selection through genetic algorithms for classification purposes. In: Proceedings of the 10th IASTED International Conference on Artificial Intelligence and Applications, AIA 2010, pp. 6–11 (2010)
5. Cateni, S., Colla, V., Vannucci, M., Vannocci, M.: Prediction of steel hardenability and related reliability through neural networks. In: 12th IASTED International Conference on Artificial Intelligence and Applications AIA 2013, pp. 169–174 (2013)
6. Colla, V., Reyneri, L.M., Sgarbi, L.M.: Neuro-wavelet parametric characterization of jominy profiles of steels. Integr. Comput. Aided Eng. **7**(3), 217–228 (2000)
7. Colla, V., Vannucci, M., Allottay, B., Malvezziy, M.: Estimation of train speed via neurofuzzy techniques. In: Mira, J., Alvarez, J.R. (eds.) Artificial Neural Nets Problem Solving Methods. Lecture Notes in Computer Science, vol. 2687, pp. 497–503. Springer, Berlin (2003)
8. Cybenko, G.: Approximation by superpositions of a sigmoidal function. Math. Control Sig. Syst. **2**(4), 303–314 (1989)

9. Doane, D., Kirkaldy, J.: Hardenability Concepts with Application to Steel: Proceedings of a Symposium Held at the Sheraton-Chicago Hotel, Oct. 24–26, 1977. Metallurgical Society of AIME (1978)

10. Li, Y.W., Yuan, T., Han, J.J., Zhu, J.F.: The application of rbf neural network in the complex industry process. In: 2012 International Conference on Machine Learning and Cybernetics (ICMLC), vol. 2, pp. 454–459 (2012)

11. Malvezzi, M., Toni, P., Allotta, B., Colla, V.: Train speed and position evaluation using wheel velocity measurements. In: IEEE/ASME International Conference on Advanced Intelligent Mechatronics, AIM, vol. 1, pp. 220–224 (2001)

12. Reyneri, L.M.: Unification of neural and wavelet networks and fuzzy systems. IEEE Trans. Neural Netw. 10(4), 801–814 (1999)

13. Valentini, R., Colla, V., Vannucci, M.: Neural predictor of the end point in a converter. Revista de Metalurgia (Madrid) 40(6), 416–419 (2004)

14. Vannucci, M., Colla, V.: Novel classification method for sensitive problems and uneven datasets based on neural networks and fuzzy logic. Appl. Soft Comput. J. 11(2), 2383–2390 (2011)

15. Wilamowski, B.: Suitability of fuzzy systems and neural networks for industrial applications. In: 2012 13th International Conference on Optimization of Electrical and Electronic Equipment (OPTIM), pp. 1–7 (2012)

The Importance of Variable Selection for Neural Networks-Based Classification in an Industrial Context

Silvia Cateni and Valentina Colla

Abstract Data pre-processing plays an important role in data mining for ensuring good quality of data especially dealing with industrial datasets. This work presents an exemplar case study for the prediction of the inclusions population in steel products, which demonstrates the importance of variable selection to obtain satisfactory classification accuracy and to achieve a deep understanding of the phenomenon under consideration. A novel variable selection approach has been applied for selecting the variables which mainly affect the target, preliminary to the design of the classifier. Five different classifiers have been designed and applied and the obtained results are presented, compared and discussed.

1 Introduction

Data pre-processing is a fundamental phase in order to build an efficient and reliable Neural Network (NN) as well as for the application of other techniques belonging to the wider class of Artificial Itelligence (AI). In this step, data are analysed in order to both extract the proper inputs and construct consistent training and test data [14]. On the other hand, experimental data which come from real-life processes such as an industrial context, often require pre-processing preliminary to their exploitation for empirical modelling and model validation [12]. The present paper describes how a real dataset coming from the flat steel products manufacturing is exploited in order to predict and classify particular surface defects named inclusions, whose presence is strictly related to the quality of such products. Actually, the higher the inclusion number, the lower the quality. It is thus of utmost importance to correctly predict the occurrence of such defects even before the rolling stage in order to both operate suitable countermeasures to avoid their formation and to address the semi-manufactured products where their presence is mostly likely toward a compatible customer or

S. Cateni · V. Colla (✉)
Scuola Superiore Sant' Anna, TeCIP Institute, Via Alamanni 13D, 56010 Pisa, Italy
e-mail: colla@sssup.it

S. Cateni
e-mail: s.cateni@sssup.it

© Springer International Publishing Switzerland 2016
S. Bassis et al. (eds.), *Advances in Neural Networks*, Smart Innovation,
Systems and Technologies 54, DOI 10.1007/978-3-319-33747-0_36

destination. In fact, depending on the product final application, the presence of such kind of defects can be either a simple drawback or a cause for product rejection.

Even more important is to understand which among a series of possible factors is mostly affecting the occurrence of this kind of defects. Information is obviously available from standard operating practice but still a deeper knowledge can be extracted from the available data in order to solve uncertainties and keep into strict control those process parameters which have the greater influence on the inclusions formation. This is an exemplar case where a particular pre-processing operation, i.e. *variable selection* can be a support for data mining and knowledge discovery.

In the present application, a binary classification problem is formulated where a surface defect that has been detected during the cold rolling stage must be classified as inclusion or not on the basis of a series of chemical and process parameters concerning the manufacturing stages proceeding to the rolling itself. Variable selection is applied to select the most relevant input variables for the classifiers through an innovative combined approach. Afterwards five binary classifiers are designed to this purpose and their accuracy on a real dataset is evaluated.

The paper is organised as follows: Sect. 2 a overviews the problem of inclusions; in Sect. 3 a general overview of the adopted variable selection approach preliminarily to the classification stage is described. The performances of the different classifiers are presented and compared in Sect. 4 in order to put into evidence how their efficiency is affected by the variable selection. Finally in Sect. 5 some concluding remarks are provided.

2   The Problem of Inclusions Detection

The industrial application which is addressed in this paper refers to the classification of a particular kind of surface defects on flat steel products, which are named *inclusions*. An *inclusion* is a very small particle of external material inside the steel matrix [9], whose presence generates a surface defect when the material is rolled (see an example in Fig. 1).

Inclusions in steel could be classified considering two criteria: their origin or their size [8, 20]. According to their origin, inclusions can be *exogenous* or *endogenous*, while considering their size, they can be classified as *micro* or *macro* inclusions. Micro/macro inclusions could be both exogenous and endogenous. The occurrence of micro inclusions is mainly due to reaction between oxygen and other additions in liquid steel; the nature of their formation is growth or nucleation and their dimension is lower than 5 μm. Macro inclusions, that are bigger than 25 μm are mainly formed in tundish due to chemical reactions, re-oxydation, slag entrapment or refractory erosion. In order to improve the so–called steel *cleanliness*, the presence of non-metallic oxide inclusions must be limited and their morphology, composition and size distribution must be controlled in all stages of the steel production [13, 15, 16, 18].

Fig. 1 An inclusions-related defect on the surface of a flat steel product

In the last years car producers increase their request for quality and reliability of the steel coils provided by the steelmakers in order to satisfy the increasing demand for weight reduction and advanced design. To face this challenge, the steelmakers can exploit a valuable and powerful tool called Automatic Surface Inspection System (ASIS), which is able to capture defects present on the surface of flat steel products at the end of the process. Most of the ASIS exploit a lighting system to lighten the strip surface while a set of cameras capture an image of the product surface. These images are elaborated in order to classify the detected surface defects. As the rolling line speed is very high and the coil length can reach 2–3 km, many images are processed in a limited time and this facts limits the reliability of the on-line classification. To this aim ASIS must be suitable trained in order to achieve reliable classification performances, especially with reference to some kinds of defects which are not very frequent or too similar to each other, although deriving from different causes. Inclusions are among those kinds of defects which are frequently misclassified. A possible solution to this problem is off-line post processing of the images: an approach based on Fuzzy Inference Systems has been proposed in [1] to correctly classify two type of surface defects, related to two different kinds of inclusions, namely the so-called *Large Populations* and the *rolled in*.

The task of the present work is different, we want to classify surface defects as inclusions or not depending on variables related to the chemistry of the steel or the processing stages which proceed the cold rolling.

3 Proposed Method

The developed procedure for defects classification is composed of three main steps: the variable selection, the classification procedure and the evaluation of the goodness of the classification: in the following a subsection is devoted to each of these steps.

3.1   Variable Selection

This task has been widely investigated and is the most relevant part of research works where datasets include a large amount of potential input variables for prediction or classification tasks. In many real world problems the variable selection is indispensable especially when the number of variables is considerable with respect to the available number of measurements. The reduction of the variables set can be performed using two different approaches: feature extraction and feature selection [3, 4]. *Feature extraction* performs a transformation of the original feature set, while *variable selection* selects a subset from the original features reducing the dimension of data without transforming the data, but just pointing out the variables that mostly affect a given process/phenomenon [5]. Variable selection techniques are capable to solve a popular paradox in classification tasks, which occurrs when the classification accuracy does not improve if more input variables are considered. On the contrary, when irrelevant or redundant features are considered, the accuracy of the classifier tends to decrease.

In literature three different type of feature selection approaches can be found: *filters*, which are independent on the learning machine used for the classification; *wrappers*, that deal with the learning algorithm as a black box; *embedded approaches*, which perform feature selection as part of the learning phase [2] which complete the design of the classifier. The most obvious for features selection consists in the analysis of all combinations of variables, also called *exhaustive search* (which belongs to the wrapper cathegory) or brute force approach but its computational time complexity is exponential, thus it is not affordable when the number of potential input variables is high.

In this paper a hybrid feature selection method [6] is applied. Firstly a combination of filter approaches is used in order to reduce the number of input variables and then the exhaustive search technique is performed in a reasonable time in order to obtain a sub-optimal subset of features. During the first step 3 filters are independently applied to evaluate the importance of each potential input feature by provide it with a normalized score s ($0 \leq s \leq 1$). A threshold decisor, fixed to the mean score, is applied to selected the reduced feature dataset, which is afterwards subjected to an exhaustive search. The final subset is pointed out which provides the best classifier accuracy.

3.2   Classification

The reduced dataset is divided into two parts: 75 % of the data are used for the classifier training, while remaining 25 % are used for the tests. Attention must be paid to preserve the original proportion among data belonging to the two classes. The dataset is used as input for five different classifiers: a Bayesian Classifier (BC), a Decision Tree (DT) [7], a Support Vector Machine (SVM) [19], a Multi-Layer Perceptron

(MLP) [10] and a Self-Organising Map (SOM) [11]. All these kinds of classifiers are very popular and widely used in literature and are applied here in their standard form, which is not depicted in detailed. The final aim here is to assess the importance of the preliminary variable selection stage by exploiting the validation dataset to calculate the performance of the classifiers. The 10-cross validation method has been used in evaluation phase and the mean performance is considered.

3.3 Accuracy Evaluation

The performance of the classifier is evaluated through the following indices:

- True Positive (TP): percentage of correctly classified samples of the unitary class;
- True Negative (TN): percentage of correctly classified samples of the null class;
- False Positive (FP): percentage of null samples incorrectly classified in the unitary class;
- False Negative (FN): percentage of unitary samples incorrectly classified in the null class.

The traditional accuracy definition provides the effectiveness of classifier by evaluating the probability of correct classification regardless the class. This is not an appropriate measure for imbalanced datasets [34–37], which are very frequent in real contexts, especially when unbalance is high. A good index is the so–called *Average Accuracy*, also called *Balanced Classification Rate* (BCR) [17]. which considers the balance between the two classes and is defined as follows:

$$BCR = 0.5 * [TP/(TP + FN) + TN/(TN + FP)] \qquad (1)$$

where the ratios TP/(TP + FN) and TN/(TN + FP) are also called *sensitivity* and *specificity*, respectively.

4 Experimental Results

In this paper a real industrial dataset has been exploited for the above-depicted analysis. Each defect is associated to the number of the coil where it is located and, subsequently, to the cast from which the coil is originated as well as to the chemical and process variables referring to that cast. In particular, the following 18 variables are available:

1. Reheating Temperature
2. Blowed Oxygen (mn^3)
3. Aluminum content (kg)
4. duration of the reheating process (min)

Table 1 Classification results

Classifier	BCR 18 vars.	BCR 4 vars.
BC	0.59	0.78
DT	0.57	0.72
SVM	0.58	0.78
MLP	0.53	0.77
SOM	0.57	0.76

5. Oxygen in ppm measured at the treatment
6. O_2 in the converter (mn^3)
7. Silicon variation (difference between Si during continuous cast and Si in ladle)
8. Nitrogen variation
9. Phosphorous variation
10. mean casting speed line 1 [cm/min]
11. mean casting speed line 2 [cm/min]
12. Total number of slabs
13. Total weight of slabs
14. Tapping temperature
15. Ladle temperature
16. Treatment temperature
17. Casting Temperature
18. Tundish Temperature

The variables selection procedures selects 4 variables, i.e. No 2, 8, 10 and 15, which means that the Blowed Oxygen, the Nitrogen content in the liquid steel, the first speed of the casting and the ladle temperature are the parameters which mostly affect the defects classification as inclusion. Table 1 shows results in terms of BCR by considering all the variables and after applying the variable selection procedure: the accuracy of all the classifiers is dramatically increased by the application of the proposed variable selection procedure, which proves the efficiency of this approach.

All the classifiers achieve results which are considered compatible with an industrial application, however the BC and the SVM-based classifiers slightly outperform the other ones.

This is also a clear example where the application of a variable selection procedure has not only the beneficial effecto on the design of efficient classifiers (based on NNs as well as on other kind of AI-based techniques), i.e. here classifiers which are actually capable of recognizing the occurrences of inclusions. Also the knowledge of the phenomenon under consideration can be confirmed and/or improved by a joint application of variable selection and NNs: firstly the variables which have the greatest effect are highlighted. Further indications can also be obtained if one considereds e.g. the percentage of detected inclusion as a function of each variable, such as in Fig. 2: it clearly appears that the lowest the Blowed Oxygen the highest the percentage of inclusions, which means that oxygen blowing have a beneficial effect in limiting the presence of inclusions.

Fig. 2 Percentage of
inclusions—variable 2

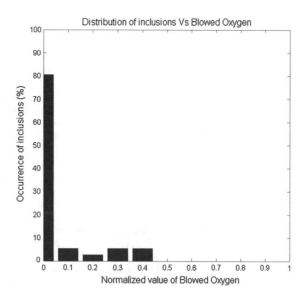

This application therefore provides an exemplar case where NNs can be applied not only as a *black box* which accomplish a task without providing the designer with further insight into the system of phenomeno under consideration, which is a very common criticisms and objection against the application of NNs in the industrial context. On the opposite, in the present case a suitable combination of variable selection and NNs (or other techniques) not only supports correct defect identification but also helps the designer to consolidate its knowledge on the phenomenon under consideration and potentially give hints on suitable countermeasures to limit defects formation.

5 Conclusions

This paper treats an exemplar industrial application related to the classification of particular surface defects of flat steel products, the inclusions, on the basis of parameters related to the steel chemistry and to the processes preceeding the cold rolling stage, i.e. those processes where the phenomenon of inclusions formation actually occurs and could be eventually controlled and corrected. The performance achieved on the tested industrial problem and the results put into evidence the advantages which can be drawn from the use of a variable selection procedure before applying AI-based methods, including NNs. The future perspectives within this field include the use of other datasets in order to further demonstrate the effectiveness of the proposed variable selection approach.

References

1. Borselli, A., Colla, V., Vannucci, M., Veroli, M.: An image inference system applied to defect detection in flat steel production. IEEE World Congr. Comput. Intell. **1**, 1–6 (2010)
2. Cateni, S., Colla, V., Vannucci, M.: General purpose input vvariable extraction: a genetic algorithm based procedure give a gap. In: Proceedings of the 9th International Conference on Intelligence Systems design and Applications ISDA'09, pp. 1278–1283 (2009)
3. Cateni, S., Colla, V., Vannucci, M.: Variable selection through genetic algorithms for classification purpose. In: Proceedings of the 10th IASTED International Conference on Artificial Intelligence and Applications, AIA 2010, pp. 6–11 (2010)
4. Cateni, S., Colla, V., Vannucci, M.: A genetic algorithm-based approach for selecting input variables and setting relevant network parameters of som-based classifier. Int. J. Simul. Syst. Sci. Technol. **12**(2), 30–37 (2011)
5. Cateni, S., Colla, V., Nastasi, G.: A multivariate fuzzy system applied for outliers detection. J. Intell. Fuzzy Syst. **24**(4), 889–903 (2013)
6. Cateni, S., Colla, V., Vannucci, M.: A hybrid feature selection method for classification purposes. In: Proceedings on UKSim-AMSS 8th European Modelling Symposium on Computer Modelling and Simulation, EMS 2014, At Pisa (Italy), pp. 39–44 (2014)
7. Duda, R., Hart, P.: Pattern Classification and Scene Analysis. Wiley, New york (1973)
8. ECSC supported project: Quality improvement for continuously cast steel products by swirling flow strategies. Technical report, Contract number 7210-PR (211)
9. Fageth, R., Allen, W., Jager, U.: Fuzzy logic classification in image processing. Fyuzzy Syst. **82**(3), 265–278 (1996)
10. Fausett, L.: Foundaments of Neural Networks. Prentice Hall, New York (1994)
11. Haykin, S.: Self Organizing Maps in Neural Networks—A Comprehensive Foundation, 2nd edn. Prentice Hall (1999)
12. Komperod, M., Hauge, T., Lie, B.: Preprocessing of experimental data for use in model building and model validation. In: The 49th Scandinavian Conference on Simulation and Modeling (2008)
13. Kwon, Y.J., Zhang, J., Lee, H.G.: A cfd based nucleation-growth-removal model for inclusion behaviour in a gas agitated ladle during molten steel deoxidation. ISIJ Int. **38**, 6 (2008)
14. Lopes, N., Ribeiro, B.: A data pre-processing tool for neural networks (dptnn) use in a moulding injection machine. In: Proceedings of the Second World Manufacturing Congress Durham, pp. 357–361. England (1999)
15. Sheng, D.Y., Soder, M., Alexis: Most relevant mechanism of inclusion growth in an induction-stirred ladle. Scand. J. Matall. **31**, 210–220 (2002)
16. Sheng, D., Soder, M., Jönsson, P.: J. Scand. J. Metall. **31**, 134–147 (2002)
17. Sokolova, M., Lapalme, G.: A systsystem analysis of performance measures for classification tasks. Inf. Process. Manage. **4**, 427–437 (2009)
18. Taguchi, K.: Complex deoxidation equilibria in molten iron by aluminium and calcium. ISIJ Int. **45**, 1572–1576 (2005)
19. Vapnik, V.: Statistical Learning Theory. Wiley, New York (1998)
20. Zhang, L., Thomas, B.G., Wang, X., Cai, K.: Evaluation and control of steel cleanliness review. In: 85th Steelmaking Conference Warrendale, PA, pp. 431–452 (2002)

Non Linear Time Series Analysis of Air Pollutants with Missing Data

Giuseppina Albano, Michele La Rocca and Cira Perna

Abstract This paper investigates the jointly use of local polynomials and feedforward neural networks for estimating the probability of exceedance of the daily average for PM_{10} in the presence of missing data. In contrast to other approaches focusing on some assumption on the distribution of PM_{10}, the reconstruction of the unobserved time series is obtained by using a procedure involving two nonparametric steps based on the estimation of the trend-cycle and of the superimposed nonlinear stochastic component of the series. By using Neural Network Sieve Bootstrap, the probability to overcross the limit established by the European Union for PM_{10} is evaluated at the dates where time series shows missing values. An application to real data is also presented and discussed.

Keywords PM_{10} · Missing data · Neural networks · Resampling scheme

1 Introduction and Background

Air quality monitoring is carried out to detect any significant pollutant concentrations which may have possible adverse effects on human health. Usually, the main indicators to measure air quality are: *(i)* NO_2 (nitrogen dioxide) that is a prominent air pollutant and it is heavier than air so, generally its presence is in the ground; *(ii)* CO (carbon monoxide) that is toxic to humans when encountered in concentrations above about 35 ppm, although it is also produced in normal animal metabolism in low quantities; *(iii)* $O3$ (ozone): above concentrations of about 100 ppb damage mucous and respiratory tissues in animals, and also tissues in plants so it makes ozone a potent respiratory hazard; *(iv)* $PM_{2.5}$ and PM_{10} (particle pollution, also known as

G. Albano (✉) · M. La Rocca · C. Perna
Department of Economics and Statistics, University of Salerno, Fisciano, SA, Italy
e-mail: pialbano@unisa.it

M. La Rocca
e-mail: larocca@unisa.it

C. Perna
e-mail: perna@unisa.it

© Springer International Publishing Switzerland 2016
S. Bassis et al. (eds.), *Advances in Neural Networks*, Smart Innovation,
Systems and Technologies 54, DOI 10.1007/978-3-319-33747-0_37

"particulate matter", less than $2.5\,\mu m$ in diameter $(PM_{2.5})$ and between 2.5 and $10\,\mu m$ in diameter (PM_{10})).

In the following we focus on the concentration of PM_{10} in the air since the International Agency of Research on Cancer and the World Health Organization designate airborne particulates a Group 1 carcinogen. Particulates are the deadliest form of air pollution due to their ability to penetrate deep into the lungs and blood streams unfiltered, causing permanent DNA mutations, heart attacks, and premature death (see [4]). In 2013, a study involving 312944 people in nine European countries revealed that there was no safe level of particulates and that for every increase of $10\,\mu g/m^3$ in PM_{10}, the lung cancer rate rose 22 % (see [8]).

The European Union has developed an extensive body of legislation which establishes health based standards and objectives for a number of pollutants in air. The admitted limits for PM_{10} established by European Union are: $40\,\mu g/m^3$ (Yearly average) and $50\,\mu g/m^3$ (daily average (24-h)). The allowed number of exceedances per year is 35.

However, such analysis is made complex by the frequently large proportions of observations missing from the data due to machine failure, routine maintenance, changes in the siting of monitors, human error, or other factors. Incomplete datasets may lead to results that are different from those that would have been obtained from a complete dataset.

One approach to solve incomplete data problems is the adoption of imputation techniques. In a study of methods for imputation of missing values in air quality datasets, Junninen et al. in [6] generated three randomly simulated missing data patterns for evaluating the methods in different missing data conditions. Blended data patterns in different proportions were constructed for examining the methods in a way that reflected the heterogeneity of the air quality datasets. The patterns were simulated with 10 and 25 % missing data. Noor et al. in [7] concentrate on replacing the real missing item of annual PM_{10} monitoring data using three methods that are linear, quadratic and cubic interpolation, assuming that the data are fitted to a lognormal distribution. A similar work was made in [11] for the Weibull distribution. Recently, Junger et al. in [5] presented an imputation-based method that is suitable for multivariate time series data, which uses the EM algorithm under the assumption of normal distribution. A different approach is developed in [9] in which a nonparametric estimator of the regression function with correlated errors is proposed when observations are missing in the response variable.

Actually, for the time series of PM_{10}, in which generally missing values can be observed, it becomes crucial to investigate the probability to observe an exceedance of the limit established by European Union for each day in which a missing value is observed.

The aim of this paper is to estimate such a probability rather than to estimate the missing value for the daily average of PM_{10}. This becomes very interesting since, generally, when a missing value is present it is assumed to be under the threshold of $50\,\mu g/m^3$.

This topic has been treated in environmental literature and a lot of approaches have been proposed based on extreme value theory (see, for example, [10] for an overview). An alternative more general approach for modeling exceedances is via methodology in time series and spatial statistics for probability distribution functions and quantile (see, for example, [1]).

Our approach combines the use of local estimator of a cycle-trend component of an environmental time series and a neural network Sieve bootstrap for the non linear component obtained after detrending the original time series.

The paper is organized as follows. In Sect. 2 the novel approach to estimate exceedance probability is proposed and discussed; It has been applied on a real data set of air pollutants measured through PM_{10} in Salerno area. These series have an high percentage of missing data, so it is plausible that they hide crossing values of the admitted limit. In Sect. 3, the data and some descriptive statistics are presented along with a discussion of the main results. Some remarks close the paper.

2 A Bootstrap Scheme to Estimate Exceedance Probability

Let us consider the model for the response variable Y_t, representing the daily average of an air pollutant at day t:

$$Y_t = m_t + Z_t + \varepsilon_t, \qquad t = 1, \dots, n. \tag{1}$$

In (1) m_t is a deterministic trend-cycle and Z_t is a generally nonlinear stochastic process in $\mathbb{R}$. The error term ε is an unobserved random process with zero mean and finite variance σ_ε^2. In the following we assume that ε_t is a white noise.

The response variable Y_t ($t = 1, \dots, n$) can have missing data. To check whether an observation is available or not, a new variable δ_t is introduced into the model as an indicator of the missing observations. Thus, $\delta_t = 1$ if Y_t is observed, and zero if Y_t is missing.

The imputed time series can be obtained by the following procedure:

1. Estimate the cycle-trend component m_t by using a local polynomial estimator (namely $\hat{m}_t$) based on the approach suggested in [9] and reconstruct the time series as

$$\tilde{Y}_t = \delta_t Y_t + (1 - \delta_t)\hat{m}_t$$

The procedure leads to a consistent estimator for time series with missing data and correlated errors.

2. Estimate the stochastic component Z_t in (1) by using a neural network obtaining $\hat{Z}(t)$. Precisely, compute the residuals $R_t = Y_t - \tilde{Y}_t$ and estimate a feedforward neural network with p input neurons and hidden layer size r on the time series R_t.

3. Compute the values

$$\hat{\varepsilon}_t = \tilde{Y}_t - \hat{m}(t) - \hat{Z}(t).$$

4. Use the NN-Sieve bootstrap (see [2, 3]) to get a bootstrap replicate of $\hat{Z}(t)$, namely Z_t^*.
5. Obtain a bootstrap estimate of Y_t as

$$Y_t^* = \delta_t Y_t + (1 - \delta_t)(\hat{m}_t + Z_t^*). \tag{2}$$

The probability to observe an exceedance for Y_t of a fixed limit c can be expressed as

$$EP := P_*(Y_t^* > c). \tag{3}$$

where P_* is the probability distribution induced by the resampling scheme.
6. To estimate EP, as usual, a Monte Carlo approach is implemented. B replicates of Z_t^*, namely $Z_{1t}^*, Z_{2t}^*, \ldots, Z_{Bt}^*$, $t = 1, 2, \ldots, n$ are generated and, plugging into (2), get the corresponding values $Y_{1t}^*, Y_{2t}^*, \ldots, Y_{Bt}^*$.
7. Estimate EP by

$$\widehat{EP} = \frac{1}{B} \sum_{i=1}^{B} I(Y_{it}^* > c) \tag{4}$$

where $I(\cdot)$ is the indicator function.

The proposed procedure combines a local estimator for the trend-cycle $m(t)$ and a feedforward neural network as an estimator of the component $Z(t)$. The first choice is justified in order to have flexible local structures which are not influenced by missing values outside the estimation window. The latter choice appears to be necessary since the detrended time series might show a non linear structure due both to the intrinsic characteristic of the data and to the cycle-trend estimation step. Indeed, the difficult task of selecting tuning parameters in that step might possibly induce neglected nonlinearities in the detrended series.

3   Application to PM_{10} Time Series from Salerno Area

The data used for this analysis is particulate matter (PM_{10}) data (measured in $\mu g/m^3$) on a time-scale of one per day (daily averaged) from 1 January 2011 to 31 December 2014 in Salerno. In Salerno there are three monitoring stations, but on the website of ARPAC (Agenzia Regionale per la Protezione Ambientale Campania) daily data of PM_{10} only from two stations are available (SA21 and SA22). The data present a very high percentage of missing values (41 % for SA21 and 30 % for SA22). Table 1 shows descriptive statistics (in mg/m^3) for the data. In Fig. 1 the PM_{10} concentrations reported on the website of ARPAC are shown for SA21 and SA22. It is evident that in the both the series a lot of consecutive missing values are present especially in the first part of the observation period.

Table 1 Descriptive statistics on PM_{10} time series data

Station	% missing	n	Min	Q_1	Q_2	Mean	Q_3	Max
SA21	0.41	863	6	20	30	35.33	44	171
SA22	0.30	1016	6	26	33	37.35	43	277

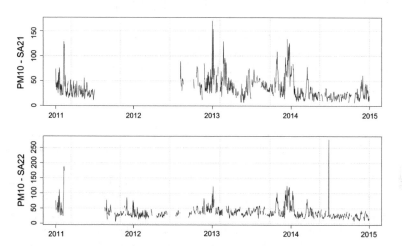

Fig. 1 Time plot of PM_{10} for the stations SA21 (*top*) and SA22 (*bottom*) from 1 January 2011 to 31 December 2014

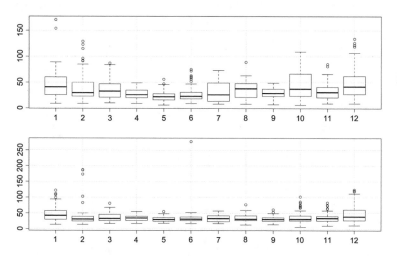

Fig. 2 Monthly boxplots of PM_{10} for the stations SA21 (*top*) and for SA22 (*bottom*) from 1 January 2011 to 31 December 2014

In the following to avoid problems of conditional heteroscedasticity (see Fig. 2) the log-transform of the data has been considered. Hence, let Y_t be the logarithm of PM_{10} at the day t.

In the first step of the proposed procedure, a local estimator for the cycle trend component of Y_t has been used. In this context, the smoothing parameter has been chosen by using a cross validation approach. On the residuals $R_t = Y_t - \tilde{Y}_t$, in order to check their non linearity, the Terasvirta neural network test has been performed. For the two time series SA21 and SA22 the test statistics are, respectively, $\chi^2 = 16.98$ and 38.50 and the p-values are $= 2 \cdot 10^{-4}$ and $4.3 \cdot 10^{-9}$. However, in both the cases, the null hypothesis of linearity is rejected.

Then, a feedforward neural network model has been estimated; again, the parameters p and r have been chosen by using a cross validation approach. Moreover, in the bootstrap scheme B has been fixed to 4999. Figures 3 and 4 report the time plot of PM_{10} in which the missing values are estimated by using the suggested procedure along with the estimation of the exceedance probability of the limit of $50\,\mu g/m^3$ established by European Union.

In particular, in both the figures the top plot shows the reconstructed time series of $\log(PM_{10})$ in which the missing values are estimated by using the mean of Y_{it}^*, $i = 1, \ldots, B, t = 1, \ldots, n$.

The bottom plot shows the estimation of EP. Clearly, when the value of PM_{10} at day t is observed, $\widehat{EP} = 0$ if the values is lower than $\log(50)$, otherwise $\widehat{EP} = 1$. When the value of PM_{10} at time t is missing, the estimate of EP is obtained through (4). Table 2 reports the number of exceedances per year of the limit $50\,\mu g/m^3$ for PM_{10} computed on the original and on the reconstructed time series. It is assumed that an exceedance is occurred if $\widehat{EP} > 0.5$; otherwise, if $\widehat{EP} \leq 0.5$, it is assumed that the missing value is under the threshold of $50\,\mu g/m^3$. When missing values are present the reported number of exceedances appear to be heavily underestimated if no reconstruction of the series is implemented.

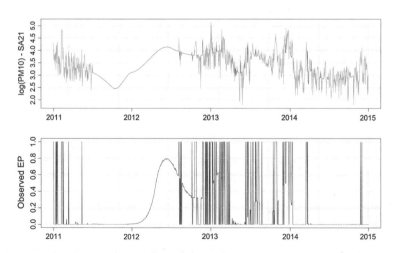

Fig. 3 Recostructed time series of PM_{10} for the station SA21 (*top*) and estimated exceedance probability of the limit $50\,\mu g/m^3$ (*bottom*)

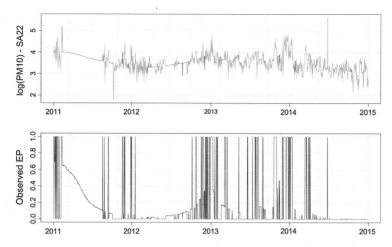

Fig. 4 Recostructed time series of PM_{10} for the station SA22 (*top*) and estimated exceedance probability of the limit $50\,\mu g/m^3$ (*bottom*)

Table 2 Number of exceedances per year of the limit $50\,\mu g/m^3$ for PM_{10} at the stations SA21 and SA22

SA21			SA22	
Year	Original	Recostructed	Original	Recostructed
2011	18	18	35	100
2012	33	141	30	30
2013	99	116	77	77
2014	15	15	26	26

4 Concluding Remarks

In this paper the problem of estimating the probability of exceedance of the daily average for PM_{10} in the presence of missing data is investigated. The novel procedure presents the advantage, with respect other procedure in the same field, to be applicable for a wide class of time series since it involves two nonparametric steps. In the first the estimation of the trend-cycle is obtained by using a local polynomial approach while the residual from the detrended time series is modeled through a feedforward neural network. Then, the exceedance probability, that is the probability to overcross the limit established by the European Union for PM_{10}, is evaluated at the dates where time series shows missing values by using a sieve bootstrap technique based on neural models. An application of the novel procedure to PM_{10} time series from Salerno area is also presented and discussed. The procedure appears to be effective in estimating the exceedance probability when missing data are present and it allows a more realistic estimates of the yearly number of crossing of the EU limits. However, the proposed procedure is still under study through an extended simulation experiment in order to evaluate its performance when different values of the tuning parameters are selected. Finally, a comparison with existing methodologies will be also part of further research studies.

Acknowledgments The authors would like to thank the anonymous reviewers for their valuable comments and suggestions to improve the quality of the paper. The authors also acknowledge partial financial support from the Italian MIUR Grant PRIN-2010, prot. 2010J3LZEN 005.

References

1. Draghicescu, D., Ignaccolo, R.: Modeling threshold exceedance probabilities of spatially correlated time series. Electron. J. Stat. **3**, 149–164 (2009)
2. Giordano, F., La Rocca, M., Perna, C.: Forecasting nonlinear time series with neural network sieve bootstrap. Comput. Stat. Data Anal. **51**, 3871–3884 (2007)
3. Giordano, F., La Rocca, M., Perna, C.: Properties of the neural network sieve bootstrap. J. Nonparametr. Stat. **23**, 803–817 (2011)
4. Health—Particulate Matter—Air & Radiation—US EPA. Epa.gov. 17 Nov 2010
5. Junger, W.L., Ponce de Leon, A.: Imputation of missing data in time series for air pollutants. Atmos. Environ. **102**, 96–104 (2015)
6. Junninen, H., Niska, H., Tuppurrainen, K., Ruuskanen, J., Kolehmainen, M.: Methods for imputation of missing values in air quality data sets. Atmos. Environ. **38**, 2895–2907 (2004)
7. Noor, M.N., Al Bakri, A.M.M., Yahaya, A.S., Ramli, N.A., Fitri, N.F.M.Y.: Estimation of missing values in environmental data set using interpolation technique: fitting on lognormal distribution. Aust. J. Basic Appl. Sci. **7**(5), 336–341 (2013)
8. Raaschou-Nielsen, O., et al.: Air pollution and lung cancer incidence in 17 European cohorts: prospective analyses from the European Study of Cohorts for Air Pollution Effects (ESCAPE). Lancet Oncol. **14**(9), 813–822 (2013)
9. Pèrez-Gonzàlez, A., Vilar-Fernàndez, J. M., Gonzlez-Manteiga, W.: Asymptotic properties of local polynomial regression with missing data and correlated errors Ann. Inst. Stat. Math. **61**, 85–109 (2009)
10. Piergorsch, W.W., Smith, E.P., Edwards, D., Smith, R.L.: Statistical advances in environmental science. Stat. Sci. **13**(2), 186–208 (1998)
11. Yahaya, A.S., Ramli, N.A., Ahmad, F., Nor, N.M., Bahrim, M.N.H.: Determination of the best imputation technique for estimating missing values when fitting the weibull distribution. Int. J. Appl. Sci. Technol. **1**(6), 278–285 (2011)

Part VI
Intelligent Cyber-Physical
and Embedded Systems

Making Intelligent the Embedded Systems Through Cognitive Outlier and Fault Detection

Manuel Roveri and Francesco Trovò

Abstract Intelligent embedded systems represent a novel and promising generation of embedded systems which are able to interact and adapt to the environment in which they operate. When these embedded systems operate in real environmental conditions, the presence of faults or outliers could perturb the acquired datastreams, hence degrading the performance of the envisaged intelligent application. The proposed solution for outlier and fault detection in intelligent embedded systems is meant to promptly detect variations in the statistical behaviour of the acquired datastreams, distinguishing among the occurrence of an outlier, a fault or model bias and activating (whenever possible) suitable mitigation actions. Following a fully cognitive approach, the proposed solution relies on the ability to model and exploit the temporal and spatial relationships present in the acquired datastreams and operate without requiring any a-priori information about the system or the possibly occurring outliers/faults. Experimental results on synthetic and real datasets show the effectiveness of the proposed solution.

Keywords Intelligent embedded systems · Cognitive fault detection and diagnosis · Cognitive outlier detection and diagnosis

1 Introduction

In the recent years there has been a steadily increasing interest in the design and development of intelligent embedded systems, which are embedded systems able to interact and adapt to the environment in which they operate [7]. Relevant application scenarios of such new generation of embedded systems are intelligent sensor

M. Roveri (✉) · F. Trovò
Dipartimento di Elettronica, Informazione e Bioingegneria,
Politecnico di Milano, Milano, Italy
e-mail: manuel.roveri@polimi.it

F. Trovò
e-mail: francesco1.trovo@polimi.it

© Springer International Publishing Switzerland 2016
S. Bassis et al. (eds.), *Advances in Neural Networks*, Smart Innovation,
Systems and Technologies 54, DOI 10.1007/978-3-319-33747-0_38

networks, smart buildings and smart cities, and intelligent objects for the Internet-of-Things (IoT), just to name a few.

The underlying assumptions of all these application scenarios are the availability and the correctness of acquired data. Unfortunately, when the embedded systems operate in (possibly harsh) real-working conditions, the acquired streams of data could suffer from incorrect or perturbed values induced by the presence of faults or outliers affecting the embedded electronics or the sensors. Hence, to maintain the application performance of the envisaged intelligent embedded system, such unlucky events must be promptly detected and (whenever possible) suitably mitigated.

Fault and outlier detection mechanisms have been widely studied in the related literature (e.g., see [11, 12] for comprehensive reviews) and applied in several application domains [13]. Traditional approaches generally require a-priori information about the system or the possibly occurring perturbations (e.g., faults or outliers) to be effective. Unfortunately, such an assumption is hard to be satisfied in complex and large-scale real-working scenarios. To overcome this constraint, in the recent years a new generation of detection and diagnosis systems has been developed under the umbrella of "cognitive" systems. The main characteristics of these systems are the ability to autonomously learn a model of the system in nominal conditions and the possible faults during the operational life (through suitably-defined machine learning mechanisms) and the capability to exploit the temporal and spatial relationships present in the acquired datastreams. Relevant examples of cognitive solutions for fault detection and fault isolation/identification can be found in [3, 4, 6, 8–10, 14, 15].

The aim of this paper is to suggest a novel solution for cognitive outlier and fault detection in intelligent embedded systems. Following a fully cognitive approach, the basis of the suggested solution is the ability to autonomously learn the statistical behaviours of the streams of data and their relationships in nominal conditions. In more detail, the proposed solution aims at promptly detecting variations w.r.t. the learned nominal conditions, distinguishing among the presence of an outlier, a fault or model bias (induced by the uncertainty in the modelling of the statistical behaviours of relationships) and activating (whenever possible) a suitable data-reconstruction mechanism to reconstruct the perturbed data. The main characteristics of the proposed solution for cognitive outlier and fault detection in intelligent embedded systems are the following:

- the joint use of cognitive outlier and fault detection mechanisms for the inspection of acquired datastreams;
- the ability to distinguish between outliers, faults and model bias;
- the ad-hoc activation of a data-reconstruction mechanism whenever an outlier perturbed the stream of data.

To the best of our knowledge, the proposed solution is the first attempt to jointly address the problem of fault and outlier detection following a fully cognitive approach. The paper is organized as follows: Sect. 2 describes the problem formulation, while Sect. 3 details the proposed solution for cognitive outlier and fault detection in intelligent embedded systems. Experimental results are presented in Sect. 4, while conclusions are finally drawn in Sect. 5.

2 Problem Formulation

Let us consider an intelligent embedded system endowed with a multi-sensor board equipped with a set of n sensors, i.e., $S = \{s_1, \ldots, s_n\}$. These sensors could be homogeneous, hence measuring the same physical quantity (e.g., two temperature sensors), or heterogeneous, hence measuring different but related physical quantities (e.g., a temperature and a humidity sensor). Moreover, let $x_i(t) \in \mathbb{R}$ be the measurement acquired by sensor s_i at time t and $\mathbf{x}_i = \left(x_i(t)\right)_{t=1}^{\infty}$ be the ith stream of data. Faults and outliers, which induce a perturbation in acquired data, are here formalized as an additive model (the extension to the multiplicative model is straightforward):

$$x_i^p(t) = x_i(t) + k_i(t)P_i(t)$$

where $x_i^p(t)$ is the perturbed value of the measurement $x_i(t)$ and $k_i(t)$ and $P_i(t)$ model the time-profile (i.e., the time evolution and duration) and the signature (i.e., the effect) of perturbation p affecting sensor s_i at time t, respectively.

In this paper we consider two general families of time profile $k_i(t)$. The first one models the presence of *outliers* $\mathcal{O}$, where the dynamic of the perturbation is faster than acquisition/processing rate, thus inducing an instantaneous spurious perturbation affecting data. Hence, outliers could be modelled by considering $k_i(t)$ as a Bernoulli random variable of parameter γ_i, where γ_i models the probability of having an outlier at the i-th sensor at each time instant. The second one models *faults* $\mathcal{F}$, here represented as a steady variation in the statistical behaviour of acquired datastreams. Fault could be permanent, transient or intermittent according to the characteristics of $k_i(t)$ as described in [12].

3 The Proposed Solution

Following a fully cognitive approach, the proposed solution relies on an initial measurements set $Z_N = (\mathbf{x}^T(1), \ldots, \mathbf{x}^T(N))$, where $\mathbf{x}^T(t) = (x_1(t), \ldots, x_n(t))$, which is assumed to be outlier-free and fault-free, to model the statistical behaviour of the intelligent embedded system in nominal conditions. Afterwards, during the operational life, discrepancies w.r.t. the trained nominal conditions have to be promptly detected and suitably mitigated by differentiating among the occurrence of an outlier, a fault or model bias. The first step of the proposed solution refers to the characterization of the temporal and spatial relationships present in Z_N through the learning of a dependency graph $\mathcal{G} = (V, E)$, where $V = S$ is the set of sensors of the intelligent embedded system and E is a set of directed edges connecting sensors defined

Fig. 1 Example of
dependency graph for a
network with $|S| = 4$
sensors. The set of sensor is
$S = \{s_1, \ldots, s_4\}$ and the set
of functional relationship is
$E = \{e_{21}, e_{32}, e_{41}, e_{42}\}$

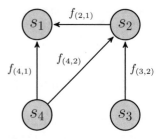

as follows: $e_{ij} = (s_j, s_i) \in E$ iif a relationship $f_{(\mathbf{x}_j, \mathbf{x}_i)}$ between datastreams $\mathbf{x}_i$ and $\mathbf{x}_j$
exists. To learn $\mathcal{G}$, we apply the Granger-based theoretical framework presented in
[5] to the set Z_N. There, relationships between datastreams are assessed through the
theoretically-grounded Granger causal dependency test. An example of dependency
graph is shown in Fig. 1. Once $\mathcal{G}$ has been learned, the proposed solution is able to
set up the outlier and fault detection phases as follows: the set Z_N of the available
measurements is partitioned into T_M and V_{N-M}, where they are formed by the first M
and the last $N - M$ columns of Z_N, respectively ($Z_N = [T_M, V_{N-M}]$). Here, T_M is used
as training set for both the outlier and the fault detection techniques, while V_{N-M} is
used as a validation set.

3.1 Cognitive Outlier Detection and Isolation

For each sensor $s_i \in V$, we learn a non-linear Multiple Input-Single Output (MISO)
model M_i^{MISO} (e.g., a Nonlinear AutoRegressive with eXogenous input -NARX-
model) on T_M, where $\mathbf{x}_i$ is the output and the set of streams $\{\mathbf{x}_j, j \neq i$ s.t. $\exists e_{ij} \in E\}$ are
considered as inputs. In addition to the n MISO models, we learn $|S|$ non-linear Sin-
gle Input-Single Output (SISO) models $M_{i,j}^{SISO}$ modelling the relationships between
a couple of datastreams $(\mathbf{x}_i, \mathbf{x}_j)$ (where $| \cdot |$ is the cardinality operator). Obviously,
$M_{i,j}^{SISO}$ is estimated iif $e_{ij} \in E$. As described in the sequel, the MISO models are used
to detect the occurrence of an outlier, while the SISO ones isolate the sensor in which
the outlier occurred.

　　More specifically, the n MISO models, whose aim is to capture the temporal
and spatial dependencies present among all the streams of data, are the basis of the
first-layer triggering mechanism to detect outliers. In fact, by inspecting the residual
$r_i(t) = x_i(t) - \hat{x}_i(t)$, where $\hat{x}_i(t)$ is the predicted output of M_i^{MISO} model at time t, we
detect the occurrence of an outlier in s_i when (see [11]),

$$r_i(t) < \hat{\mu}_i^{MISO} - \lambda_i^D(t)\sigma_i^{MISO} \vee r_i(t) > \hat{\mu}_i^{MISO} + \lambda_i^D(t)\sigma_i^{MISO}, \tag{1}$$

where $\hat{\mu}_i^{MISO}$ and σ_i^{MISO} represent the mean and the standard deviation of r_i computed on V_{N-M}, respectively, and $\lambda_i^D(t)$ represents a confidence parameter for the

detection of outlier at the i-th sensor at time t. We emphasize that Eq. (1) is a quite straightforward solution for outlier detection [11] but, here, the main idea is that the confidence parameter $\lambda_i(t)$ can vary over time in response to the outlier detection and the reconstruction of an outlier as described below.

Once the outlier has been detected by relying on r_i as per Eq. (1), we need to isolate the sensor (among the output sensor and the input sensors of M_i^{MISO}) in which the outlier occurred. Even though the MISO models are able to accurately model the entire system and detect if a change in the functional relationship structure occurred, they does not provide any information on the location where this change occurred. To achieve this goal we rely on $\mathcal{G}$ and we inspect the SISO models $M_{i,j}^{SISO}$ where the output is $\mathbf{x}_i$ as follows (similarly to Eq. (1)):

$$r_{i,j}(t) < \hat{\mu}_{i,j}^{SISO} - \lambda_i^I(t)\sigma_{i,j}^{SISO} \vee r_{i,j}(t) > \hat{\mu}_{i,j}^{SISO} + \lambda_i^I(t)\sigma_{i,j}^{SISO}, \qquad (2)$$

where $r_{i,j}(t) = x_i(t) - \hat{x}_{i,j}(t)$ is the residual between $x_i(t)$ and the predicted output $\hat{x}_{i,j}(t)$ of the $M_{i,j}^{SISO}$ model, $\hat{\mu}_{i,j}^{SISO}$ and $\sigma_{i,j}^{SISO}$ represent the mean and the standard deviation of $r_{i,j}(t)$ computed on V_{N-M}, respectively, and $\lambda_i^I(t)$ represents a confidence parameter for the isolation of outliers in the i-th datastream at time instant t (which could be, in principle, equal or less than $\lambda_i^D(t)$). We emphasize that all these SISO models share the output stream $\mathbf{x}_i$ and let v be the number of SISO models that detected a discrepancy according to Eq. (2), one of the following three situations arises:

- $v \geq 2$ (i.e., more than one SISO model detected a discrepancy): we can safely consider the output sensor s_i as the sensor in which the outlier occurred;
- $v = 1$ (i.e., just one SISO model $M_{i,\bar{j}}^{SISO}$ detected a discrepancy): we can safely claim that the outlier occurred in the input sensor $s_{\bar{j}}$;
- $v = 0$ (i.e., none of the SISO models confirmed the presence of a discrepancy): we can associate the detection of the M_i^{MISO} model to a model bias.

In the first two cases we confirmed the presence of an outlier and isolated it among all the sensors involved in the M_i^{MISO} model. Let s_h be the isolated sensor, we can mitigate the occurrence of the outlier at time t by replacing the perturbed sample $x_h(t)$ with its prediction $\hat{x}_h(t)$ provided by the M_h^{MISO} model. In the third case, i.e., model bias, no mitigation actions are needed.

We are aware that the introduction of a reconstructed value in the dataset might introduce an unforeseen uncertainty in the outlier detection phase of subsequent samples. Hence, once an outlier has been detected at time $\bar{t}$ and isolated at sensor s_h, the parameters $\lambda_h^D(t)$ of Eq. (1) and $\lambda_h^I(t)$ of Eq. (2) are temporarily increased to take into account the presence of reconstructed values:

$$\lambda_h^K(t) = \max\left\{\bar{\lambda}_h^K, \bar{\lambda}_h^K \cdot \Gamma \cdot \tau^{t-\bar{t}}\right\}, \qquad (3)$$

where $K \in \{D, I\}$, $\bar{\lambda}^K$ is the reference user-defined value of the outlier confidence parameter $\lambda_h^K(t)$, Γ is the peak increase following the outlier detection and isolation and τ is a decay rate.

3.2 Cognitive Fault Detection

Once spurious samples have been removed in the outlier detection phase, the fault detection phase can be applied by inspecting the statistical behaviour of relationships in $e_{ij} \in E$ over time. For this purpose, we rely on the Hidden Markov Model(HMM)-based cognitive fault detection mechanism presented in [3] applied to each couple of datastream $(\mathbf{x}_j, \mathbf{x}_i)$ s.t. $e_{ij} \in E$.

The peculiarity of such an approach is the presence of a HMM-Change Detection Test (HMM-CDT) able to work in the parameter space of linear time-invariant predictive models (modeling the functional relationships corresponding to edges in E) to detect the occurrence of a fault in one of the sensors of the intelligent embedded system. In more detail, for each considered relationship $f_{(\mathbf{x}_j, \mathbf{x}_i)}$, we estimate over time the parameters of a linear Single-Input Single-Output Autoregressive with Exogenous (ARX) model built on batches of data extracted from the available datastreams. The parameters referring to the training set T_M are used to train the HMM, while a change detection threshold is automatically computed on the validation set V_{N-M} according to the HMM-CDT parameter C. Then, during the operational life, the likelihood of the HMM measures the statistical compatibility of estimated ARX parameters over time w.r.t. the trained HMM. When the likelihood decreases below the automatically-defined threshold a change in the relationship is detected. Afterwards, as described in [4], the dependency graph $\mathcal{G}$ is used to discriminate between fault, change in the environment and model bias. When a fault is detected, the application of the intelligent embedded system is informed about the event and suitably-defined reaction mechanisms is activated (e.g., downgrading the running application to operate without the faulty sensor, or setting up a virtual sensor to replace the faulty one).

4 Experimental Section

We evaluated the effectiveness of the proposed solution on a synthetically generated non-linear dataset and on a real-world dataset acquired by a sensing unit of the rock-collapse forecasting system described in [1, 2]. We compare the proposed solution (denoted by COFD) with the method where no outlier or fault detection mechanisms have been adopted (denoted by B and considered here as a baseline), with the method where only the outlier detection and mitigation phase are considered (denoted by COD), and with the method encompassing only the HMM-CDT (denoted by CFD) without any outlier detection/mitigation phase. We consider the following figures of merit (when available) averaged over the total number of experiments:

- *mse*: mean square error between the (possibly reconstructed) data and the generated data without any perturbation, from the beginning of the test set till a fault detection (if provided by the method) has occurred;
- *FN*: fault detection false negatives rate, i.e., the ratio between the number of experiments where the algorithm does not provide a fault detection even though the fault is present and the total number of experiments;

- *FP*: fault detection false positives rate, i.e., the ratio between the number of experiments where the algorithm provides a fault detection before the actual fault occurred and the total number of experiments;
- *DD*: fault detection delay, i.e., the number of instants $t' - t_{init}$ needed to provide a detection (at time t') of the fault once it has occurred at t_{init} .

Synthetic Dataset In the synthetic dataset, we model an intelligent embedded system whose dependency graph has the following characteristics $|S| = 4$ and $|E| = 4$ as described in [5] and where the data generating process is:

$$x_i(t) = \sin\left\{ \theta_{ij}^T \left[x_i(t-1), x_i(t-2), x_j(t-1), x_j(t-2) \right] \right\} + \eta(t),$$

where θ_{ij} is a randomly generated parameter vector and $\eta(t) \sim \mathcal{N}(0, 0.05)$. A sequence of 3881 samples was generated and the first $N = 1841$ are used for the training and validation phases, while the remaining ones constitute the test set for the detection phases. In the interval $[1842; 2861]$ we injected outliers uniformly, with probability $\gamma_i \in \{0.01, 0.05\}$, whose signature is equal to $P_i(t) = M\Delta_i$, where $M \in \{0.2, 0.4, 0.8\}$ and $\Delta_i = \max_t x_i(t) - \min_t x_i(t)$. In the interval $[t_{init}, t_{end}] = [2862; 3881]$ we injected two different kinds of abrupt faults: an additive one **A**, i.e., $x_i^p(t) = x_i(t) + 0.2\Delta_i$, for $t \geq t_{init}$ and a stuck-at one **S**, i.e., $x_i^p(t) = x_i(t_{init})$, for $t \geq t_{init}$. In the considered solutions the confidence of α_g of the dependency graph learning procedure was set to 0.01, while for the outlier detection and isolation phases we set $\Gamma = 2$, $\tau = 0.7$, $\bar{\lambda}_i^D = 3$ and $\bar{\lambda}_i^I = 2$ $\forall i$, and the parameter C of the HMM-CDT was set to 4.1 (the estimation procedure of the threshold was conducted as in [5]). We average the results over 100 independent runs.

The results are presented in Table 1, where we can see from the *mse* values of the baseline (B) and the algorithm with solely the outlier mitigation procedure (COD) that COFD is able to effectively reduce the error induced by outliers and the fault. By comparing the *mse* of the methods employing the HMM-CDT (i.e., CFD and COFD), the proposed method is able to effectively reduce the error induced by the injected outliers. This is particularly evident for high magnitudes of outliers that appear with high probability, e.g., see the *mse* results in the case $\gamma_i = 0.05$ and $M = 0.8$. As expected the detection delays provided by COFD are larger than the one provided by CFD (due to the automatic correction of faulty data as if they were outliers), but we emphasize its ability to significantly reduce FP (w.r.t. CFD) throughout different outlier magnitudes and probabilities.

Rock-collapse Forecasting System The sensing unit of the considered rock-collapse forecasting system is endowed with four sensors (internal/external temperature sensors and two clinometer sensors). The dataset, which is made available to the scientific community,[1] is composed by 1500 samples acquired in year 2011 with a sampling period of 5 minutes. The dataset includes a real stuck-at fault in the interval $[t_{init}, t_{end}] = [1291; 1316]$. We considered the first $N = 1000$ data for training ($M = 800$) and we injected outlier with $\gamma_i = 0.05$ and $P_i(t) = 0.8\Delta_i$. The

---

[1]The dataset can be downloaded from http://roveri.faculty.polimi.it/software-and-datasets.

Table 1 Results for synthetic and real datasets for different outlier Θ and fault F profiles

Θ		F	mse				DD		FP		FN	
			B	COD	CFD	COFD	CFD	COFD	CFD	COFD	CFD	COFD
Synthetic	$\gamma_i = 0.01$	A	0.00446	0.00338	0.00071	0.00054	78.2	99.5	0.226	0.242	0.024	0.016
	$M = 0.2$	S	0.00399	0.00260	0.00037	0.00029	56.5	88.5	0.226	0.242	0	0
	$\gamma_i = 0.05$	A	0.00484	0.00357	0.00079	0.00074	74.8	108.2	0.537	0.276	0.016	0.016
	$M = 0.2$	S	0.00410	0.00269	0.00060	0.00044	60.1	85.7	0.537	0.276	0.000	0.000
	$\gamma_i = 0.01$	A	0.00541	0.00365	0.00093	0.00058	72.9	98.3	0.439	0.220	0.016	0.016
	$M = 0.4$	S	0.00410	0.00261	0.00062	0.00027	57.1	90.9	0.439	0.220	0.000	0.000
	$\gamma_i = 0.05$	A	0.00701	0.00385	0.00183	0.00060	48.4	91.9	0.886	0.244	0.000	0.024
	$M = 0.4$	S	0.00483	0.00264	0.00181	0.00029	57.3	95.2	0.886	0.244	0.000	0.000
	$\gamma_i = 0.01$	A	0.01029	0.00379	0.00213	0.00068	64.7	159.4	0.871	0.218	0.008	0.056
	$M = 0.8$	S	0.00506	0.00254	0.00185	0.00026	61.8	92.2	0.871	0.218	0.000	0.000
	$\gamma_i = 0.05$	A	0.01467	0.00349	0.00824	0.00078	43.0	160.5	0.992	0.200	0.000	0.080
	$M = 0.8$	S	0.00914	0.00250	0.00824	0.00034	68.0	93.5	0.992	0.200	0.000	0.008
Real	$\gamma_i = 0.05$ $M = 0.8$		2.09033	0.05318	10.88364	0.07873	N.a.	45.0	1	0	0	0

experimental results related to this real dataset are presented in Table 1 (averaged over 10 experiments). Here we can see that the behaviour of the *mse*s is in line with the one provided in the synthetic experiments. By resorting to the proposed COFD we are able to provide a positive detection, while a traditional CFD method provides only false positive detections.

5 Conclusions

The aim of this paper was to suggest a novel solution for outlier and fault detection in intelligent embedded systems. The proposed solution relies on a fully cognitive approach, hence being able to exploit temporal and spatial relationships present in acquired datastreams and without requiring any a-priori information to operate. The novelties of the proposed solution reside in the ability to promptly detect perturbations in the acquired streams of data, the capability to distinguishing among outliers, faults and model bias and the ability to reconstruct data affected by outliers. The effectiveness of the proposed solution has been tested on both synthetic and real datasets.

References

1. Alippi, C., Camplani, R., Galperti, C., Marullo, A., Roveri, M.: An hybrid wireless-wired monitoring system for real-time rock collapse forecasting. In: Mobile Adhoc and Sensor Systems (IEEE MASS), 2010. pp. 224–231. IEEE (2010)
2. Alippi, C., Camplani, R., Galperti, C., Marullo, A., Roveri, M.: A high-frequency sampling monitoring system for environmental and structural applications. ACM Trans. Sens. Netw. (TOSN) 9(4), 41 (2013)
3. Alippi, C., Ntalampiras, S., Roveri, M.: An hmm-based change detection method for intelligent embedded sensors. In: The 2012 International Joint Conference on Neural Networks (IJCNN), pp. 1–7. IEEE (2012)
4. Alippi, C., Ntalampiras, S., Roveri, M.: A cognitive fault diagnosis system for distributed sensor networks. IEEE Trans. Neural Netw. Learn. Syst. **24**(8), 1213–1226 (2013)
5. Alippi, C., Roveri, M., Trovò, F.: Learning causal dependencies to detect and diagnose faults in sensor networks. In: 2014 IEEE Symposium on Intelligent Embedded Systems (IES), pp. 34–41, Dec 2014
6. Alippi, C., Roveri, M., Trovò, F.: A self-building and cluster-based cognitive fault diagnosis system for sensor networks. IEEE Trans. Neural Netw. Learn. Syst. **25**(6), 1021–1032 (2014)
7. Alippi, C.: Intelligence for Embedded Systems. Springer (2014)
8. Cen, Z., Wei, J., Jiang, R.: A gray-box neural network-based model identification and fault estimation scheme for nonlinear dynamic systems. Int. J. Neural Syst. **23**(06) (2013)
9. Demetriou, M., Polycarpou, M.: Incipient fault diagnosis of dynamical systems using online approximators. IEEE Trans. Autom. Control **43**(11), 1612–1617 (1998)
10. Farrell, J., Berger, T., Appleby, B.D.: Using learning techniques to accommodate unanticipated faults. Control Syst. IEEE **13**(3), 40–49 (1993)
11. Hodge, V., Austin, J.: A survey of outlier detection methodologies. Artif. Intell. Rev. **22**(2), 85–126 (2004)

12. Isermann, R.: Fault-Diagnosis Systems: An Introduction from Fault Detection to Fault Tolerance. Springer (2006)
13. Isermann, R.: Fault-Diagnosis Applications: Model-Based Condition Monitoring: Actuators, Drives, Machinery, Plants, Sensors, and Fault-Tolerant Systems. Springer (2011)
14. Roveri, M., Trovò, F.: An ensemble of hmms for cognitive fault detection in distributed sensor networks. In: Artificial Intelligence Applications and Innovations. Springer (2014)
15. Trunov, A., Polycarpou, M.: Automated fault diagnosis in nonlinear multivariable systems using a learning methodology. IEEE Trans. Neural Netw. **11**(1), 91–101 (2000)

A Study on the Moral Implications of Human Disgust-Related Emotions Detected Using EEG-Based BCI Devices

Beatrice Cameli, Raffaella Folgieri and Jean Paul Medina Carrion

Abstract This paper reports results obtained from a set of experiments aiming to compare the behaviour and cerebral rhythms in response to the vision of images related to different types of disgust: core, animal nature and moral. The approach combines Information Technology methods and cognitive technologies (specifically Brain Computer Interfaces) as well as behavioural study methods, especially referring to FACS. The presented experiment has been designed with the intent to define the use of images instead of videos in eliciting a disgust response from participants. A sample of individuals has been involved in a process during which they received visual stimuli based on disgust. The stimuli concerned both core/animal nature disgust and moral disgust. The obtained results show the interesting effects of the reaction of participants, abstracted from freeze framed images, as well as EEG correlates.

Keywords Emotional decoding · Cognitive science · Disgust · Brain computer interface · Synchronization analysis · FACS

The names of the authors are in alphabetic order since each made a significant contribution to the research reported.

B. Cameli · J.P.M. Carrion
Cognitive Science and Decision Making, Università degli Studi
di Milano, Milan, Italy

R. Folgieri (✉)
Department of Philosophy, Università degli Studi di Milano, Milan, Italy
e-mail: raffaella.folgieri@unimi.it

© Springer International Publishing Switzerland 2016
S. Bassis et al. (eds.), *Advances in Neural Networks*, Smart Innovation,
Systems and Technologies 54, DOI 10.1007/978-3-319-33747-0_39

1 Introduction

The human capacity to recognize emotions is considered a fundamental innate activity for social communication and survival [1].

The aim of this work is to analyse, by a behavioural and EEG (ElectroEncephaloGraphy) point of view, the emotion of disgust. We chose disgust as the most interesting emotions to be studied in this particular experimental setup, because it is not only elicited as a mean of survival in its most basic features, but it has a major social function, as a bond against what is perceived socially repugnant.

Disgust is the emotion of repulsion: it is studied as a basic emotion since a taste or a smell can trigger it. Its fundamental role in an evolutionary perspective made the study of disgust fall into a domain of emotional behaviour observable not only in Humans, but in most mammals [4]. An experience of disgust includes the motivation of getting away or getting rid of the repulsive entity, often followed or coupled with the desire to wash off, remove or purify any residual physical contact made with the entity. Very specific behavioural and emotional patterns are associated with disgust, from changes in facial expression and mood, to nausea and vomiting [8].

Nevertheless, investigations in cognitive neuroscience have shown that human disgust is based on emotional and behavioural features that can qualify it as a moral emotion [12].

Disgust can be categorized into different categories: the most elementary forms of disgust are distaste and core disgust. They are triggered by an offensive taste or smell, while more complex and exclusive of humans subcategories of disgust are interpersonal and moral disgust, that occur in the social domain.

Animal nature disgust relates to animal nature and mortality of humans, and is triggered by violations of the body envelope, sex, poor hygiene and, of course, death.

Interpersonal disgust is elicited by people, in a way to avoid direct or in indirect physical contact with strangers, probably an adaptive mechanism to avoid the risk of contracting infections and diseases, and to maintain social order.

Moral disgust relates essentially to protecting the social order and as a mean for socialization. It is triggered by some moral offences (Table 1).

In this study, we investigate whether the vision of a core/animal nature and moral disgust-related image could impact both behaviour (facial expression) and brain rhythms, collected by a BCI (Brain Computer Interface) device, namely an Emotiv Epoc headset. Similar studies [10] have been conducted using fMRI (functional Magnetic Resonance Imaging) technology or using videos showing the disgusting action in its making [6]. In this study the moral disgusting action is either shown in its results or freezed framed, making it necessary for the subject to actively place it in the correct frame of the referring context. We considered images instead of videos to verify if also a cognitive abstraction of an action could elicit the emotion of disgust, focusing exclusively on the visual stimulus and not on an action, intrinsic in a video.

The paper is organized as follows: after the context description, in the second paragraph we explain the experimental setup that is the technology used to collect

Table 1 Categories of disgust

Categories of disgust
Distaste: general elicitation present in mammals in response to "bad" tastes
Core disgust: triggered by animals and their products if perceived as potential foods • Food related • Body and animal waste product
Animal nature disgust: elicited by the animal nature and mortality of humans • Sex related • Poor hygiene • Violation of the body envelope • Death and body decay
Interpersonal disgust: contact with individuals perceived as sub-humans • Strangeness • Disease • Misfortune • Moral taint
Moral disgust: related to the protection of the social status quo. Triggered by contact with people perceived as different or dangerous the s.q (e.g. homosexuals, people from a different ethnic or cultural background, criminals).

EEG brain rhythms, the method applied for the behavioural and the synchronization analysis and the sample of individuals participating in the experiment. In paragraph three we will present and discuss the results, while in the last paragraph four we will trace our conclusions and future research directions.

2 Experimental Setup

2.1 Participants

Eleven volunteers (6 females and 5 males) attending or recently graduated from the University of Milan, participated in the study. Participants age was between 21 and 19 years old. All volunteers agreed to participate and to have their picture taken twice a second. All the subjects weren't informed of the aim of the study, but were debriefed afterwards.

2.2 Materials and Methods

The primary aim of our study consists in comparing results from behavioural analysis and brain rhythms responses to either core/animal nature and moral disgust. In order to pursue this goal, we analysed the cortical correlates of the disgust emotion by the use of a Brain Computer Interface.

The reliability of commercial non-invasive BCI devices and the lower cost of these EEG-based systems, compared to other brain imaging techniques, such as fMRI or high-density EEG, determined the increasing interest in their application in different Research fields, also thanks to the portability of the equipment. This last feature makes BCI devices particularly suited for experiments involving virtual [7] and real [9] situations and a larger number of subjects, especially when evaluating emotional or cognitive response of individuals. In fact, during EEG measures, anxiety induced by invasive devices could influence the emotive response of individuals. At the opposite, BCI devices, being projected also for entertainment applications, are particularly comfortable and allow the collection of data in a more ecological fashion. Commercial BCI [1] devices consist in a simplification of the medical EEG equipment, communicating an EEG response to stimuli by WiFi connection, allowing people to feel relaxed, to reduce anxiety and to move freely in the experimental environment, acting as in the absence of the BCI devices. Among the different BCIs proposed by commercial Companies, two small-sized, inexpensive devices are currently largely in use in the scientific and in the entertainment communities: the Emotiv Epoc [3] and the Neurosky MindWave [11]. In our study, we used the Emotiv Epoc headset, equipped with fourteen wet sensors, positioned in different areas of the scalp. For the aim of our study, we considered data from the frontal area of the scalp, involved in visual cognitive processes. The contribution of the frontal lobes, in fact, is fundamental in our research, since we are interested in analysing higher cognitive brain activity, mainly recordable in frontal regions, rather than elementary stimulus related activity. BCI devices collect several cerebral frequency rhythms: the Alpha band (7–14 Hz), related to relaxed awareness, meditation, contemplation; the Beta band (14–30 Hz), associated with active thinking, active attention, solving practical problems; the Delta band (<4 Hz), frontal in adults, posterior in children with high amplitude waves, detected during some continuous attention tasks [2]; the Theta band (4–7 Hz), usually related to emotional stress, such as frustration and disappointment; the Gamma band (30–80 Hz), generally related to cognitive processing of multi-sensorial stimuli (Basar 2012). The BCI device communicates the EEG response to stimuli by Wi-Fi or Bluetooth connection, allowing people to move freely in the experimental environment, thus reducing anxiety.

We noticed that the signal transmitted by the device is already filtered to remove the 50 Hz frequency band related to electric power equipment in our country. Moreover, it detects eye blinks, thus avoiding a specific filtering work on this kind of artifacts. Other motion related activities should be detected with ad hoc signal analysis, that it has not been considered, since from a visual inspection the signal in general did not show these typical features.

We improved the signal quality of the Emotiv Epoch using a bipolar spatial filtering, referencing one electrode the one other electrode.

To improve the interpretation of cognitive and behavioural processes, we combined the use of FACS (Facial Action Coding System) [5], statistics and

Information Technology (specifically signal analysis) methods. In fact, though the main objective of the study consists in investigating the impact of disgusting messages on cognitive performances, we also aimed to verify how EEG detection through BCI device can improve the analysis and the interpretation of cognitive processes through the combined approach of cognitive science and information technology methods. The analysis of the collected data was divided into three stages: (1) the behavioural data analysis, based on the findings deriving from the questionnaire submitted to participants; (2) the EEG signal analysis, performed on the data collected from individuals participating in the experiments, using the correlation analysis between two coupled brain rhythms and the synchronization index; (3) the comparison of the results from the two approaches to match (or not) the respective findings.

To manage the connection with the device, and to synchronize the recording of EEG data with the computer task, we used specifically implemented Java code (Processing[1]).

For the data analysis, we used different procedures ad hoc implemented using R^2 software statistical environment. Particularly, the adopted R procedure (manmely HHT) we used, solves any problem due to the not monotonic phase returned by the Hilbert transform. For details, see original description of the HHT procedure (http://srl.geoscienceworld.org/content/84/6/1074.full).

2.3 Procedure

To test the similarities between core/animal nature and moral disgust on the subjects we set up three different experimental conditions. The subjects were presented with a core or animal nature disgust stimuli, a neutral, grey frame and a moral disgust stimulus. The grey frame is neutral because it does not elicit any emotions, as for the colour as for the absence of stimulus images. The sequence of stimuli was repeated 3 times.

The moral disgust stimuli were chosen based on pre-existing literature [5, 12]. Twice per second, the computer camera was programmed to shoot a picture of the subject while they were observing the stimuli.

In order to obtain genuine facial expressions in response to the stimuli, the participants were not informed on what they were about to see, but were debriefed after the experiment ended.

Participants were seated in a comfortable position, at a distance of fifty centimetres from the computer screen, wearing the BCI device. They were asked to avoid talking or moving during the observation of the stimuli, in order to avoid the

---

[1]https://processing.org/.

[2]http://www.r-project.org/.

influence of Electromyography (EMG) signal in the collected data. The only task assigned to users was to stay focussed on observing the images.

The images chosen for the experiment have been the following:

- Image one: infected tonsils with visible pus (core disgust)
- Image two: a beaten woman, with a bruised face (moral disgust)
- Image three: a wounded foot (animal nature disgust)
- Image four: a homeless man ferociously attacked (moral disgust)
- Image five: a wounded back, with the spinal cord showing (animal nature disgust)
- Image six: violence against an old woman (moral disgust)

2.4 Data Analysis

We want to recall that Evoked Potentials (EP) or Evoked Fields (EF) indicate the variations in EEG or MEG signals associated with a sensorial stimulus. More generally, Event-Related Potential (ERP) or Event Related Field (ERF) stand for the variations induced by an event (a sensorial stimulus, a motor act or an endogenous event). ERP/Fs are the result of an adjustment of the phase of the cerebral rhythms related to the event and of an increment of the power of the signal. For this reason, in the EEG signal analysis, we also calculated a synchronization index. Considering a single electrode side, the phase coherence time can be easily revealed by inspecting the analytic phase. In fact, irregularities in analytic phase plots reveal the so-called phase slips, occurring when the phase is "reset". Phase coherence time lapse can last 100 ms up to a few seconds, and then new phase locked oscillations arise. Around a stimulus onset, the phase coherence interval is shorter. This phenomenon can be observed by computing the analytic phase of single filtered bands. To perform the phase analysis, we applied the Hilbert transform of the EEG signal collected by the single electrode. In this work we are interested in calculating the synchronization index between each band couple.

Given phase's $\phi^1(t)$ and $\phi^2(t)$ of two signal's frequency bands, we get the phase difference $\phi^2(t) - \phi^1(t) = \phi^{1,2}(t)$ and a synchronization index, as in the following formula:

$$g_{12}^2 = \langle cos(\phi^{1,2}(t)) \rangle^2 + \langle sin(\phi^{1,2}(t)) \rangle^2 \tag{1}$$

where brackets $\langle \rangle$ denote the average of the computed cos and sine values. The index g_{12} range is [0, 1], where 1 represents the perfect phase synchronization and 0 stays for the absence of phase synchronization. The presence of phase synchronization in cortical activity reveals neurons firing in-phase, corresponding to the neurons co-operation for perceptual or cognitive tasks [13]. In our experimental setup we collected one signal from a specific topographic position so we could explore phase synchronization index between different wave bands, revealing the kind of functional activity of the brain.

On the collected data, we calculated also the correlation index. The analysis performed on the collected data was aimed to detect the presence of common effects (disgust elicitation) on the participants. We considered the brain rhythms Alpha, Beta and Theta and their combinations, particularly: theta versus lowAlpha; theta versus highAlpha; theta versus lowBeta; theta versus highBeta; lowAlpha versus lowBeta; lowAlpha versus highAlpha; highAlpha versus lowBeta; highAlpha versus highBeta.

We focused on the presence of significant levels of correlation and synchronization for the six different visual stimuli for all the participants. Two baselines were defined for each subject, below which the data were discarded as non-significant.

The baseline was built through the calculation of correlation and synchronization during the vision of the neutral stimuli, thus making it possible to bestow the excess of the significance level to the vision of the disgust-related stimuli.

On a total of 11 participants, it was considered that the emotional effect could be considered if at least 5 of them presented a significant correlation and synchronization index.

For each participant the first step was to assess and calculate a correlation index for the disgust related stimuli for the combination of the considered rhythms (Alpha, Beta, Theta). Then those brain rhythms were further subdivided into couples, and for each couple calculations were made in order to obtain a synchronization index for each of the disgust-related stimuli.

The following step consisted in considering those neutral stimuli that alternated the disgust-related stimuli throughout the whole duration of the experiment: it was necessary to figure out the baseline for both correlation and synchronization for each participant, so to establish if they actually experienced an emotional reaction during the vision of the target disgust images. From this baseline, two more selections of data were made: first, were excluded those data that did not present a correlation and a synchronization level higher than the baseline; secondly the situations (related either to stimuli and to brain rhythms) which did not present a significant level both for correlation and discrimination were also discarded.

3   Results and Discussion

Collected data have been divided into stimulus-image related epochs. The epoched data were, then, considered in the following analysis.

Table 2 illustrates, vertically, the rhythm combinations considered, while horizontally the sequential disgust-related stimuli that were presented to the participants (images from one to six). The table shows how many individuals presented

Table 2 Significant correlation and synchronization levels for paired rhythms and stimuli

Frequency of significant correlation and synchronization levels for paired rhythms and stimuli						
Paired rhythms	1st	2nd	3rd	4th	5th	6th
theta versus lowAlpha	3	5	2	7	7	5
theta versus highAlpha	6	5	3	3	4	4
theta versus lowBeta	2	5	3	3	5	5
theta versus highBeta	2	7	1	3	4	7
lowAlpha versus lowBeta	5	5	3	4	4	7
lowAlpha versus highAlpha	6	4	3	5	4	2
highAlpha versus lowBeta	5	4	5	4	4	8
highAlpha versus highBeta	5	4	6	2	4	6

simultaneously a correlation and synchronization level higher than their correspondent baseline values.

Results show that stimuli 1 and 6 had a greater effect on the considered brain rhythms. The higher crossed involvement level was recorded with the stimulus 6, while the strongest rhythms combination, present in eight subjects was the pair Low Alpha–Theta, followed by the combination low Alpha–low Beta, present crosswise in 7 of the participants. Generally, the combination Low Alpha–Theta was dominant in the majority of the stimuli, being present in either core, animal nature and moral disgust related images.

To confirm the findings, data in the following Table 3 show the correlation indices related to the brain rhythms couple lowAlpha versus theta, while Table 4 show the corresponding synchronization indices.

Table 3 The correlation indices for the brain rhythms couple lowAlpha–theta for each subject and image

lowAlpha versus theta						
Subject	img 1	img 2	img 3	img 4	img 5	img 6
1	0.867886	**0.521222**	0.249079	**0.488216**	**0.815683**	0.231436
2	0.099919	−0.34318	0.056528	−0.09191	**0.979994**	0.331599
3	−0.05419	0.173223	−0.04331	**0.857738**	**0.782544**	0.016156
4	0.157697	**0.423884**	0.600283	**0.840045**	0.21476	**0.700796**
5	0.24812	**0.815112**	0.139878	0.143301	**0.866526**	0.588761
6	−0.32521	0.360061	0.054704	**0.66627**	**0.838608**	**0.946766**
7	0.827053	0.925656	−0.00982	−0.29816	−0.40613	**0.060964**
8	−0.18252	0.315769	0.12435	**0.710008**	0.768373	0.174461
9	−0.16643	−0.20734	−0.08597	**0.782573**	**0.641046**	0.196369
10	0.557261	0.790038	0.506167	**0.924878**	0.053997	**0.719287**
11	−0.28845	−0.7229	0.412449	−0.33565	−0.69591	**0.658835**

Table 4 The synchronization indices for the brain rhythms couple lowAlpha–theta for each subject and image

lowAlpha versus theta						
Subject	img 1	img 2	img 3	img 4	img 5	img 6
1	0.742892	**0.523592**	0.476043	**0.506265**	**0.926214**	0.624047
2	0.563289	0.198139	0.674353	0.273915	**0.931137**	0.580984
3	0.285354	0.395217	0.350354	**0.709949**	**0.580349**	0.699399
4	0.642046	**0.280144**	0.4643	**0.596356**	0.385927	**0.539331**
5	0.610222	**0.796978**	0.490566	0.775072	**0.805532**	0.753448
6	0.52676	0.583856	0.5281	**0.683405**	**0.657567**	**0.870306**
7	0.490859	**0.792947**	0.592375	0.247033	0.259808	**0.391169**
8	0.094745	0.556784	0.422513	**0.857078**	**0.895583**	0.684119
9	0.320089	0.51938	0.513217	**0.880059**	**0.787146**	0.308538
10	0.891502	**0.915413**	0.744418	**0.575906**	0.639669	**0.570224**
11	0.42287	0.54993	0.460743	0.408512	0.355483	**0.713635**

In the combination theta versus highAlpha we detected transversal effects in images 1 and 2. The maximum synchronization index is achieved by the subject 10 with the first image, with a value of 0.931807, and a corresponding correlation index of 0.838136.

In the combination theta versus lowBeta we had transversal effects in images 2, 5 and 6, while in the couple theta–highBeta for the images 2 and 6. In lowAlpha versus lowBeta we detected the effects in images 1, 2 and 6, while in lowAlpha versus highAlpha we obtained the results for the images 1 and 4. Concluding, in highAlpha versus lowBeta the involved images have been 1, 3 and 6, while in highAlpha versus highBeta we had transversal effects in images 1, 3 and 6.

In conclusion, even if we need to consider the limited number of participants to the experiments and of the adopted images, the results are promising and suggest that high levels of correlation and synchronization of low Alpha and Theta brain

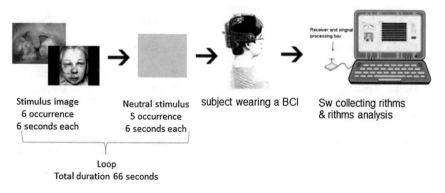

Stimulus image
6 occurrence
6 seconds each

Neutral stimulus
5 occurrence
6 seconds each

subject wearing a BCI

Receiver and singnal processing box

Sw collecting rithms
& rithms analysis

Loop
Total duration 66 seconds

Fig. 1 The experiment set-up

rhythms in response to different disgust related stimuli. This results, obtained on disgust elicitation are supported by the subjects' answers on the questionnaires and confirms the authors' expectation about detecting the emotion of disgust in response to an image instead of a video, since this correlation was present in either core and animal disgust related images and moral disgust related stimuli. Taken on a whole, this result might be particularly true for well educated participants raised in a western society, while moral disgust related stimuli could be perceived as non-repulsive by people from a different ethnic or social background.

4 Conclusions

In this study, we investigate the correlations between facial expressions and brain rhythms in response to different disgust-related stimuli with the use of a non-invasive BCI technology.

As discussed in the specific paragraph, the behavioural analysis showed that the task was well understood by participants, and that feeling moral disgust towards a frozen framed image has very few differences if compared to feeling the same emotions by observing the action taking place. We can hence assume that the cognitive abstraction made by participants in this study can be interpreted as a valid mean for further investigations in the field.

References

1. Allison, B.Z., Wolpaw, E.W., Wolpaw, J.R.: Brain-computer interface systems: progress and prospects. Expert Rev. Med. Devices **4**(4), 463–474 (2007)
2. Aston-Jones, G., Cohen, J.D.: Adaptive gain and the role of the locus coeruleus-norepinephrine system in optimal performance. J. Comput. Neurol. **493**(1), 99–110 (2005)
3. Başar, E.: A review of alpha activity in integrative brain function: fundamental physiology, sensory coding, cognition and pathology. Int. J. Psychophysiol. (2012)
4. Darwin, C.: The Expression of the Emotions in Man and Animals, 3rd edn. Oxford University Press, New York (1998)
5. Ekman, P., Friesen W.V.: Unmasking the Face: A Guide to Recognizing Emotions from Facial Clues. Malor Books (2003). Reprint edition
6. Ekman, P., Friesen, W.V.: Facial Action Coding System: A Technique for the Measurement of Facial Movement. Consulting Psychologists Press, Palo Alto (1978)
7. Folgieri, R., Zichella, M.: A BCI-based application in music. Theor. Pract. Comput. Appl. Entertain. **10**(3) (2012)
9. Haidt, J.: The moral emotions. In: Davidson, R.J., Scherer, K.R., Goldsmith, H.H. (eds.) Handbook of Affective Sciences, pp. 852–870. Oxford University Press, Oxford, UK (2003)
8. Krause, C.M., Sillanmaki, L., Koivisto, M., Saarela, C., Haggqvist, A., Laine, M., Hamalainen, H.: The effects of memory load on event-related eeg desynchronization and synchronization. Clin. Neurophysiol. **111**(11), 2071–2078 (2007)

10. Moll, J., et al.: The moral affiliations of disgust: a functional MRI study. Cogn. Behav. Neurol. **18**(1), 68–78 (2005)
11. Nakamura, T., Sakolnakorn, O.P.N., Hansuebsai, A., Pungrassamee, P., Sato, T.: Emotion induced from colour and its language expression. In: Proceedings of Interim Meeting of the International Colour Association, pp. 29–36 (2004)
12. Rozin, P., Fallon, A.E.: A perspective on disgust. Psychol Rev. **94**, 23–41 (1987)
13. Varela, F., et al.: The brainweb: phase synchronization and largescale integration. Neture Rev. Neurosci. **2**, 229–239 (2011)

Learning Hardware Friendly Classifiers Through Algorithmic Risk Minimization

Luca Oneto and Davide Anguita

Abstract Conventional Machine Learning (ML) algorithms do not contemplate computational constraints when learning models: when targeting their implementation on embedded devices, restrictions are related to, for example, limited depth of the arithmetic unit, memory availability, or battery capacity. We propose a new learning framework, i.e. Algorithmic Risk Minimization (ARM), which relies on the notion of stability of a learning algorithm, and includes computational constraints during the learning process. ARM allows to train resource-sparing models and enables to efficiently implement the next generation of ML methods for smart embedded systems. Advantages are shown on a case study conducted in the framework of Human Activity Recognition on Smartphones, on which we show that effective and computationally non-intensive models can be trained from data and implemented on the destination devices.

1 Introduction

Embedded systems are nowadays playing an important role in the exploration of novel alternatives for the retrieval of information directly from the users and the surrounding environment [9]. Embedding smart models, trained from empirical data through Machine Learning (ML) algorithms, on such devices could target a wide range of new applications that benefit from the device processing and sensing capabilities. However, creating models, which can be effectively implemented on embedded system, is not a trivial task: complex calculations can be either impossible to implement on the target device (e.g. due to the absence of Floating-Point Units—FPUs), or incompatible with some requirements (e.g. in terms of battery life for stand-alone mobile nodes) [5, 11].

L. Oneto (✉) · D. Anguita
DIBRIS, University of Genoa, Via Opera Pia 13, 16145 Genoa, Italy
e-mail: luca.oneto@unige.it

D. Anguita
e-mail: davide.anguita@unige.it

© Springer International Publishing Switzerland 2016
S. Bassis et al. (eds.), *Advances in Neural Networks*, Smart Innovation,
Systems and Technologies 54, DOI 10.1007/978-3-319-33747-0_40

Learning effective classifiers, to be implemented on smart embedded systems, is thus a key task and, for this reason, we propose to run the learning phase offline (e.g. on a conventional PC), thus we deal with the identification of hardware-friendly models to be deployed on the destination embedded system. In other words, we mainly target energy efficiency of the learned model, so that the latter requires fewer system resources for its operation, if compared to a conventional implementation. Several works dealt with similar objectives in the past [1, 5, 13]: however, they typically rely on post-processing existing (real-valued) models, so to adapt them to the embedded system. In this work, we aim at proposing an entire novel learning framework, where limitations in terms of energy efficiency can be contemplated from the beginning. In this way, we are able to select the best model given the constraints. Moreover limiting the amount of information exploited while learning models (e.g. limited number of bits in sparse classifiers [17]) leads also to a beneficial regularization effect, allowing to learn simpler models which even outperform more complex ones and are "lighter" to implement. In particular, we will show that increasing the amount of information, used for learning, leads only to run the risk of overfitting data.

For this purpose, while typically model learning relies on the well-known Structural Risk Minimization (SRM) pillar [19], the objective of this work is accomplished by defining the Algorithmic Risk Minimization (ARM) framework, where: (i) spotlights are moved from classes of models to algorithms, with beneficial outcomes on the overall performance of the trained model, as also underlined in [16]; (ii) learning simpler classifiers is enabled by contemplating only a limited number of bits for representing the data, as well as by favoring the sparsity of the trained model (i.e. by reducing the number of features contemplated when running the classifier on the destination device) [3]. In this framework, a new learning procedure, based on the notion of Stability [7, 18], is introduced and exploited. Advantages of this new approach are shown on a case study conducted in the framework of Human Activity Recognition on Smartphones (HARoS) both in terms of accuracy and energy efficiency of the learned models.

2 Algorithm Risk Minimization

Let us consider the conventional binary classification framework where $\mathcal{X} \in \mathbb{R}^d$ and $\mathcal{Y} \in \{\pm 1\}$ are, respectively, the input and the output space. We consider a set of n labeled i.i.d. samples $S_n : \{z_1, \dots, z_n\}$, where $z_{i \in \{1, \dots, n\}} = (x_i, y_i)$, sampled from an unknown distribution μ. A learning algorithm $\mathcal{A}_{\mathcal{H}}$, characterized by a set of hyperparameters $\mathcal{H}$, maps S_n into a function $f : \mathcal{A}_{(S_n, \mathcal{H})}$ from $\mathcal{X}$ to $\mathcal{Y}$. In particular, $\mathcal{A}$ choses $f \in \mathcal{F}$ where $\mathcal{F}$ is generally unknown and depends on $\mathcal{H}$. The accuracy of $\mathcal{A}_{(S_n, \mathcal{H})}$ in representing μ is measured with reference to a loss function $\ell(\mathcal{A}_{(S_n, \mathcal{H})}, z) :$ $\mathcal{Y} \times \mathcal{Y} \to [0, \infty)$. The purpose of any learning procedure is to minimize the generalization error: $R(\mathcal{A}_{(S_n, \mathcal{H})}) = \mathbb{E}_z \ell(\mathcal{A}_{(S_n, \mathcal{H})}, z)$ [7, 20]. This task can be accomplished by comparing the performance of a set of algorithms $\{\mathcal{A}^1, \mathcal{A}^2, \dots, \}$ or, given algorithm

$\mathcal{A}$, different configurations of the hyperparameters $\{\mathcal{A}_{\mathcal{H}_1}, \mathcal{A}_{\mathcal{H}_2}, ...\}$. This process is usually performed offline, and allows defining the function $\hat{f}$ that will be finally deployed on the destination device. Note that, since μ is generally unknown, we have to resort to an empirical estimator $\hat{R}(f)$ of $R(f)$ (e.g. the empirical error [20] or the Leave One Out error [7]).

According to Statistical Learning Theory (SLT) [20], a learning process consists in aprioristically selecting an appropriate set of hypothesis spaces of increasing size (or class of functions) $\mathcal{F}_1 \subseteq \mathcal{F}_2 \subseteq ...$ and, then, in choosing the most suitable model from it, based on the available data. Unfortunately $\mathcal{F}$ if generally unknown making the SRM framework hard to implement. Consequently, previous works [7, 16, 18] suggested to study the following quantity:

$$R(f) \leq \hat{R}(f) + |R(f) - \hat{R}(f)|, \quad f = \mathcal{A}_{(S_n, \mathcal{H})}. \tag{1}$$

The analysis above does not require $\mathcal{F}$ to be known, while only the algorithm $\mathcal{A}$ must be defined. This approach allows to only take into account those functions that can be actually trained by the algorithm (based on the available samples). The bound of Eq. (1) suggests to revise the conventional SRM framework, so that the focus is moved from sequences of classes of functions to sets of algorithms: we define then the Algorithmic Risk Minimization (ARM) framework. In particular, we consider different algorithms $\{\mathcal{A}^1, \mathcal{A}^2, ...\}$ and, for each algorithm $\mathcal{A}^i$, a different configuration of its hyperparameters $\{\mathcal{H}_1^i, \mathcal{H}_2^i, ...\}$. Then, ARM suggests to select the algorithm and its hyperparameters, such that the bound of Eq. (1) is minimized, and the best model is identified:

$$\mathcal{A}^*_{(S_n, \mathcal{H})} = \arg \min_{\mathcal{A}_{\mathcal{H}} \in \{\mathcal{A}^1_{\mathcal{H}_1^1}, \mathcal{A}^1_{\mathcal{H}_2^1}, ..., \mathcal{A}^2_{\mathcal{H}_1^2}, \mathcal{A}^2_{\mathcal{H}_2^2}, ...\}} \{\hat{R}(\mathcal{A}_{(S_n, \mathcal{H})}) + \hat{D}(f)\}. \tag{2}$$

where $\hat{D}(f) = |R(\mathcal{A}_{(S_n, \mathcal{H})}) - \hat{R}(\mathcal{A}_{(S_n, \mathcal{H})})|$. This approach leads to general benefits in model selection [16]. Unfortunately $D(f)$ cannot be computed with the data since μ is unknown so, in order to study $D(f)$ required by the methodology defined by Eq. (2), we make use of the notion of Stability [7, 18]. Stability aims at upper bounding $D(f)$. For this purpose, we define a modified training set $S_n^{\backslash i}$, where the ith element is removed: $S_n^{\backslash i} : \{z_1, ..., z_{i-1}, z_{i+1}, ..., z_n\}$. The Leave-One-Out (LOO) error is used as empirical estimator of the generalization error [7, 16]: $\hat{R}(f) = \hat{R}_n^{loo}(\mathcal{A}_{(S_n, \mathcal{H})}, S_n) = \frac{1}{n} \sum_{i=1}^n \ell(\mathcal{A}_{(S_n^{\backslash i}, \mathcal{H})}, z_i)$. The deviation of the generalization error from the LOO error is analyzed:

$$\hat{D}(f) = \hat{D}^{loo}(\mathcal{A}_{(S_n, \mathcal{H})}, S_n) = \left| R(\mathcal{A}_{(S_n, \mathcal{H})}) - \hat{R}_n^{loo}(\mathcal{A}_{(S_n, \mathcal{H})}, S_n) \right|. \tag{3}$$

In [16], under some mild conditions, it is proved that the following Stability bound, holding with probability $(1 - \delta)$:

$$R(\mathcal{A}_{(S_n,\mathcal{H})}) \leq \hat{R}^{\text{loo}}(\mathcal{A}_{(S_n,\mathcal{H})}, S_n) + \sqrt{\frac{2}{2n\delta} + \frac{6}{\delta}\left(\hat{H}^{\text{loo}}(\mathcal{A}_{(S_{\frac{\sqrt{n}}{2}},\mathcal{H})}, S_{\frac{\sqrt{n}}{2}}) + \sqrt{\frac{\log(\frac{2}{\delta})}{\sqrt{n}}}\right)}, \qquad (4)$$

where

$$\hat{H}^{\text{loo}}(\mathcal{A}_{(S_{\frac{\sqrt{n}}{2}},\mathcal{H})}, S_{\frac{\sqrt{n}}{2}}) = \frac{8}{\sqrt{n^3}}\sum_{i,j,k=1}^{\frac{\sqrt{n}}{2}}\left|\ell(\mathcal{A}_{(\check{S}^k_{\frac{\sqrt{n}}{2}},\mathcal{H})}, \check{z}^k_j) - \ell(\mathcal{A}_{(\check{S}^k_{\frac{\sqrt{n}}{2}})^{\backslash i},\mathcal{H})}, \check{z}^k_j)\right| \qquad (5)$$

and $\check{S}^k_{\sqrt{n}/2} : \left\{z_{(k-1)\sqrt{n}+1}, \ldots, z_{(k-1)\sqrt{n}+\sqrt{n}/2}\right\}, \check{z}^k_j : z_{(k-1)\sqrt{n}+\sqrt{n}/2+j}, \forall k \in \{1, \ldots,$
$\sqrt{n}/2\}$. Note that the bound of Eq. (4) takes into account only empirical quantities. By plugging the bound of Eq. (4) into the procedure of Eq. (2) we get a fully empirical ARM procedure.

3 Training Hardware-Friendly Classifiers

In this section we consider models $f(x) = w^T x$, where $w \in \mathbb{R}^d$, thus defining linear classifiers in the space $\mathcal{X}$ [20]. Our target is to learn models which can be represented by using a limited number of bits κ. We can then define the set of numbers, which can be represented with κ bits [4]: $\mathbb{B}_\kappa = \{-2^{\kappa-1} + 1, \ldots, 0, \ldots, 2^{\kappa-1} - 1\}$. Then, the spaces of inputs and of weights are: $\mathcal{X} \in \mathbb{B}_\kappa^d$ and $w \in \mathbb{B}_\kappa^d$. Moreover, we also target to learn models, able to exploit only a limited subset of features. In particular, we can fix a maximum percentage of inputs to exploit: $\sum_{i=1}^d [w_i \neq 0] \leq d\zeta$. Then, we introduce the following set: $\mathbb{B}_\kappa^{d\zeta} = \{\mathbb{B}^d : \text{ only } d\zeta \text{ elements } \neq 0\}$. Finally, in order to avoid overfitting issues we have to introduce a regularization effect [12]. For this purpose first step consists in introducing a smooth loss function, the Hinge Loss function $\ell_\xi(f,z) = \max[0, 1 - yf(x)]$, which is a convex upper bound of the hard loss function [20]. This allow us to define our resource efficient training algorithm.

$$\min_w \sum_{i=1}^n \max[0, 1 - yf(x)] \quad \text{s.t.} \quad \|w\|_1 \leq 2^{\kappa-1}\omega, \ w \in \mathbb{B}_\kappa^{d\zeta} \qquad (6)$$

where $\|w\|_1 = \sum_{i=1}^d |w_i|$ and ω is an hyperparararameter that balances the effects of regularization. Note that $\|w\|_1$ is used in place of $\|w\|_2^2 = \sum_{i=1}^d w_i^2$, as the Manhattan-norm is known to allow reducing the number of $w_{j\in 1,\ldots,d} \neq 0$ [14]. Note also that κ and ζ are hyperparameters to tune during the model selection phase: several combinations are tested, and the one that allows obtaining the best model is finally chosen. Eventually, values or ranges of κ and ζ can be imposed by physical constraints (e.g. depth of the Arithmetic Logic Unit—ALU—on the destination device, availability of memory, etc.). Unfortunately Problem (6) is computationally intractable, since it contemplates combinatorial constraints. Consequently, we propose an algorithm, able to find approximate solutions to Problem (6). We give up dealing with combinatorial constraints, by simply coping with a convex relation of the problem and looking for the nearest integer solution that meets the constraints: these tricks have shown to lead to effective solutions in similar problems [1, 5]. For this purpose, let us define $\mathcal{N}(v, \mathcal{V})$ as the nearest neighbor of v in the set $\mathcal{V}$. Then, the learning algorithm can be defined as:

$$w = \mathcal{N}(w^+ - w^-, \omega \mathbb{B}_\kappa^{d\zeta}) : \quad w^+, w^- = \arg\min_{w^+, w^-, \xi} \mathbf{1}_n^T \xi \tag{7}$$

$$\mathbf{1}_d^T(w^+ + w^-) \le 2^{\kappa-1}\omega, \ YX(w^+ - w^-) \ge \mathbf{1}_n - \xi, \ \xi \ge \mathbf{0}_n, \ w^+, w^- \ge \mathbf{0}_d$$

where $X = [x_1 | \dots | x_n]^T$ and Y is a diagonal matrix with $Y_{i,i} = y_i \ \forall i \in \{1, \dots, n\}$. Moreover, be a a constant and e an integer, then a_e is a vector of e elements all equal to a. Note that Problem (7) is a Linear Programming problem: then, it can be effectively and easily solved through many state of the art methods [8, 10], by iteratively neglecting variables that have no influence on the learned model (i.e. the ones for which $w_i = 0$). The hyperparameters of Problem (7) are $\mathcal{H}^L = \{\kappa, \zeta, \omega\}$, namely: the number of bits κ, the sparsity of the solution ζ, and the regularization hyperparameter ω. The learning algorithm, defined by Problem (7), is then: $\mathcal{A}^L_{\{\kappa, \zeta, \omega\}}$. Note that, in order to tune $\{\kappa, \zeta, \omega\}$ the ARM procedure defined in the previous section can be exploited. Since $w \ne 0$ only for few features and $w \in \mathbb{B}_\kappa^d$, implementing the model $f(x) = w^T x$ even on a resource-limited device is straightforward, as it only necessitates a fixed-point unit.

4   Results on HARoS and Discussion

Smartphones are nowadays playing an important role in the exploration of novel alternatives for the retrieval of information directly from the users. When dealing with applications of ML approaches on smartphones, energy efficiency is a key target [9], and is currently the main limitation of these mass-marketed devices: battery draining is an issue, and we have to deal with it in order to realize effective smartphone apps with added value for the final users. As a consequence, we exploit the Human Activity Recognition on Smartphones (HARoS) dataset [2, 6] in order to test the performance of the learning procedures, proposed in the previous section, under different perspectives: in particular, accuracy and energy efficiency will be concerned.

Since the dataset consists of six classes and, in this work, we deal with binary problems, a One-vs-One (OvO) approach is applied. We use $n \in \{25, 100, 225, 400, 625, 900\}$ data, randomly sampled, to create the training set S_n, while the remainder data are exploited as test set $\mathcal{T}$ to verify the performance for the selected model on an independent dataset (note that $S_n \cap \mathcal{T} = \emptyset$): the Hard Loss Function, which counts the number of misclassifications, is used to compute $\hat{R}_\mathcal{T}$, i.e. the error performed on $\mathcal{T}$. We search for ω and γ in the range $[10^{-6}, 10^3]$ among 30 values equally spaced in a logarithmic scale. The number of bits is selected in the set $\kappa = \{1, 2, 4, 8, 16, 32, \infty\}$, where $\kappa = \infty$ is related to the use of double precision floating-point arithmetic. Moreover, the sparsity of the solution is tested for $\zeta = \{0.5, 1\}$. All the procedures are replicated 30 times, randomly sampling different training sets, in order to guarantee that statistically relevant results are derived.

Table 1 presents the error $\hat{R}_{\mathcal{T}}$ (in percentage) performed on the test sets by the learned classifiers when the ARM is exploited: in particular, the values of ζ and κ, selected by the model selection procedure, are also reported. In order to evaluate whether the performance of ARM models is valuable or not, we also perform a test set model selection, where hyperparameters are optimized in order to minimize the error on $\mathcal{T}$ (namely, to obtain the smallest error rate on the test sets): note that this procedure is not feasible in practice, as $\mathcal{T}$ is obviously unknown in real applications, and is performed here only for benchmarking purposes. The corresponding results are shown in Table 2: by comparing them with the ones in Table 1, it is straightforward noting that the ARM learning procedure allows selecting very efficient models. In order to investigate how the ARM works in practice we report in Table 3 the Stability value for the algorithm $\hat{H} = \hat{H}^{\text{loo}}(\mathcal{A}^{\text{L}}_{(S_{\sqrt{n}/2},\{\kappa,\zeta,\omega\})}, S_{\sqrt{n}/2})$, given the tuple of hyperparameters used. All values are shown by varying n and the hyperparameters ζ and κ: for each combination, ω is optimized and the corresponding statistical measure is shown. It is worth noting that the Stability decreases as more complex algorithms are chosen and the Stability decreases while increasing the cardinality of the training set. As a final issue, we deal with the estimation of the advantage, in terms of computational resources usage, of exploiting hw-friendly models with respect to conventional ones. In particular, we consider the best classifiers selected with the ARM procedure, and we average two indexes of performance over the different OvO binary problems: (i) the number of Predictions Per Second (PPS) and (ii) the Battery Life in Hours (BLH). Moreover, we compare them with a conventional floating-point implementation: the values are shown in Table 4. These results are obtained by implementing the models on the Samsung Galaxy S II smartphone, where we have realized both fixed-point and floating-point (32 bits) procedures, in accordance with the selected values of κ. It is worth highlighting the large difference between the rates obtained using the fixed-point representation and the results achieved by floating-point arithmetic. Concerning battery consumption with the floating-point and fixed-point representations, the experiment consisted of continuously running the HAR smartphone application and measuring the battery discharging time from a fully charged state down to a minimum level of 10 %. We have found that the average battery life is increased up to 100 % when a fixed-point 8-bit application is running instead of a floating-point one. These results are highly dependent on the exploited hardware and operating system; however, they allow showing the improvements that can be achieved thanks to the proposed approach, and are a good indicator of the benefits that this method can offer for saving battery life and the possibility of being integrated into devices for everyday life. In current scenarios, even small savings in battery consumption make a big difference in deciding whether or not to use a mobile app: this is the case where HAR applications are required to deliver activity information to other higher-level decision applications (e.g. phone apps for

Table 1 Error on the test set $\hat{R}_T$ performed by the models, selected in the ARM framework

Prob.	n	ζ	κ	$\hat{R}_T$	n	ζ	κ	$\hat{R}_T$	n	ζ	κ	$\hat{R}_T$	n	ζ	κ	$\hat{R}_T$	n	ζ	κ	$\hat{R}_T$	n	ζ	κ	$\hat{R}_T$
1vs2	25	1.0	32	16.9	100	1.0	16	8.7	225	1.0	4	6.3	400	1.0	8	6.0	625	1.0	8	6.0	900	1.0	8	5.0
1vs3		1.0	8	16.5		1.0	16	10.0		1.0	8	9.7		1.0	8	9.2		1.0	16	8.3		1.0	8	6.2
1vs4		1.0	4	1.9		1.0	4	0.6		1.0	8	0.2		1.0	4	0.1		1.0	4	0.1		1.0	8	0.1
1vs5		1.0	16	1.6		1.0	4	0.3		1.0	8	0.2		1.0	8	0.2		1.0	4	0.1		1.0	4	0.1
1vs6		1.0	8	2.9		1.0	8	2.6		1.0	2	0.8		1.0	2	0.0		1.0	4	0.0		1.0	4	0.0
2vs3		1.0	4	19.7		1.0	16	9.0		1.0	8	6.6		1.0	8	6.6		1.0	16	6.6		1.0	16	4.0
2vs4		1.0	8	3.4		1.0	8	0.8		1.0	1	0.5		1.0	8	0.2		1.0	8	0.2		1.0	8	0.2
2vs5		1.0	8	2.1		1.0	8	0.5		1.0	2	0.1		1.0	4	0.1		1.0	4	0.1		1.0	8	0.0
2vs6		1.0	8	2.3		1.0	4	0.9		1.0	1	0.3		1.0	8	0.0		1.0	1	0.0		1.0	1	0.0
3vs4		1.0	2	1.8		1.0	8	0.8		1.0	8	0.2		1.0	8	0.1		1.0	8	0.1		1.0	8	0.1
3vs5		1.0	2	0.9		1.0	2	0.2		1.0	4	0.0		1.0	2	0.0		1.0	8	0.0		1.0	4	0.0
3vs6		1.0	4	0.8		1.0	16	0.5		1.0	2	0.0		1.0	4	0.0		1.0	4	0.0		1.0	2	0.0
4vs5		1.0	8	29.5		1.0	8	17.8		1.0	16	13.7		1.0	8	9.7		1.0	16	8.7		1.0	8	8.1
4vs6		1.0	4	2.0		1.0	4	0.3		1.0	8	0.0		1.0	2	0.0		1.0	8	0.0		1.0	16	0.0
5vs6		1.0	8	1.6		1.0	2	0.0		1.0	8	0.0		1.0	2	0.0		1.0	8	0.0		1.0	4	0.0

Table 2 Lowest error rate $\hat{R}_T$, obtained by optimizing the hyperparameters to minimize the error on T for benchmarking purposes

Prob.	n	ζ	κ	$\hat{R}_T$	n	ζ	κ	$\hat{R}_T$	n	ζ	κ	$\hat{R}_T$	n	ζ	κ	$\hat{R}_T$	n	ζ	κ	$\hat{R}_T$	n	ζ	κ	$\hat{R}_T$
1vs2		1.0	8	16.9		1.0	16	8.7		1.0	4	6.3		1.0	8	6.0		1.0	8	6.0		1.0	8	5.0
1vs3		1.0	8	16.5		1.0	16	10.0		1.0	8	9.7		1.0	4	9.2		1.0	4	8.2		1.0	8	6.2
1vs4		1.0	1	1.7		1.0	2	0.6		1.0	2	0.2		1.0	1	0.1		1.0	4	0.1		1.0	1	0.1
1vs5		1.0	2	1.6		1.0	4	0.3		1.0	2	0.2		1.0	4	0.1		1.0	4	0.1		1.0	4	0.1
1vs6		1.0	1	2.5		1.0	4	2.6		1.0	1	0.8		1.0	1	0.0		1.0	1	0.0		1.0	1	0.0
2vs3		1.0	4	19.7		1.0	2	8.8		1.0	4	6.5		1.0	4	6.4		1.0	8	6.5		1.0	16	4.0
2vs4		1.0	2	3.4		1.0	8	0.8		1.0	4	0.5		1.0	8	0.2		1.0	8	0.2		1.0	8	0.2
2vs5	25	1.0	4	2.1	100	1.0	2	0.5	225	1.0	4	0.1	400	1.0	2	0.1	625	1.0	4	0.1	900	1.0	4	0.0
2vs6		1.0	1	2.3		1.0	1	0.7		1.0	1	0.3		1.0	1	0.0		1.0	1	0.0		1.0	1	0.0
3vs4		1.0	1	1.7		1.0	4	0.8		1.0	1	0.2		1.0	2	0.1		1.0	4	0.1		1.0	4	0.1
3vs5		1.0	2	0.9		1.0	1	0.2		1.0	4	0.0		1.0	2	0.0		1.0	1	0.0		1.0	8	0.0
3vs6		1.0	4	0.8		1.0	1	0.5		1.0	1	0.0		1.0	1	0.0		1.0	1	0.0		1.0	1	0.0
4vs5		1.0	2	29.5		1.0	8	17.8		1.0	8	13.7		1.0	8	9.7		1.0	8	8.7		1.0	8	8.1
4vs6		1.0	1	2.0		1.0	8	0.3		1.0	1	0.0		1.0	4	0.0		1.0	2	0.0		1.0	2	0.0
5vs6		1.0	2	1.4		1.0	1	0.0		1.0	1	0.0		1.0	1	0.0		1.0	1	0.0		1.0	1	0.0

Table 3 Stability $\hat{H}$ by varying n, κ, and ζ when $A^L_{(\kappa,\zeta,\omega)}$ is used (we report the result just for $n \in \{225, 625\}$ because of space constraints)

n	225														625													
ζ	1.0							0.5							1.0							0.5						
κ	1	2	4	8	16	32	∞	1	2	4	8	16	32	∞	1	2	4	8	16	32	∞	1	2	4	8	16	32	∞
1vs2	12	10	8.7	8.7	8.6	8.6	8.6	14	12	10	10	10	10	10	8.6	6.0	4.7	4.5	4.5	4.5	4.5	10	7.3	5.7	5.5	5.5	5.5	5.5
1vs3	12	9.6	8.2	7.8	7.9	7.9	7.9	15	11	10	9.6	9.7	9.7	9.7	14	13	11	11	11	11	11	17	16	13	13	13	13	13
1vs4	1.7	1.6	1.6	1.5	1.5	1.5	1.5	2.1	2.0	2.0	1.9	1.9	1.9	1.9	0.3	0.2	0.2	0.2	0.2	0.2	0.2	0.3	0.3	0.2	0.2	0.2	0.2	0.2
1vs5	1.0	1.1	1.1	1.1	1.1	1.1	1.1	1.3	1.4	1.3	1.3	1.3	1.3	1.3	0.3	0.2	0.2	0.2	0.2	0.2	0.2	0.3	0.2	0.2	0.2	0.2	0.2	0.2
1vs6	1.6	1.4	1.5	1.5	1.5	1.5	1.5	1.9	1.7	1.8	1.9	1.9	1.9	1.9	0.4	0.4	0.3	0.3	0.3	0.3	0.3	0.5	0.5	0.4	0.4	0.4	0.4	0.4
2vs3	12	11	10	10	10	10	10	15	13	12	12	12	12	12	13	11	10	10	10	10	10	16	13	13	12	12	12	12
2vs4	1.5	1.5	1.6	1.6	1.6	1.6	1.6	1.8	1.8	2.0	2.0	2.0	2.0	2.0	9.7	7.6	1.7	1.4	1.4	1.4	1.4	11	9.3	2.1	1.8	1.8	1.8	1.8
2vs5	1.0	0.8	0.8	0.9	0.9	0.9	0.9	1.3	0.9	1.0	1.1	1.1	1.1	1.1	0.2	0.2	0.2	0.2	0.2	0.2	0.2	0.2	0.2	0.2	0.2	0.2	0.2	0.2
2vs6	2.1	2.0	2.1	2.0	2.0	2.0	2.0	2.5	2.5	2.5	2.5	2.5	2.5	2.5	0.7	0.7	0.7	0.8	0.8	0.8	0.8	0.9	0.9	0.9	0.9	0.9	0.9	0.9
3vs4	1.9	1.7	1.6	1.6	1.6	1.6	1.6	2.3	2.1	1.9	1.9	1.9	1.9	1.9	9.7	8.2	2.7	1.7	1.7	1.7	1.7	11	9.9	3.2	2.0	2.0	2.0	2.0
3vs5	1.1	1.0	0.9	0.9	0.9	0.9	0.9	1.3	1.2	1.1	1.1	1.1	1.1	1.1	15	13	1.6	0.3	0.3	0.3	0.3	18	16	1.9	0.4	0.4	0.4	0.4
3vs6	2.1	2.0	2.0	2.0	2.0	2.0	2.0	2.5	2.4	2.5	2.5	2.5	2.5	2.5	0.8	0.8	0.7	0.7	0.7	0.7	0.7	1.0	0.9	0.9	0.9	0.9	0.9	0.9
4vs5	27	24	20	20	20	20	20	33	29	24	24	24	24	24	23	19	17	16	16	16	16	28	23	21	20	20	20	20
4vs6	5.1	4.7	4.6	4.5	4.5	4.5	4.5	6.2	5.7	5.6	5.5	5.5	5.5	5.5	1.1	0.9	0.9	0.9	0.9	0.9	0.9	1.3	1.2	1.1	1.1	1.1	1.1	1.1
5vs6	4.6	3.8	3.8	3.7	3.8	3.8	3.8	5.6	4.7	4.7	4.6	4.6	4.6	4.6	1.3	1.1	1.1	1.1	1.1	1.1	1.1	1.6	1.3	1.3	1.3	1.3	1.3	1.3

Table 4 PPS and BLH averaged over the best models, selected in the ARM framework, against the indexes obtained by floating-point (32-bit) models

	ARM models	Floating-point models		ARM models	Floating-point models
PPS	2100	230	BLH	230	110

maintaining a healthy lifestyle through HAR [15]), thus implying sharing system resources. A general aim is to build a device able to operate at least during a full day, so that the battery recharges can occur during the night time.

References

1. Alba, E., Anguita, D., Ghio, A., Ridella, S.: Using variable neighborhood search to improve the support vector machine performance in embedded automotive applications. In: IEEE International Joint Conference on Neural Networks (2008)
2. Anguita, D., Ghio, A., Oneto, L., Parra, X., Reyes-Ortiz, J.L.: Energy efficient smartphone-based activity recognition using fixed-point arithmetic. J. Univers. Comput. Sci. **19**, 1295–1314 (2013)
3. Anguita, D., Ghio, A., Oneto, L., Ridella, S.: A support vector machine classifier from a bit-constrained, sparse and localized hypothesis space. In: International Joint Conference on Neural Networks (2013)
4. Anguita, D., Ghio, A., Oneto, L., Ridella, S.: Smartphone battery saving by bit-based hypothesis spaces and local rademacher complexities. In: International Joint Conference on Neural Networks (2014)
5. Anguita, D., Ghio, A., Pischiutta, S., Ridella, S.: A support vector machine with integer parameters. Neurocomputing **72**(1), 480–489 (2008)
6. Bache, K., Lichman, M.: UCI Machine Learning Repository (2013). http://archive.ics.uci.edu/ml
7. Bousquet, O., Elisseeff, A.: Stability and generalization. J. Mach. Learn. Res. **2**, 499–526 (2002)
8. Boyd, S., Vandenberghe, L.: Convex Optimization. Cambridge University Press (2009)
9. Cook, D.J., Das, S.K.: Pervasive computing at scale: transforming the state of the art. Pervasive Mob. Comput. **8**, 22–35 (2012)
10. Dantzig, G.B.: Linear Programming and Extensions. Princeton University Press (1998)
11. Ghio, A., Pischiutta, S.: A support vector machine based pedestrian recognition system on resource-limited hardware architectures. In: Research in Microelectronics and Electronics Conference PRIME (2007)
12. Ivanov, V.: The Theory of Approximate Methods and Their Application to the Numerical Solution of Singular Integral Equations. Springer (1976)
13. Lesser, B., Mücke, M., Gansterer, W.: Effects of reduced precision on floating-point SVM classification accuracy. Proc. Comput. Sci. **4**, 508–517 (2011)
14. Meinshausen, N., Bühlmann, P.: Stability selection. J. R. Stat. Soc. Ser. B (Stat. Methodol.) **72**(4), 417–473 (2010)
15. Nicholas, L., Mashfiqui, M., Mu, L., Xiaochao, Y., Hong, L., Shahid, A., Afsaneh, D., Ethan, B., Tanzeem, C., Andrew, C.: Bewell: a smartphone application to monitor, model and promote wellbeing. In: IEEE International ICST Conference on Pervasive Computing Technologies for Healthcare (2012)

16. Oneto, L., Ghio, A., Ridella, S., Anguita, D.: Fully empirical and data-dependent stability-based bounds. IEEE Trans. Cybern. (2014). doi:10.1109/TCYB.2014.2361857
17. Oneto, L., Ghio, A., Ridella, S., Anguita, D.: Learning resource-aware models for mobile devices: from regularization to energy efficiency. Neurocomputing (2015)
18. Poggio, T., Rifkin, R., Mukherjee, S., Niyogi, P.: General conditions for predictivity in learning theory. Nature **428**(6981), 419–422 (2004)
19. Shawe-Taylor, J., Bartlett, P.L., Williamson, R.C., Anthony, M.: Structural risk minimization over data-dependent hierarchies. IEEE Trans. Inf. Theor. **44**(5), 1926–1940 (1998)
20. Vapnik, V.N.: Statistical Learning Theory. Wiley (1998)

A Hidden Markov Model-Based Approach to Grasping Hand Gestures Classification

Anna Di Benedetto, Francesco A.N. Palmieri, Alberto Cavallo and Pietro Falco

Abstract Gesture recognition is a hot topic in research, due to its appealing applications in real-life contexts, from remote control to assistive robotics. In this paper we focus on grasping gestures recognition. This kind of gestures is particularly interesting, because it requires not only analyzing hand trajectories, but also fingers position and fingertip forces, of utmost importance in manipulation tasks. We used a discrete HMM-based model for gesture recognition. Input codebooks for the model are gesture elementary phases, obtained through a LLS-regression segmentation algorithm, and feature vectors representing hand position over time.

Keywords Human hand · Manipulation · Grasping · Gesture recognition · Gesture segmentation · Dataglove · Human-computer interaction · Linear least squares · Hidden Markov models

1 Introduction

Gestures are a form of non-verbal communication, expressing feelings, ideas and messages by means of visible bodily actions. In particular, the interpretation of hand gestures using automated systems has many real-life applications, e.g. sign

A. Di Benedetto (✉) · F.A.N. Palmieri · A. Cavallo · P. Falco
Dipartimento di Ingegneria Industriale e dell'Informazione,
Seconda Università degli Studi di Napoli (SUN), Aversa, Italy
e-mail: anna.dibenedetto@unina2.it

F.A.N. Palmieri
e-mail: francesco.palmieri@unina2.it

A. Cavallo
e-mail: alberto.cavallo@unina2.it

P. Falco
e-mail: pietro.falco@unina2.it

© Springer International Publishing Switzerland 2016
S. Bassis et al. (eds.), *Advances in Neural Networks*, Smart Innovation,
Systems and Technologies 54, DOI 10.1007/978-3-319-33747-0_41

Fig. 1 Typical gesture recognition workflow [9]

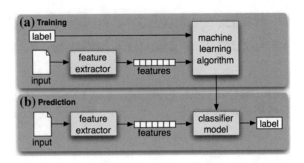

language recognition, assistive robotics, remote control and interaction with video games. The typical workflow leading to a gesture classifier is depicted in Fig. 1.

Researchers mainly focused on sign language [1, 2] and hand gesture trajectory [3, 4] recognition. When the input device is a 2D camera, appearance-based hand models and edge detection algorithms are used to identify hand shape [3–5]. In other works hand movements in 3D space are acquired by means of wired gloves and magnetic field trackers [1, 2, 6]. Cartesian coordinates [7] and angles [3], plus velocity and acceleration [8] are common features used to track hand trajectory in sign language applications (e.g. detecting the end of a word when the hand stops). In appearance-based approaches, also hand shape and skin color are considered [4, 5]. When the input device is a wired glove, instead, wrist position and orientation and joint angles [10] or fingertips position [6] are acquired. Feature vectors are often obtained from raw data after a pre-processing step involving Fourier transform [5, 7], PCA [1, 2, 8] or rotation-invariant transformations.

Algorithms used for training and classification include neural networks [11], SVMs [2], BNs [12], and HMMs [1, 3–8, 13]. In case of discrete HMMs, K-means clustering can be used to obtain codebooks [4, 6, 8].

In this paper we will present our experimental results related to hand gesture recognition. We followed the workflow of Fig. 1 to develop a system able to recognize gestures acquired while grasping different objects. Grasping gestures are similar: all the actions have the same elementary phases (rest, approaching, grasping, releasing and again rest) and similar movements of fingers are performed while approaching the object. The ability of recognizing apparently similar gestures using few representative parameters can be surely seen as an index of effectiveness of an algorithm; we were able to classify three different categories of grasping gestures with a high recognition rate, using a limited training set.

An innovative point in our work is the inclusion of fingertip forces in gesture dataset. Fingertip forces represent the interaction between hand and grasped object, crucial in manipulation tasks; despite this, in full-body motion applications forces exchanged between body and external environment are too often not taken in account.

2 Dataset

Our dataset was extracted at Karlsruhe Institute of Technology, Germany, during a previous work. In particular we considered 30 sample gestures, acquired while grasping a bottle, a plate, and unscrewing the cap of a bottle (10 realizations per gesture). A CyberGlove II, a Polhemus FASTRAK, and Pressure Profile FingerTPS were used to measure respectively angular finger position, hand trajectory and force reactions on fingertips, 32 signals in total. Values obtained were then mapped on a skeletal model of the hand, shown in Fig. 2. For details, see [14].

Data were arranged in a matrix, $\mathbf{X}$, where each column is a different signal si coming from one of the data glove sensors. The kth row $\mathbf{x}(k)$, instead, represents a time sample of the gesture as the concatenation of multiple vectors:

$$\mathbf{x}(k) = [\mathbf{p}(k), \mathbf{o}(k), \mathbf{f}_1(k), \mathbf{f}_2(k), \mathbf{f}_3(k), \mathbf{f}_4(k), \mathbf{f}_5(k), \mathbf{ff}(k)] \tag{1}$$

where:

$\mathbf{p} = [x, y, z]$	$\in R^3$	Wrist position [m]
$\mathbf{o} = [qx, qy, qz, qw]$	$\in R^4$	Wrist orientation—unit quaternions [adim.]
$\mathbf{f_i} = [p0_i, p1_i, p2_i, a_i]$	$\in R^4$	Phalanxes flexion and abduction for ith finger [rad]
$\mathbf{ff} = [f_1, f_2, f_3, f_4, f_5]$	$\in R^5$	Forces on fingertips [N]

Fig. 2 Skeletal hand model (courtesy of KIT)

z y
x
world axes

3   Proposed HMMs Structure

Our classifier is based on HMMs [15]. A HMM is a double stochastic process: on top there is a Markov chain, leading to a sequence of hidden states we cannot access. In each hidden state, observable emissions are produced, according to a certain probability distribution. In a discrete HMM $\lambda(\mathbf{T}, \mathbf{E}, \pi)$, this distribution is a *pmf* over a finite set of values. The parameters of our model are:

- Hidden state space $S = \{s_1, s_2 \ldots s_N\}$, where $|S| = N$;
- Emission space $V = \{v_1, v_2 \ldots v_M\}$, where $|V| = M$;
- Initial state probability distribution π, where $\pi_i = P(S_1 = s_i)$, $1 \le i \le N$;
- State transition matrix $\mathbf{T}$, with $t_{ij} = P(S_{n+1} = s_j | S_n = s_i)$, $1 \le i, j \le N$;
- Emission matrix, $\mathbf{E}$, where $e_{jk} = P(V_n = v_k | S_n = s_j)$, $1 \le j \le N$, $1 \le k \le M$.

The idea leading to the choice of HMMs to model gestures in our dataset is intuitive: when grasping an object, all the gestures follow the same elementary phases, like "rest-close-grasp-open-rest". What really differentiates one gesture from another is the fact that, in each of these phases, the postures the hand assumes are different, depending on the object to grasp. As stated in [5], we can think to gesture phases as something related to some sort of "mental state" (the desire of doing an elementary action, like opening the hand), and to hand postures as the translation of this mental state into an observable sequence of movements, different for each object and for each realization of the same gesture. In a few words, phases are the states of our model, while hand postures constitute the emissions.

In the following sections we will first focus our discussion on how to split a gesture in phases (also called segments). Then this information will be used to train our HMMs.

4   Proposed Gesture Segmentation Algorithm

Segmenting a gesture means splitting it into its elementary phases. We propose an algorithm based on Linear Least Squares prediction [16].

At the beginning of each segment, our predictor computes its coefficients and "learns" the signal characteristics, by taking into account the behavior of the matrix $\mathbf{X}$ in a time window of given length. Then predictions start; predicted value and actual value are compared at each time sample. As long as the signal behavior is similar to that observed during learning, estimated value and actual value will be close to each other. If the behavior changes, prediction error will grow, until a certain threshold is exceeded; this means a new segment has begun. Coefficients are recalculated on a new window at the beginning of the new segment, in order to adapt to the new trend. Predictions will then restart and continue until the error grows again, and so on, until the end of the time samples. Notice that hand signals are, by their nature, correlated; just think of the difficulties we have to move middle and ring fingers independently. It is useful to include in the calculation of the ith signal also information about all other signals.

Let us remember that each column of $\mathbf{X}$ is a signal $\mathbf{s_i}$, and let us define:

- N (total number of signals);
- M (width of the time window used for learning at the start of each segment);
- W (number of past samples of each signal considered for the prediction);
- θ (prediction error threshold).

Predictor structure. A new row of the matrix $\mathbf{X}$ is computed at time instant k as linear combination of the W past values of each signal with the predictor coefficients:

$$\hat{\mathbf{x}}(k) = \left[\mathbf{s}_1^T(k-1:k-W)\,\mathbf{s}_2^T(k-1:k-W)\ldots\mathbf{s}_N^T(k-1:k-W)\right]\mathbf{C} \qquad (2)$$

$\mathbf{C}$ is a NW $\times$ N matrix, whose ith column contains the NW coefficients needed for the prediction of the single signal $\mathbf{s_i}(k)$. The colon notation represents a time interval.

Learning. At the start of each segment the predictor is trained with M examples; this means computing $\mathbf{C}$ by inverting the following matrix equation:

$$X_M = S_M C \qquad (3)$$

The M rows of $\mathbf{X_M}$ and $\mathbf{S_M}$ are obtained considering the (2) in M different time instants of the learning window. In our case, $\mathbf{C}$ is calculated from the (3) by using Moore-Penrose pseudoinverse since M < NW.

Prediction error. A new segment starts when

$$E(k) = \frac{\|\hat{x}(k) - x(k)\|}{\|x(k)\|} > \theta \qquad (4)$$

5 Feature Extraction and Definition of the Codebooks

The next step is to extract feature vectors from motion. Feature vectors are the basis from which we will derive the discrete codebooks for states (segments) and emissions (hand postures) of our HMMs:

States. A feature vector for a segment (that is a submatrix of $\mathbf{X}$) is obtained by vectorizing it. Notice that not all segments have the same time duration, so we time-interpolated the data in order to have vectors of the same length for all segments.

Emissions. A hand posture is characterized by wrist position, finger bending, and fingertip forces. A feature vector for a hand posture thus is simply a row of $\mathbf{X}$, that is the hand posture in a discrete-time instant k.

The components of our feature vectors can assume infinite values in R; in order to obtain discrete codebooks, we must operate a vector quantization. A K-Means clustering algorithm was used to separately cluster states and emission feature vectors extracted from gestures in training set. The number of clusters, K, must be chosen depending on the desired quantization granularity; the aim is to synthesize data with a few representative elements, while still maintaining a distinction between different hand postures, or different gesture phases.

6 Experimental Results

6.1 Building the Classifier

Segmentation. Each element of the training set was segmented using the proposed LLS algorithm, with $N = 32$, $M = 10$, $W = 2$, $\Theta = 0.25$.

Figure 3 shows segmentation result for a realization of the gesture "grasp a bottle".[1]

Clustering. Discrete codebooks were obtained from feature vectors as following:

States. We chose $K = 5$ as desired number of clusters for the gestures "grasp an empty bottle" and "grasp a small plate". These correspond roughly to the phases "rest", "open", "adjust force", "grasp", "close". For the gesture "unscrew cap" we instead chose $K = 6$; in fact this gesture is more complex, since it includes unscrewing and repositioning of the hand. The initial codebook has been chosen taking K representative vectors, one for each elementary phase of a gesture.

Emissions. Here K must be chosen depending on the desired quantization granularity: a smaller K leads to a more compressed data representation. A value of $K = 20$ was enough to obtain a balance between compression and accuracy in data representation. We also performed a more compressive clustering with $K = 10$. The initial codebook was chosen at random. In Fig. 4, the gesture "grasp a bottle" is represented as a sequence of cluster centroids from K-Means clustering on emissions ($K = 20$).

Training. At this point we needed to train three HMMs, one for each gesture, computing **T**, **E**, π. Without knowing hidden state paths in the training set, we would have used the Baum-Welch algorithm. However, thanks to the segmentation operation, we made explicit the state sequence for all training gestures, so we could estimate **T** and **E** simply computing the relative frequency of transitions and emissions. π is deterministic, since all of our gestures start from the same state (rest).

---

[1]Gesture data is heterogeneous (angles, trajectories, forces). To give an idea of algorithm results to the reader, however, we need to represent all signals on the same plot. We will omit measurement units for simplicity in this and all following figures showing hand signals.

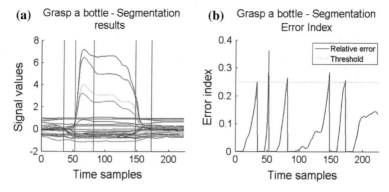

(a) Grasp a bottle - Segmentation results

(b) Grasp a bottle - Segmentation Error Index

Fig. 3 Segmentation results (**a**) and error index (**b**) for the gesture "grasp a bottle"

Fig. 4 Quantized gesture "grasp a bottle"

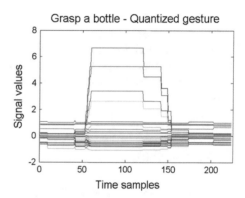

Grasp a bottle - Quantized gesture

We estimated the parameters for our HMMs using the emission codebook with $K = 20$. Then we trained other three HMMs using the emission codebook with $K = 10$.

6.2 Test Results

A twofold cross-validation has been carried out to test our system: the first fold was obtained splitting our dataset D (30 samples) into two halves, S_1 (training set) and S_2 (test set); each half contains 5 samples per gesture. The HMM for a gesture was trained using the corresponding 5 samples taken from S_1, while the tests have been done using the samples in S_2, unknown to the classifiers. The second fold was obtained simply inverting the roles of S_1 and S_2. Each test set element was translated into a sequence of symbols from the emission codebook using a minimum distance criterion: each row of **X** (hand posture) was assigned to the symbol corresponding to the less distant cluster centroid for emissions.

Table 1 Test results

		First fold					Second fold				
Test set		1	2	3	4	5	6	7	8	9	10
K = 20	Bottle	✓	✓	✓	✓	✓	✓	✓	✓	✓	✓
	Plate	✗	✓	✓	✓	✓	✓	✓	✓	✓	✓
	Unscrew cap	✓	✓	✓	✓	✓	✓	✓	✓	✓	✓
	Recognition rate	93.3 %					100 %				
		Average recognition rate = 96.7 %									
K = 10	Bottle	✓	✓	✓	✓	✓	✓	✓	✓	✓	✓
	Plate	✓	✗	✓	✓	✓	✓	✓	✓	✓	✓
	Unscrew cap	✓	✓	✓	✓	✓	✓	✓	✓	✓	✓
	Recognition rate	93.3 %					100 %				
		Average recognition rate = 96.7 %									

The decision criterion is, as usual with HMMs, based on log-likelihood: assign the input emission sequence O to the gesture whose model λ_i holds the highest logarithmic production probability, $\log(P(O|\lambda_i))$.

We tested both the systems with K = 20 and K = 10. Test results are shown in Table 1.

7 Conclusion

In this paper we proposed a LLS-based gesture segmentation algorithm and an HMM-based method for classification of grasping gestures. We obtained a recognition rate of about 97 %, despite a training set made up of just 5 samples per model; only one action has been mispredicted in both cases, also using a reduced codebook of only 10 representative postures. For each model, we had to save only $S \times S$ transition probabilities, $S \times E$ emission probabilities, $E + S$ cluster centroids; with few information we could encapsulate the characteristics of a complex gesture.

The classifier retained its performances also with the reduced codebook; furthermore, algorithms to solve HMMs are fast. This is a great advantage in terms of space and time complexity and shows how effective HMMs are for our case-study.

HMMs trained making explicit the state sequences are also able to operate segmentation: since hidden states correspond to segments, given a sequence of emissions we can estimate the underlying state path using the Viterbi algorithm.

We did not take in consideration inputs that do not match any of the modeled gestures; any input is always assigned to the class with the largest production probability for that input. A possible solution to this issue is proposed in [13], where an additional "non-gesture" model was designed to capture meaningless gestures.

The work done so far constitutes the basis for applications in many fields regarding human-computer interaction such as assistive robotics, where automated systems could recognize the actions of mobility-impaired people and help them in their everyday life.

References

1. Bashir, F., et al.: HMM-based motion recognition system using segmented PCA. In: IEEE International Conference on Image Processing, 2005. ICIP 2005, vol. 3. IEEE (2005)
2. Li, C., Kulkarni, P.R., Prabhakaran, B.: Segmentation and recognition of motion capture data stream by classification. Multimedia Tools Appl. 35(1), 55–70 (2007)
3. Elmezain, M., et al.: A hidden markov model-based isolated and meaningful hand gesture recognition. Int. J. Electr. Comput. Syst. Eng. 3(3), 156–163 (2009)
4. Yoon, H.-S., et al.: Hand gesture recognition using combined features of location, angle and velocity. Pattern Recogn. 34(7), 1491–1501 (2001)
5. Chen, F.-S., Fu, C.-M., Huang, C.-L.: Hand gesture recognition using a real-time tracking method and hidden Markov models. Image Vis. Comput. 21(8), 745–758 (2003)
6. Ekvall, S., Danica, K.: Grasp recognition for programming by demonstration. In: Proceedings of the 2005 IEEE International Conference on Robotics and Automation, 2005. ICRA 2005. IEEE (2005)
7. Yang, J., Xu, Y., Chen, C.S.: Gesture interface: modeling and learning. In: 1994 IEEE International Conference on Robotics and Automation, 1994. Proceedings. IEEE (1994)
8. Kong, W.W., Ranganath, S.: Automatic hand trajectory segmentation and phoneme transcription for sign language. In: 8th IEEE International Conference on Automatic Face and Gesture Recognition, 2008. FG'08. IEEE (2008)
9. Bird, S., Klein, E., Loper, E.: Natural language processing with Python. O'Reilly Media, Inc. (2009)
10. Kim, J.-S., Jang, W., Bien, Z.: A dynamic gesture recognition system for the Korean sign language (KSL). IEEE Trans. Syst. Man Cybern. Part B: Cybern. 26(2), 354–359 (1996)
11. Murakami, K., Taguchi, H.: Gesture recognition using recurrent neural networks. In: Proceedings of the SIGCHI Conference on Human Factors in Computing Systems. ACM (1991)
12. Suk, H.-I., Sin, B.-K., Lee, S.-W.: Hand gesture recognition based on dynamic Bayesian network framework. Pattern Recogn. 43(9), 3059–3072 (2010)
13. Iba, S., et al.: An architecture for gesture-based control of mobile robots. In: 1999 IEEE/RSJ International Conference on Intelligent Robots and Systems, 1999. IROS'99. Proceedings, vol. 2. IEEE (1999)
14. Cavallo, A., Falco, P.: Online segmentation and classification of manipulation actions from the observation of kinetostatic data. IEEE Trans. Human-Mach. Syst. 44(2), 256–269 (2014)
15. Rabiner, L.: A tutorial on hidden Markov models and selected applications in speech recognition. Proc. IEEE 77(2), 257–286 (1989)
16. Haykin, S.: Adaptive Filter Theory. Prentice-Hall, Englewood Cliffs, NJ (1986)

A Low Cost ECG Biometry System Based on an Ensemble of Support Vector Machine Classifiers

Luca Mesin, Alejandro Munera and Eros Pasero

Abstract The electrocardiogram (ECG) found many clinical applications. Recently, it was proposed as a promising technology also for biometric applications, i.e., to recognize a subject within a group of known people. For such an application, the accuracy of classical ECG clinical recordings is usually not needed, but the measurement procedure should be fast, robust and cheap. We developed an embedded wearable system for recording one-lead ECG from the wrists of a person. The system was used to record data from 10 subjects. Data were pre-processed to reduce the noise content. Then fiducial points were detected and used to train an ensemble of support vector machines to identify a person among the group. Mean classification accuracy was higher than 95 % if a single heartbeat was considered and higher than 98 % if 3 consecutive heartbeats were used, choosing by majority. The system is fast (a few seconds are needed), not invasive and can be used either standalone or together with other identification techniques to increase the safety level.

Keywords Biometry · Electrocardiogram (ECG) · Support vector machine · Intelligent sensors · Embedded system

1 Introduction

Biometry found important security applications [1]. Different data may be used for biometric purposes, for example DNA [2], face [3], ears [4], iris [5], fingerprints [3]. However, some external physiological data like fingerprint, iris and face are easy to mimic or to fake and some internal data like DNA require expensive technology to be processed. On the other hand, an internal physiological data like electrocardiogram

L. Mesin (✉) · A. Munera · E. Pasero
Department of Electronic and Telecommunication (DET), Politecnico di Torino,
Corso Duca degli Abruzzi 24, 10129 Turin, Italy
e-mail: luca.mesin@polito.it

E. Pasero
e-mail: eros.pasero@polito.it

© Springer International Publishing Switzerland 2016
S. Bassis et al. (eds.), *Advances in Neural Networks*, Smart Innovation,
Systems and Technologies 54, DOI 10.1007/978-3-319-33747-0_42

(ECG) can be easily recorded and processed and cannot be faked. Indeed, individual differences, such as chest geometry or heart position and size reflect in unique characteristics in the ECG, which can then be used as a biometric attribute. For this reason, ECG is an emerging novel biometrics for human identification [6].

Different biometric methods based on ECG have been introduced [6]. One-channel ECG (single lead) is the most used acquisition system for this application [7], as it is easy to record. Electrodes were placed in different positions, e.g., on the chest [8, 9], in the standard position making the RA-LA Eindhoven triangle [10], between thumbs and indices [11].

The recorded signal is usually very noisy, so that a pre-processing is needed: a band-pass filter or a pre-emphasis filter [8] is used to remove cross-talk from contracting muscles and power line interference [9–11].

Then, different features can be extracted from the ECG, giving rise to different classification techniques, that can be broadly distinguished into two main categories [7]: those based on fiducial points [6, 9] or not [8, 12, 14] (e.g., using the Short Time Fourier Transform [16], Wavelet Transform [17], chaos extractor [18], Pulse Active Width [19]). The extracted features can be further processed to reduce their redundant information, e.g., by principal components analysis (PCA, [13]) or independent component analysis (ICA, [15]).

Also classification can be performed in many different ways [7]: artificial neural networks [8], linear discriminant analysis [9], template correlation [12], normalized Euclidean distance [14], normalized Gaussian log likelihood [14], Support Vector Machine [22] and others [7]. Classification has been done with different purposes, as heartbeat detection [21] and biometric recognition in different physiological conditions [23].

In this work, we acquired the ECG by a custom made system based on wrist sensors connected to a smart-phone by Bluetooth. The recorded data were then processed by a classifier to recognize the person of interest.

2   Methods

The following steps were considered to develop the identification system: ECG acquisition, signal pre-processing (to reduce the measurement noise), fiducial points extraction and classification.

2.1   Acquisition System

ECGs were recorded by a custom made acquisition system [24, 25], using three electrodes placed on the wrists of the subject as shown in Fig. 1a. The placement of the electrodes corresponds to Lead One of the Eindhoven triangle. A picture of the device is shown in Fig. 1b. The signal was amplified (with gain 1000) and

(a)   **(b)**

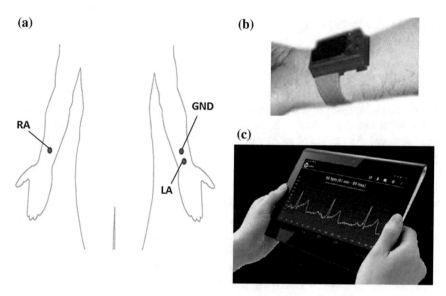

Fig. 1 Acquisition system. **a** Placement of electrodes. **b** ECG sensor. **c** Android App for ECG trace recording and visualization

sampled at 1 kHz. The device includes a bandpass filter (with cutoff frequencies of 10 and 170 Hz) and a Driven Right Leg Circuit to reduce common mode interference. The system transmits via Bluetooth the data to a Tablet, where an Android App is used to visualize the ECG trace (as shown in Fig. 1c). It is powered by a 3.6 V–240 mAh lithium-Polymer battery and the consumption is about 0.05 mW for 20 s of acquisition.

2.2 Experiments

We recorded ECGs from 10 healthy subjects in rest conditions (3 women and 7 men, aged from 22 to 28 years). Nine ECG traces were recorded from each subject. Each ECG trace had a duration of 10 s. The traces were split into 3 groups, each composed of 3 of them, that were used to prepare the training, validation and test sets used to build and test the classifier, as detailed below.

A single additional healthy subject was studied also under fatiguing conditions, in order to test the robustness of our solution to a variation of the heart rate: 5 acquisitions were acquired at rest and 4 during a fatiguing task, inducing an increase of the heart rate of about 100 %.

2.3   ECG Processing

2.3.1   Pre-processing

The raw ECG is often rather noisy and contains distortions of various origins. Visual analysis of our ECGs indicated that a pre-processing was needed to perform three major tasks: baseline drift correction, frequency-selective filtering and signal enhancement. As a result of some empirical trials, the following combination of pre-processing methods was selected [26].

Baseline drift correction was implemented using a detrending method for heart rate variability (HRV) analysis [27]. The method is based on the smoothness priors approach and operates like a time-varying FIR high pass filter.

An adaptive bandstop filter was applied to remove power-line noise (with adaptation constant equal to 1.5), adapting to possible changes of power-line interference.

Finally, we used a lowpass filter (Butterworth filter, passband corner frequency 40 Hz, stopband corner frequency 60 Hz, passband ripple 0.1 dB, stopband attenuation 30 dB) and smoothing (moving average based on 5 samples), to reduce residual high frequency noise.

2.3.2   Fiducial points estimation

After noise reduction, R peaks were detected using the Pan-Tompkins algorithm [28]. The detected heartbeats were then aligned with respect to the R peaks and truncated by a window of 0.8 s.

Possible outliers (due to signal truncation) were removed by imposing a correlation over 0.6 between the heartbeats and the average waveform. An example of pre-processing is shown in Fig. 2.

The fiducial points, shown in Fig. 2c, were then estimated. The P, Q, S and T positions were computed by finding local minima (in the case of Q and S) or maxima (P and T) in specific regions, using the R points detected in the pre-processing step as reference. The position of all points (P, Q, S and T) with respect to R peaks and the amplitude of the ECG trace in such points normalized with respect to the R peak were considered as the 8 features characterizing each waveform.

Typically, the heart rate of a normal sinus rhythm is 60–100 bpm. However, it can widely change, as it is significantly affected by many factors: stress, exercise, shock or body chemistry, which change also the morphology of the ECG. Moreover, even in relaxed conditions, there is a physiological HRV, reflecting the control by the autonomous nervous system [29]. To compensate for these effects, we considered the following formula to correct the QT interval [30]

$$QTc = QT + 0.154 * (1000 - RR) \tag{1}$$

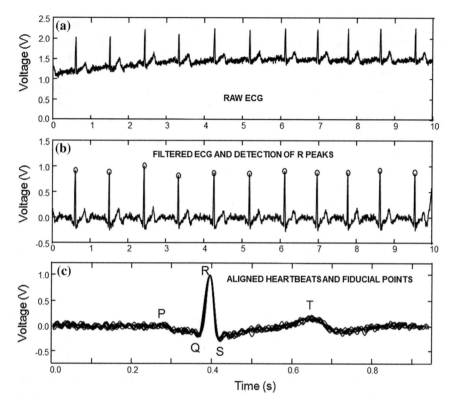

Fig. 2 Example of recorded signals. **a** Raw signal. **b** Filtered signal. **c** ECG waveforms aligned and indication of the fiducial points

where RR is the interval between two consecutive R peaks, QT is the delay between the beginning of the Q and the end of the T waves and QTc is the corrected QT value. As indicated in the Discussion Section, a robust estimation of the QT interval was not possible, so that it was approximated as the time delay between the Q and T points. All positions of the fiducial points within the QT interval (thus, R, S and T, neglecting the delay between the beginning of the Q wave and the Q point) were rescaled accordingly.

2.4 Classification by SVMs

Each ECG heartbeat was characterized by a set of 8 features derived from the fiducial points, as detailed in the previous section. In order to identify the heartbeat of a specific subject, different SVMs [31] were trained (using the training set) to perform a binary classification (indicating if the heartbeat belonged to the subject or not). As the classes were not linearly separable, the input space was mapped into a feature space using a polynomial kernel, with order finely tuned to increase the classification performances (specifically, the order of the polynomial corresponding to the best average accuracy among the subjects on the validation set was selected out of the range 2–5; the optimal order was 4). All possible combinations of features were considered: all possible choices of single feature, pairs, triplets, ... till using all 8 features. In this way, a subject specific exhaustive search among all 255 possible SVM classifiers was performed. The classification performance of each of these SVMs was tested on the validation set. The 9 classifiers with best generalization on the validation set were selected and used to perform a classification on the test set, choosing by majority the final output.[1]

The performances of the classification were tested for each subject in terms of accuracy, sensitivity and specificity, defined as

$$accuracy = \frac{TP + TN}{TP + TN + FP + FN} \tag{2}$$

$$sensitivity = \frac{TP}{TP + FN} \tag{3}$$

$$specificity = \frac{TN}{TN + FP} \tag{4}$$

where TP, TN, FP and FN indicate true positives, true negatives, false positives and false negatives, respectively.

3 Results

The classification performances are shown in Table 1 for the 10 subjects at rest, in the cases in which a single or 3 heartbeats were used for the classification (when more heartbeats were used, the final classification was chosen by majority). Classification performances always increased if more heartbeats were considered. All possible 6 permutations of training, validation and test sets were considered, showing the mean accuracy, sensitivity and specificity obtained.

The mean number of features included in the best classifiers was 4.58. The features most included in the best classifiers were the amplitude of the S and T fiducial points (included in the 72 % and 61 % of the best classifiers, respectively) and the

---

[1]The number of classifiers to be used was chosen among the odd numbers in the range 3–11 (only odd numbers were considered in order not to have ambiguous conditions for the majority choice). The number of classifiers with best generalization performances was chosen. We also considered the possibility of using a weighted average of the classifications, with weights proportional to the accuracy of the classifiers in the validation set. However, the performances were not statistically different with respect to the choice by majority, so that, for simplicity, the latter method was used.

Table 1 Performances of the optimal SVM-based classifiers for the identification of each subject (mean and standard deviation, in round brackets), considering 1 heartbeat (left) or the major classification among 3 consecutive heartbeats (right)

	Classification based on 1 heartbeat			Classification based on 3 heartbeat		
Subject	Accuracy (%)	Sensitivity (%)	Specificity (%)	Accuracy (%)	Sensitivity (%)	Specificity (%)
1	92.2(6.4)	50.0(40.7)	96.8(2.7)	95.6(5.8)	58.4(46.7)	99.7(0.5)
2	94.9(2.9)	66.4(16.5)	98.5(1.5)	98.2(1.9)	73.5(22.9)	99.9(0.1)
3	94.3(2.8)	69.7(24.0)	96.9(2.8)	98.6(1.4)	77.1(33)	99.5(1.0)
4	95.6(3.5)	79.0(29.0)	98.5(1.4)	97.6(3.5)	79.5(31.9)	99.9(0.3)
5	96.6(4.6)	77.9(30.9)	98.7(1.8)	98.6(2.7)	81.4(39.9)	99.8(0.2)
6	98.8(0.7)	98.4(1.7)	98.9(0.8)	100(0.0)	100(0.0)	100(0.0)
7	98.9(0.6)	96.4(3.7)	99.2(0.9)	100(0.1)	99.5(1.3)	100(0.0)
8	98.0(1.1)	90.3(9.6)	98.8(0.7)	100(0.0)	100(0.0)	100(0.0)
9	99.9(0.2)	99.1(2.1)	100(0.0)	100(0.0)	100(0.0)	100(0.0)
10	98.9(0.7)	93.4(4)	99.4(0.5)	100(0.0)	100(0.0)	100(0.0)

The percentage values were rounded to the first decimal digit

positions of the P and S waves (included in the 65 % and 61 % of best classifiers, respectively).

In order to test the effect of an increase of the heart rate, the ECG of an additional subject was recorded also during fatiguing exercises (as stated in the Methods section). A single representative test was considered, using 3 recordings at rest as training set, 1 recording at rest and 2 under fatigue for validation and the last acquisitions (again 1 measurement at rest and 2 under fatigue) for the test set. The classifier was trained to recognize such an additional person out of the group of 10 subjects considered previously. The following performances were achieved, proving that the method is not much sensitive to a variation of heart rate: accuracy 99.12 %, sensitivity 96.3 %, specificity 99.5 %.

4 Discussion

An innovative wireless and low cost system was used to acquire ECG traces from the wrists and a classifier was implemented to recognize the subject among a group of known people. The classification was based on a set of SVMs trained to identify a subject based on the fiducial points extracted from her/his ECG.

The classification reached high scores, even using a few heartbeats to perform the identification (with different recognition performances for different subjects, indicating that some of them had more discriminating ECG traces). The performances increased by considering more heartbeats.

Only a few fiducial points were considered, as some ECG traces did not allow a robust estimation of others, like the beginning/end points of P, R and T waves. However, depending on the specific ECG trace and on the future improvements of the acquisition system, more features could be extracted and used to increase the classification performances.

An off-line implementation was here considered, but we can expect that the classification of a fully embedded system can be fairly fast. Indeed, the time needed for recognition depends on the number of heartbeats considered (chosen on the basis of the needed accuracy). For example, considering 3 heartbeats, 3–4 s are needed for the recording; the computation of R peaks and fiducial points is not computationally intensive (it is based on simple thresholding and extremal points localization in predefined regions around the R peak); a negligible time is necessary for the classification (the SVMs are trained off-line and their mere application can be performed fast).

Acknowledgments This work has been partly funded by the Italian MIUR OPLON project and supported by the Politecnico of Turin NEC laboratory.

References

1. Jain, A.K., Ross, A., Prabhakar, S.: An introduction to biometric recognition. IEEE Trans. Circ. Syst. Video Technol. **14**(1), 4–20 (2004)
2. Soltysiak, S., Valizadegan, H.: DNA as a Biometric Identifier. Computer Science and Engineering Department. Michigan State University (2008)
3. Hong, L., Jain, A.: Integrating faces and fingerprints for personal identification. IEEE Trans. Pattern Anal. Mach. Intell. **20**(12), 1295–1307 (1998)
4. Yuan, L., Mu, Z., Xu, Z.: Using Ear Biometrics for Personal Recognition in Advances in Biometric Person Authentication. Lecture Notes in Computer Science, vol. 3781, pp. 221–228 (2005)
5. Daugman, J.: Recognising Persons by Their Iris Patterns: Advances in Biometric Person Authentication. Springer, Berlin (2005)
6. Biel, L., Pettersson, O., Philipson, L., Wide, P.: ECG analysis: a new approach in human identification. IEEE Trans. Instr. Meas. **50**(3), 808–812 (2001)
7. Odinaka, I., et al.: ECG biometric recognition: a comparative analysis. IEEE Trans. Inf. Forensics Secur. **7**(6), 1812–1824 (2012)
8. Loong, J.L.C., et al.: A new approach to ECG biometric systems: a comparative study between LPC and WPD systems. World Acad. Sci. Eng. Technol. **68**, 759–764 (2010)
9. Israel, S.A., et al.: ECG to identify individuals. Pattern Recogn. **38**(1), 133–142 (2005)
10. Boumbarov, O., Velchev, Y., Sokolov, S.: ECG personal identification in subspaces using radial basis neural networks. In: IEEE International Workshop on IDAACS: Technology and Applications, IDAACS 2009 (2009)
11. Chan, A.D.C., et al.: Wavelet distance measure for person identification using electrocardiograms. IEEE Trans. Instr. Meas. **57**(2), 248–253 (2008)
12. Fatemian, S.Z., Hatzinakos, D.: A new ECG feature extractor for biometric recognition. In: 16th International Conference on Digital Signal Processing (2009)
13. Wang, Y., et al.: Analysis of human electrocardiogram for biometric recognition. EURASIP J. Adv. Sign. Proc. **2008**, 19 (2008)

14. Plataniotis, K.N., Hatzinakos, D., Lee, J.K.M.: ECG biometric recognition without fiducial detection. In: IEEE Biometric Consortium Conference (2006)
15. Ye, C., Coimbra, M.T., Vijaya Kumar, B.V.K.: Investigation of human identification using two-lead electrocardiogram (ECG) signals. In: IEEE 2010 Fourth IEEE International Conference on BTAS (2010)
16. Abo-Zahhad, M., Ahmed, S.M., Abbas, S.N.: Biometric authentication based on PCG and ECG signals: present status and future directions. Sig. Image Video Process. **8**(4), 739–751 (2014)
17. Belgacem, N., et al.: ECG based human authentication using wavelets and random forests. Int. J. Cryptogr. Inf. Secur. (IJCIS) **2**, 1–11 (2012)
18. Chen, C.-K., Lin, C.-L., Chiu, Y.-M.: Individual identification based on chaotic electrocardiogram signals. In: IEEE 6th IEEE Conference on ICIEA (2011)
19. Safie, S.I., Soraghan, J.J., Petropoulakis, L.: ECG biometric authentication using Pulse Active Width (PAW). In: IEEE Workshop on Biometric Measurements and Systems for Security and Medical Applications (BIOMS) (2011)
20. Saechia, S., Koseeyaporn, J., Wardkein, P.: Human identification system based ECG signal. IEEE Region 10 TENCON (2005)
21. Jiang, W., Kong, S.G.: Block-based neural networks for personalized ECG signal classification. IEEE Trans. Neural Netw. **18**(6), 1750–1761 (2007)
22. Da Silva H.P., Plcido, H., et al.: Finger ECG signal for user authentication: usability and performance. In: IEEE 6th International Conference on BTAS (2013)
23. Sidek, K.A., Khalil, I., Smolen, M.: ECG biometric recognition in different physiological conditions using robust normalized QRS complexes. Comput. Cardiol. (CinC), 97–100 (2012)
24. Caffarelli, F.: ECG Wearable Device For Biometric Recognition, Master thesis in Electronic Engineering, Politecnico of Turin, Italy (2013)
25. http://www.neuronica.polito.it/ecg.aspx
26. Nemirko, A.P., Lugovaya, T.S.: Biometric human identification based on electrocardiogram. In: Proceedings of XII-th Russian Conference on Mathematical Methods of Pattern Recognition. MAKS Press, Moscow, pp. 387–390 (2005)
27. Tarvainen, M., Ranta-aho, P., Karjalainen, P.: An advanced detrending method with application to HRV analysis. IEEE Trans. Biomed. Eng. **49**, 172–175 (2001)
28. Pan, J., Tompkins, W.J.: A real-time QRS detection algorithm. IEEE Trans. Biomed. Eng. **32**, 230–236 (1985)
29. Li, N., Cruz, J., Chien, C.S., Sojoudi, S., Recht, B., Stone, D., Csete, M., Bahmiller, D., Doyle, J.C.: Robust efficiency and actuator saturation explain healthy heart rate control and variability. Proc. Natl. Acad. Sci. USA **111**(33), E3476–E3485 (2014)
30. Sagie, A., Larson, M.G., Goldberg, R.J., et al.: An improved method for adjusting the QT interval for heart rate (the Framingham heart study). Am J Cardiol. **70**, 797–801 (1992)
31. Cortes, C., Vapnik, V.N.: Support-vector networks. Mach. Learn. **20**, 273–297 (1995)

Part VII
Reconfigurable- Modular- Adaptive- Smart Robotic Systems for Optoelectronics Industry: The White'R Instantiation

A Regenerative Approach to Dynamically Design the Forward Kinematics for an Industrial Reconfigurable Anthropomorphic Robotic Arm

Oliver Avram and Anna Valente

Abstract Reconfigurable industrial robots have tremendous potential to provide broad adaptive capacity to fast changes in market requirements. Nevertheless the reconfigurability concept adds complexity to the modeling of one of the classical but not less important problems in robot manipulators, which is forward kinematics. This paper presents an automated procedure to determine the kinematic parameters of a reconfigurable 6DOF industrial robot as a first step towards the design of a complete reconfigurable control architecture. The efficacy of the proposed approach has been demonstrated on an existing prototype of a reconfigurable anthropomorphic robotic arm developed for the optoelectronic industry.

Keywords Robotic kinematics · Automated frame assignment · Industrial reconfigurable robot

1 Introduction

Reconfigurable robots are versatile, more adaptive and present numerous advantages over their conventional counterparts in the light of a highly changing production environment [1, 2]. Nevertheless, the number of reconfigurable robots prototyped for industrial applications is rather limited [3–5]. Valente [6] tackles some aspects hindering the wide industrial acceptance of reconfigurable robots and proposes an integrated design approach to support the robot designer in the assessment of the manipulator capabilities and functionalities over time. A reconfigurable robot must

O. Avram (✉) · A. Valente
SUPSI ISTePS - Institute of Systems and Technologies for the Sustainable
Production, Galleria 2, Manno 6928, Switzerland
e-mail: oliver.avram@supsi.ch

A. Valente
e-mail: anna.valente@supsi.ch

© Springer International Publishing Switzerland 2016
S. Bassis et al. (eds.), *Advances in Neural Networks*, Smart Innovation,
Systems and Technologies 54, DOI 10.1007/978-3-319-33747-0_43

437

be able to satisfy various production tasks with minimal additions and modifications. Following upon the assembly of a new configuration, the robot must be fully operative in the shortest time. However a robot controller only works if the robot configuration and its kinematic and dynamic functionality are known a priori. In consequence, the development of an automated model for the forward kinematics computation is an important first step on the way towards the development of a real-time reconfigurable control platform.

One of the common approaches used to automatically generate the kinematics of reconfigurable robot is based on the Denavit-Hartenberg (D-H) methodology and consists in generating a mapping between modular design variables and the robot configuration in terms of D-H notation. Automated kinematic modeling techniques by addressing the individual kinematic description of the modular units and a kinematic chain assembly technique were proposed by several authors [7–10]. Djuric et al. [11] carried out the development of a global kinematic model able to generate all possible kinematic configurations of any n-DOF reconfigurable machine. They defined the D-H parameters as variables capable to represent and satisfy the kinematic properties of any configuration while considering both rotational and translational joints.

Another common approach addressed the definition of the modular reconfigurable robot in terms of a series of dyads. Pan et al. [12] defined the initial pose of every dyad based on screw theory and subsequently computed the kinematics based on product-on-exponential formulation. Chen et al. [13, 14] employed a graph based representation scheme termed Assembly Incidence Matrix to represent the ever changing configurations and the formulation of the kinematics was also based on local representation of the product-of-exponential formula.

This paper concentrates on the forward kinematics computation of an anthropomorphic robot arm performed automatically with respect to every new configuration.

This is facilitated by a number of rules concerning the individual frame assignment on the modular units of the robot, their storage in a dedicated repository and a recursive algorithm to kinematically describe a new configuration without any manual intervention. The rest of this paper is organized as follows: Sect. 2 gives an overview of the modeling framework, Sect. 3 describes the main steps in the model development and Sect. 4 discusses the implementation aspects. The simulation outcomes are discussed in Sect. 5 and finally Sect. 6 gives the conclusion and future work.

2   Modeling Framework Overview

A generic reconfigurable kinematic module represents a forward step towards a reconfigurable architecture having the capability to accept either new modules or new applications without restoring from scratch. This article covers exclusively the generic formulation of the forward kinematics for a prototype industrial robot arm conceived to meet a wide spectrum of production requirements of significant relevance for the optoelectronics industry.

In order to develop a kinematic model compliant with the norms of the future reconfigurable industry, there is a need to firstly introduce the reconfigurable aspects of the prototype robotic arm. The reconfigurable robotic arm consists of pluggable modules, joints, links and end effectors respectively, which can be of different shapes and sizes. A dedicated library stores the geometrical characteristics of the modules and robot configuration variations can be produced by using different kinematic chain topologies.

The modeling framework is depicted in Fig. 1 with the main focus on the reconfigurable kinematics. The type and sequence of robotic modules within the configuration that suits the best a production scenario is fixed and provided as an input by a model developed a priori, the robot configuration model. The kinematic and dynamic performance will be assessed by dedicated models. The control strategy will be designed based on the outcomes of these two models.

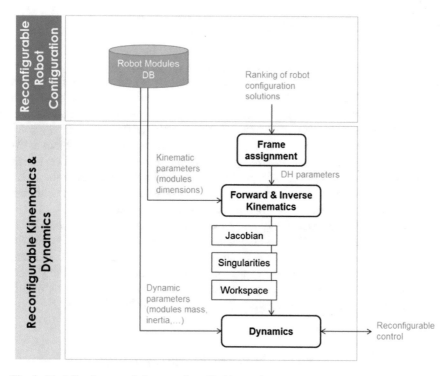

Fig. 1 Modeling framework for reconfigurable kinematics

3   Model Development

In order to generate various kinematic configurations for the 6DOF anthropomorphic robot arm, the D-H parameters are used to represent the kinematics of all possible configurations. These parameters are defined as variables whose values will be determined specifically for every new configuration after the complete characterization of the reference frames along the kinematic chain carried out by employing concepts of line geometry and vector algebra.

The input required to perform the forward kinematics consists of:

- the description of the fundamental kinematic chain and the best robot configuration for the production task;
- the number and types of constituent joints and links;
- the robot base frame;
- the reconfigurable parameters describing the relative orientation of consecutive joint axes.

3.1   Frame Assignment on Individual Robot Modules

The first step of the kinematic modeling is the proper assignment of coordinate frames to the robot modules. Each coordinate system is orthogonal and the axes obey the right-hand rule. Generally, the reference frames are assigned on each module in a way that enables an easier description of the motion between consecutive joints and the characterization of the geometry of the links.

A convenient way to assign the input and output frames is to place them at the extremity of the links on the assembly ports to facilitate the alignment between their z-axes and the z-axes of both adjacent joints. This will allow expressing the links' length or links' offset as the distance between the z-axes along the common normal or the x-axes respectively of the two aforementioned joint frames.

Two types of links respectively will be considered. The PL type connects joints with parallel axes and the PP type connects joints with intersecting axes. For the PL link, the frames have the same orientation and are placed with the z-axis coincident with the axis of the assembly pins and the origins placed on the two assembly axes. The x axes are constructed as perpendiculars from each origin to the axis of reference pins and they must be oriented along the length of the link (Fig. 2).

3.2   Frame Assignment Along the Kinematic Chain

Based on the kinematic chain of a specific robot configuration, the next activity is to perform the assembly of the modules with respect to the frames assigned

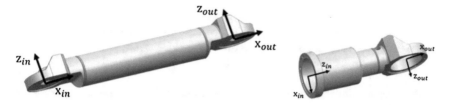

Fig. 2 Input/output frame assignment on PL-type (*left*) and PP-type (*right*) links

individually to each module and then to identify the initial location of the origins of joints' frames from the base to the end effector of the kinematic chain. The relative offset between the origins of two consecutive joints is determined by the length of the link connecting them and represented by the transformations L_{T1}, L_{T2} or L_{TCP} relating the two fixed coordinate frames located on each end of the link:

$$L_{T1} = Trans(ll_x, 0, 0) \tag{1}$$

$$L_{T2} = Trans(0, 0, ll_z) \tag{2}$$

$$L_{TCP} = Trans(0, 0, ll_{z\_TCP}) \tag{3}$$

where ll_x is the length of the PL type link along x axis, ll_z is the length of the PP type link along the z-axis and ll_{z_TCP} is the length of the robot's end effector.

The initial location of the coordinate frames are expressed one by one in zero reference position and the orientation of the joint axes is established based on the sequence of the modules and a set of reconfigurable parameters uniquely describing the chosen kinematic chain.

3.3 Forward Kinematics

For the considered anthropomorphic arm with spherical wrist the location of the origins of the reference frames is subject to the following assignment: on the arm side the first and second joint origins coincides and on the wrist side the joints 4, 5 and 6 have also the same location. The extraction of the D-H parameters (joint offset d_i, link length a_i and twist angle α_i) is a process performed recursively from the base to the last link which cannot be initiated before revisiting the initial joint origins and altering their location if required by a specific robot configuration.

In order to describe the complete transformation of the arm from the base to the end effector all the transformations between consecutive joints must be described. Each homogenous transformation can be determined based on the generic transformation matrix:

$$
{}_{i}^{i-1}A = \begin{bmatrix} \cos\theta_i & -\sin\theta_i & 0 & a_i \\ K_{Ci}\sin\theta_i & K_{Ci}\cos\theta_i & -K_{Si} & -d_iK_{Si} \\ K_{Si}\sin\theta_i & K_{Si}\cos\theta_i & K_{Ci} & d_iK_{Ci} \\ 0 & 0 & 0 & 1 \end{bmatrix}
$$

This formulation allows an easier assessment of various reconfigurations by varying the reconfigurable parameters K_{Si} and K_{Ci} rather than developing unique models for each configuration [11]. The forward kinematics of the reconfigurable model can be calculated using the multiplication of all the homogenous matrices from the base frame to the end effector frame.

4   Implementation

The reconfigurable kinematic model, implemented in Matlab, is constructed progressively joint by joint according to the generic algorithm flowchart described in Fig. 3. The sequence of the joints and links in the kinematic chain as well as their ID's are provided by the robot arm configurator module. The information about the initial frame assignment on each module for a specific configuration is retrieved by querying a MySQL database designed as an inventory of all available robotic modules.

The initial location of the origins are computed by summation of the link lengths taking into account an initial pose of the robot as shown in Fig. 4.

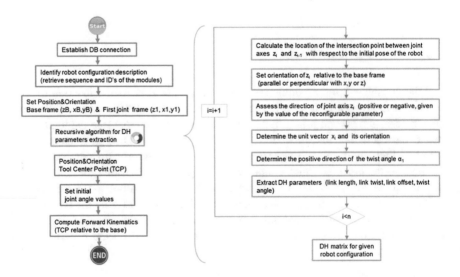

Fig. 3 Recursive algorithm for automated computation of the forward kinematics

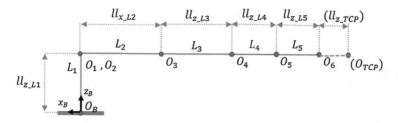

Fig. 4 Initial representation of the joints' origins in the kinematic chain

The next step addresses the orientation of the joint axis z_i (the origin O_i previously assigned is considered as being a point on the axis z_i) with respect to the base coordinate frame. At this stage, the orientation is assessed in termed of parallelism and three cases are possible: $z_i \| x_B$, $z_i \| y_B$ or $z_i \| z_B$.

For instance if $z_i \| x_B$, $z_{i-1} \| y_B$ and $O_i - O_{i-1} = (0, 0, z)$ where z represents the offset between the two origins along z_B, then $z_i \perp z_{i-1}$ in O_{i-1}. Otherwise, if $z_{i-1} \| x_B$ and $z_i \| y_B$ then $z_i \perp z_{i-1}$ in O_i. However, there are cases when the orientation and orthogonality between the last joint axis of the arm (i.e. z_3) and the first joint axis of the wrist of the robot (i.e. z_4) will lead to a coincidence between O_3 and O_4, if the logic stated above is applied. Therefore, in order to ensure that $O_4 = O_5 = O_6$ the corresponding equations pertaining to the location of the origins must be adapted as follows:

$$O_4 = O_5 = O_6 = O_4 + L_4 = (-(ll_{x_{L2}} + ll_{z_{L3}} + ll_{z_{L4}}), 0, ll_{z_{L1}}) \tag{4}$$

$$O_{TCP} = O_5 + L_5 + L_{TCP} = (-(ll_{x_{L2}} + ll_{z_{L3}} + ll_{z_{L4}} + ll_{TCP}), 0, ll_{z\_L1}) \tag{5}$$

If $z_{i-1} \| z_B$ and $z_i \| z_B$ then, obviously, $z_i \| z_{i-1}$. If the origin O_i is translated along a single axis (either x_B or y_B) with respect to O_i then the D-H parameter $a_i = O_i - O_{i-1}$. The determination of the orthogonality/parallelism between a joint axis z_i and the reference frame of the base is followed by the identification of the orientation of the joint with respect to the previous joint axis z_{i-1}. This is controlled by the reconfigurable parameters K_{si} and K_{ci}, defined as the sinus and cosines of the twist angles.

Once the matrix of the z-axis unit vectors is determined every orthogonal reference frame can be fully represented if the corresponding x and y axes are known. Their unit vectors can be estimated by employing concepts of line geometry and vector algebra.

The last step is the extraction of the D-H parameters by evaluating the following situations: (a) if $z_i \perp z_{i-1}$, $z_i \| y_{i-1}$ then $x_i \| x_{i-1}$ and $d_{DHi} = O_i - O_{i-1}$ and $a_i = 0$, (b) if $z_i \perp z_{i-1}$ and $z_i \| x_{i-1}$ then $d_{DHi} = 0$ and $a_i = 0$, (c) if $z_i \| z_{i-1}$ then $a_i = O_i - O_{i-1}$ and $d_{DHi} = 0$.

The values of the D-H parameters previously determined together with the values of the reconfigurable parameters are input into the generic homogenous transformation matrix introduced in Sect. 3.3. This matrix describes the transformation of coordinates between two consecutive frames.

5 Simulation and Discussion

The developed model derives the forward kinematics with respect to every new configuration and can display the results in either numerical or symbolical form. The sequence and id's of the robotic modules, the sequence and values of the reconfigurable parameters, the base location and the initial pose of the links modules relative to the base characterize the input information of each robot configuration. Following upon a database query to collect the geometric information of the robot modules, the initial location of the origin axes and the z, x and y unit vectors describing the joint reference frames are iteratively determined and arranged in matrix form:

$$
O_{16} = \begin{bmatrix} 0 & 0 & 0 \\ 0 & 0 & 92 \\ 400 & 0 & 92 \\ 645 & 0 & 92 \\ 793 & 0 & 92 \\ 963 & 0 & 92 \end{bmatrix} \quad
Z_{16} = \begin{bmatrix} 0 & 0 & 1 \\ 0 & -1 & 0 \\ 0 & -1 & 0 \\ -1 & 0 & 0 \\ 0 & -1 & 0 \\ 1 & 0 & 0 \end{bmatrix} \quad
DH_m = \begin{bmatrix} 0 & 0 & 92 & 0 \\ 1.57 & 0 & 0 & 0 \\ 0 & 400 & 0 & 0 \\ 1.57 & 0 & 393 & 0 \\ 1.57 & 0 & 0 & 0 \\ 1.57 & 0 & 0 & 0 \end{bmatrix}
$$

The joint reference frames, initially located in the origin points given by the matrix O_{16}, are reassigned by considering that $O_1 = O_2$ and $O_4 = O_5 = O_6$. Next, the D-H matrix is generated based on the D-H parameters recursively extracted with respect to the new location of the origins. Any change in the robot kinematic

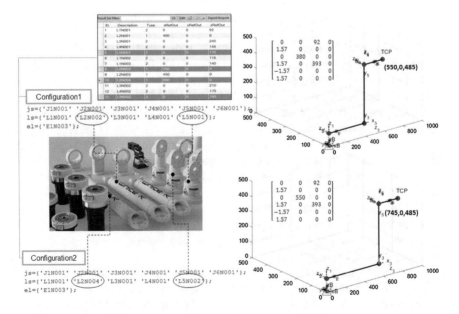

Fig. 5 Alternative robot arm configurations and forward kinematics computation

configuration, triggered by the need to meet specific production requirements, is automatically handled by the model and the extraction of the D-H parameters and computation of the D-H matrix are adapted accordingly.

For instance, Fig. 5 pictures a visual extract of the modules library and the description of two robot configurations. With respect to Configuration 1, Configuration 2 implements two longer versions for the link 2 and 5.

As can be seen in the above figure, for the same pose of the robot, the forward kinematics computation generated automatically a new location of the tool center point (TCP) by considering the geometric dimensions of the alternative links.

6 Conclusions and Future Work

The proposed model enables the automated extraction of D-H parameters and computation of forward kinematics for alternative configurations of an anthropomorphic robotic arm. Initially, an individual frame assignment is defined with respect to the robot modules and the information related to the robot configuration variations is stored in a dedicated library. Once the joint frames are assigned with respect to the chosen kinematic chain, the recursive extraction of the D-H parameters can be initiated. The kinematic model build upon the set of D-H parameters identified for the analyzed configuration is used to compute the forward kinematics.

The automated generation of the kinematic parameters is a first step towards the development of a real-time reconfigurable software platform enabling the fast adaptation of a robot architecture to evolving production requirements without having them programmed in an explicit manner. Future work will address the development of the inverse kinematics and dynamics.

Acknowledgments The research has been partially funded by FP7 EU Project "white'R white room based on Reconfigurable robotic Island for optoelectronics".

References

1. Koren, Y., Heisel, U., Jovane, F., Moriwaki, T., Pritschow, G., Ulsoy, G., Van Brussel, H.: Reconfigurable manufacturing systems. CIRP Ann. Manufact. Technol. **48**(2), 527–540 (1999)
2. Yim, M., Shen, W., Salemi, B., Rus, D., Moll, M., Lipson, H., Klavins, E., Chirikjian, G.: Modular self-reconfigurable robot systems. IEEE Robot. Autom. Mag. **14**(1), 43–52 (2007)
3. Bruzzone, L., Razzoli, R., Acaccia, G.: A modular robotic system for industrial applications. Assem. Autom. **28**, 151–162 (2008)
4. Xie, Z., Zhang, Q., Wang, B., Liu, Y.: HIT-ARM I high speed dexterous robot arm. In: Proceedings of the 2012 IEEE International Conference on Robotics and Biomimetics, Guangzhou (China) (2012)

5. Xijian, H.,Yiwei, L., Li, J., Hong, L.: Design and development of A 7-DOF humanoid arm. In: Proceedings of the 2012 IEEE International Conference on Robotics and Biomimetics, Guangzhou (China) (2012)
6. Valente, A.: Reconfigurable industrial robots—an integrated approach to design the joint and link modules and configure the robot manipulator. In: 3rd IEEE/IFToMM International Conference on Reconfigurable Mechanisms and Robots, 20–22 Jul 2015, Beijing, China
7. Kelmar, L., Khosla, P.: Automatic generation of forward and inverse kinematics for a reconfigurable modular manipulator system. In: Proceedings of the IEEE International Conference on Robotics and Automation (ICRA'88), Apr 1988, pp. 663–668
8. Bi, Z.M., Zhang, W.J., Chen, I.-M., Lang, S.Y.T.: Automated generation of the D-H parameters for configuration design of modular manipulators. J. Robot. Comput. Integr. Manufact. **23**(5), 553–562 (2007)
9. Rajeevlochana, C.G., Saha, S.K., Kumar, S.: Automatic extraction of DH parameters of serial manipulators using line geometry. In: The 2nd Joint International Conference on Multibody System Dynamics, May 29–June 1 2012, Stuttgart, Germany
10. Benhabib, B., Zak, G., Lipton, M.G.: A generalized kinematic modeling method for modular robots. J. Robot. Syst. **6**(5), 545–571 (1989)
11. Djuric, A.M., Al Saidi, R., ElMaraghy, W.: Global kinematic model generation for n-DOF reconfigurable machinery structure. In: 6th Annual IEEE Conference on Automation Science and Engineering Marriott Eaton Centre Hotel Toronto, Ontario, Canada, 21–24 Aug 2010
12. Pan, X., Wang, H., Jiang, Y., Yu, C.: Research on kinematics of modular reconfigurable robots. In: Proceedings of the 2011 IEEE International Conference on Cyber Technology in Automation, Control, and Intelligent Systems, 20–23 Mar 2011, Kunming, China
13. Chen, I.-M., Yang, G.: Configuration independent kinematics for modular robots. In: Proceedings of the 1996 IEEE International Conference on Robotics and Automation Minneapolis, Minnesota, Apr 1996
14. Chen, I.-M., Yang, G., Yeo, S.H.: Automatic modeling for modular reconfigurable. In: Cubero, S. (ed.) Robotic Systems—Theory and PracticeIndustrial Robotics: Theory, Modeling and Control. ISBN: 3-86611-285-8

An ANN Based Decision Support System Fostering Production Plan Optimization Through Preventive Maintenance Management

Marco Cinus, Matteo Confalonieri, Andrea Barni and Anna Valente

Abstract Besides pursuing the economic goals of low costs and high profits, companies are becoming more and more aware of the environmental and social impact of their actions. Companies striving for the integrated optimization of environmental and economic perspectives within their production processes, need to be supported by tools helping to understand the effects of the decision making process. In this context, this paper describes the Artificial Intelligence developed for a Decision Support System (DSS) which enables the early identification of problems occurring on manufacturing. The decision making process beneath the DSS starts from the aggregation of production lines sensors data in Key Performance Indicators (KPI). The data are then processed by means of an Artificial Neural Networks (ANN) based knowledge system which enables to suggest preventive maintenance interventions. The proposed maintenance activities, elaborated throughout a scheduling engine, are integrated within the weekly production schedule, according to the selected optimization policy.

Keywords Decision support system · Production line · Preventive maintenance

M. Cinus (✉) · M. Confalonieri · A. Barni · A. Valente
SUPSI-DTI, University of Applied Sciences and Arts of Southern Switzerland,
Manno, Switzerland
e-mail: marco.cinus@supsi.ch

M. Confalonieri
e-mail: matteo.confalonieri@supsi.ch

A. Barni
e-mail: andrea.barni@supsi.ch

A. Valente
e-mail: anna.valente@supsi.ch

© Springer International Publishing Switzerland 2016
S. Bassis et al. (eds.), *Advances in Neural Networks*, Smart Innovation,
Systems and Technologies 54, DOI 10.1007/978-3-319-33747-0_44

447

1 Introduction

The only two good reasons to stop a production line are: (1) performing regularly scheduled maintenance or (2) equipment failure. Thus, the decisional process associated with maintenance activities on manufacturing lines becomes a critical activity, especially when considering the repercussions of a production line stop can have on the whole plant. Performing timely and necessary maintenance is indeed critical to preventing failures that may result in costly production interruptions, but relying on a fixed schedule may result in higher than necessary costs for both parts and labor. In many large-scale plant-based industries, the costs related to the maintenance of such level of availability account as much as 40 % of the operational budget. Moreover, up to one-half of these maintenance costs can be considered wasted by the application of ineffective maintenance management methods [1]. The dominant reason for this ineffective use of maintenance expenditures is the lack of factual data that quantifies when and what kind of maintenance is needed to maintain, repair or replace critical machinery, equipment and systems within a plant [1]. In particular, the identification of anomalous machine behavior not directly ascribable to failures and the reconstruction of degradation patterns supporting preventive maintenance are severely complex.

The maintenance and production management related issues become of particular interest if considering that machine energy consumption and consequently the related emissions, increase while machine approach to a degraded behavior [2]. Thus providing an efficient preventive maintenance, not only allows avoiding unnecessary stops of the production line, but also can save energy and lowering the emissions of machinery. In this context, the application of tools supporting condition based maintenance has dramatically reduced non-value added maintenance by eliminating the need to unnecessarily shutdown equipment for maintenance checks. Predictive maintenance has become a key part of the modern maintenance department and more and more companies are taking on board these technologies in order to maximize the reliability of their equipment by detecting failures well in advance. Some failure modes cannot be designed out (i.e., mechanical bearings are here to stay, electrical panels will always be an integral part of any system), but if failures can be detected early, the maintenance team can plan the work in an organized manner [3].

Despite the state of the art progresses in the application of sensors and tools able to track machine behavior and performances, organizations are still lacking in a structured and reliable approach for collecting and analyzing equipment performance, executed maintenance tasks, failure history or any of the other data that could, and should be used to plan and schedule tasks that would prevent premature failures, extend the useful life of critical plant assets, and reduce their life cycle cost. Model-based maintenance decision support systems are thus needed to achieve high productivity and cost effectiveness of the overall system [4, 5].

In this paper we are going to explore the current State-of-the-Art in the field of Decision Support Systems (Sect. 2), present the tool developed inside the project (Sect. 3), focus on the Artificial Intelligence of the tool (Sect. 3.1), present the achieved result (Sect. 4) and the final considerations about the work (Sect. 5).

2 SOA

Decision Support Systems (DSS) are interactive computer-based systems intended to help decision makers in using communication technologies, data, documents and knowledge or models, to identify and solve problems, perform decision processes and make decisions [6]. Such a general definition of DSS reflects the great number of different uses that these systems have, this also confirmed by literature reports coming from different fields like management science [7, 8], project management [9], finance [10], logistics [11] and scheduling [12].

Manufacturing plants are currently facing three challenges which can be tackled by the adoption of DSS in the management phase. First of all, a large amount of data, the so called big-data, arises more and more in the industry. The management of these data ranges from collection and selection to analysis and manipulation. The alternatives decisions amongst which the decision maker must make his choice made can range from a few to a few thousand. A DSS can help in the process by narrowing the possibilities down to a reasonable number. Computers can evaluate alternatives, especially where the alternatives can be put into numerical terms. Even when this is not the case, the computer can assist the decision-maker in presenting the alternatives in a form that facilitates the decision [6, 7] developed a DSS framework integrating business and engineering decisions aimed at maximizing the profit and the product quality of an oil refinery at the same time. This development showed that the second challenge which concerns the modern manufacturing plants is the nature of data and decision criteria, which are mostly heterogeneous, coming from different sources and areas and most of the time presented and stored in an unstructured manner. In manufacturing plants there are different levels, and scopes, for a decision, each of it with a different time horizon. This decision type strongly influences the DSS data management dynamics. This can be seen as the third challenge.

With specific reference to the manufacturing applications, the DSS solutions proposed in the scientific literature mostly involve the adoption of mathematical optimization tools based on different programming methods or artificial intelligence knowledge-based algorithms [8]. Some of them also merge the analytical formulation with discrete simulation-based techniques so to realize off-line testing of the solutions before actually deploying them in the shop-floor. A manufacturing methodology for non-destructive, real-time process optimization and quality certification has been developed by [9]. A DSS has been developed with the purpose of improving the performances of a high frequency welding process for pipes and tubes milling. The developed DSS assisted the operator by suggesting decision made upon real time data coming from the process, merged with quality control information. It indicates the corrective action to be taken, by including both quantitative models and qualitative analyses. This allowed the operator to react quickly and on time adjusting the process parameters. In the meanwhile, the DSS acted as an information system, especially by collecting and formalizing data, for the higher level manages, helping also them in the decisional process. Knowledge-driven DSSs are able to suggest or recommend actions based upon knowledge that has been gathered using artificial intelligence in data analysis or other statistical tools as, for instance, case-based reasoning and Bayesian networks. Data mining techniques are strictly connected to this category being them able to find hidden patterns and relationships between data in a database [10]. In the modeling problems where it is necessary the DSS capability of predicting behaviors and patterns starting from datasets, artificial neural networks (ANN) come instrumental.

Artificial Neural Networks (ANN, NN), in particular the standard multilayer feed forward networks with as few as one hidden layer using arbitrary squashing functions, are capable of approximating any Borel measurable function, from one finite dimensional space to another, to any desired degree of accuracy [11]. This results in NN to be defined as "Universal Approximators". In [12] is stated that NN works well due to (1) their ability to spot and capture regularities within patterns; (2) the capability of working with dataset of high dimensionality; (3) the ability to map relationships which are difficult to be mapped with conventional approaches. As stated in [13], in accuracy tests against other approaches, NN always have a high score. The main advantages of NN, always according to [12], are the following: (1) Ability to extract hardly noticeable trends and patterns; (2) adaptive learning based on data; (3) self-organization, the network itself creates the data representation; (4) real time operation, execution and training can be done in parallel to speed up the computation time. Being black-boxes is one of the main critics for the NN, they are good at predicting outputs but they lack the ability to fully understand the underlying model. This can be view as the first drawback. The second one is their susceptibility to over fitting, meaning that a network with a high learning capacity trained on a dataset with too few samples leads to network which perfectly predicts the training set, but poorly evaluates on the test set. When a net performs this way it is usually said to be over fitted and as consequences it is not able to generalize well.

Limiting to manufacturing field, the use of ANNs to support critical and strategic decisions is documented in [14]. Their study shows that a supervised learning method as probabilistic neural network has the best classification capabilities compared with other artificial intelligence approaches in selecting the plant location for a clothing manufacturing factory. The problem here is given by the manifold of macro-environmental factors (economical, social, political, and legal) combined with other operative factors as customers, competitors' suppliers.

3 The Developed Tool

The work described in this paper has been carried out under the FP7 Factory-Ecomation project. The developed hardware, software and methodological interventions are tested and validated in the two project industrial demonstrators, namely IKEA Portugal and Brembo Poland. The project involved the development of a Decision Support System (DSS) able to predict machine failures and spotting problems in order to help the Decision Maker in planning the preventive maintenance, minimizing costs and maximizing its benefits.

The Factory-Ecomation DSS has been built based on three main blocks, which are showed in Fig. 1. Those modules are the Profiler, the Assesser and the Optimizer. The first one is responsible for the data collection and the assessment of the facility status and behavior. This module is composed by two sub-modules, the "DSS Indicators" which responsibilities are data acquisition from the shop floor devices including sensors, PLCs and SCADA systems. The second sub-module being the "DSS KPIs", filters and processes the collected data and organize them in human understandable KPIs.

The second module is the Assesser which determines the recovery strategy to be implemented systematically in the factory so to ensure the persistent accomplishment of the goals. The third module is the Optimizer, which through the scheduling of the preventive maintenance tasks and the ordinary jobs, tries to find the best solution to improve the schedule.

A more extensive and in depth analysis of the Factory-Ecomation DSS can be found in Confalonieri et al. [15].

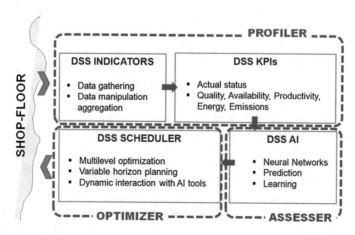

Fig. 1 DSS main building blocks

3.1   Artificial Intelligence

This tool adaptability need, in fact, has pushed forward the search for an approach which could be adjusted firstly to our two industrial and, in parallel, on further industrial cases.

Moreover, from an analysis of the IKEA line and of all the gathered data, an impressive quantity of data has been recognized, i.e.: the line is composed from 24 productive resources, each of it having 9 to 12 different parameters. Considering that all these parameters have a role in the detection of degradation pattern, the problem dataset has undoubtedly a high dimensionality. Last but not least, the gathered data is also huge in quantity, since the production line has been monitored continuously for more than 9 months, causing us to search for an approach which could lead to low computation time in order to be useful on the operational day by day management in the shop floor.

A matching between the recognized needs, (adaptability to different models, the high dimensionality of data and the great quantity of gathered data) and the advantages/disadvantages of NN showed that all the needs were covered by the advantages and the drawbacks could be limited by training techniques and by a correct choice of the dimension of training, testing and validation set.

4   Test and Results

The dataset used has been constructed from the sensor and performance log files of the IKEA line of about 9 months of work. These data were classified using information coming from the maintenance log file of the same work period. The resulting data set was composed by about 10'000 samples, each composed by an input vector of 248 elements. Since the data was of different nature (e.g.: idle times, number of failures, mean time between failures, energy consumption, dust emissions and absorbed dust emissions) the dataset has been normalized. In the first case a min-max normalization has been applied and in the second case a Gaussian normalization.

The NN were trained to classify which is the faulty machine, without taking into account which type of failure this machine had.

The selected NN, is a feed forward neural network implemented using the Pybrain [16] library, with a single hidden layer, based on the sigmoid activation function, with a soft-max output layer.

In order to achieve valid test results, the data has been divided in training (80 %) and test (20 %) sets. In order to evaluate the generalization capabilities of the trained network a first series of test has been carried out using a fivefold cross validation of the training set and tested with the test set every 5 epochs of training. These training had a number of hidden neurons ranging between 75 and 125, chosen experimentally (Figs. 2, 3, 4 and 5). The achieved results are presented in

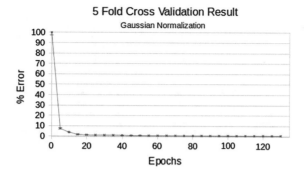

Fig. 2 Gaussian normalization, fivefold cross validation 105 hidden nodes

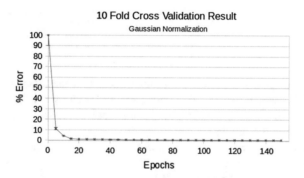

Fig. 3 Gaussian normalization, tenfold cross validation 135 hidden nodes

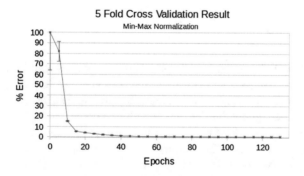

Fig. 4 Min-max normalization, fivefold cross validation 95 hidden nodes

Fig. 2 and in Fig. 4, for both type of normalization). A second series of training has been carried out using a tenfold cross validation and with a number of hidden neurons ranging between 100 and 200. The achieved results are visible in Figs. 3 and 5.

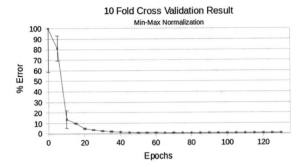

Fig. 5 Min-max normalization, tenfold cross validation 140 hidden nodes

Table 1 Decision tree results

Algorithm	% Correctly classified
J48	93.59
Hoeffding tree	93.70

In order to be able to evaluate the achieved result, at the moment, there are no experts in IKEA which are able to correctly identify a faulty machine based on the data registered on the logs, they actually have to look at the resource failure sensor to detect which is the failing resource.

As a simple benchmark, also decision trees generated from the most known algorithms have been tested on the same dataset using a tenfold cross validation. The trees were generated using the WEKA data mining suite [17]. Results are presented in Table 1.

5   Conclusions

In this paper, a DSS supporting decision making process within manufacturing plants has been described with focus on the Neural Network used to predict the anomalies occurring on the production lines. The tool in general is expected to give a number of advantages at different levels within the production plant with multiple positive effects on the whole manufacturing activity: (I) integration of human and machines based decisions within the production management process; (II) capturing the streaming dynamics of the production choices across the shop-floor in order to early identify incoming machine failures; (III) the ability of controlling only those machines that show the beginning of a malfunctions and update the production schedule consequently.

The analysis and early spotting of incoming machine failure is carried out using a Feed Forward Neural Network, with a single sigmoid activated hidden layer and a softmax output layer. The preliminary results showed a very good performance on the identification of the faulty machine with a success higher than 98 % correct identified faulty resource. The NN performed well against some simple benchmark, which only reached scores of about 94 %.

Next steps would be the training and consequent validation of a new Neural Network which goal is to identify both the machine and the type of incoming failure. The NN will then be trained in both industrial demonstrators and validated on site.

Acknowledgements The presented research activities have received funding from the European Union Seventh Framework Programme (FP7/2007-2013) under grant agreement n° 314805 (Factory-ECOMATION).

References

1. Eti, M., Ogaji, S., Probert, S.: Reducing the cost of preventive maintenance (PM) through adopting a proactive reliability-focused culture. Appl. Energy **83**, 1235–1248 (2006)
2. Jihong, Y., Dingguo, H.: Energy Consumption Modeling for Machine Tools After Preventive Maintenance, Industrial Engineering and Engineering Management. IEEM), Macao (2010)
3. Liggan, P., Lyons, D.: Applying predictive maintenance techniques to utility systems. Pharm. Eng. **31**, 8–16 (2011)
4. van Houten, F., Tomiyama, T., Salomons, O.: Product modelling for model-based maintenance. CIRP Ann. **47**, 123–128 (1998)
5. Lung, B., Veron, M., Suhner, M., Muller, A.: Integration of maintenance strategies into prognosis process to decision-making aid on system operation. CIRP Ann. **54**, 5–8 (2005)
6. Wallach, E.G.: Understanding decision support systems and expert systems. Irwin (1994)
7. Hu, W., Almansoori, A., Kannan, P., Azarm, S., Wang, Z.: Corporate dashboards for integrated business and engineering decisions in oil refineries: an agent-based approach. Decis. Support Syst. 729–741 (2012)
8. Choi, S., Kang, S.: An integrated platform for distributed knowledge-based production scheduling. IEEE, 111–116 (2005)
9. Swanepoel, K.: Decision support system: real-time control of manufacturing processes. J. Manuf. Technol. Manage. 68–75 (2004)
10. Klein, M., Methlie, L.: Knowledge-based Decision Support Systems with Applications in Business. Wiley (1995)
11. Hornik, K.: Multilayer feedforward networks are universal approximators. Neural Netw. **2**, 359–366 (2003)
12. Ciobanu, D., Vasilescu, M.: Advantages and disadvantages of using neural network for predictions. Ovidius University Annals, Economic sciences series **13**, 444–450 (2013)
13. Smith, S.J., Berson, A., Thearling, K.: Building Data Mining Applications for CRM. McGraw-Hill (1999)
14. Au, K., Wong, W., Zeng, X.: Decision model for country site selection of overseas clothing plants. Int. J. Adv. Manuf. Technol. 408–417 (2006)
15. Confalonieri, M., Barni, A., Valente, A., Cinus, M., Pedrazzoli, P.: An AI based decision support system for preventive maintenance and production optimization in energy intensive manufacturing plants. in: International Conference on Engineering, Technology, and Innovation (2015)
16. Schaul, T., Bayer, J., Wierstra, D., Sun, Y., Felder, M., Sehnke, F., Rueckstiess, T., Schmidhuber, J.: PyBrain. J. Mach. Learn. Res. **11**, 743–746 (2010)
17. Hall, M., Frank, E., Holmes, G., Pfahringer, B., Reutemann, P., Witten, I. H.: The WEKA data mining software: an update. SIGKDD Explor. **11**(1) (2009)

Gaussian Beam Optics Model for Multi-emitter Laser Diode Module Configuration Design

Hao Yu, Giammarco Rossi, Yu Liu, Andrea Braglia and Guido Perrone

Abstract The fiber laser market has rocketed by more than 10 % in the last decades and as, a consequence, multi-emitter laser diode modules, which constitutes the fiber laser pump sources have seen a sharp increase in sales. Until now, the design process for a high coupling efficiency laser diode module is still mainly based on commercial ray tracing simulation tools and a trial and error approach. Such a design methodology lowers designers' productivity, extends the development cycle and increases R&D cost. Nowadays more and more customers need customised manufacturing to meet their specific requirements which means that R&D and production functions must respond immediately to customers' input by providing reliable and mature solutions ready for volume production. To overcome the flows of conventional design, we have developed a model based on Gaussian beam optics that has been validated by an extensive set of experiments. This model is used to design the multi-emitter laser diode modules that will be produced in the white'R Island.

1 Introduction

Nowadays the market of high-power fiber laser surges up. The producers of fiber laser have to meet the requirements of the market fast enough. However the design tools for multi-emitter laser diode module, which is used as the pump source of the high power fiber laser, are still mainly developed by classical optics that is not accurate since the laser beam behaves differently from the light generated by spontaneous emission. Hence it is argent to develop an model for evaluating the performance and the possibility of the multi-emitter laser diode module design.

H. Yu (✉) · Y. Liu · G. Perrone
Department of Electronics and Telecommunications, Politecnico di Torino,
24 Corso Duca degli Abruzzi, 10129 Torino, Italy
e-mail: hao.yu@polito.it

G. Rossi · A. Braglia
OPI Photonics s.r.l, 3 Via Conte Rosso, 10129 Torino, Italy
e-mail: grossi@opiphotonics.com

© Springer International Publishing Switzerland 2016
S. Bassis et al. (eds.), *Advances in Neural Networks*, Smart Innovation,
Systems and Technologies 54, DOI 10.1007/978-3-319-33747-0_45

A multi-emitter module is composed of a number of single emitters or bars whose output beams are combined by using the well-known "beam stacking approach" where an optical system arranges side by side the beams and then focus them into a multi-mode fiber. The output power of the fiber laser rises as long as the number of laser diodes inside the laser diode module goes up or the coupling efficiency increases.

The beam parameter product (BPP) is widely utilized to specify the beam quality of a laser beam. It has long been established that BPPs are 'optical invariants' that cannot be improved upon by optical systems. It is therefore obvious that the BPPs of the diode/stack would determine the best fiber BPP possible. For this purpose, the following equation has been widely used [1]:

$$BPP_{dia} = (BPP_{slow}^2 + BPP_{fast}^2)^{1/2} \qquad (1)$$

where BPP_{dia} is the beam of the laser diode in the diagonal direction, BPP_{fast} and BPP_{slow} denote the BPPs in the fast axis (FA) and slow axis (SA) direction. Previous literature states that for the fiber coupling to be efficient, BPP_{dia} has to be smaller than the BPP_{fiber} of the fiber. The proof for Eq. 1 cannot be found in published literatures and in [2] the author proved the other equation:

$$BPP_{dia} = BPP_{slow} + BPP_{fast} \qquad (2)$$

In our work, we found Eq. 2 is the minimum for the BPP of laser diode stack before the entrance of fiber and depends on the choice of optical components. It can be achieved if and only if fast axis collimator (FAC), slow axis collimator (SAC), FA focusing lens and SA focusing lens are rightly chose. In this paper, we develop a model for calculating the minimal diagonal BPP of laser diode stack.

2   Model Description

For the sake of simplicity but without losing generality, we assume that:

1. The beam of laser diode is a rectangular and the divergences in both directions are constants.
2. The beam radius is constants and is identical to the waist radius, because the well-collimated beam has extremely small divergences in two directions.
3. Each laser diode is incoherent respect to others and there is no interference.
4. All lenses are aberration-free.

Since the laser diode beam is not exactly a Gaussian beam, but it contains higher order Gaussian modes [3]. The M^2 factor is used in the expression for waist, the $1/e^2$ intensity beam radius at z, becomes then:

$$w(z) = w_0[1 + (\frac{M^2 \lambda z}{\pi w_0^2})^2]^{1/2} \tag{3}$$

where z is the distance from the waist of the laser beam, λ is the wavelength, w_0 is the $1/e^2$ intensity beam waist radius, $w(z)$ is the $1/e^2$ intensity beam radius at z.

The effective focal length of the collimating lens usually is much greater than $\pi w_0^2/M^2\lambda$, which will be shown later, consequently Eq. 3 yields:

$$w(f_{col}) \approx \frac{M^2 \lambda}{\pi w_0} f_{col} = \theta f_{col} \tag{4}$$

where f_{col} is the effective focal length (EFL) of the collimating lens, θ is the half divergence of the beam at the $1/e^2$ intensity.

After being collimated, the beam passes through a focusing lens and concentrates at the focal plane. The spot size is given by [4]:

$$w_{spot} = \frac{M^2 \lambda}{\pi w_{in}} f_{foc} \tag{5}$$

where w_{in} is the beam waist radius at the lens, f_{foc} is the EFL of the focusing lens.

We have assumed that the divergence of the collimated beam is negligible and the beam waist radius is a constant after collimating. Substituting Eq. 4 into Eq. 5 proves:

$$w_{FA} = \frac{M^2 \lambda}{\pi w(f_{FAC})} f_{foc_{FA}} = \frac{BPP\, f_{foc_{FA}}}{\theta_{FA}\, f_{FAC}} \tag{6}$$

where f_{FAC} is the EFL of the FAC, $f_{foc_{FA}}$ is the EFL of the FA focusing lens and θ_{FA} is the half divergence of the FA.

The half divergence of the focused beam is:

$$\theta_{foc_{FA}} = \frac{w(f_{FAC})}{f_{foc_{FA}}} = \frac{\theta_{FA} f_{FAC}}{f_{foc_{FA}}} \tag{7}$$

Let's now consider multi-emitter module whose N laser diodes are distributed along FA, the spot waist radius remains the same but the N collimated beams can be seen as a unique beam of width equal to $N * w(f_{col})$. Hence the divergence of the focused beam changes to:

$$\theta_{foc_{FA}} = N \frac{w(f_{FAC})}{f_{foc_{FA}}} = N \frac{\theta_{FA} f_{FAC}}{f_{foc_{FA}}} \tag{8}$$

Assume there are N_{FA} laser diodes along FA direction and N_{SA} along SA direction. The beam waist radius of this N_{FA} by N_{SA} laser diode stack at the focal plane can be expressed as:

$$w_{dia} = [(\frac{BPP_{FA}}{\theta_{FA}} \frac{f_{foc_{FA}}}{f_{FAC}})^2 + (\frac{BPP_{SA}}{\theta_{SA}} \frac{f_{foc_{SA}}}{f_{SAC}})^2]^{1/2} \tag{9}$$

where BPP_{FA} is the BPP of the FA, BPP_{SA} is the BPP of the SA, where θ_{FA} is the half divergence of the beam along the FA at the $1/e^2$ intensity, θ_{SA} is the half divergence of the beam along the SA at the $1/e^2$ intensity, $f_{foc_{FA}}$ is the effective focal length of the FA focusing lens, $f_{foc_{SA}}$ is the effective focal length of the SA focusing lens, f_{FAC} is the effective focal length of the FA collimating lens, f_{SAC} is the effective focal length of the SA collimating lens.

The divergence of the the the focused beam will be:

$$\theta_{dia} = [(N_{FA} \frac{\theta_{FA} f_{FAC}}{f_{foc_{FA}}})^2 + (N_{SA} \frac{\theta_{SA} f_{SAC}}{f_{foc_{SA}}})^2]^{1/2} \tag{10}$$

By multiplying Eqs. 9 and 10, the BPP of the diagonal can be obtained as follow:

$$
\begin{aligned}
BPP_{dia} =& [(\frac{BPP_{FA}}{\theta_{FA}} \frac{f_{foc_{FA}}}{f_{FAC}})^2 + (\frac{BPP_{SA}}{\theta_{SA}} \frac{f_{foc_{SA}}}{f_{SAC}})^2]^{1/2} \bullet \\
& [(N_{FA} \frac{\theta_{FA} f_{FAC}}{f_{foc_{FA}}})^2 + (N_{SA} \frac{\theta_{SA} f_{SAC}}{f_{foc_{SA}}})^2]^{1/2} \\
=& [(N_{FA} BPP_{FA})^2 + (N_{FA} BPP_{SA} \frac{\theta_{FA} f_{FAC} f_{foc_{SA}}}{\theta_{SA} f_{SAC} f_{foc_{FA}}})^2 + \\
& (N_{SA} BPP_{SA})^2 + (N_{SA} BPP_{FA} \frac{\theta_{SA} f_{SAC} f_{foc_{FA}}}{\theta_{FA} f_{FAC} f_{foc_{SA}}})^2]^{1/2}
\end{aligned} \tag{11}
$$

3   Experimental Setup

In the experiment, we measure two parts. The first part is to measure the characteristics of the laser diode Model SEC9-915-01 by II-VI Laser Enterprise. The second is to verify our model. The laser diode in the experiment works at 8 A current, approximating to 8 W, in order to keep the temperature below 30 °C.

3.1   Near-Field Measurements

The experimental setup to measure the near field of a single laser diode is shown in Fig. 1. Firstly, the beam is collimated by the F1 lens (Thorlabs Molded Glass Aspheric Lens C330TMD-C, EFL = 3.1 mm and NA = 0.68). Secondly, the most of the beam passes though the wedge prism (Thorlabs Round Wedge Prisms PS814-C) and is absorbed by the damper. Only a small amount of the beam is reflected. There are

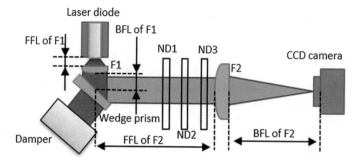

Fig. 1 Near-field measurement setup

three reasons to adopt the wedge prism: attenuating the beam, preventing the laser diode from the damage due to the back reflection of the back surface and avoid the ghost image. Thirdly, the attenuated beam transits three neutral density filters (Thorlabs Mounted Absorptive Neutral Density Filters) in orders, namely ND1 (NE20A), ND2 (NE30A) and ND3 (NE40A). Finally, the beam is focused by the F2 lens (Thorlabs Mounted Achromatic Doublets AC254-100-C-ML, EFL = 100 mm) and then is imaged onto the CCD camera (Lumenera Lw230M - 2.0 Megapixel Monochrome Camera Module, 7.1 mm × 5.4 mm active area with a resolution of 1616 × 1216). Distances between each surface is also illustrated in Fig. 1, which FFL (Front Focal Length) of F1 = 1.8 mm, BFL (Back Focal Length) of F1 = 2.1 mm, FFL of F2 = 104.3 mm (the thickness of three filters has been considered) and BFL of F2 = 90.3 mm.

After obtaining the image from the CCD camera, the size of the active region can be calculated by the following expression:

$$w_0 = \frac{f_1}{f_2} d_{mess} \tag{12}$$

where f_1 is the EFL of F1, f_2 is the EFL of F2 and d_{mess} is the distance measure on the image.

The area of the image is 0.0528 mm × 3.0096 mm. The EFL of F1 is 3.1 mm and the EFL of F2 is 100.1 mm and the converted values are shown in (Table 1).

Table 1 Near-field characteristics

Parameter	Value
FA emitter width @ $1/e^2$ intensity (um)	1.6
SA emitter width @ $1/e^2$ intensity (um)	93.3

3.2 Far-Field Measurements

The far-field is measured by the setup illustrated in Fig. 2. The setup is almost the same as the setup used to measure the near-field except replacing the F1 lens with a larger EFL lens (Thorlabs Molded Glass Aspheric Lens C240TME-C, EFL = 8 mm and NA = 0.5, FFL = 5.9 mm and BFL = 7.9 mm) and inserting the F3 lens (Thorlabs Mounted Achromatic Doublets AC254-50-C-ML, EFL = 50 mm) between the F2 lens and the CCD camera. The distance between the F3 lens and the CCD camera is the BFL of F3, 40.9 mm.

When the measurement is done, the half divergences of two axes can be calculated by the equation as follow:

$$\theta = \frac{f_2}{f_1 f_3} \frac{d_{mess}}{2} \tag{13}$$

where f_3 is the EFL of F3.

By substituting the measurements from CCD camera to Eq. 12 with the EFL of F1 = 8 mm, the EFL of F2 = 100.1 mm and the EFL of F3 = 49.8 mm, the divergences of FA and SA are able to be computed, which is shown in Table 2.

3.3 Laser Diode Stack Measurements

Until now we don't have a stack for holding laser diodes and collimators, a machine for moving and clamping lenses and a oven for soldering laser diodes. We have to

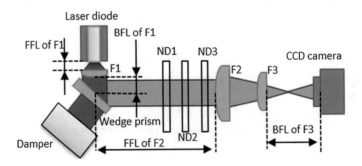

Fig. 2 Far-field measurement setup

Table 2 Far-field characteristics

Parameter (deg)	Value
FA half divergence @ $1/e^2$ intensity	22.8
SA half divergence @ $1/e^2$ intensity	6.6
FA FWHM divergence	25.3
SA FWHM divergence	8.5

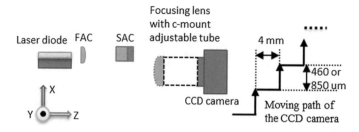

Fig. 3 Approach to measure laser diode stack by using a single laser diode

use other approach to mimic the behavior of laser diode stack. Figure 3 shows the schematic of this method. The approach can be divided into 7 steps:

1. Keeping the FAC (Ingeneric FAC-08-600, EFL = 600 um and NA = 0.8 or Ingeneric FAC-06-1100, EFL = 1100 um and NA = 0.6) fixed.
2. Adjusting the SAC (FISBA OPTIK SAC 12000, EFL = 12 mm and NA = 0.15) and the laser diode until the beam is well collimated.
3. Changing the position of focusing lens so that the focal plane locates on the CCD exactly.
4. Moving the CCD camera along the edge of the stairs and shooting at the beginning of each stair. The CCD camera is fixed on a platform consisting of 3-axis (Newport M-UTM50PP.1, M-UTM150PP.1 and M-UTM150CC.1) which is controlled by Newport MM4006 8-axis controller.
5. Removing the focusing lens and moving the CCD camera to the position of the front surface of the focusing lens.
6. Repeating the 4th step.
7. Using Matlab to load all images and combining them into two images: one is the beam on the entrance of the focusing lens and the other one is the spot at the focal plane.
8. Calculating the BPP of the beam.

The results shows in the Tables 3 and 4.

Table 3 Comparison of 6 combined beams between the measurement and the calculation (FAC-06-1100)

Item	Measurement	Calculation	Error (%)
BPP (mm * mrad)[a]	11.2	11.7	4.3
BPP (mm * mrad)[b]	11.6	11.7	0.9
BPP (mm * mrad)[c]	11.9	11.7	4.3

[a]Focusing lens of EFL = 100 mm
[b]Focusing lens of EFL = 50 mm
[c]Focusing lens of EFL = 8 mm

Table 4 Comparison of 6 combined beams between the measurement and the calculation (FAC-08-600)

Item	Measurement	Calculation	Error (%)
BPP (mm * mrad)^a	8.0	8.2	2.4
BPP (mm * mrad)^b	8.2	8.2	0
BPP (mm * mrad)^c	8.4	8.2	2.4
BPP (mm * mrad)^d	11.5	11.3	1.8
BPP (mm * mrad)^e	11.8	11.3	4.4
BPP (mm * mrad)^f	12.0	11.3	6.2

^{a} Focusing lens of EFL = 100 mm, 6 combined beams
^{b}Focusing lens of EFL = 50 mm, 6 combined beams
^{c}Focusing lens of EFL = 8 mm, 6 combined beams
^{d}Focusing lens of EFL = 100 mm, 10 combined beams
^{e}Focusing lens of EFL = 50 mm, 10 combined beams
^{f}Focusing lens of EFL = 8 mm, 10 combined beams

4   Conclusion

The model of introducing the impact of FAC, SAC and focusing lens has been developed. The comparison between the measured BPP and the calculation fully proves the validation of the model and providing a fast way to evaluate the performance of a design. The future work is to further develop the model with the influence of focal shift and the distance between laser diodes and focusing lens.

This project sponsored by the China Scholarship Council (CSC).

References

1. Rodrguez-Vidal, E., Quintana, I., Etxarri, J., Azkorbebeitia, U., Otaduy, D., Gonzlez, F., Moreno, F.: Optical design and development of a fiber coupled high-power diode laser system for laser transmission welding of plastics. Opt. Eng. **51**, 124301 (2012)
2. Wang, Z., Drovs, S., Segref, A., Koenning, T., Pandey, R.: Fiber coupled diode laser beam parameter product calculation and rules for optimized design. In: SPIE LASE, (International Society for Optics and Photonics, 2011), p. 791809
3. Sun, H.: Laser Diode Beam Basics, Manipulations and Characterizations. Springer Science & Business Media (2012)
4. NelsoN, C., Crist, J.: Predicting laser beam characteristics. Laser Technik J. **9**, 3639 (2012)

Flexible Automated Island for (Dis) Assembly of a Variety of Silicon Solar Cell Geometries for Specialist Solar Products

Sarah Dunnill, Ivan Brugnetti, Marco Colla, Anna Valente, Harry Bikas and Nikolaos Papakostas

Abstract The demand for non-standard, custom design, solar products is rapidly increasing with a growing number of companies wanting to incorporate sustainable energy solutions into their products. The FP7 funded white'R project aims to move away from the current manual assembly processes by developing a new automated manufacturing tool, capable of tabbing and stringing a wide variety of different size and shape solar cells. The island will have the capability of scanning incoming solar cells to be (dis)assembled and associate them to a number of tasks to be executed. It will also have the intelligence to automatically recognise, select and configure the proper "Plug&Produce" (P&P) equipment to be used for the operations. To date, a reference model has been created, detailing current production processes, considered products and foreseen equipment, in order to support the configurable implementation of the island. Designs for the fixturing and storage systems along with the end effectors have been produced.

Keywords Solar module · Solar cell · PV module · Manufacturing and processing · Automatic assembly

S. Dunnill
Solar Capture Technologies, Albert Street NE24 1LZ, Blyth, UK
e-mail: sd1@solarcapturetechnologies.com

I. Brugnetti · M. Colla (✉) · A. Valente
SUPSI-ISTePS, Galleria 2, 6928 Manno, Switzerland
e-mail: marco.colla@supsi.ch

I. Brugnetti
e-mail: ivan.brugnetti@supsi.ch

A. Valente
e-mail: anna.valente@supsi.ch

H. Bikas · N. Papakostas
Laboratory for Manufacturing Systems & Automation (LMS),
University of Patras, 26500 Patras, Greece
e-mail: bikas@lms.mech.upatras.gr

N. Papakostas
e-mail: papakost@lms.mech.upatras.gr

© Springer International Publishing Switzerland 2016
S. Bassis et al. (eds.), *Advances in Neural Networks*, Smart Innovation,
Systems and Technologies 54, DOI 10.1007/978-3-319-33747-0_46

465

1 Introduction

In 2014, the total installed capacity for solar photovoltaics (PV) amounted to around 177GW [1], a 65 % increase from the previous year and a significant increase from just 5 years ago in 2008 when it was 15.795GW [2]. This increase is mainly due to advancements in solar technology and manufacturing processes resulting in a decrease in price, where currently the average cost of a standard flat-plate solar module is only USD $0.548 per Watt [3]. The increased popularity of solar has resulted in a growing number of companies wanting non-standard, custom design solar solutions incorporated into their products.

While there have been advancements in the manufacturing technology for standard flat-plate solar modules, the available technology for non-standard solar modules remains limited, resulting in high costs for specialist products. Solar Capture Technologies (SCT) is a UK manufacturer of custom design solar modules and systems. Each solar module designed by SCT uses different size and shape solar cells depending on customer requirements. The tabbing and stringing assembly process used by SCT to connect the solar cells within a module is at present entirely manual. Currently, there is no flexible automated system commercially available to string cells that vary widely in size and shape.

The key objective of the FP7 funded white'R project[1] is to develop a new automated manufacturing tool, (based on the combination of fully automated, self-contained 'white room' modules) for tabbing and stringing solar cells.

2 State of the Art

The white'R manufacturing island aims to move away from the current manual and rigid assembly processes by developing a new automated manufacturing tool, capable of tabbing and stringing a wide variety of different size and shape solar cells. White'R is a self-contained white room consisting of a multi-robotic island made up of intelligent, cooperative, fully automated "Plug&Produce (P&P)" modules that presents a dual level of reconfigurability and reusability. The reconfigurability refers to the capability of the island, as a whole, to match the evolving variable family of solar cells. The P&P modules interact with each other to accomplish multi-level optimization strategies related to both assembly and disassembly processes. The island will have the capability of scanning incoming solar cells to be (dis)assembled and associate them to a number of tasks to be executed. It will also have the intelligence to automatically recognize, select and configure the proper (P&P) equipment to be used for the operations. Each single module of the island, from the end-effectors to the robotic arm, is designed and built as a

[1]Funded by the 7th Framework Programme (FP7), FoF.NMP.2013-2: Innovative Re-Use of Modular Equipment Based on Integrated Factory Design. Project reference no. 609228.

self-declaring intelligent module capable of interacting with other island entities. The same capabilities, considered at a higher level, make the island as a whole a P&P solution as it is a compact white room that can be easily integrated in the shop-floor where pre-existing equipment is operating.

The automation of the tabbing and stringing process would significantly increase production capabilities and decrease the cost of custom design solar modules. The technical objectives of the white'R system are a 50 % reduction of cost compared to current production systems and 30 % set-up and ramp-up times reduction by self-adaptive reconfigurability.

3 Framework Description

All white'R features and capabilities are derived from the need to manufacture a range of complex products. The creation of a detailed reference model is necessary to identify the production tasks and requirements expected from the P&P modules. To create this reference model a detailed company profile is created outlining the company structure and organisation, the family of products, current production processes and pre-existing equipment. This information is instrumental to address the configuration and restrictions of the P&P modules and the island's integration into the shop-floor, as well as its efficient ramp up and management over time.

3.1 Products and Operations

The gathered information needs to be recorded in such a way that it is readable by both human and machine. Considering this, all parts and operations to be managed by the white'R island are clearly defined and coded. The complex shape products to be managed by the SCT white'R island are the individual strings of solar cells (for use within a solar module). These are described as a set of multiple components from the single elementary component, to an intermediate assembly, up to the final product. The components are described as 'Entities' which are assigned an Entity Identification code and complete name. This enables all users and software tools to refer uniquely to the specific component. The entities are linked to the reference family of products so are subsequently linked to the company infrastructure. Each entity is associated to a description of its function in the overall product and is related to a number of characteristics. This includes the physical and operational characteristics, the lifecycle, the sustainability and the production characteristics.

Following the outline of the production requirements, information related to the operations for the manufacture of the reference family of products is gathered. The operations are a combination of current manual assembly operations with possible robotic assembly steps. This information refers respectively to physical, time and space transformations. Any operation can be associated to a number of possible control operations such as visual or electrical tests. As with the information about the products, the operations are described in a way that can be gradually developed over time to fit evolving or new products.

A list of production tasks resulting from the mapping of the operations to the various entities contributing to the string manufacturing is then created, with exclusive reference to the production tasks to be performed within the white'R island. Different production tasks are created from varying combinations of operations and entities. From this, the production processes to be executed can be built by combining some of the production tasks. The order of these tasks may be technologically imposed or left free as a design choice. These production processes can be similarly combined to build specific string designs.

Through this comprehensive company infrastructure and production facility assessment it was identified that the SCT white'R island would manage both the tabbing and stringing of solar cells of variable sizes and the repair of defective strings.

3.2 White'R Island Design

The primary purpose of the P&P modules required for the SCT white'R island (as identified by the reference model) is the automated tabbing and stringing of a large variety of solar cell geometries. This includes those currently known by the manufacturer along with possible new families related to future orders. The assembly and disassembly processes require the integration of several technologies in one production solution.

The white'R island design challenge consists of selecting the best set of resources to satisfy the production requirements outlined in Sect. 3.1 during the whole system lifecycle with the maximum expected profit. The considered resources are robots and/or operational machines (referred to in this section as operational resources) equipped with the related end effectors, transporters, load/unload stations, fixturing and storage. All considered resources are either performing technological processes such as soldering and cutting or are dedicated to mechanical assembly. An outline of the tasks to be performed by the main components is shown in Table 1. Fixtures are assumed to be easily moveable among machines by proper automated handling systems and can be passive or active.

The cost of the potential resources selected in the design phase depends on their architecture and performance. The system design process can start from green field or can regard an existing production solution to be retrofitted to match new requirements (brown field design). Along with the set of selectable resources, the system designer needs information about the demand and the feasible process plans to manufacture the products. In particular, the demand is characterized by a set of products with their demand volumes and technological features. Volumes and product features can evolve over time, e.g. products may be modified, new products

Table 1 Outline of the main components of the SCT white'R island and descriptions of the tasks to be performed

Resource	Description
Modular robot	Robot responsible for transportation of cells and tab wire as well as soldering of them
Cells storage	Location of the solar cells required for the assembly process
Tab wire feeder	Device for dispensing tabbing wire
Fixturing	Reconfigurable device used for supporting the assembly of the cells. After assembly it also supplies current to the solar cell string to perform the luminescence test
Conveyor belt	Transport system responsible for carrying the complete string outside of the robotic cell
Electroluminescence camera	Camera optimized for the electroluminescence test

can be introduced and demand volumes can change. Moreover, demand volumes can also be subject to mid-term variability. Information regarding the technological requirements of the products to be manufactured determines the selection of resource types, while information regarding production volumes drives the choice of the number and productivity of resources.

The output of the design challenge consists of the selection of resources and process plans to be implemented in a system configuration. Since the production problem evolves over the planning horizon, it may be necessary to properly reconfigure the system. A set of possible reconfigurations can be defined according to the future outcomes of the random variables affecting the problem. Therefore, the output of system design should be a capacity plan with an initial system configuration, a set of possible reconfigurations (e.g. acquisition/dismissing of resources) and possible related changes of process plans.

4 Results

4.1 Fixturing System

Within the white'R island the primary purpose of the fixturing system is to securely and accurately mount the individual solar cells, allowing for support during the machining and assembly operations. It will also serve to reduce working time by allowing quick set-up, and by smoothing the transition form part to part, reducing the complexity of the process.

Owing to the unique requirement for flexibility, this system must be easily adaptable for a range of cell geometries. Considering this, a modular vacuum fixture system was chosen. As outlined in Fig. 1, this system presents the possibility to accommodate one or multiple sliding bars, offering a low complexity and low cost

Fig. 1 Flexible fixturing
system design

solution. This allows for quick and easy adaptation, not only in the current range of
products but also for any future variants or new products, as it is not limited to
fixed, predefined positions of the components.

The sliding bars ensure correct cell and tab placement while supporting the cells
and allowing soldering tool access. Every cell, and consequently, string of cells,
requires at least three bars in order to be fixed safely: the tabbing wires of the cells
are positioned on the central bar while the two lateral bars guarantee the stress
suffered from the string soldering operations is distributed along the cell along
while providing a secure fixturing by means of a special array of vacuum cups. The
bars are set to mirror the cell dimensions according to the cell size. The vacuum
system provides a small, controllable clamping force without damaging the fragile
cells. This system is only activated by contact of the cell, so that the number of
vacuum cups utilized always matches the currently installed cells. Integrated con-
tacts within the bars also provide the required electrical connection for electrolu-
minescence testing of the strings produced. This helps to identify any cracked or
damaged cells that could result in the failure of the assembled string therefore
reducing the overall failure rate.

4.2 Storage

The storage of both the cells required for the assembly process and the strings
produced is an important component within the white'R island. The cells and
strings must be easily transported while also being supported in order to prevent any
damage or breakages.

A mask solution with specific spacing for individual products provides a low
cost solution that is easy to manufacture and replace for future variants and new
products. Mounted on a modular carousel, the mask acts as a loading device and
storage for cells, allowing high positioning accuracy. Through careful design,
masks can be used for more than one family of products, further reducing costs.
Unloading of the tabbed strings of cells is to be carried out by means of a conveyor
belt.

Fig. 2 End effector design
consisting of two modules:
the body and the tips

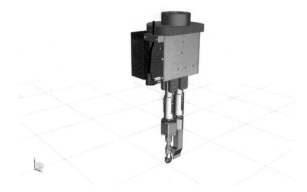

4.3 End Effectors

The operational requirements of the end effectors for the tabbing and stringing of solar cells are the handling and bonding. The handling steps include the picking, placing and holding of the cells while the bonding involves the soldering of the tabbing wire to the cell.

The end effectors consist of two modules, the body and the tips, as shown in Fig. 2. The body is equipped with a rotational motor generating linear movements to be transferred to the tips. It is also equipped with standard interfaces which enable the adoption of multiple tips to exploit the manipulation of multiple operations at the same time. They can manipulate either the cells or the strings of multiple dimensions by relying upon a vacuum system. They can also enable the soldering of strings to connect pairs of cells, thus concurrently applying heat and pressure on the string material only from one side (the top) of the cell.

The end effector design presents two major advantages. It permits a considerable reduction of the manufacturing times by avoiding the intermediate working step of tabbing cells (including soldering operation on the back of the cells), which in turn requires continuous set-up changes and tabbing wire scraps.

On the other hand, the interchangeable tips and the multiple technologies integrated on them allow an enhanced capacity of matching a large family of products while respecting the fragility constraints of the parts. The flexible tips can manipulate and manufacture variable cell sizes and different tips can be nested to the same tool body, if necessary, to accomplish new sets of operations.

4.4 Robotic Systems

The end effectors are manipulated by a robotic arm whose structure is designed to be modular and reconfigurable. The arm is conceived as a serial sequence of joints and links whose dimensions and performance vary across the position they cover on the robotic chain. The robot actuation and control system is physically located in the modules so that every module is an autonomous entity embedding the motors, drives and sensors. The standard interface of these modules and reconfigurability of the robot guarantee the possibility to generate multiple robotic arm configurations with variable reaching and degrees of freedom, by adding or subtracting modules. The selected arm configuration will be communicated to the robot controller that will adapt the related logics and settings. This reconfigurability feature is essential for the adaption to the needs of the customized solar module manufacturing sector, i.e. a large variety of solar cells with varying sizes, geometries and production demands.

5   Conclusions

The white'R island consists of a combination of components (robots, end effectors, transport, handling and tooling systems) which are conceived as "Plug&Produce" mechatronic sub modules, properly configured to match the PV production requirements.

A reference model has been produced, outlining the components and processes to be managed by the white'R island in a way that is clearly readable by both human and machine. The products and operations to be managed by the island have been outlined and a list of production tasks created. The flexible aspect of the island was considered and the model was designed in a way that allows it to evolve over time to reflect new or modified products.

Designs for the fixturing system, end effectors and robotic systems have been proposed taking into account the wide range of cell geometries possible. The fixturing system design compromises three adjustable sliding bars, with an inbuilt vacuum system, for a cell or string of cells that can be set to match the cell dimensions. A storage system that accommodates both the cells and strings has been developed, allowing high positioning accuracy while removing the need for separate (un)loading equipment.

The end effector has interchangeable tips, enhancing the capability to match a large family of products. The use of standard components also enhances the reusability of the system for different products, while keeping costs down. The proposed robotic arm can be adapted to have varying reaching and degrees of freedom. This reconfigurability is essential for adaptions according to the custom string designs.

Ongoing development work includes a detailed description of the robotic architecture and performance and the development of sensing systems for the inline monitoring and testing of the strings constructed. A number of adaptive process planning policies which adjust the process flow to match changes to the production requirements (new products, variants or production volumes) as well as equipment maintenance interventions will also be developed.

Acknowledgments The authors would like to thank the staff of Solar Capture Technologies and all those involved in the white'R project for their support. This research was conducted within the white'R project, funded by the European Commission under the Grant Agreement no. 609228 in the Seventh Framework Programme.

References

1. Green Tech Media: http://www.greentechmedia.com/articles/read/The-Global-Solar-PV-Market-Hit-177GW-in-2014-A-Tenfold-Increase-From-2008
2. Clean Technica: http://cleantechnica.com/2014/04/13/world-solar-power-capacity-increased-35-2013-charts/
3. Energy Trend: http://pv.energytrend.com/pricequotes.html

Analysis of Production Scenario and KPI Calculation: Monitoring and Optimization Logics of White'R Island

Manuel Lai and Simona Tusacciu

Abstract The purpose of the analysis is to identify the main variables that affect the production efficiency of the white'R island. In order to provide a detailed framework of the machine performances, a life cycle assessment study was undertaken, by collecting relevant information within a tool, in the form of MS Excel spreadsheet, properly developed and named White'R Energy Assessor Tool (WEATool) and thus handled to calculate a preliminary estimation of the energy consumption of the machine. The WEATool defines the white'R performance using a list of KPIs, very significant parameters that examine the economic, energy and environmental features of the island. The WEATool can be used during the design phase to provide a preliminary evaluation of the environmental impact; during the production planning such as a decision making tool to reduce the energy consumption and the production costs; and finally, during the demonstration phase to collect the data of the measurement campaign.

1 Introduction

White'R concept comes up from the need to reduce manufacturing costs of opto-electronics components which production process requires a large number of manual operations, such as multi-emitter laser diodes and photovoltaic cells. To achieve the goal three main fronts were analysed: labour cost, investment cost and energy consumption.

A suitably developed high-efficiency modular robot, able to perform currently manual operations, will ensure a drastic reduction of labour cost and cycle time production, thus increasing the productivity of the manufacturing system. The lack

M. Lai
IRIS S.r.l., Corso Unione Sovietica 612/21, Turin 10135, TO, Italy
e-mail: manuel.lai@irissrl.org

S. Tusacciu (✉)
IRIS S.r.l., Corso Unione Sovietica 612/21, Cercenasco 10135, TO, Italy
e-mail: simona.tusacciu@irissrl.org

© Springer International Publishing Switzerland 2016
S. Bassis et al. (eds.), *Advances in Neural Networks*, Smart Innovation,
Systems and Technologies 54, DOI 10.1007/978-3-319-33747-0_47

of personnel within the industrial area will minimize workspaces by reducing investment, operating and maintenance costs, which are very significant entries in the economic balance of clean rooms. Finally, the modularity and reconfigurability of the product on one hand facilitates the optimization analysis in order to maximize process efficiency of operating cycles while reducing energy consumption, on the other hand will ensure to the island a high level of customizability, making it more adaptable to future market requirements.

To certify the performances of the island and thereby increase the chances of results dissemination, a proceeding to achieve a Green Label was started during the initial phase of the project. Achieve a Green Label means characterize the product in terms of environmental, economic and social aspects, by highlighting the sustainability of the white'R island compared to the best available technologies.

To evaluate the whole life cycle of a product, the European Union's Ecodesign Directive 2009/125/EC establishes a framework to set mandatory ecological requirements for energy-using and energy-related products. Following the Ecodesign mandate, the Fraunhofer IZM Institute [1] has developed the Methodology for the Ecodesign of Energy-related Products (MEErP) [2], whose purpose was analyse materials and efficiency indicators to seek out suitable and robust assessment criteria compliant with Ecodesign requirements. The study offers also the MEErP Ecoreport Tool, a calculation template able to provide a Life Cycle by assessing the consumption of resources (such as water, coolant ...), the consumption of electricity and heat, the production of non-recyclable waste and air pollutant emissions.

2 White'R Life Cycle Assessment

Life-cycle assessment (LCA, also known as life-cycle analysis, ecobalance, and cradle-to-grave analysis) is a technique to assess environmental impacts associated with all the stages of a product's life, from cradle to grave (i.e., from raw material extraction through materials processing, manufacture, distribution, use, repair, maintenance and disposal or recycling). The aim of LCA is to compare the full range of environmental effects assignable to products and services, by quantifying all inputs and outputs of material flows and assessing how these material flows affects the environment (Key Performance Indicators).

To analyze white'R life cycle, according to MEErP, significant environmental aspects related to product design must be identified and classified by the following life cycle phases:

1. raw material selection and use;
2. manufacturing;
3. packaging, transport, and distribution;
4. installation and maintenance;

5. use;
6. end-of-life, (the state of a product having reached the end of its first use until its final disposal).

MEErP describes, for each phase, which environmental aspects must be assessed. Among these ones, white'R methodology will consider:

1. predicted consumption of materials and energy;
2. anticipated emissions to air, water or soil;
3. anticipated pollution through physical effects such as noise, vibration, radiation, electromagnetic fields;
4. expected generation of waste material;
5. possibilities for reuse, recycling and recovery of materials and/or of energy (Directive 2002/96/EC);
6. use of substances classified as hazardous to health and/or the environment according to Council Directive 67/548/EEC.

The MEErP EcoReport Tool, contains a powerful method to assess energy consumption and other several indicators all along the whole process chain. Taking into account the white'R features, the existing tool can be considered comprehensive concerning beginning of life and end of life of each module. The use phase modelling section is adequate to the environmental assessment scope but too general for a device optimization. For this lacking we decided to set up a new calculation tool, which can be integrated throughout the design phase, named "White'R Energy Assessor Tool" which has a close connection to the MEErP EcoReport Tool. On the one hand, EcoReport will be used to gather information about environmental indicators related to the beginning and the end of life, on the other hand the White'R Energy Assessor Tool will describe with more details the usage of the machine by analyzing the production life of it.

3 White'R Energy Assessor Tool

The White'R Energy Assessor Tool (WEATOOL) is a tool in the form of a MS Excel spreadsheet, developed to assess quantitatively the environmental impact of white'R island's use phase, integrating product and process description related to the two demo application: assembly of multi-emitter diodes and PV panels. It includes:

- an interface with MEErP tool, used to asses economic, environmental and energy impact of the beginning of life and end of life of white'R modules and island;
- a specific new model to assess the economic, environmental and energy impact of the middle of life (usage) of white'R modules and island;
- a specific new model to assess and compare different scenarios in terms of KPIs.

All WEATool modules, models and interfaces are developed within MS Excel including several VBA routines to automate the extraction of data from MEErP and to improve the user friendliness of user interfaces. At this stage WEATool has been customized for both industrial use cases (multi-emitter diodes and PV panels) including bill of materials, bill of processes, KPIs, scenarios and MEErP data.

3.1   Input Data: Description of Product

To describe the white'R use phase, we decided to begin by defining the various components that the island will be able to produce. Since two distinct types of market products will be manufactured, two separate versions of the WEATool were developed. To perform a whole description of each case study, two kind of data must be collected:

- Information regarding the operation of the machine concerning the use phase thus listing Production Tasks, Processes and Use cases;
- Information related to Modules and Submodules that physically compose the island.

The details of operating processes of the machine are defined within three different level of detail:

- Production Tasks: elementary operations which describe each individual action that the machine is able to make (e.g. "Active alignment of Mirror");
- Processes: sets of Production Tasks (e.g. "Mounting Chip On Carrier");
- Use Cases: sets of Processes. They indicate the different items available in the white'R island production mix (e.g. "Single emitter assembly" or "Multi emitter repairing").

The connections between the different kind of data (e.g. how each process is involved in each use case) and information concerning working devices find place within the tool structure.

To calculate environmental, technical and economic, KPIs WEATool provides a process that involves the following steps:

1. Analysis of the Bill Of Materials of several devices composing the island;
2. Calculation of environmental KPIs related to the beginning of life and end of life through the MEErP tool;
3. Analysis of the production scenarios and comparison with the state of the art and calculation of environmental KPIs related to middle of life (compiling the MEErP tool with data processed by WEATool in the production scenarios);
4. Calculation of technical and economic KPIs.

3.2 Input Data: Bill of Material

To carry out the analysis of the life cycle of the island the better way to work was departing from the investigation of Sub-modules life cycle, the reasons are mainly two: the first one is that the components are manufactured or purchased by different partners, thus information come from different sources; the second one is that within a research project it is necessary that data system implementation is flexible enough to adapt itself the progress of the project without suffering distortions in the structure, in this way the replacement of some components can be managed individually without issues.

For the examination of beginning and of life (BOL) and end of life (EOL) it is required the knowledge of materials associated with the construction of each Sub-module; instead for the middle of life (MOL) investigation it's required the description of processes (use phase).

In this phase it's therefore crucial to focus attention on the bill of materials; for this reason a form called "BOM file" ("Bill Of Material") was sent to each partner of the consortium, it is structured as follows:

- Product data: name and description of the sub-module;
- List of components which compose the sub-module and its weight in grams;
- Identification of the category and sub-category of material for each component.

The material category can be selected among the options provided by MEErP tool: Bulk plastics, Technical Plastics, Ferrous Metal, Non Ferrous metal, Coating, Electronics, Miscellaneous. For each category several sub-categories are available.

The BOM files are collected in a specific folder that WEATool analyzes to identify any changes to the white'R Bill Of Materials. The information contained in the various files are methodically placed inside the MEErP, providing environmental KPIs such as emissions of carbon dioxide, the resources used (water, refrigerant) waste (recyclable and non-recyclable), energy consumption, etc. The KPIs are provided separately for the BOL and EOL and extrapolated by WEATool to process them within scenarios analysis spreadsheets.

3.3 Production Scenarios

Since that one of the objectives of the white'R energy assessor tool is the process optimization, the analysis was carrying out comparing different production scenarios, which provide an overview of several possible configurations of the machine.

For each scenario, the user can modify some variables: productive years, working shift and their duration, yearly working days, saturation, technical efficiency, machine inactivity, production mix. The collected information are introduced to the MEErP tool for the calculation of middle of life environmental KPIs. Each scenario is composed by three different sections: energy consumption, cost assessment and environmental data.

The first one provides analysis of cycle time and related energy consumption, the second one allows to evaluate investment, maintenance, operating and disposal costs and the third one collect the set of environmental KPIs of each sub-modules calculated by the MEErP, split by three life cycle phases (BOL, MOL, EOL).

Each scenario can be compared with a spreadsheet containing information of state of the art technologies.

3.4   Key Performance Indicators

To achieve a high-level of process optimization, some technical KPIs were defined, such as time cycle of a product, energy consumption per unit (kJ/unit) and total energy consumption per year (kJ/year), energy consumption in standby mode (kJ/year and kJ/h), etc. Finally to assess the economic sustainability of white'R island, the WEATool provides also some KPIs useful to evaluate the impact of investment cost to the productivity, as well as maintenance cost and labor cost.

The proposed KPIs are the unit cost (€/unit) calculated for different production mix, the percentage of unit cost given by the labor cost, the percentage of yearly cost derived by the auxiliary system of the island, not directly involved in the production process (the most significant operating cost of diodes is done by the operating cost of the white room devices used to keep clean the working area), the reconfiguration cost, analyzed in a specific scenario and finally a set of KPIs useful to compare white'R island scenarios with the state of the art production system.

4   Conclusions

At this phase of the project, the WEATool provide guidelines for industrial partner such as decision making instrument, by picking best technical choices like the identification of materials or the size of robotic island. When the prototype will be ready to work, a campaign of measurements will be made by populating database with real measured values, in order to develop logical optimization scenarios identifying the most economically efficient and environmentally sustainable solutions. Finally, after results validation, the calculated KPIs will be used to submit the request for Green Label.

References

1. http://www.izm.fraunhofer.de/
2. http://www.meerp.eu/

Standardization and Green Labelling Procedure Methodology Developed for the Achievement of an Eco-Label for White'R Island

Manuel Lai, Giulia Molinari, Maurizio Fiasché and Rossella Luglietti

Abstract In order to certify the positive environmental impact emerging from white'R results, a procedure to apply for a Green Labelling was defined and applied [1]:

Product Description for white'R Island
white'R island lifecycle description
Preliminary prediction of white'R KPIs
Choice between mono/multi criteria and third party/self-certificate Eco label
Assessment of white'R KPIs
Scenario selection and white'R island optimization to maximise KPIs
Certification of environmental performance
Achievement of white'R Green Label
At this stage of the project, the first 4 steps of white'R green labelling methodology have been progressed and type III environmental declaration was chosen. At the end of the project, it will be possible facing a Life Cycle Assessment to achieve an Environmental Product Declaration (EPD) [2]
To that purpose, all the data necessary for the achievement of an EPD are being collected/calculated and a first draft of the EPD document is being progressed.

1 Introduction

This paper presents the eco labelling procedure that is being adopted in white'R project to achieve an Eco label for the product, a robotized white room.

M. Lai · G. Molinari (✉)
IRIS S.r.l., Via Borgata San Rocco 19, Cercenasco, TO, Italy
e-mail: giulia.molinari@irissrl.org

M. Lai
e-mail: manuel.lai@irissrl.org

M. Fiasché · R. Luglietti
Department of Management, Economics and Industrial Engineering,
Politecnico di Milano, Milano, Italy

© Springer International Publishing Switzerland 2016
S. Bassis et al. (eds.), *Advances in Neural Networks*, Smart Innovation,
Systems and Technologies 54, DOI 10.1007/978-3-319-33747-0_48

481

White'R Green Labelling Methodology is based on the general principles from ISO14020 Environmental labels and declarations and its purpose is to investigate all relevant information that might cause potential negative effects on the environment analysing the Life cycle flow of the product under discussion and comparing it with the current situation.

Since white'R ecolabel procedure uses indicators and methods proposed by EcoDesign directive, particularly the Eco Report Tool, a "white'R Energy Assessor Tool" has been setup in the form of a MS Excel spreadsheet. white'R energy assessor tool can calculate relevant KPIs from white'R machine's bill of materials, bill of processes and scenarios; it also implements the MEErP [3] EcoReport Tool 2011 made available by the European Commission in the framework of the Eco-Design directive.

2 Green Labelling Standardization Framework

Green Labels are statements that certify, through the communication of verifiable information, accurate and not misleading environmental aspects of products or services in order to promote the supply and demand for those products and services that cause less damage to the environment, helping to encourage a process of continuous environmental improvement driven by the market.

Obtain a Green Label means carry out a life cycle analysis of the product that allows to achieve different kind of benefits:

- environmental, minimizing the product impact in terms of emissions, energy and resources consumption,
- economic, by analysing and optimizing processes in order to reduce production costs,
- and social, by identifying white'R as an environmentally safe product, realized by using eco-friendly materials and processes and consequently improving the competitiveness on the market and the possibilities of an effective dissemination of the results.

2.1 Green Labels Classification

The International Organization for Standardization (ISO) [4] has identified three broad types of Eco label that follow the principles listed into ISO 14020:

- type I (ISO 14024) is a voluntary, multiple-criteria based, third party program based on life cycle considerations;
- type II (ISO 14021) is a self-certificate program and single-criteria made by manufacturer;

- type III (ISO 14025) is a third party program that includes the environmental impacts during the entire life cycle flow, the methodology used is the Life Cycle Assessment (LCA, ISO 14044).

2.2 The Environmental Product Declaration (EPD)

The Environmental Product Declaration (EPD) that has been chosen for white'R island is a type III ecolabel. The objectives of type III environmental declarations (defined into ISO 14025) are to provide LCA-based environmental related information about the product in order to assist users and purchasers to make informed comparison between products and to provide information for assessing the environmental impacts of products over their life cycle.

Creating an Environmental Product Declaration includes the following steps:

1. Find/create relevant PCR (Product Category Rules) document for the product category
2. Perform LCA study based on PCR
3. Compile environmental information into the EPD reporting format
4. Verification and certification
5. Registration and publication

The most important part of the EPD procedure is the LCA study; it evaluates the environmental impacts investigating the resources and energy consumption, and all the emissions into the system (air, water and soil) during the life cycle flow.

3 White'R Green Labelling Methodology

The green labelling procedure to be adopted for white'R island starts analysing the Life cycle flow of the product under discussion, comparing it with the current situation. The Life Cycle of white'R island includes the beginning of life, that is the design of the machine, the middle of life, which is the usage phase, and the end of life that takes into account the recycling and dismantling phases, including the benefits coming from recycling and reuse.

The aim of this green labelling study consists in the investigation of the relevant information analysed during the case study analysis. Particularly, it has been examined the entire existing production of optoelectronics diodes and photovoltaic cells to highlight the requirements and the possible environmental implications.

3.1 White'R Green Labelling Procedure Step List

The green labelling procedure, as mentioned before, has been divided in different steps in accordance with the goal of the project and the evolution of data collection.

Basing on the here above described guidelines a clear straightforward procedure has been set up, identifying for each step scopes, inputs, outputs and tools to be used.

Step 1 Product Description for white'R island: bill of materials

The first step of the methodology aims to describe white'R island in terms of: bill of materials, list of devices and modules being implemented and list of operational functions that the island will be required to perform (starting with the tasks related to multi-emitter diodes and PV panels assembly). This step and will give, as output, a list of devices, tasks, processes and use cases.

The first step is the definition and evaluation of input and output in terms of resources required during the island production. All the data shall be accurate and verifiable in accordance with the ISO 14020 principles. This step has been implemented after the design phase, when all relevant information about the island requirements has been defined.

Step 2 White'R island lifecycle description

In the second step, the lifecycle from cradle to grave of each item/device/module has been described, collecting details on: raw materials, use of recycled material, use, maintenance and end-of-life management.
White'R island life cycle has been divided into three phases, as mentioned before: beginning of life, middle of life and end of life.

Step 3 Preliminary prediction of white'R environmental indicators on life cycle basis

The scope of the third step of the methodology is to give a first estimation of all the relevant KPIs (Key Performance Indicators), including data related to each sub-module and device of the whole system and give, as output, white'R environmental indicators and the contribution from each individual lifecycle phase.

Step 4 Choice between mono/multi criteria and third party/self-certificate Eco-label

Basing on the analysis of STEP3 data, it was possible to make a choice between mono and multi-criteria certifications and checking on the possibility of a suitable third party to validate that certification.
The achievement of one of the three ISO certification methodology depends on data quality and collection. The methodology used and the completeness of collected data allowed to apply for type III certification.

Step 5 Assessment of white'R KPIs relevant to criteria selected under step4, basing on Best Available Techniques

Step 5 gives a certified assessment of all the relevant key performance indicators including data related to each sub-module and device of the whole system, on the base of a scenario integrating the best available technologies. White'R environmental indicators will be calculated using a tool (WEATool—Whiter Energy Assessor Tool) created specifically for this purpose, based on the MEErP Tool, which is a calculation template in the form of a spreadsheet, based on Eco-design Directive and used to analyze the entire life cycle of an energy-related product providing information on environmental impacts, energy and resources consumption and cost assessment.

Assessed environmental indicators are being elaborated under the holistic KPI framework, and will be exploited from other ongoing research, in order to guarantee a tangible economic and industrial sustainability of the environmentally optimal solution.

Step 6 Scenario selection and white'R island optimization, including Best not yet Available Techniques, to maximise KPIs

Once identified the environmental implications on white'R life cycle, the scope of the sixth step is to investigate a performance optimization, including alternatives and not yet available techniques. Once environmental indicators have been assessed, they will be elaborated under the holistic KPI framework in order to guarantee an economic and industrial sustainability of the environmentally optimal solution.

Step 7 Certification of environmental performance

Basing on the analysis of Step 6 data collection, on the choice made on Step 4 (green label type) and on the specific procedure defined by the suitable third party selected in Step 4, it follows the preparation of the documentation necessary for a green label certification. At the end of the procedure, the interested parties are invited to verify the data recollected and the results, in accordance with the principles of ISO 14020.

Step 8 Achievement/publication of White'R Green Label

After the achievement of a green label for white'R island, the publication shall be available for consultation to interested parties.

4 Green Label Achievement

White'R green labelling procedure final goal is to obtain an international and structured certification, which might show the benefits of automatic production system, compared with a manual one. The optimal choice will be to achieve an EPD.

4.1 Life Cycle Assessment Study (LCA)

In order to achieve an Environmental Product Declaration it is necessary to perform a complete life cycle assessment study. Life Cycle seeks to identify possible improvements to products in the form of lower environmental impacts and reduced use of resources across all life cycle stages. This begins with raw material extraction and conversion, then manufacture and distribution, through to use phase. It ends with re-use, recycling of materials, energy recovery and ultimate disposal.

Data selected for the LCA depend on the goal and scope of the study. Such data may be collected from the production sites associated with the unit process within the system boundary, or may be obtained or calculated from other sources. In practice, all data may include a mixture of measured, calculated or estimated data. Inputs may include: use of mineral resources, emission to air and water, energy consumption, etc. What is important is that all the data are accurate and verifiable in accordance with the ISO 14020 principles.

The data that should explicitly specified in the EPD document includes all the data collected during the LCA study based on the PCR document for the product category and are divided into 7 main categories:

- specification of manufacturing company, that includes information on environmental management system of the manufacturing company;
- specification of the product, that includes the description of the machine and its main components and technical information such as energy consumption and productivity:
- material content declaration;
- collected data about the core module of the product (from raw materials to assembling);
- collected data about the upstream module of the product (production of raw materials);
- collected data about the downstream module of the product, that includes both the use phase (production phase) and the end of life (from disassembly to disposal/recycle) of the machine.

4.2 Implementation of White'R Green Labelling Methodology for the Achievement of Type III Environmental Declaration

The first action that was undertaken in order to achieve an environmental product declaration for white'R island is to find a relevant PCR (Product Category Rules). PCRs are documents that define the rules and requirements for EPDs of a certain product category. It already exists a Product Category that matches white'R island's description: category 44918, which is about machines and apparatus used principally for the manufacture of semiconductors devices. The corresponding PCR (449) describes in detail which performance related information (KPIs) are needed for the life cycle assessment study that must be included in the EPD document.

After performing the life cycle assessment study for white'R island prototype using the WEATool, following the instructions included in the 449 PCR document, all the environmental related information collected/calculated has to be compiled in the final EPD document. The reporting format of an EPD shall include the following five parts: 1. Programme-related information 2. Product-related information 3. Environmental performance-related information 4. Additional environmental information 5. Mandatory statements.

Once compiled the EPD document it is essential for the market acceptance of EPD that the data and other environmental information given are considered reliable and trustworthy. The underlying data, the data handling and the EPD itself should therefore be subject for an independent verification and official registration.

5 Conclusions

At this stage of the project, the first four steps of the methodology have been progressed, while the assessment of white'R KPIs and the scenario selection for white'R island optimization are being undertaken. In order to achieve an EPD for white'R island after the end of the project, next steps include compiling the information into the EPD reporting format and formally submitting the EPD document.

References

1. ISO14020 Environmental Labels and Declarations—General Principles: http://www.iso.org/iso/catalogue_detail?csnumber=34425
2. http://www.environdec.com/
3. http://ec.europa.eu/enterprise/policies/sustainablebusiness/ecodesign/methodology/index_en.html
4. http://www.iso.org/iso/home.html

A Novel Hybrid Fuzzy Multi-objective Linear Programming Method of Aggregate Production Planning

Maurizio Fiasché, Gaia Ripamonti, Francesco G. Sisca, Marco Taisch and Giacomo Tavola

Abstract In this work a novel fuzzy multi-objective linear programming (FMOLP) method based on hybrid fuzzy inference systems is proposed for solving the general framework of integration of self-contained assembly unit in a fuzzy environment where the product price, unit cost of not utilization of resources, work force level, production capacity and market demands are fuzzy in nature. The proposed model attempts to minimize total production costs, maximizing the shop floor resources utilization and the profits, considering inventory level, and capacity. Pareto solutions optimization is computed with different techniques and results are presented and discussed with interesting practical implications.

Keywords Aggregate production planning · Hybrid fuzzy system · APP · Fuzzy Multi-objective programming

1 Introduction

Nowadays world markets are driven by demand for personalization, products life cycles shrinking and customers grown awareness. The manufacturing systems are driven towards the deployment of reconfigurable machinery and robots in order to support mass customized highly personalized products and fast reactions to variability of market demands.

In despite of the worldwide trends sectors as the ones producing high precision content devices for industrial and professional application, where often dimensions of the devices to be produced and handled are very small are still hugely employing manual activities in their production processes.

Markets push industries to deliver customized products often characterized by a high rate of complexity, high variants in relation to product geometry, technological

M. Fiasché (✉) · G. Ripamonti · F.G. Sisca · M. Taisch · G. Tavola
Department of Management, Economics and Industrial Engineering,
Politecnico di Milano, Milano, Italy
e-mail: maurizio.fiasche@polimi.it

© Springer International Publishing Switzerland 2016
S. Bassis et al. (eds.), *Advances in Neural Networks*, Smart Innovation,
Systems and Technologies 54, DOI 10.1007/978-3-319-33747-0_49

features and performance ranges, frequent features evolution over time to match the market dynamics. All of these aspects are coupled often with low production volumes. These type of products are called in literature as HMLV, acronym that stands for High Mix Low Volume [1].

This paper has been developed within the scope of the white'R FP7 European project, which aims to deliver a self- contained automatic island based on reconfigurable and modular infrastructure for the assembly/disassembly of optoelectronic devices. This new self-contained high automated reconfigurable assembly technology tool brings along the need to develop a management framework to sustain the integration of the island into an existing shop floor and to investigate models and techniques to solve the decisional problem of production planning and scheduling.

In this paper it is proposed a fuzzy multi-objective linear programming model based on the management integration framework proposed in [2] in order to solve the Aggregate Production Planning (APP) decision problem of a shop floor integrating self-contained production island white'R producing HMLV optoelectronics.

The study is organized as follow: In Sect. 2 it is started analyzing the issue of the integration of a self-contained automated assembly station in a pre-existing shop floor. Afterwards in Sect. 3 it is briefly discussed the general management framework of [2] while in Sect. 4 it is proposed a novel fuzzy multi-objective linear programming model solving the aggregate production planning decision problem presented in the general framework. In Sect. 5 numerical example is presented and a development method is discussed and finally, in Sect. 6 conclusions are inferred and the important topics for the future research discussed.

2   White'R Integration in a Pre-existing Shop-Floor

The re-configurable assembly island white'R is located in a way that the material flow goes along with the technological cycle. Consequently, from the preparation of raw material in the white'R storage till the final assembly operations of the final product, a flow shop is constituted. The flow shop may be internal white'R, or constituted by the island and one or more complement manual assembly processes (Fig. 1), downstream (a) or upstream (b), or both (c).

The different integration scenarios are showed in Fig. 2.

The integration may be basically described through two different decision levels: the way white'R will be employed to assemble products (Use choices) and the way it will be integrated in the already existing production environment (Integration choices) (b). Regarding the use choices, white'R may be used for delivering in output mono-product batches (batches of only one product family in which it would be assembled just one item or multiple items), or multi-products batches (batches of more than one family of products in which two further options may be followed such as one item for each product family or more than one item for each family of products).

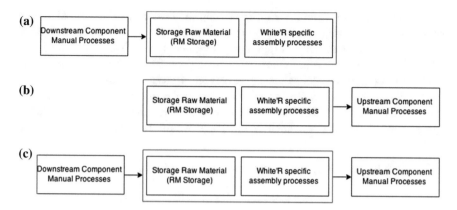

Fig. 1 The different integration scenarios

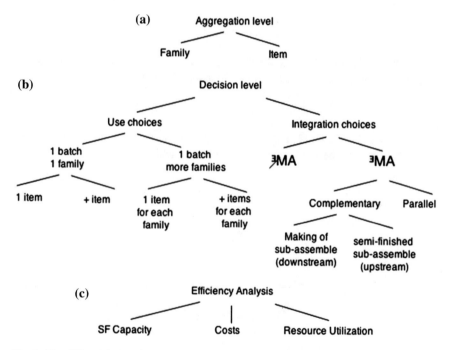

Fig. 2 The different integration scenarios

Regarding the Integration choices, it may be assumed that:

(1) White'R substitutes wholly the previous resources concentrating within itself all the production/assembly phases (^{3}/MA stands for "don't exist Manual Assembly").

(2) White'R supports the existing resources (in parallel to an existing independent assembly unit or along with downstream or/and upstream component assembly resources).

The integration of white'R within an existing shop-floor may be analyzed through different drives: aggregation level (a) and Efficiency (c). Moreover the integration may focus on product families (aggregate level) or items (disaggregate level) (a). The different integration scenarios may also be considered through an efficiency analysis through Shop Floor Capacity, Costs and Resource Utilization (c).

The different Capacity Scenarios associated to the different integration choices have been discussed in [2]. Briefly, in this specific production environment after the integration of the self-contained island there would be four different capacity scenarios:

(1) White'R used along with other assembly units (in flow).

$$C_{ass.shoflloor} = C_{min} = Tc_{lj}. [pcs/hrs]*Av_{lj}[hrs/day] \qquad (1)$$

(2) White'R replaces all the assembly units previously carried on in the shop-floor:

$$C_{ass.shofloor} = C_{wrAL} \qquad (2)$$

(3) White'R is used besides the assembly units previously carried on in the shop-floor concurring to fulfill the demand of the same family of products:

$$C_{ass.shoflloor} = C_{maAL} + C_{weAL} \qquad (3)$$

(4) White'R is dedicated to a specific family of products while a manual assembly line assemble the products not assembled by white'R:

$$\sum_{j=1}^{m-1} c = C_{maAL} \qquad (4)$$

$$\sum_{j=m}^{n} c = C_{wrAL} \qquad (5)$$

$$C_{ass.shoflloor} = C_{maAL} J = 1 \ldots (m-1) \qquad (6)$$

$$C_{ass.shofloor} = C_{wrAL} J = m \ldots n \qquad (7)$$

3   The Management Integration Framework

In this section it will be discussed and extended the production planning and scheduling framework presented in [2] whose aim is to support the management of a company employing white'R visualizing the production planning and scheduling decisional problem and implementing proper Linear Programming (LP) models to solve it (see Sect. 5).

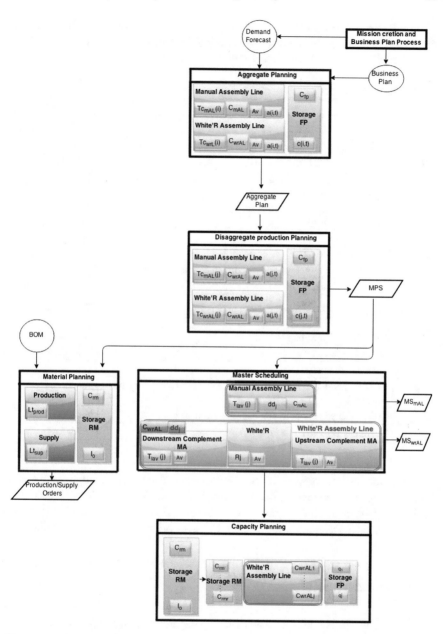

Fig. 3 Parametric framework

The plane framework presented and discussed in [2] it has been here developed parametrically in order to visualize preliminary parameters to consider for developing proper techniques for solving production planning, scheduling, material requirement and capacity planning problems (Fig. 3).

The nature of the parameters considered goes along with each block of the diagram representing the Production Planning System (PPS). Therefore they have been thought in four different categories below showed:

APP and DPP parameters		
T_c	Cycle time	[min * unit]
C	Capacity	[units/day]
Av	Availability	[%]
C_{fp}	Finished product capacity	[units/month]
a	Set-up cost	[euro/setup]
C	Holding inventory cost	[euro/months * units]
Master scheduling parameters		
T_{lav}	Working time	[min * unit]
dd_j	Due date for item j	[day]
R_j	Self-contained re-configurable assembly island throughput	[units * day]
Material planning parameters		
Lt_{prod}	Production lead time	[days]
Lt_{sup}	Supply lead time	[days]
C_{rm}	Raw material maximum capacity	[units/day]
I_0	Beginning inventory	[units/day]
Capacity planning parameters		
C_{wrrw}	Self-contained assembly unit storage raw material capacity for material k	[units]
Cwr_j	Self-contained assembly unit capacity for item j	[units/day]
q_j	Quantities assembled of item j	[units/day]

The parametric framework in this work presented has been thought as the preliminary step to implement a linear programming model to tackle APP and DPP decisional problems. Thus, in the next session it will be investigated a Fuzzy Multi Objective Linear Programming for the APP level of the PPS system described.

4 Problem Formulation

In this section, the model formulation for aggregate planning is presented.
The signs ~ and ^ indicates the fuzzy constraint and parameter.

4.1 Indices

Here below are listed the indices used for decision variables and parameters.

t	Period	1, ..., T
I	Product family	1, ..., N
l	Line	1, ..., L
j	Item	1, ..., J

4.2 Decision Variables

The decision variables of the model presented in this work are:

$\hat{q}_{ilt}$	Assembly units of product family i in line l in period t	[units]
H_{it}	Inventory level of product family in period t	[units]
s_{ilt}	Binary variable of set up for family i in line l in period t	[0; 1]

4.3 Parameters

The parameters of the model presented in this work are:

$\hat{z}_{il}$	Unit assembly cost for product family i in line l	[euro/unit]
$\hat{a}_{il}$	Set-up cost for family i in line l	[euro/set-up]
$\hat{c}_l$	Unit inventory cost for product family i	[euro/stocked unit]
$\hat{p}_{il}$	Average unit price for family i	[euro/unit]
$\hat{b}_{il}$	Availability of the line l for the family i at period t	[%]
$\hat{g}_{il}$	Average assembly time of family i in line l	[min/unit]
C_{sf}	Shop floor assembly capacity	[units/day]
C_{fp}	Finished product inventory capacity	[units/month]
$\hat{D}_{it}$	Demand of product family i at period t	[units]

4.4 Objective Functions

This work develops an optimal production plan to satisfy the forecasted demand on a mid-term horizon basis. This plan determines inventory levels, assembly quantities and number of set-ups for each period t of the planning horizon T in order to minimize overall assembly costs, maximize profit and the utilization of the shop-floor resources.

The objective functions are as follows:

$$\min \sum_{i=1}^{N} \sum_{l=mAL}^{wrAl} \sum_{t=1}^{T} (\hat{z}_{il} * \hat{q}_{ilt} + \hat{a}_{il} * st_{il}) + \sum_{i=1}^{N} \sum_{t=1}^{T} H_{it} * \hat{c}_i \tag{8}$$

This objective is to minimize overall assembly costs.

$$\max \sum_{i=1}^{N} \sum_{l=mAL}^{wrAl} \sum_{t=1}^{T} [(\hat{p}_{il} * \hat{q}_{ilt}) - (\hat{z}_{il} * \hat{q}_{ilt})] \tag{9}$$

This objective is to contribute to profit.

$$\max \sum_{i=1}^{N} \sum_{l=mAL}^{wrAl} \sum_{t=1}^{T} (g_{il} * b_{ilt} * \hat{q}_{il}) / (g_{il} * \hat{q}_{ilt}) \tag{10}$$

This objective is to maximize the shop-floor utilization level.

4.5 Constraints

For each period, the following constraints apply:

The demand satisfaction constraint

$$\sum_{i=1}^{N} \hat{q}_{ilt} + H_i \stackrel{=}{=} \hat{D}_{it}$$

The inventory capacity constraint

$$\sum_{i=1}^{N} H_{it} \leq C_{fp}$$

The Shop Floor Assembly capacity constraint

$$\sum_{i=1}^{N} \hat{q}_{ilt} \leq C_{sf}$$

The constraint on allocation of assembly lines to specific families

$$q_{itl} = 0 \, for \, l = mAL \, and \, i = 3$$

The non negativity constraint on decision variables

$$q_{itl}, H_{it} \geq 0$$

The constraint on the boolean nature of set-up variable

$$s_{ilt} \in \{0, 1\}$$

5 Model Development

5.1 FMOLP Solving Method

The proposed FMOLP model is constructed using the piecewise linear membership function of Hannan [3] to represent the fuzzy goals of the Decision Maker (DM) in the MOLP model, together with the minimum operator of the fuzzy decision-making of Bellman and Zadeh [4].

Moreover, the original fuzzy MOLP problem can be converted into an equivalent crisp LP problem and it is easily solved by the standard simplex method. There are different fuzzy goal programming models [4–7].

The important differences among these models result from the types of membership functions and aggregation operators they apply. In general, aggregation operators can be roughly classified into three categories: intersection, union, and averaging operators [8].

The minimum operator is preferable when the DM wishes to make the optimal membership function values approximately equal or when the DM feels that the minimum operator is an approximate representation. The application of the aggregation operator to draw maps above the maximum operator and below the minimum operator may be important in some practical situations.

Alternatively, averaging operators consider the relative importance of fuzzy sets and have the compensative property so that the result of combination will be medium. The g-operator [9], which yields an acceptable compromise between empirical fit and computational efficiency, seems to be the convex combination of the minimum and maximum operators [10]. Zimmermann [8] pointed out that the following eight important criteria must be applied selecting an adequate aggregation

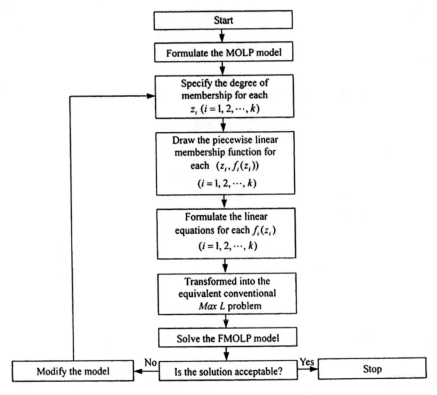

Fig. 4 Block diagram of the interactive FMOLP

operator-axiomatic strength, empirical fit, adaptability, numerical efficiency, compensation, range of compensation, aggregating behavior, and required scale level of membership function.

In order to solve the FMOLP problem involving fuzzy parameters it has been adopted the algorithm proposed by Wang et Liang [11]. The Block diagram of the interactive FMOLP is represented in Fig. 4.

Step 1   Formulate the original fuzzy MOLP model for the considered problem

Step 2   Given the minimum acceptable membership level, α, and then convert the fuzzy inequality constraints with fuzzy available resources (the right-hand side) into crisp ones using the weighted average method

Step 3   Specify the degree of membership for several values of each objective function

Step 4   Draw the piece-wise linear membership functions for each objective function

Step 5   Formulate the piece-wise linear equations for each membership function

Step 6   Introduce an auxiliary variable, thus enabling the original fuzzy multi-objective problem to be aggregated into an equivalent ordinary LP form using the minimum operator

Table 1 Comparison among methods tested

	Hybrid MCDM-EMO	NSGA-II	NBIm
f1	3500	40000	300000
f2	29500	30000	250000
f3	55000	103000	70000
f4	20000	90000	86000

Step 7 Solve the ordinary LP problem, and execute the interactive decision process. If the decision maker is dissatisfied with the initial solutions, the model must be adjusted until a set of satisfactory solutions is derived

5.2 The Multi-objective Linear Programming Associated—Results and Comments

The approach proposed above has been tested on a synthetic data-set built from simulations models of the white'R environment (made in SIMIO®). Mathematical programming and Evolutionary algorithms have been used, as well as the evolutionary multi-objective optimization (EMO) algorithms for solving the MOLP associated. Thus Normal Boundary Intersection (NBI) [12], Modified Normal Boundary Intersection (NBIm) [13], Normal Constraint (NC), [14, 15] Successive Pareto Optimization (SPO) [16] and for the seconds Non-dominated Sorting Genetic Algorithm-II (NSGA-II) [17] and Strength Pareto Evolutionary Algorithm 2 (SPEA-2) [18]. Also a hybrid approach incorporating Multi-Criteria Decision Making (MCDM) approaches into EMO algorithms as a local search operator and to lead a DM to the most preferred solution(s). Here the local search operator is mainly used to enhance the rate of convergence of EMO algorithms [19].

A comparison on several running (fi) is highlighted in Table 1. Here the Hybrid approach can achieve the same subset Pareto solutions faster than other techniques and when there is not a solution the search stop faster than other techniques. The other methods, performs worst on almost all our tests for our database.

6 Conclusions

In order to develop a model for production planning in a production environment of HMLV products employing white'R it is important to outline white'R main functions so that a wide description framework can be drawn. Furthermore identifying all different scenarios of integration is the second step in order to pinpoint all different cases in terms of production capacity of the shop-floor. The production planning system developed along with the integration of the white'R has been described by parameters so that the further model implementation for APP resolution was easier. The FMOLP developed on the basis of the management frameworks proposed presented fuzzy variables and fuzzy constraints because the production environment is fuzzy in nature. Along with this characteristic has been used an algorithm where an important step was the de-fuzzyfication of the FMOLP proposed. A comparison of different methods has been run and interesting results have been reported. In the future authors would propose some EMO algorithm with quantum Principle [20, 21] in the method proposed above.

References

1. Mahoney, R.M.: High-Mix Low-Volume. Hewlett Packard (1997)
2. Fiasché, M., Ripamonti, G., Sisca, F.G., Taisch, M., Valente, A.: Management integration framework in a shop-floor employing self-contained assembly unit for optoelectronic products. In: Research and Technologies for Society and Industry Leveraging a better tomorrow (RTSI), 2015 IEEE 1st International Forum, pp. 569–578. Turin (2015). doi: 10.1109/RTSI.2015. 7325159
3. Hannan, E.L.: Linear programming with multiple fuzzy goals. Fuzzy Sets Syst. 6, 235–248 (1981)
4. Bellman, R.E., Zadeh, L.A.: Decision-making in a fuzzy environment. Manage. Sci. 17, 141–164 (1970)
5. Zimmerman, H.-J.: Fuzzy programming and linear programming with several objective functions. Fuzzy Sets Syst. 2, 209–215 (1978)
6. Leberling, H.: On finding compromise solutions in multicriteria problems using the fuzzy min-operator. Fuzzy Sets Syst. 6, 105–118 (1981)
7. Sakawa, M.: An interactive fuzzy satisfiying method for multiobjective linear programming problems. Fuzzy Sets Syst. 28, 114–129 (1988)
8. Zimmermann, H.J.: Fuzzy Set Theory and ITS application. Kluwer, Boston (1996)
9. Zimmermann, H.J., Zysno, P.: Latent connectives in human decision making. Fuzzy Sets Syst. 4, 37–51 (1980)
10. Zimmermann, H.-J.: Fuzzy linear programming. In: Gal, T., Greenberg, H.J. (eds.) Advances in Sensitivity Analysis and Parametric Programming, pp. 15.1–15.40. Kluwer, Boston (1997)
11. Wang, R.-C., Liang, T.-F.: Application of fuzzy multi-objective linear programing to aggregate production planning. Comput. Ind. Eng. 46, 17–41 (2004)
12. Das, I., Dennis, J.E.: Normal-boundary intersection: a new method for generating the pareto surface in nonlinear multicriteria optimization problems. SIAM J. Optim. 8(3), 631 (1998). doi:10.1137/S1052623496307510
13. Motta, R.S., Afonso, S.M.B., Lyra, P.R.M.: A modified NBI and NC method for the solution of N-multiobjective optimization problems. Struct. Multi. Optim. (2012). doi:10.1007/s00158-011-0729-5
14. Messac, A., Ismail-Yahaya, A., Mattson, C.A.: The normalized normal constraint method for generating the Pareto frontier. Struct. Multi. Optim. 25(2), 86–98 (2003). doi:10.1007/s00158-002-0276-1
15. Messac, A., Mattson, C.A.: Normal constraint method with guarantee of even representation of complete Pareto frontier. AIAA J 42(10), 2101–2111 (2004). doi:10.2514/1.8977
16. Mueller-Gritschneder, D., Graeb, H., Schlichtmann, U.: A successive approach to compute the bounded pareto front of practical multiobjective optimization problems. SIAM J. Optim. 20(2), 915–934 (2009). doi:10.1137/080729013
17. Deb, K., Pratap, A., Agarwal, S., Meyarivan, T.: A fast and elitist multiobjective genetic algorithm: NSGA-II. IEEE Trans. Evol. Comput. 6(2), 182 (2002). doi:10.1109/4235.996017

18. Zitzler, E., Laumanns, M., Thiele, L.: SPEA2: improving the performance of the strength pareto evolutionary algorithm, Technical Report 103, Computer Engineering and Communication Networks Lab (TIK), Swiss Federal Institute of Technology (ETH) Zurich (2001)
19. Sindhya, K., Deb, K., Miettinen, K.: A local search based evolutionary multi-objective optimization approach for fast and accurate convergence. In: Parallel Problem Solving from Nature—PPSN X. Lecture Notes in Computer Science, vol. 5199. p. 815 (2008). doi:10.1007/978-3-540-87700-4_81
20. Fiasché, M.: A quantum-inspired evolutionary algorithm for optimization numerical problems. In: ICONIP 2012, Part III, LNCS, vol. 7665, pp. 686–693. (Part3) (2012). doi:10.1007/978-3-642-34487-9_83
21. Fiasché, M., Taisch, M.: On the use of quantum-inspired optimization techniques for training spiking neural networks: a new method proposed. In: Smart Innovation, Systems and Technologies, vol. 37, pp. 359–368 (2015). doi:10.1007/978-3-319-18164-6_35

A Production Scheduling Algorithm for a Distributed Mini Factories Network Model

M. Seregni, C. Zanetti and M. Taisch

Abstract Distributed Manufacturing is considered as one of the modern pervasive production paradigm spreading as a response to demand for green and customized products with low cost and fast delivery time. Mini factory seems to effectively overcome challenges posed by the modern business environment (Reichwald et al. The Practical Real-Time Enterprise, Springer, Berlin, pp. 403–434,2005, [1]). However, design of the Mini factory network has to consider several inner and external variables to reach high performances. Then, in this paper authors analyze how products demand volume impact on the size and the configuration, i.e. typologies of Mini-factory, of the Mini factory network. To do that, an EFUNN adapted for this application has been implemented. Results show an accuracy of over 90 % for running with 3 different MFs used (Triangular, Trapezoidal, Gaussian), with a constraints of 2 possible configuration number of mini-factories range. In conclusion, this model seems to be an accurate tool to predict the best network architecture, given market demand. to be satisfied.

Keywords Distributed manufacturing systems · Mini factory · Production scheduling · Neural networks

1 Introduction

According to a recent Eurostat survey [2], European furniture industry accounts for 130.000 companies, which raised 95 billions of Euros turnover in 2010. They mainly consist in local, small enterprises, and they employee about 1,04 millions

M. Seregni (✉) · C. Zanetti · M. Taisch
Politecnico di Milano, Piazza Leonardo Da Vinci 32, 20133 Milan, Italy
e-mail: marco.seregni@polimi.it

C. Zanetti
e-mail: cristiano.zanetti@polimi.it

M. Taisch
e-mail: marco.taisch@polimi.it

© Springer International Publishing Switzerland 2016
S. Bassis et al. (eds.), *Advances in Neural Networks*, Smart Innovation,
Systems and Technologies 54, DOI 10.1007/978-3-319-33747-0_50

people [2]. This data highlights the great relevance of this sector for the EU economies. Moreover, Europe plays a leading role within the worldwide furniture industry, as far as its influence on fashion and design are concerned. However, the recent economic crisis, while affecting Western Europe Countries since 2009, has raised the level of global competition, causing an on-going production transfer from EU countries to low-labour cost countries. On the other hand, customers' expectations have been rapidly increasing in terms of tailoring degree and environmental sustainability, especially among young and middle age people [3].

To this extent, European enterprises are required to be more and more innovative in each business aspect, e.g. product, manufacturing processes and organization.

While bearing in mind the above mentioned challenges, a production system based on a Mini-Factories network could represent an eventual effective solution [4], which could ensure highly personalized products while coping with the highly fragmented EU local markets [5].

As far as the 'Mini-Factory' definition is regarded, in this paper it is referred to a small-scale production unit based on a single, general purpose CNC machine, which is supported by a design configurator, aiming at manufacturing customized furniture.

In the second section, the state of the art of Mini-Factory model for the furniture industry has been presented, together with the logical concept and the Production System Model.

The networking of the Mini-factories is presented in the next section, thus introducing the concept of a Macro factory organization, which provides supporting activities for a Mini-Factories network.

A basic assumption has been investigated in this paper, namely to verify both the effectiveness and the efficiency of the Mini-factories within a distributed network. Based on the proposed Production System Model (i.e. PSM), a specific algorithm was needed to pursue the best compromise between Mini-Factory utilization ratio and number of Mini-factories against the change in the selling mix. In addition, assumptions were made on the difference among Mini-factories facilities (e.g. aged equipment with lower efficiency) and localization (e.g. high labour cost countries).

Finally, limitations in the scope and future outlooks are emphasized in order to provide hints for further research steps.

2 Mini-Factory

2.1 Mini Factory Concept

Reichwald, Stotko, and Piller defined in 2005 a Mini-Factory as "[...] a designed scalable, modular, geographically distributed unit that is networked with other units of this type" [1]. While taking into consideration both their assumptions and the information gathered through the above mentioned surveys and case study analysis and interviews, the following basic principles have been determined as essential elements of a Mini-factory:

- Reconfigurability/Modularity, defined as the ability of a manufacturing or assembly system to switch with minimal effort and delay to a particular family through the addition or removal of functional elements [6]
- Change-over ability, intended as a single machine which performs operations on a known work piece or subassembly with low effort and delay [6]
- Customer proximity, i.e. in terms of location (Urban location) and provided products/services (Configuration, Sales, After-sales)

In addition, the localization as close as possible to the Customers, enables the growth of repeating purchase, hence exploiting the economies of integration [7]. According to Zaeh and Wagner this distributed instantiation of Mini-Factories could also have economic returns in terms of decreased logistic costs and delivery time [8].

Therefore, a Mini-Factory network represents an opportunity for synergies among enterprises who intend to take advantage of common task coordination, such as procurement, HR administration and logistics, sharing of information, etc.

As such, it mainly consists of one of the several models included in "Distributed Manufacturing" (DM) concept [9]. This term has been defined, among others, by Kühnle in 2009 as "...geographically dispersed manufacturing locations of one enterprise". Furthermore, Distributed Manufacturing Systems may be defined as "... class of manufacturing systems, focused on the internal manufacturing control and characterized by common properties: autonomy, flexibility, adaptability, agility and decentralization [10]".

In this regard the Mini-factory network consists of a central command unit, which carries out support to the network of Mini-Factories, i.e. standardized raw materials procurement, operators' training and basic product features definition, while keeping proximity to the Customers its main aim.

2.2 A Sample of Mini-Factory for the Furniture Industry

An operational model of Mini-Factory production system has been developed on the basis of case studies, surveys and one-to-one interviews.

The outcomes of these activities have constituted the knowledge background, i.e. best practices and performance monitoring, to develop a model of Mini-Factory specifically thought for furniture industry.

The re-localization of manufacturing activities within urban contexts, the downsizing of production systems and the integration/hybridization among production, design and distribution, in terms of processes and locations, represent three emerging trends, which define a potential new paradigm in the field of the industrial production. In this regard, a small scale production system, properly located within a shopping mall or a Do-It-Yourself center, is able to directly interact with the customers, while being close to them in terms of design expectations, quality, environment, costs an delivery time. In addition, a recent survey by the Wall Street Journal [11] revealed that the vacancy rate is up to 8 % in regional malls in United States and similar trends are expected in Europe, due to steady increase of on-line market places. This opportunity results in renting cost reduction and in an incentive to relocate small production centers as close as possible to the customers (i.e. CTC).

The scope of this paper is focused on woodworking Mini-Factories only, even though findings might be extended to further industrial sectors, while bearing in mind regulatory, technological, operational and social constraints.

Product Definition

A robust scientific literature has explored the relationship between design and mass customization, thus highlighting the need for special attention on both the product structure and the product aesthetics [12]. These have been reflected into two general design rules, which are based on the maximum reduction of hardware and the design of new standard parts and components facilitating the development of modular furniture. Both trends could lead to a 'smarter' influence on the Mini-Factory production system model, reducing the warehouse surface and complexity, while meeting the customers' expectations.

It is very important to take into account that furniture integrate a part of a larger and interactive customer experience, which strongly integrates design, production and distribution. Consumer involvement essentially consists in matching his requirements and technological constraints trough a parametric configurator software that both professionals and unskilled users can easily use.

As far as product logic is regarded, three main criteria should be adopted:

- pieces of furniture should be flat-pack based, ready to be assembled and easy to be disassembled;
- packaging consumable and instructions shall be included;
- common kinds of wooden panels: (laminated) chipboard, (laminated) plywood and (varnished) MDF should be used only.

With regards to product families the research scope has been limited to living rooms and children bedrooms, which represent general purpose environment. Further products may be developed on occurrence, bearing in mind relevant product/process requirements and constraints.

Operations Management

A Mini-Factory is both a production unit and a distribution and contact point, which provides finished products to the customers without intermediaries. Hence, CTC Mini-Factory is divided into two areas, i.e. "the Shop" and "the Factory. Its PSM has been conceived as a single CNC machine, directly connected to a design configurator, aiming at producing customized furniture. Both upstream and downstream processes have been taken into account, as depicted in the functional schema (Fig. 1).

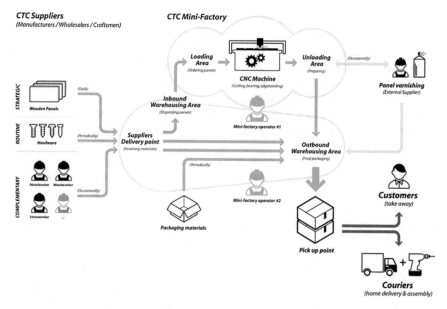

Fig. 1 Mini-Factory production system model functional schema

Suppliers' network is essentially a set of woodworkers, wholesalers and artisans, who work in cooperation with the Mini-Factory, thus providing raw materials and consumables. They have been classified in three categories on the basis of the purchase volume and customization required:

- Strategic: High volume and High Customization (e.g. wooden panels)
- Routine: High volume and Low Customization (e.g. hardware parts and components supplier);
- Complementary: Low volume, high customization (e.g. Iron Workers, Glass Workers, etc.).

As far as strategic suppliers are concerned standardization of purchased panels is essential to reduce the management complexity and not-value-added storing areas. A "pre-nesting" activity has to be performed, thus heuristically establishing panels mix for manufacturing a specific piece of furniture. By analyzing the local panel markets, the highest range of products are usually guaranteed by wholesaler within 24–48 h, thus limiting inventory.

Routine suppliers typically deliver hardware and components, which are referred to as joints, brackets, fitting and accessories. The favorite procurement strategy is based on on-gong replenishment policies, such as on-line delivery or consignment stock, providing that an agreement can be reached with a large manufacturer/wholesaler.

Complementary suppliers (e.g. Iron Workers, Glass Workers, Decorators, Marble workers, etc.) provide those materials are can be integrated into the piece of furniture, thus meeting specific customers' demands on occurrence. The following complementary suppliers have been identified:

Since these raw materials are not standardized, a purchase-on-demand policy should be pursued, to reduce space renting costs and obsolescence. In terms of location, interviews with the Woodworkers showed that it is preferable to rely on local Suppliers, to reduce time to market and to enable tailored solutions.

Mini-Factory operations include all the activities necessary to manufacture finished and pre-packed furniture components, starting from standard panels and consumables.

A single general purpose CNC machine, served by an anthropomorphic robot, is capable of performing a whole series of tasks, such as cutting, boring, edge banding and nesting, in order to transform wooden panels into pieces of furniture. The operator should be monitoring the whole process and he should be assigned the following specific tasks: sorting out, loading and positioning the stack of panels in front of the machine, packaging the pieces of furniture after machining into a cardboard box.

Bearing in mind that long term production planning is difficult to achieve for a Mini-Factory, production scheduling deals with a design-to-order (i.e. DTO) approach, which is essentially based on two key decisions:

1. Assignment Problem (Routing): timed assignment of production orders to machines according to their availability (not necessary for a stand-alone machine);
2. Production sorting (Sequencing): definition of the orders processing sequences on the basis of priority criteria.

The objective of the scheduling process within a Mini-Factory is therefore to avoid delays and to minimize production swarfs. The latter represents a challenge and it has been furtherly analyzed, to take proper directions. Some simple assumptions have been made for this analysis, based on the interviews with SMEs:

- 20 % of the swarfs amount is irrecoverable
- Operator average wage: 35.000 Euros per year
- Mini-Factory annual production; about 50 tons.
- Overheads: about 20 % of total costs.

Due to the significant space renting costs and management of small pieces of furniture, disposal of scraps is the most convenient policy for low level of scraps, rather than reusing or selling back swarfs, thanks to the low loss of recoverable material compared to management and warehousing costs.

The delivery process is extremely "lean", different solutions have been analyzed and a SWOT analysis (see Table 1) has been carried out accordingly; it has been designed only a small area for storing finished products, which is intended to contain a quantity of orders equal to an average daily production stock plus a safety stock. End users may decide to pick up their own furniture by themselves or, alternatively, they may rely on an external courier at an average extra charge of 7 %

on sale price (within a range of 100 km), according to interviews with woodworker SMEs. In house solution may represent an alternative for great volumes.

Organizational Model and ICT Tools

Working within a mall means that employees performing work affecting either the product or the service delivered by the Mini-Factory shall be complying with the shopping center requirements; moreover, they must be competent on the basis of appropriate education, training, skills and experience.

The main roles of a Mini-Factory have been identified as follows [13]:

- CNC Operators, who essentially take care of operations.
- Shop Operators, who represent the interface to the Customers on the front desk; they are responsible for managing the production scheduling
- Manager, who is responsible for the Mini-Factory organization and for the profit center.

Due to high availability of panel wholesalers and from interviews with Woodworkers, it was determined that a limited warehouse is necessary for a Mini-Factory, to contain 2–3 days production panels and relevant accessories.

A proposal for a rational layout has been developed as a result of workspace minimization, thus reducing unintended material transfer.

Both layout and warehouse should be reviewed in compliance with the shopping mall requirements and policies.

As far as ICT is concerned, the following tools are needed to properly support the PSM: a Scheduler, a Supply Chain Manager and an Inventory Manager (Fig. 2).

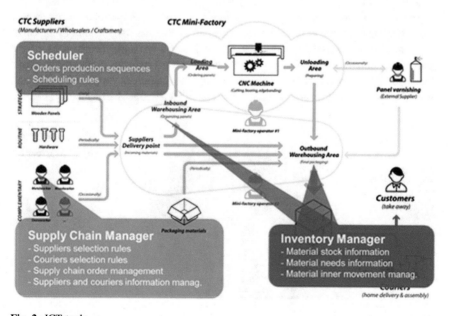

Fig. 2 ICT tools

Different ICT tools are available on the market and the software selection is out of scope as far as this paper is concerned. Nevertheless, typical features have been determined, to identify an open source tool, which could prove to be as flexible as it is required by a Mini-Factory [14].

To identify the specific management software a two steps analysis has been performed, dealing with non functional and functional requirements [15].

3 Mini-Factory Network

The production system model of a CTC Mini-Factory can be extended setting-up a scenario characterized by a network of Mini-Factories distributed in a region or in a country. The network concept needs a second-tier organization Macro-factory for supporting coordination among Mini-Factories and exploiting economies of scale in some activities [16, 17]. The Macro-Factory is both a physical and an organizational entity that provides vital support and information to several Mini-Factories, without participating in the manufacturing activities. It is mainly conceived and intended for support activities, such as: long-term planning and forecasting, order management, contracts with suppliers, maintenance, technical support, installations, logistics, etc. Single Mini-Factories can access to these shared activities, thus exploiting economies of scale.

The introduction of this second level reduces the activity in charge of single Mini-Factories, thus focusing resources on higher added value tasks, such as the CRM (Customer Relationship Management) and the Customer oriented processes. The consequent reduction of the processes those are locally implemented, positively affects the standardization of the business model; in other words, the simplification of the Mini-Factory model increases the repeatability level and effectiveness of the process instantiation. Finally, with reference to the business model of a single Mini-Factory, procurement, transport, human resources management, and supervision-coordination activities could be centralized in the Macro-factory (Fig. 3).

Fig. 3 Macro-factory activities

4 Production Scheduling Analysis

4.1 Hypothesis

The data set synthetically built is done as a collection of Mini-factory networks, seen as nodes of a graph. An in depth analysis should conducted also with other data collection about market requirements and targets, with overall strategy of each factory and of the whole network, and with other features of the system. In this paper we want to answer only two research questions:

- How changes in Total Production Volume and Mix affect Mini-factory network configuration?
- How changes in Total Production Volume and Mix affect Utilization Rate of Mini-factories within the network?

Thus, a classification of the configuration with the percentage of utilization is to identify. Thus in our network we have 2 output linked together: the Utilization Ratio (UR), the ratio between the requested production and the available production, and the configuration with different number and types of mini-factory units.

4.2 Analysis

We used a hybrid neural system, an Evolving Fuzzy Neural Networks (EFUNN) model. Fuzzy neural networks are connectionist structures that can be interpreted in terms of fuzzy rules [17–19]. Fuzzy neural networks are Neural Networks (NN), with all the NN characteristics of training, recall, adaptation, and so on, whereas neuro-fuzzy inference systems are fuzzy rule-based systems and their associated fuzzy inference mechanisms that are implemented as neural networks for the purpose of learning and rules optimization. The evolving fuzzy neural network (EFuNN) presented here is of the modified type, for adapting the network structure to problem space created.

EFuNNs have a five-layer structure (Fig. 4). Here nodes and connections are created/connected as data examples are presented [19]. An optional short-term memory layer can be used through a feedback connection from the rule (also called case) node layer. The layer of feedback connections could be used if temporal relationships of input data are to be memorized structurally. The input layer represents input variables. The second layer of nodes (fuzzy input neurons or fuzzy inputs) represents fuzzy quantisation of each input variable space.

For example, two fuzzy input neurons can be used to represent 'small' and 'large' fuzzy values. Different membership functions (MF) can be attached to these neurons. The number and the type of MF can be dynamically modified. The task of the fuzzy input nodes is to transfer the input values into membership degrees to which they belong to the corresponding MF. The layers that represent fuzzy MF are

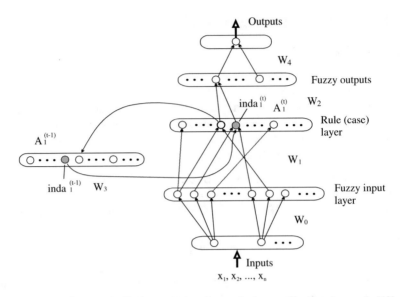

Fig. 4 An EFuNN with a feedback connection that works in an online learning mode [19]

optional, as a nonfuzzy version of EFuNN can also be evolved with only three layers of neurons and two layers of connections.

The third layer contains rule (case) nodes that evolve through supervised and/or unsupervised learning. The rule nodes represent prototypes (exemplars, clusters) of input–output data associations that can be graphically represented as associations of hyperspheres from the fuzzy input and the fuzzy output spaces. Each rule node r is defined by two vectors of connection weights, W1(r) and W2(r), the latter being adjusted through supervised learning based on the output error. The fourth layer of neurons represents fuzzy quantisation of the output variables, similar to the input fuzzy neuron representation. Here, a weighted sum input function and a saturated linear activation function is used for the neurons to calculate the membership degrees to which the output vector associated with the presented input vector belongs to each of the output MFs. The fifth layer represents the values of the output variables. Here a linear activation function is used to calculate the defuzzified values for the output variables.

The analysis of our DB shows that the EFuNN evolving procedure leads to a similar local incrementally adaptive error and EFuNNs allow for rules to be extracted and inserted at any time of the operation of the system thus providing knowledge about the problem and reflecting changes in its dynamics. In this respect the EFuNN is a flexible, incrementally adaptive, knowledge engineering model.

One of the advantages of EFuNN is that rule nodes in EFuNN represent dynamic fuzzy-2 clusters, but this could be useful for our problem to choose in a dynamic way the configuration of mini-factories.

Generally speaking, despite the advantages of EFuNN, there are some difficulties when using them:

(a) EFuNNs are sensitive to the order in which data are presented and to the initial values of the parameters.
(b) There are several parameters that need to be optimised in an incrementally adaptive mode. Such parameters are: error threshold, number, shape, and type of the membership functions; type of learning; aggregation threshold and number of iterations before aggregation, etc. It is possible to approach these problems choosing a suited optimizer like a genetic algorithm (GA).

4.3 Outcomes

We used a custom implementation of EFUNN already published in [20–24] modified for our particular application and we obtained an accuracy of over 90 % for running with 3 different MFs used (Triangular, Trapezoidal, Gaussian), with a constraints of 2 possible configuration number of mini-factories range [0–3].

5 Conclusion

The definition of a Mini-Factory Production System Model (i.e. PSM), as presented in the state of the art literature, has been reviewed and integrated, to provide SMEs with a business model as a close as possible to Customers, in terms of product design, quality and delivery time.

Core processes and development choices have been discussed together with requirements for organizational aspects and ICT tools.

Strategic, tactic and operational results have been dealt with, thus highlighting the opportunities for a stand alone Mini-Factory and the synergies provided by a network of Mini-Factories coordinated by a Macro-factory.

Finally, the PSM might be extended to further industrial sectors, providing that regulation, technological, environmental and social issues are properly coped with.

A framework for classification and for decision support system has been described and proposed in previous sections, its application in this business case has been very useful for defining the manufacturing network size and configuration, i.e. mini factory typologies, according with the product demand volume.

Acknowledgments The work presented here is part of the project "CTC- Close To Customer"; this project has received funding from the European Union's Seventh Framework Programme for research, technological development and demonstration under grant agreement no FoF. NMP.2013-6 608736—CTC.

References

1. Reichwald, R., Stotko, C. M., & Piller, F.T.: Distributed mini-factory networks as a form of real-time enterprise: concept, flexibility potential and case studies. In: The Practical Real-Time Enterprise, pp. 403–434). Springer, Berlin (2005)
2. http://ec.europa.eu/eurostat/statistics-explained/index.php/Manufacture_of_furniture_statistics_-_NACE_Rev._2
3. CBI.: The domestic furniture market in the EU (2007)
4. Seregni, M., Opresnik, D., Zanetti, C., Taisch, M., Voorhorst, F.: Mini factory: a successful model for european furniture Industry?. In: Advances in Production Management Systems. Innovative and Knowledge-Based Production Management in a Global-Local World (pp. 571–578). Springer, Berlin (2014)
5. Arbeit und Leben Bielefeld e.V. (DGB/VHS).: European Sector Monitor of the wood/furniture industry (2009)
6. Wiendahl, H.P., ElMaraghy, H. a., Nyhuis, P., Zäh, M.F., Wiendahl, H.H., Duffie, N., Brieke, M.: Changeable manufacturing—classification, design and operation. CIRP Ann.—Manufact. Technol. **56**(2), 783–809 (2007). http://doi.org/10.1016/j.cirp.2007.10.003
7. Piller, F.T. Economies of Interaction and Economies of Relationship: Value Drivers in a Customer Centric Economy, pp. 1–15. Brisbane (2002)
8. Zäh, M.F., Wagner, W.: Planning Mini-Factory Structures for the Close-to-Market Manufacture of Individualized Products. In: Proceedings of the MCPC, p. 3 (2003)
9. Mourtzis, D., Doukas, M.: Decentralized manufacturing systems review: challenges and outlook. In: Robust Manufacturing Control, pp. 355–369. Springer, Berlin (2013)
10. Kühnle, H. (ed.).: Distributed manufacturing: paradigm, concepts, solutions and examples. Springer Science & Business Media (2009)
11. http://www.wsj.com/articles/SB10001424052702303417104579543760893597806
12. http://p2pfoundation.net/Mass_Customization
13. Pedrazzoli, P., Cavadini, F. A., Corti, D., Barni, A., Luvini, T.: An innovative production paradigm to offer customized and sustainable wood furniture solutions exploiting the mini-factory concept. In: Advances in Production Management Systems. Innovative and Knowledge-Based Production Management in a Global-Local World pp. 466–473. Springer, Berlin (2014)
14. Wang, S., Wang, H.: a survey of open source enterprise resource planning (ERP) Systems. Int. J. Bus. Inf. **9**(1) (2014)
15. Benlian, A., Hess, T.: Comparing the relative importance of evaluation criteria in proprietary and open-source enterprise application software selection–a conjoint study of ERP and Office systems. Inf. Syst. J. **21**(6), 503–525 (2011)
16. Marcotte, F., Grabot, B., Affonso, R.: Cooperation models for supply chain management. Int. J. Logistics Syst. Manag. **5**(1), 123–153 (2009)
17. Rudberg, M., West, B.M.: Global operations strategy: coordinating manufacturing networks. Omega **36**(1), 91–106 (2008)
18. Furuhashi T., Hasegawa T., Horikawa S., et al.: An adaptive fuzzy controller using fuzzy neural networks. In: Proceedings of Fifth International Fuzzy Systems Association World Congress, pp. 769–772. IEEE (1993)
19. Lin, C.T., Lee, C.S.G.: Neural Fuzzy Systems: a Neuro-Fuzzy Synergism to Intelligent Systems. Prentice-Hall, Inc., Upper Saddle River (1996). ISBN 0-13-235169-2
20. Fiasché, M.: A quantum-inspired evolutionary algorithm for optimization numerical problems. In: LNCS, ICONIP 2012, Part III, vol. 7665, pp. 686–693 (2012). doi:10.1007/978-3-642-34487-9_83
21. Kasabov, N.: Evolving connectionist systems: the knowledge engineering approach. Springer Science & Business Media. M. Fiasché, A quantum-inspired evolutionary algorithm for optimization numerical problems. In: LNCS: 2012, ICONIP 2012, Part III, vol. 7665, pp. 686–693 (2007). doi:10.1007/978-3-642-34487-9_83

22. Fiasché, M., Verma, A., Cuzzola, M., Morabito, F.C., Irrera, G.: Incremental–adaptive–knowledge based–learning for informative rules extraction in classification analysis of aGvHD. In: IFIP Advances in Information and Communication Technology, Engineering Applications of Neural Networks. vol. 363, pp. 361–371. Springer, Berlin (2011). doi:10.1007/978-3-642-23957-1_41

23. Fiasché, M., Verma, A., Cuzzola, M., Iacopino, P., Kasabov, N., Morabito, F.C.: Discovering diagnostic gene targets and early diagnosis of acute GVHD using methods of computational intelligence over gene expression data. In: Networks–ICANN 2009. Artificial Neural Networks—ICANN 2009. Part II, LNCS 5769/2009, Springer, Berlin, pp. 10–19 (2009). doi:10.1007/978-3-642-04277-5_2

24. Verma, A., Fiasché, M., Cuzzola, M., Iacopino, P., Morabito, F.C., Kasabov, N.: Ontology based personalized modeling for type 2 diabetes risk analysis: an integrated approach. LNCS 5864(2), 360–366 (2009). doi:10.1007/978-3-642-10684-2_40

A Multi-horizon, Multi-objective Training Planner: Building the Skills for Manufacturing

Marta Pinzone, Paola Fantini, Maurizio Fiasché and Marco Taisch

Abstract Fast technological progress and the dynamics generated by economic, ecological and societal mega-trends are putting manufacturing companies under strong pressure to radically change the way in which they operate and innovate. Consequently, human work in manufacturing is also visibly changing and the need to up-skill and re-skill workers is rapidly increasing in order to sustain industry competitiveness and innovativeness as well as employability. Life-long training is therefore crucial to achieve the vision of a successful and socially sustainable manufacturing ecosystem in which industry, society and individuals can thrive together. In this paper we present a novel *training planner*, which provides worker-specific training recommendations based on workers' knowledge, skills and preferences, job content and allocation statistics, and factory demands. Multiple objectives, related to economic and social performances are taken into account.

Keywords Training · Manufacturing · Worker-centric · Mathematical model · MOLP · Multi-objective · Multi-horizon · Quantum evolutionary algorithm

M. Pinzone (✉) · P. Fantini · M. Fiasché · M. Taisch
Politecnico di Milano, Department of Management,
Economics and Industrial Engineering, piazza Leonardo da Vinci 32,
20100 Milan, Italy
e-mail: marta.pinzone@polimi.it

P. Fantini
e-mail: paola.fantini@polimi.it

M. Fiasché
e-mail: maurizio.fiasche@polimi.it

M. Taisch
e-mail: marco.taisch@polimi.it

© Springer International Publishing Switzerland 2016
S. Bassis et al. (eds.), *Advances in Neural Networks*, Smart Innovation,
Systems and Technologies 54, DOI 10.1007/978-3-319-33747-0_51

1   Introduction

Life-long training is crucial to sustain industry competitiveness and innovativeness while building socially sustainable and successful manufacturing ecosystems in which industry, society and individuals can thrive together [1].

On the one hand, a lack of training poses a threat to the employability of a large proportion of the labor force [2]. On the other hand, without paying attention to the knowledge and skills of their staff, manufactures cannot achieve and sustain successful performances in the long-term [3–5].

Training is essential to support the up-skilling and re-skilling of adult and older workers, who increasingly need to update and broaden their competences in order to keep pace with rapid technological changes [6].

Training is also of paramount importance for attracting and developing young workers [7], since in future it will be even harder to find highly skilled workers in order to fill in the positions left vacant because of the retirement of baby-boomers [8].

Extant research on training has explored a variety of approaches and methods in order to identify training requirements, prioritize them and develop optimal training plans for human capital development [9, 10].

Notwithstanding the attention paid to training, previous research primarily focused on optimizing production goals and/or cost objectives [11] and overlooked the importance of incorporating the different perspectives of the company and the workers in order to determine optimal training plans [11–13].

Consequently, recent research studies have called for the adoption of a more "worker-centric" perspective in planning training in manufacturing, according to which "the human dimension is a key cornerstone of the factory of the future" [14–16].

In this paper, we aim at contributing to this stream of the literature and advancing both research and practice by providing a multi-objective optimization problem which takes into consideration the trade-off between (i) closing the gaps in workers' knowledge and skills, (ii) satisfying workers' training preferences, and (iii) minimizing the training cost sustained by the company.

Our contribution will help managers decide optimal "worker-centric" plans for training/re-training according to (i) the knowledge and skill level needed to achieve production objectives, (ii) the social performance goals at the company as well as at the individual level, and (iii) the cost of training.

According to its objectives, the paper is structured as follows. In the next section, we provide an overview of extant research on training in manufacturing. Subsequently, we introduce the formal description and mathematical modeling of the training problem, considering both the company and the worker perspective. We then provide a case based numerical example and discuss preliminary results. Finally, conclusions and suggestions for future research are outlined.

2 Literature Background

Assessing worker needs for training and planning for future activities aimed at building human capital are key managerial tasks which have attracted increasing research attention in the last years (e.g., [9, 17–19]).

So far studies have primarily addressed the question of how much training workers should receive and/or how many different competences workers should learn in order to achieve performance objectives [19].

In this respect, [20] developed four integer programming models and optimal solution approaches for creating training plans. Each model is characterized by a different objective function, namely (i) minimizing the total cost of training; (ii) maximizing the flexibility of the workforce; (iii) minimize the total time required for training; (iv) optimizing the trade-off between minimizing the total cost of training and maximizing the flexibility of the workforce [20].

Slomp et al. [21] proposed an integer programming model to select workers to be cross-trained for particular machines, considering trade-offs between training costs and the workload balance among workers in a manufacturing cell.

Azizi and Liang [22] presented a mathematical programming model that determines the training schedule in order to minimize the total costs of the company, including training costs.

Liu et al. [23] developed a multi-objective mathematical model to simultaneously achieve task-to-worker training plans and worker-to-*seru* assignment plans, in order to reconfiguring a *seru* production system.

As a main drawback, so far studies have neglected the importance of taking into account the perspective and willingness of workers receiving training in formulating their mathematical models [14, 15].

In this regards, notable exceptions are represented by [11, 12, 24].

Elmes et al. [24] proposed a decision support system that in the first phase generates a list of initial solutions taking into consideration a single minimizing function, namely (i) minimizing training costs, (ii) minimizing the skill gap or (iii) minimizing worker dissatisfaction with training. Then the tool allows the decision maker to choose one of these solutions and iteratively alter it according to the decision maker's preferences with respect to the other objectives [24].

Gong and Yu [11] and Li et al. [12] addressed the cross-training plan problem in cellular manufacturing, taking into consideration worker satisfaction on tasks and salary costs. The authors defined satisfaction as "the match between the worker preferred tasks and the trained tasks" [11, 12] and argued that, by considering satisfaction, the applicability of their model in real life-settings increases and the emergence of barriers to cellular manufacturing is less likely [11].

Finally, a review about a more general problem of job shop scheduling was presented by [25]. In this study the allocation of resources to machinery is approached with some interesting aspect for our work.

3   Training Planner: The Multi-objective Optimization Problem

3.1   Mathematical Model

The objective of the proposed model is to allocate training to workers in order to (i) maximize the gain in knowledge and skill, and (ii) maximize worker satisfaction in terms of training preferences met, while (iii) minimizing the training cost.

In formulating the problem, we assume that there are a series of jobs (namely a collection of tasks) which need to be performed by workers. Each worker has a particular knowledge and skill set, as well as a level of proficiency associated with each knowledge and skill. Similarly, each job is associated to a knowledge and skill set, as well as a certain level of proficiency required to complete the job at a desired quality.

If a worker is assigned to a job where her knowledge/skill level is below the level required, then a knowledge/skill mismatch is present. In this case, the worker is a candidate to get training in that knowledge/skill.

Furthermore, each worker has defined preferences regarding knowledge and skills on which she would like to be trained, and each unit increase in the level of knowledge and skill is associated to a certain training cost to be sustained.

Indices

i is the number of workers, $i = 1 \ldots I$
j is the number of jobs, $j = 1 \ldots J$
l is the characteristic, $l = 1 \ldots L$

Parameters

$\theta_1, \theta_2, \theta_3$ are the weights assigned to give preference to the evaluation criteria $(\theta_1 + \theta_2 + \theta_3 = 1)$
C_{il} = qualification of worker i in characteristic l; $C_{il} \in [0, N]$
S_{jl} = qualification in characteristic l to perform job j; $S_{jl} \in [0, N]$
$M_{ijl} = C_{il} - S_{jl} < 0$

concerning characteristic l, mismatch between the qualification hold by worker i and the qualification required in job j to which he was assigned

W_{jl} = weighting of characteristic l in job j
F_{ij} = Frequency of worker i assigned to job j
P_{il} = Preference of worker i to be trained in characteristic l
T_l = standardized cost of a unit of training in characteristic l

Variables

X_{il} = increase in characteristic l of worker i through training

Constraints

C1. Total capacity of training

$$\sum_i \sum_l X_{il} \leq A$$

C2. Bounds of the increase in characteristic l of worker i trough training

$$N \geq X_{il} \geq 0$$

C3. Threshold effect of negative mismatch to zero

$$M'_{ijl} = \begin{cases} 0, & \text{if } X_{il} + C_{il} - S_{jl} > 0 \\ X_{il} + C_{il} - S_{jl}, & \text{if } X_{il} + C_{il} - S_{jl} \leq 0 \end{cases}$$

Performance function

$$MAX\,Z = \theta_1 \Delta MISMATCH + \theta_2 WORKER\,PREFERENCE - \theta_3 TRAINING\,COST$$

$$\Delta MISMATCH = \sum_i \sum_j \sum_l \left[M'_{ijl} - M_{ijl} \right] F_{ij}\,W_{jl}$$

$$WORKER\,PREFERENCE = \sum_i \sum_l X_{il} P_{il}$$

$$TRAINING\,COST = \sum_i \sum_l X_{il} T_l$$

Time horizons

Depending on the meaning attributed to the variables and parameters, the mismatch term can be used to model different problems with different time horizons. In fact, qualification S can be defined on the basis of current jobs or of future jobs or both. Analogously training costs can be referred to short time, long time or both. In case of different time horizons, additional time indices have to be introduced in the model.

3.2 Evolutionary Strategies Adopted

In this paper we approach the problem building a synthetic dataset of 1.000 records with the 20 attributes. The structure of the database is presented in Fig. 1.

We choose to use a heuristic approach, in particular an evolutionary strategy with the use of quantum principles presented in [26], for solving the Multi-objective Linear Programming (MOLP) problem described above.

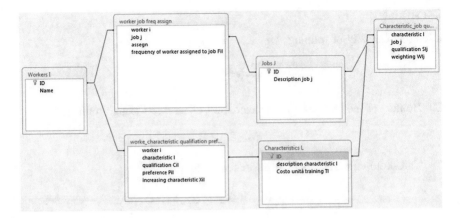

Fig. 1 Structure of the database

3.3 Quantum Theory Principles

A Quantum Evolutionary Algorithm (QEA) was proposed by [27] in 2002, which was inspired by the concept of quantum computing. According to quantum computing concept the smallest information unit in today's digital computers is represented by the quantum bit (*qbit*) [27]. *qbit* can assume the '1' or '0' states, but also a *superposition* of both states. Superposition allows the possible states to represent both 0 and 1 simultaneously based on its probability. A *qbit* state $|\Psi\rangle$ can be described as:

$$|\Psi\rangle = \alpha|0\rangle + \beta|1\rangle$$

where α and β are complex numbers defining the probability of which corresponding states is likely to appear when a *qbit* is read (measured, collapsed). $|\alpha|^2$ and $|\beta|^2$ give the probability of a *qbit* being found in state '0' or '1' respectively. Normalization of the states to unity guarantees at any time that:

$$|\alpha|^2 + |\beta|^2 = 1$$

In order to modify the probability amplitudes, *quantum gate operators* can be applied to the states of a *qbit*. A quantum gate is represented by a square matrix, operating on the amplitudes α and β in a Hilbert space, with the only condition that the operation is reversible. Such gates are: NOT-gate, rotation gate, Hadamard gate, and others [27]. General notation for an individual with several *qbit* can be defined as:

$$Q(t) = \left\{ q_1^t, \ q_2^t, \ \ldots, q_n^t \right\}$$

where n is the size of the population and q_j^t is a Q-bit individual.

Another quantum principle is the *entanglement*—two or more particles, regardless of their location, can be viewed as "correlated", undistinguishable, "synchronized", coherent. If one particle is "measured" and "collapsed", it causes for all other entangled particles to "collapse" too.

3.4 Quantum Inspired Evolutionary Algorithms

Inspired by the concept of quantum computing, Quantum inspired Evolutionary Algorithm (QiEA) is designed with a Q-bit representation, a Q-gate as a variation operator, and an observation process.

Evolutionary algorithm with Q-bit representation has a better characteristic of population diversity than any other representations, since it can represent linear superposition of states probabilistically.

QiEA maintains a population of Q-bit individuals, where Q(t) at generation t, where n is the size of population, and q_j^t is a Q-bit individual defined as

$$\begin{bmatrix} \alpha_1 & \alpha_2 & \dots & \alpha_m \\ \beta_1 & \beta_2 & \dots & \beta_m \end{bmatrix}$$

where the following holds for $i = 1, 2, \dots, m$ as described previously (i.e., the string length of the Q-bit individual, and $|\alpha|^2 + |\beta|^2 = 1, j = 1, 2, \dots, n$).

The QiEA procedure is explained in detail in [27] and results of its application are reported in the next section.

4 Preliminary Results

We apply a new variant to the QiEA that gives a better performance of classical QEA in our test [26]. We used an initial value of Han and a customization in termination criteria. We compare the algorithm with a compact Genetic Algorithm (cGA) and with a classical Particle Swarm Optimization (PSO) on 3 run for solving the MOLP presented previously on 3 clusters identified on the same database (Table 1). The QiEA always over-performs in terms of number of evaluation for finding the best solution in every test. Sometimes the PSO finds a solution not found with the QiEA, only in a 5 % of cases.

Table 1 Mean number of function evaluations for each experiment

	QiEA	cGA	PSO
f1	4000	6000	15000
f2	9500	10000	90000
f3	65000	100000	80000

5 Conclusions and Future Development

In this paper we present a novel multi-objective linear model of the training problem, considering both the company and the workers. We take into account the trade-off between (i) closing the gaps in workers' knowledge and skills, (ii) satisfying workers' training preferences, and (iii) minimizing training costs sustained by the company.

We also propose some preliminary results from the application of QiEA, a probabilistic algorithm inspired by the concept of quantum computing.

In doing so, we contribute to shed light on the development of "worker-centric" solutions for training in manufacturing, as recently called by many authors (e.g., [11, 12, 14, 16]).

The above mentioned contributions should be interpreted in light of the limitations this study has, which could be valuably addressed in future research.

First, future studies could differentiate between types of knowledge and skills (e.g. hierarchical and categorical skills [10]) and also take into consideration uncertainty.

Second, future research could distinguish between different types of training and take into account their effect on performance.

Third, future studies could combine optimization and simulation models in order to address the training problem.

Finally, in future studies the proposed model could be applied in real life situations in order to evaluate consequent implications in the short and long term.

In conclusion, despite the limitations this study has, it advances both research and practice by providing a new "worker-centric" perspective on the training problem in manufacturing.

Acknowledgments This work has been partly funded by the European Commission through Man-Made project (Grant Agreement No: FoF.NMP.2013-3 6090730). The authors wish to acknowledge the Commission for its support.

References

1. SO SMART: Socially Sustainable Manufacturing for the Factories of the Future Coordination and Support Action. http://www.sosmarteu.eu/ (2014)
2. European Commission: Education and Training Monitor 2014 (2014)

3. Guerrazzi, M.: Workforce ageing and the training propensity of Italian firms: cross-sectional evidence from the INDACO survey. Eur. J. Train. Dev. **38**(9), 803–821 (2014)
4. Mavrikios, D., Papakostas, N., Mourtzis, D., Chryssolouris, G.: On industrial learning and training for the factories of the future: a conceptual, cognitive and technology framework. J. Intell. Manuf. **24**(3), 473–485 (2013)
5. Roca-Puig, V., Beltrán-Martín, I., Segarra Cipres, M.: Combined effect of human capital, temporary employment and organizational size on firm performance. Pers. Rev. **41**(1), 4–22 (2012)
6. European Ministers for Vocational Education and Training, the European Social Partners and the European Commission: Bruges Communiqué on enhanced European Cooperation in Vocational Education and Training for the period 2011-2020 (2010)
7. European Commission: EU Skills Panorama. Focus on Advanced manufacturing (2014)
8. Economist Intelligence Unit: Plugging the skills gap: shortages among plenty (2012)
9. Tan, K., Denton, P., Rae, R., Chung, L.: Managing lean capabilities through flexible workforce development: a process and framework. Prod. Plan. Control Manag. Oper. **24**(12), 1066–1076 (2013)
10. De Bruecker, P., Van den Bergh, J., Beliën, J., Demeulemeester, E.: Workforce planning incorporating skills: state of the art. Eur. J. Oper. Res. **243**(1), 1–16 (2015)
11. Gong, J., Yu, M.: A multi-objective cross-training plan based on NSGA-II. J. Netw. **8**(10), 2310–2316 (2013)
12. Li, Q., Gong, J., Fung, R., Tang, J.: Multi-objective optimal cross-training configuration models for an assembly cell using non-dominated sorting genetic algorithm-II. Int. J. Comput. Integr. Manuf. **25**(11), 981–995 (2012)
13. Hunt, I., Brien, E., Tormey, D., Alexander, S., Mc Quade, E., Hennessy, M.: Educational programmes for future employability of graduates in SMEs. J. Intell. Manuf. **24**(3), 501–510 (2013)
14. Bettoni, A., Cinus, M., Sorlini, M. May, G., Taisch, M., Pedrazzoli, P.: Anthropocentric workplaces of the future approached through a new holistic vision. In: Advances in Production Management Systems. Innovative and Knowledge-Based Production Management in a Global-Local World, pp. 398–405. Springer, Berlin, Heidelberg (2014)
15. May, G., Maghazei, O., Taisch, M., Bettoni, A., Cinus, M., Matarazzo, A.: Toward human-centric factories: requirements and design aspects of a worker-centric job allocator. In: Advances in Production Management Systems. Innovative and Knowledge-Based Production Management in a Global-Local World, pp. 417–424. Springer, Berlin, Heidelberg (2014)
16. May, G., Taisch, M., Bettoni, A., Maghazei, O., Matarazzo, A., Stahl, B.: A new human-centric factory model, pp. 103–108. In: Procedia CIRP (2015)
17. Marentette, K., Johnson, A., Mills, A.: A measure of cross-training benefit versus job skill specialization. Comput. Ind. Eng. 937–940 (2009)
18. Thannimalai, P., Kadhum, M., Feng, C., Ramadass, S.: A glimpse of cross training models and workforce scheduling optimization. In: IEEE Symposium on Computers and Informatics (2013)
19. Nembhard, D.: Cross training efficiency and flexibility with process change. Int. J. Oper. Prod. Manage. **34**(11), 1417–1439 (2014)
20. Stewart, B., Webster, D., Ahmad, S., Matson, J.: Mathematical models for developing a flexible workforce. Int. J. Prod. Econ. **36**(3), 243–254 (1994)
21. Slomp, J., Bokhorst, J., Molleman, E.: Cross-training in a cellular manufacturing environment. Comput. Ind. Eng. **48**(3), 609–624 (2005)
22. Azizi, N., Liang, M.: An integrated approach to worker assignment, workforce flexibility acquisition, and task rotation. J. Oper. Res. Soc. **64**(2), 260–275 (2013)
23. Liu, C., Yang, N., Li, W., Lian, J., Evans, S., Yin, Y.: Training and assignment of multi-skilled workers for implementing seru production systems. Int. J. Adv. Manuf. Technol. **69**(5–8), 937–959 (2013)
24. Elmes, B., Evans, G., DePuy, G.: Multi-objective decision support system for workforce training. In: IIE Annual Conference. Proceedings (2008)

25. Gen, M., Lin, L.: Multiobjective evolutionary algorithm for manufacturing scheduling problems: state-of-the-art survey. J. Intell. Manuf. **25**(5), 849–866 (2014)
26. Fiasché, M.: A quantum-inspired evolutionary algorithm for optimization numerical problems. In: Neural Information Processing, pp. 686–693. Springer, Berlin, Heidelberg (2012)
27. Han, K., Kim, J.: Quantum-inspired Evolutionary Algorithm for a Class of Combinatorial Optimization. IEEE Trans. Evol. Comput. **6**(6), 580–593 (2002)

Use of Laser Scanners in Machine Tools to Implement Freeform Parts Machining and Quality Control

Marco Silvestri, Michele Banfi, Andrea Bettoni, Matteo Confalonieri, Andrea Ferrario and Moreno Floris

Abstract This work illustrates the results obtained integrating the most recent laser scanner within machine tools for precision machining. The main aspects addressed include an in-depth analysis of currently available devices and their test on a specially adapted measuring machine, the use of the same axis that move the tool to achieve the scan and the use of sensors that provide different measurements types combined together. Achieved results demonstrated the possibility, for the chosen measurement system, to directly measure the position of geometric features with an accuracy of less than 2 μm and to identify their position through 3D matching with an error, calculated along repeatability tests, always less than 18 μm.

Keywords Laser scanning · On-Machine inspection (OMI) · Freeform surface localization · Computer-Aided inspection planning (CAIP)

M. Silvestri (✉) · M. Banfi · A. Bettoni · M. Confalonieri · A. Ferrario · M. Floris
University of Applied Sciences and Arts of Southern Switzerland, Manno, Switzerland
e-mail: marco.silvestri@supsi.ch

M. Banfi
e-mail: michele.banfi@supsi.ch

A. Bettoni
e-mail: andrea.bettoni@supsi.ch

M. Confalonieri
e-mail: matteo.confalonieri@supsi.ch

A. Ferrario
e-mail: andrea.ferrario@supsi.ch

M. Floris
e-mail: moreno.floris@supsi.ch

M. Silvestri · M. Confalonieri
University of Parma, Parma, Italy

© Springer International Publishing Switzerland 2016   527
S. Bassis et al. (eds.), *Advances in Neural Networks*, Smart Innovation,
Systems and Technologies 54, DOI 10.1007/978-3-319-33747-0_52

1 Introduction

This work is aimed at illustrating, through industrial examples, the results obtained integrating the most recent laser scanner within machine tools for precision machining. Although this approach has been developed to meet the specific requirements of a particular machine tool, in-depth analysis of currently available devices and their test on a specially adapted measuring machine made it possible to develop some knowledge whose use can be extended to many other modern production centers with multi-axis controls.

The obtained results are, first of all, a complete implementation of the On-Machine-Inspection concept, which is one of the most important approach to face the challenges of modern manufacturing requirements [1]. In literature, few are the research projects addressing the study of self-adaptation systems based on On-Machine Measurements for the workpiece positioning.

Some of them concern the implementation of sensors to increase machining process reliability but they seldom address free form high precision machining. Zhao et al. [2] reviewed Computer-Aided Inspection Planning (CAIP) i.e. a system developed for On-Machine Inspection systems that provides direct inspection in manufacturing and quality control, vital for automated production, providing an updated and meaningful overview. They highlight how the need for more automated inspection process planning and better decision support tools increases as the complexity and variety of products increase and the product development cycle decreases. In fact, in a conventional quality control system, a machined workpiece requires to move to a coordinate measuring machine (CMM) to check its dimensional accuracy; On-Machine Inspection allows the implementation of an automated system, using the machine and the inspection device while the part is secured on the machining centre with its coordinate system intact [3], as illustrated in Fig. 1. These systems allow, as the errors occurring during machining processes are detected, to promptly correct part distortions adjusting the subsequent machining operations.

It should also be noted that the impossibility to directly measure the movement of the tool center point (TCP) is no longer an absolute obstacle to the adoption of these systems. In fact there are documented examples of applications exceeding the problem through sophisticated measurement or calibration techniques [4, 5]. Finally, On-Machine measuring systems have also been used to detect (with in-process embedded traceable measurements) and compensate (with adaptive control and self-learning) for geometrical effects of varying external and internal dimensions, such as environmental temperature and workpiece mass, as demonstrated in the European project SOMMACT [6].

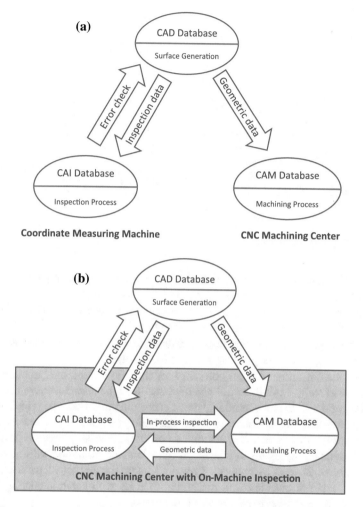

Fig. 1 Inspection processes using **a** CMM, and **b** On-Machine-Inspection [3]

2 Application Relevant Aspects

The case at hand focuses primarily on the need to accurately localize the surface to be machined in order to improve the machining accuracy. This need is characteristic of all productions involving pieces without reference surfaces (e.g. *freeform surfaces*, nowadays widely used in all engineering design disciplines) and with holding fixtures of limited accuracy.

The machining process is then defined by a sequence of operations which shall cover in the first place the calibration of the sensor, by recognizing a reference part integral with with the machine axis, and then measuring the workpiece. The system

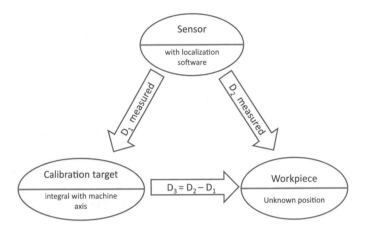

Fig. 2 The measuring principle

can thus calculate by the difference between the two measurements the position of the workpiece relative to the machine axis with the desired precision, as summarized in Fig. 2, avoiding the costs and complications necessary to reduce tolerances in the workpiece fixture and in the sensor holder.

The most important lessons learned along this activity were three: to use the same axis that move the tool to achieve the scan; to use sensors that provide measurements of different types combined together; to rethink the human-machine interaction to really exploit the full potential that this type of architecture presents. It was then possible to get something much more effective than the simple combination of a measuring station next to the machining one (result anyhow important, since it allows to not lose the reference position) because it makes possible to extend machine employing to a series of non-standard situations with a considerable economic interest as rework of repaired parts, machining of parts when the drawing file is not available and machining of parts with ceramic coatings which must be partly removed before working the metal.

The overall result can be seen as the transition from a machine tool capable of performing some specific processes only in an absolutely controlled context (receiving input parts to be machined perfectly in tolerance, mounting them in the machine through extremely precise and expensive holding fixtures and performing only the predefined tasks designed for the piece in nominal conditions) to a station with high flexibility, which can be used for functions of quality control, processing and repair in different contexts and even by personnel at different levels of training, through a system capable of treating, as Fig. 3 summarizes, the different sources of error that affect the accuracy of the work.

On the first point, the need to make dimensional measurements on free form parts at high speeds (i.e., excluding systems with touch probe), led to devote an important part of the preliminary analysis to the comparison between line laser scanners and structured light sensors (see Fig. 4). The second, costing about twice,

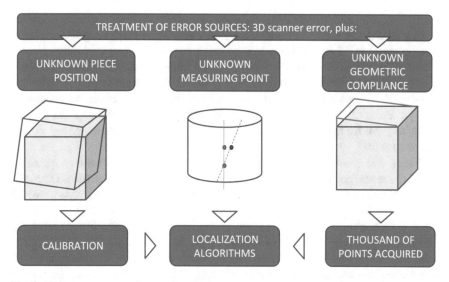

Fig. 3 Major error sources for part localization

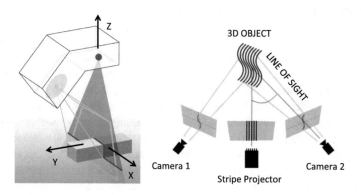

Fig. 4 Comparison between line laser scanner and structured light sensor working principle

have the obvious advantage of providing, in a single acquisition, a measure which refers to an entire area giving a three-dimensional reconstruction of the surface as a point cloud.

The alternative, represented by the line laser scanner, instead provides the profile corresponding to a single line of light and requires, to obtain a surface reconstruction, to move the scanner with obvious greater acquisition time and integration in machine problems. Initially, even by end users, seemed much more interesting to acquire "one-click" the three-dimensional reconstruction of the piece; the choice of the other solution was the result of an afterthought complex, not only motivated by the price difference of the two devices.

To perform tests, a coordinate measuring machine (CMM) has been modified setting up a double support (touch probe/laser sensor) and repeating its calibration and certification. This has allowed to create a workbench where each measurement obtained from the optoelectronic system can be immediately validated by the touch probe and which provides axis extremely stable and accurate for the scanning movement. The tests have shown that using structured light sensor has constraints (size, working distance, field of view, depth of field) which can be very limiting and more invasive for integration in machine compared to the problems related to laser scanner motion.

Figure 5 shows the used machine and the implemented modifications necessary to bring the test scenario to the problems of a real production machine. Of course each application will have different dimensions and characteristic distances which can lead to different conclusions about the integrability of a structured light sensor, but it will be anyhow true that the machine tool has already available controlled axis and that they move with sufficient accuracy for the application. The laser scanner inherent error is very low, and if the axis on which the scanner is mounted is (and it should be) accurate enough to move the tool, it will also be accurate enough to make the measurement. This allows to fully exploit the savings given by the cheaper sensor without major drawbacks in terms of other hardware needs and consequently makes the project much more interesting from an economic standpoint.

The second point to be highlighted is something peculiar to the technology adopted: a laser scanner is, simultaneously, a triangulation system that locates a geometric point in space and a system for image acquisition detecting the intensity of reflected light. Often optical systems, especially in industrial environments, are difficult to use because the lighting conditions are not the optimum ones and this affects the the operation robustness over time. Being able to combine the depth measurement, obtained by triangulation, with the image analysis allows to have simultaneously two different types of measurement that can be used in some cases to confirm the information detected, and in other to capture more informations than using only one of the two technologies. Figure 6 shows, in the image on the left, the acquisition of several reference surfaces.

Fig. 5 The CMM and the implemented modifies

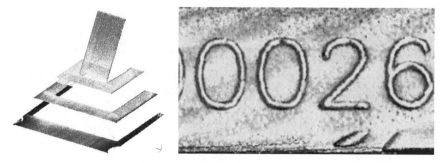

Fig. 6 Different target and role of reflected color

In this case the only relevant information is given by the position of the points found, while their color, which depends on the different degree of light reflection, is not significant. The image on the right instead shows the acquisition of an identification code, which can be obtained from the detected color using an optical character recognition (OCR) software.

Concerning the human-machine interface, the introduction of a tracking system of the piece capable of detecting large portions of geometry has allowed to eliminate the cumbersome and complex operations of identification of points to be touched with a probe. More in general, the machine will be able to modify the part-program and to adapt it to the geometrical variations of each semi-finished workpiece without the need of a full cycle through the CAM software (measurement—CAM—program generation).

3 Results and Conclusions

Laser scanner performances are evolving, thanks to the developments of lenses, signal conditioning systems and emitter sources. Regardless of this evolution, each scanner generation is available in a range of products, using different lenses, allowing to choose the best compromise between line measured width, depth of field and resolution. Currently for workpieces with a size of a few centimeters, a good compromise can be achieved with scanners which provide, on the X axis (see Fig. 4) an optical resolution of 1280 px on a line width of 25 mm, corresponding to 25 µm per pixel. This value can be improved by sub-pixel interpolation techniques up to ten times in the best reflecting conditions. Y resolution is in principle much better, but it is closely dependent on the mechanical characteristics of the axis that performs the scanning movement, not only in terms of accuracy, but also of stability and absence of vibrations. These characteristics, together to those of sensor electronics, also influence the maximum scanning speed that, in the present case, reaches 40 mm/s. Finally, the Z axis features a depth of field of about 20 mm and a resolution of 2 µm.

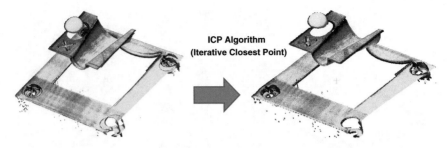

Fig. 7 ICP localization algorithm

Figure 7 shows how the system is able to superimpose two complex surfaces acquired with a given displacement, with an accuracy and a repeatability which depend on a very high number of factors.

In fact, for this purpose the characteristics of the sensor are just the first link in the system metrological chain. The actual position of the workpiece is the result of the application, to the cloud of points obtained by scanning, of a localization algorithm that, starting from a previous acquisition or from a CAD model, identifies its location in space. Point cloud matching with the CAD model showed that more than 95 % of the measured points agree with a punctual tolerance of 25 μm with the given 3D model (see Fig. 8).

Performing repeatability test consisting in repeating five times an alternate linear movement and measuring the workpiece position, the system is able to identify the

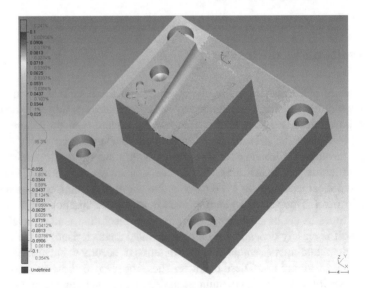

Fig. 8 An acquisition of a 3D fitted on its CAD model

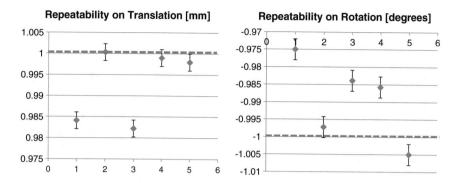

Fig. 9 Repeatability of 3D matching algorithm. The *error band* indicates sensor accuracy

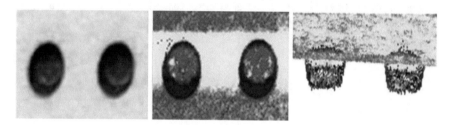

Fig. 10 Analysis of holes depth

position of geometric features with a positioning error always less than 18 μm. Similarly, an alternate rotating movement repeated five times resulted in an error always less than 0.016° (see Fig. 9).

Furthermore, the analysis of the cloud of points acquired and the color of the reflected light allows to detect defects that would not be detectable with the use of cameras only, such as the presence of holes partially occluded (see Fig. 10) or imperfect ablation of a ceramic layer.

Overall, the results confirmed the research objectives. The use of laser sensors does not allow to achieve the same level of accuracy of the touch probe to measure a single point, but the localization of complex parts with low tolerances may be far more accurate and more robust, thanks to the acquisition of millions of points describing a significant part of the surface of interest.

The analysis presented in this paper will be followed by an implementation on a particular machine tool with the aim to come to market in 2015.

References

1. Verwys, W.: Quality in aerospace turbine blade metrology. Reference guide, pp. 40–66. Aerospace Manufacturing and Design, Jan 2007
2. Zhao, F., Xu, X., Xie, S.Q.: Computer-aided inspection planning—the state of the art. Comput. Ind. **60**, 453–466 (2009)
3. Cho, M.W., Seo, T.L.: Inspection planning strategy for the on-machine measurement process based on CAD/CAM/CAI integration. Int. J. Adv. Manuf. Tech. **19**, 607–617 (2002)
4. D'Antona, G., Davoudi, M., Ferrero, R., Giberti, H.: A model predictive protection system for actuators placed in hostile environments. In: 2010 IEEE International Instrumentation and Measurement Technology Conference (I²MTC 2010) Proceedings, pp. 1602–1606 (2010)
5. D'Antona Sr, G., Davoudi, M., Ferrero, R., Giberti, H.: A model-based approach to the protection of the steering mechanism of high-power antennas placed in a nuclear fusion tokamak. Instrum. Meas. IEEE Trans. **61**(1), 55–63 (2012)
6. Silvestri, M., Pedrazzoli, P., Boër, C., Rovere, D.: Compensating high precision positioning machine tools by a self learning capable controller. In: Spaan, H., Shore, P., Van Brussel, H., Burke, T. (eds.) Proceedings of the 11th international conference of the european society for precision engineering and nanotechnology, pp. 121–124 (2011)

Author Index

Printed in the United States
By Bookmasters